TURBULENT FLUXES THROUGH THE SEA SURFACE, WAVE DYNAMICS, AND PREDICTION

NATO CONFERENCE SERIES

- I Ecology
- II Systems Science
- III Human Factors
- IV Marine Sciences
- V Air–Sea Interactions
- VI Materials Science

V AIR–SEA INTERACTIONS

Volume 1 Turbulent Fluxes through the Sea Surface, Wave Dynamics, and Prediction
Edited by A. Favre and Klaus Hasselmann

TURBULENT FLUXES THROUGH THE SEA SURFACE, WAVE DYNAMICS, AND PREDICTION

Edited by
A. Favre
Membre de l'Académie des Sciences
Institut de Mécanique Statistique de la Turbulence
Marseille, France

and

Klaus Hasselmann
Max-Planck Institut für Meteorologie
Hamburg, Germany

Published in coordination with NATO Scientific Affairs Division

PLENUM PRESS · NEW YORK AND LONDON

Library of Congress Cataloging in Publication Data

NATO Conference on Turbulent Fluxes through the Sea Surface, Wave Dynamics, and Prediction, France, 1977.
Turbulent fluxes through the sea surface, wave dynamics, and prediction.

(NATO conference series: V, Air—sea interactions; v. 1)
Includes index.
1. Eddy flux—Congresses. 2. Ocean-atmosphere interaction—Congresses. 3. Ocean waves – Congresses. I. Favre, Alexandre. II. Hasselmann, Klaus, 1931- III. Title. IV. Series.
GC190.2.N17 1977 551.4'7022 78-7339
ISBN-13:978-1-4612-9808-3 e-ISBN-13:978-1-4612-9806-9
DOI: 10.1007/978-1-4612-9806-9

SCIENTIFIC COMMITTEE

A. FAVRE, Institut de Mécanique Statistique de la Turbulence, Marseille, France
K. HASSELMANN, Max-Planck-Institut für Meteorologie, Hamburg, Germany
H. LACOMBE, Muséum National d'Histoire Naturelle, Paris, France
M. S. LONGUET-HIGGINS, University of Cambridge, Cambridge, England
E. MOLLO-CHRISTENSEN, Massachusetts Institute of Technology, Cambridge, U.S.A.

ADMINISTRATIVE DIRECTOR

F. RESCH, Institut de Mécanique Statisque de la Turbulence, Marseille, France

DIRECTORS OF THE SYMPOSIUM

A. FAVRE, Institut de Mécanique Statistique de la Turbulence, Marseille
K. HASSELMANN, Max-Planck-Institut für Meteorologie, Hamburg

LOCAL ORGANIZING COMMITTEE

A. FAVRE, M. COANTIC, and F. J. RESCH,
Institut de Mécanique Statistique de la Turbulence, Marseille

Proceedings of the NATO Conference on Turbulent Fluxes through the Sea Surface, Wave Dynamics, and Prediction held in Marseille, France, September 12–16, 1977 sponsored by the NATO Special Program Panel on Air–Sea Interactions

Softcover reprint of the hardcover 1st edition 1978

A Division of Plenum Publishing Corporation
227 West 17th Street, New York, N.Y. 10011

Foreword

A primary aim of the NATO Science Committee since its establishment in 1958 has been to further international scientific cooperation among scientists from the member countries of the Alliance. Almost all fields of basic science have received substantial support for the purposes of enhancing exchange of information, providing opportunities for advanced instruction and collaborative research. Up to 1977, more than 10,000 fellowships have been awarded to allow scientists to complete their scientific education in another country and more than 1,500 joint research projects involving scientists in at least two different countries per project have been financed by the Committee. In addition, more than 800 advanced study institutes have been held under NATO auspices which have allowed more than sixty thousand scientists to get together to hear leading scientists lecture on advanced research topics on the frontier of research of many disciplines and to discuss the most recent results of their research activities. The Science Committee also identifies specific areas as deserving special encouragement or preferential support for limited periods. A variety of mechanisms - research grants, study visits, conferences, etc. - is utilized to stimulate greater international activity. This effort is embodied in several individual programmes, each guided by a special programme panel of internationally eminent experts. Air-sea interaction is one of these programmes. Within the Air-Sea Interaction programme, many topics have been identified as being of special significance for the advancement of knowledge in this field, and this conference is well within the Programme's areas of major interest. We, on behalf of the Special Programme Panel on Air-Sea Interaction, wish to thank Professor Favre and Professor Hasselmann for having accepted the difficult task of organizing this conference, and we hope that the joint efforts of those who have experience in field work and those devoted to theoretical and experimental work in laboratories will bring a new insight into the complex phenomena of interaction between the atmosphere and the oceans.

(M. di Lullo)
Executive Officer

(J.-A. Andersen)
Chairman

Special Programme Panel on Air-Sea Interaction

Preface

One of the major goals of air-sea interaction studies is the parametrisation of the fluxes of energy, mass and momentum across the air-sea interface which represent the principal driving forces of the global atmospheric and oceanic circulation. An important second motivation is the need to predict sea state, one of the most critical environmental parameters affecting man's activities at sea. Both problems are, of course, closely interrelated. Transfers across the air-sea interface are directly affected by the dynamical properties of the moving surface itself and, conversely, the growth of waves is governed by their interaction with the turbulent atmospheric boundary layer which supports the fluxes. Finally, improved flux parametrisations including sea-state properties are of limited practical value unless accompanied by numerical wave prediction models which can be integrated in parallel with the integration of the circulation models for which the parametrisations are intended. The importance of combining flux and boundary-layer studies with wave-dynamical investigations has been generally recognised in recent years, and it therefore appeared an opportune time to bring together experts in this field in a joint symposium. In addition to theoretical and field investigations, strong emphasis has been placed on laboratory studies, which have proven particularly valuable in resolving the detailed structure of complex interaction processes. Many of the advances in air-sea interaction and wave-dynamics studies during the last decade were made possible through international cooperative projects supported by the Special Programme Panel on Air-Sea Interaction and the former Sub-Committee on Oceanography of NATO. It is therefore very appropriate that this meeting was held under the auspices of the Air-Sea Interaction Panel, and it is a pleasure for the Organizing Committee to acknowledge this support. Thanks are also due to the Local Organizing Committee for the smooth organisation and excellent facilities of the meeting and to Wolfgang Rosenthal for assistance in preparing these proceedings.

(A. Favre) (K. Hasselmann)

Contents

Section 1

FLUXES THROUGH THE AIR-SEA INTERFACE

Section 2.2: Shallow Water Waves

Section 3

WIND - WAVE INTERACTION

Section 3.2: Effects of the Waves on the Atmospheric Boundary Layer

Section 4

NUMERICAL WAVE PREDICTION MODELS

INTRODUCTORY REMARKS TO THE SESSION ON SEA-SURFACE FLUXES

Erik Mollo-Christensen

Department of Meteorology,
Massachusetts Institute of Technology
Cambridge, Massachusetts, U.S.A.

It is of course difficult indeed to make a few enlightening and nontrivial remarks to a group of active workers in the area of air-sea interaction. Rather, it devolves upon me to state a few obvious truths, to remind ourselves of a few facts and relationships that we all agree upon, but which most speakers find unnecessary to state because they are accepted or really belong in some neighboring area of science which we are not directly concerned with at this meeting. I shall be short, but will still remind you of the obvious.

First, we all realize that the fluxes of momentum, heat, latent heat, water and gases and kinetic energy across the sea surface cannot be considered as a separate problem from wave generation by wind, the problem of formation of bubbles and aerosols, the couplings between generation of disturbances by velocity shear and by convective instability. At this symposium, as is customary, the papers are arranged in convenient groupings, because it is easier to arrange them according to some kind of simple rules. But as all the other sessions at this symposium, we shall deal with the physics of air-sea interaction, and the papers are distinguished by the motivation of the authors rather than by any fundamental physics or dynamical process.

One recent theme in this session will no doubt be the intermittency of fluxes, however one choses to define intermittency or marshall the evidence that expresses the degree of intermittency. Among the manifestations of intermittency is the tendency for the probability distribution of certain variables connected with fluxes or interactions to have a non-normal probability distribution. The most common non-normal distribution we will hear mentioned is the log-normal distribution. Since the authors will not take the time to remind us of the details of the derivation of the log-normal distribution, I shall save them the trouble and do so. This will be a better use of the few minutes of my time than further pious and inspirational remarks

In order to remember where the lognormal distribution comes from, we first need to remind ourselves of the central limit theorem: A variable x, formed as the sum of many, independent, identically distributed variables Δx, will tend to have a normal distribution.

$$x = \sum_{j=1}^{N} \Delta x_j \text{ will tend to have the density}$$

$$f_x(x) = \frac{1}{\sigma\sqrt{2\pi}}\, e^{-\frac{x^2}{2\sigma^2}}$$

for large values of N.

Now, to remember the derivation of the lognormal distribution it helps to think of the following story which contains some elements of truth, but which has been improved from reality because of its didactic effectiveness. In a certain city, during the industrial revolution, which was also the age of gin consumption in England, it was discovered that the weight of the livers of cadavers was not normally, but more lognormally, distributed. This is of course logical in retrospect, as is so much of science. When a person is alive, if he has a diseased liver, the damage from drinking will be worse than if he has a normal liver. Let y be the weight of a normal liver, Δy the increment in liver size due to one more drink Δx, and let them be related as follows:

$$\Delta y = (y - a)\, \Delta x \tag{1}$$

This shows that a normal liver, of weight a, will not grow from another drink, but diseased livers are damaged proportionally to their abnormality. Rearrange, and sum to get:

$$\sum \frac{\Delta y}{(y-a)} = \sum \Delta x \tag{2}$$

The summation can be approximated by an integration:

$$\int \frac{dy}{(y-a)} = \int dx = \ln(y-a) = x \tag{3}$$

Since we assume that drinks are statistically independent, with the same norm and distribution functions, we may conclude from the central limit theorem that the total gin consumption for a person during his lifetime is normally distributed, that is, normally distributed in x-space. Therefore the logarithm of liver size is normally distributed in x (drink) -space. In liver size space (y-space) the liver size distribution will therefore be:

$$f_y(y) = \frac{1}{\sigma\sqrt{2\pi}} \frac{e^{-(\ln(x-a))^2/(2\sigma^2)}}{(y-a)} \tag{4}$$

This is the lognormal distribution function. It is generated by a process where the response to excitation is proportional to the sum of previous incremental responses. In that way, it has some similarities to an instability phenomenon, but there is a "normal" size that is a threshold value that evokes no response. Since this nonlinear process generates the lognormal distribution, it is often taken to indicate the presence of a process that is non-linear and where the response depends upon a departure from "normality".

But the lognormal distribution is of course just one of a possible large population of possible distributions caused by nonlinear response. Is is difficult to distinguish the lognormal distribution from neighboring distributions. The real difference may often be found in the values of the distribution function far away from the mean, where there will be very few observations anyway. Furthermore, if one observes lognormal distributions when averaging over each of successive time intervals, the distribution function of a sum of many lognormally distributed variables will tend to be normal under suitably restricted conditions. The occurrence of a lognormal distribution in a variable is a useful indicator of nonlinearity. If a conditionally sampled variable is lognormal, while the remainder of the variable is not, one has a hint, at least, as to where the important nonlinearity may occur. But that is really the main use of the non-normal distribution functions, as indicators of processes that may be important, hints as to where the nonlinearity may lie, and the possible separation of observations into "events" and debris of events in the past.

I am sure that possibly erroneous impressions and interpretations that I have communicated here will be rectified by the papers to be delivered in this session, and that the important considerations of selfsimilarity and similarity that I have not touched upon will be illuminated by the following speakers.

ON THE STRUCTURE OF CONVECTIVE ELEMENTS IN THE AIR NEAR THE SURFACE

J. A. Businger and S. J. S. Khalsa

Department of Atmospheric Sciences
University of Washington
Seattle, Washington U.S.A.

ABSTRACT

A review is given of the observational characteristics of convective plumes in the atmospheric surface layer. It is shown that any attempt to model these plumes must include the pressure distribution.

The presence of these convective plumes in the unstable surface layer generates a bimodal distribution in the dissipation ε as derived from the inertial subrange of the horizontal velocity. An objective method has been developed to determine when a sensor is in or out a plume on the basis of the level of dissipation. This method has been further substantiated by a probability analysis which shows that $\ell n\ \varepsilon$ is bimodally distributed. The two normal distributions are directly related to the "on" and "off" states as defined by the level of dissipation.

1. GENERAL DESCRIPTION OF STRUCTURE

It is difficult from the rather extensive literature concerning convective elements to arrive at a consistent picture of the structure of so-called plumes in the surface layer. The picture that will be presented here is biased in the direction of the authors preconceived notions and so are the references. However, it is felt that this picture is consistent with a fair number of direct observations.

A starting point is provided by the case study of Kaimal and Businger (1970). Fig. 1 provides the observations at two heights

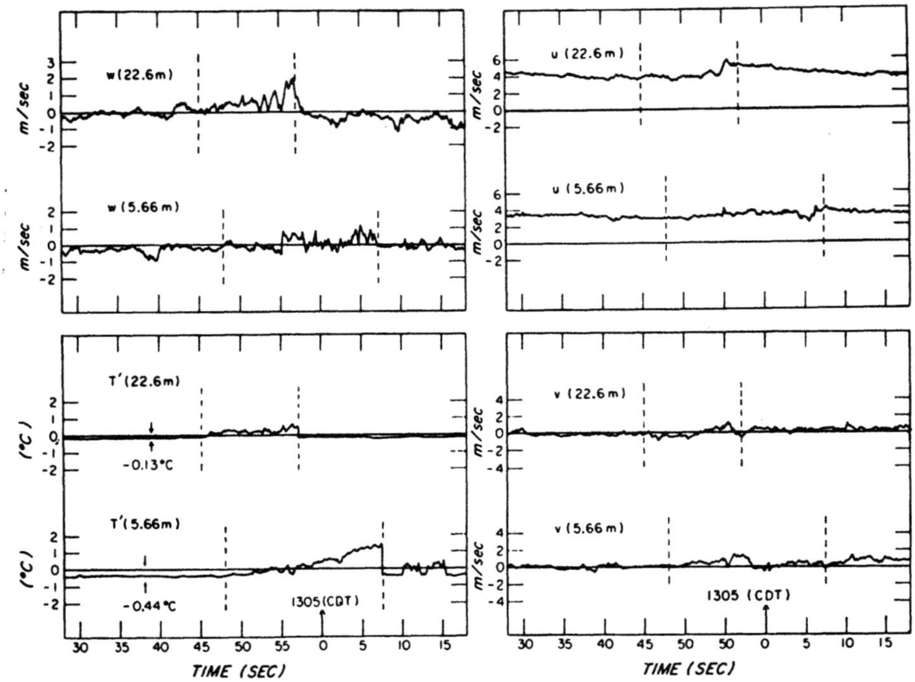

Fig. 1. Traces of u, v, and T during passage of the convective plume. (after Kaimal and Businger, 1970)

from which the case has been developed. The sharp drop in temperature (front) at the upwind side of the plume-like structure has been used as the reference point for the two levels. Fig. 2 indicates the schematic picture that can be deduced from the observations in Fig. 1. The main feature is that the slope of the front is apparently maintained and that the front itself represents the trajectory of the warm parcels of air that are swept up from the surface. The structure can move along over a horizontal surface and maintain its identity for some time. What are the dynamics that allow for this quasi steady state condition? We are not in a position to answer this question yet, but we can identify the following elements that are characteristic for the plume:

(1) The amount of heat transported upward at the lower level is about the same as at the upper level. The temperature excursion at the lower level is much greater than at the upper level. The vertical velocity on the other hand is considerably larger at the upper level.

(2) The amount of mass transported upward increases from the lower level to the upper level. Air is entrained from the sides

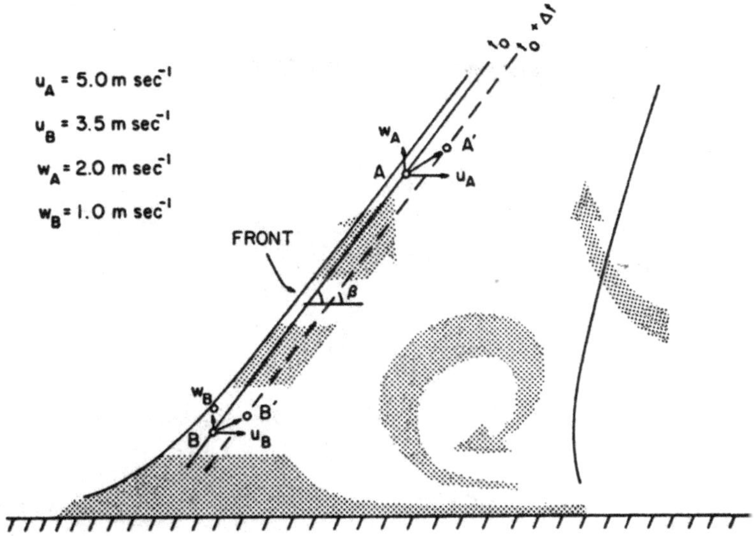

Fig. 2. Two-dimensional model of the convective plume. (after Kaimal and Businger, 1970)

into the plume at a substantial rate (at 22.6 m the mass flux is about 2.5 times the transport at 5.6 m).

(3) In Kaimal and Businger's paper also an argument is given that the vertical momentum inside the plume can only be balanced if there is a pressure gradient in the vertical to accelerate the air upward.

(4) The horizontal momentum flux, which on the average is downward, shows a marked upward contribution at the upper level close to the front (Fig. 3).

(5) The translation speed, u_t, of the plume can be compared to the average windspeed at various levels. It turns out that the plume speed is equal to the average windspeed at about 0.5 - 1 m for the case study presented in Figs. 1 and 2. Kaimal (1974) showed that the translation speed of the plume is independent of stability (Fig. 4). He found that $u_t/u_* \simeq 16$, where u_* is the friction velocity. It is of interest to find out whether this constant is the same for different roughnesses.

(6) The structure of a convective plume has some inertia because when the plume travels from a warm surface to an irrigated surface, the cool air from the irrigated area is sucked into the plume as evidenced by the appearance of inverted temperature signatures, see Fig. 5 (Businger and Frisch, 1972).

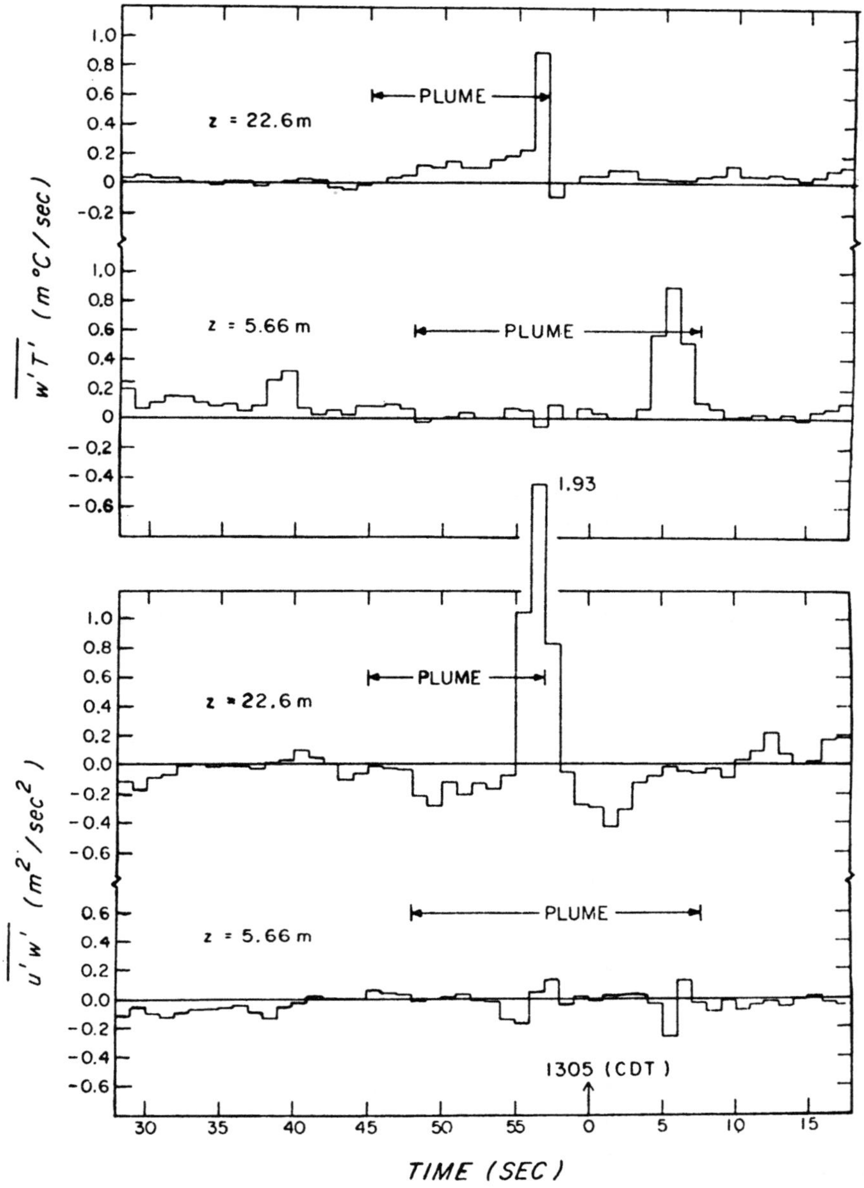

Fig. 3. Variation of heat and momentum fluxes (1-sec averages) for the period shown in Fig. 1. (after Kaimal and Businger, 1970)

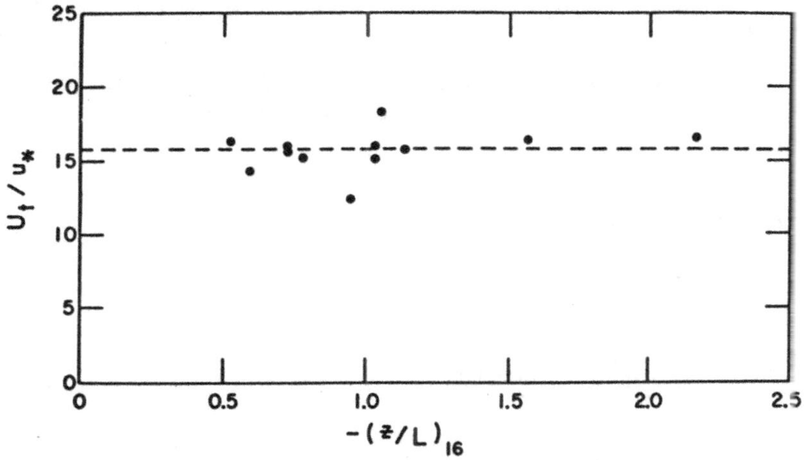

Fig. 4. Ratio of plume translation velocity to friction velocity, plotted as a function of z/L at 16 m. (after Kaimal, 1974)

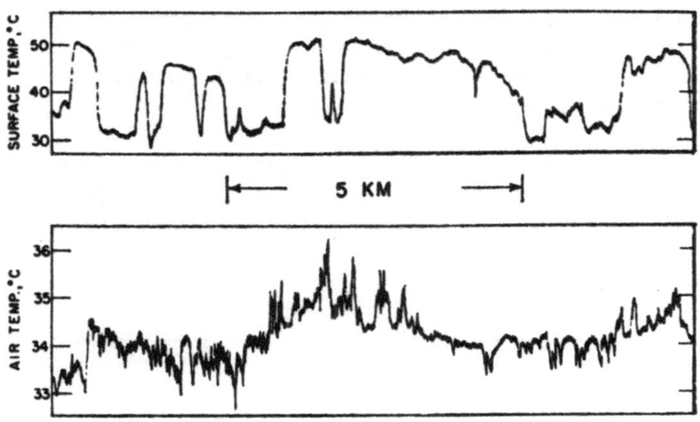

Fig. 5. An example of inverted temperature structure over a cool surface at 30 meters height. (after Businger and Frisch, 1972)

In order to describe these characteristics in a satisfactory way we need to develop a dynamical model of the plume. Pressure forces are important and form an integral part of the system. At this time all we can do is provide some simple qualitative arguments concerning the pressure forces and their consequences.

The plume structure is filled with relatively warm air. This column is lighter than the environment, therefore the hydrostatic pressure under the plume p_p will be slightly lower than outside the plume, p_e, and we may assume that

$$p_a - p_p = \Delta p \propto - g\rho' z_p \tag{1}$$

where ρ' is the density inside the plume, z_p is the height of the plume and g is the acceleration due to gravity. Furthermore we assume that

$$- \frac{\partial p'}{\partial z} \propto \frac{\Delta p}{z_p} = - g\rho'$$

or

$$- \overline{\frac{w'}{\bar{\rho}} \frac{\partial p'}{\partial z}} \propto - \frac{g}{\bar{\rho}} \overline{\rho' w'} = \frac{g}{\bar{\theta}} \overline{w'\theta'} \tag{2}$$

where w is the vertical component of velocity and θ is the potential temperature. The left hand side of (2) is the pressure transport term in the turbulent kinetic energy equation and the right hand side is the buoyancy production term. This qualitative argument suggests that the pressure transport term is proportional to the buoyancy term for well developed plume structures, which is more or less in agreement with the findings of Wyngaard and Coté (1971), see Fig. 6, where the imbalance term corresponds to the left hand side of (2).

Furthermore because the plume is bent in the direction of the wind, there will be a horizontal pressure gradient from the front to the center of the plume. This provides the parcels of air which

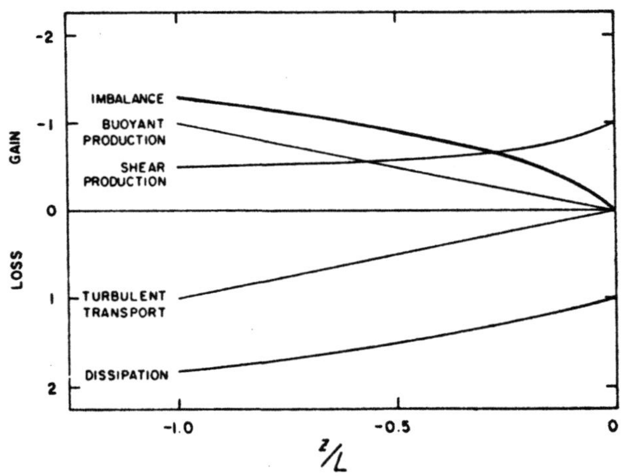

Fig. 6. Dimensionless energy budget under unstable conditions. (after Wyngaard and Coté, 1971)

rise along the front with the necessary acceleration in the direction of the mean wind, as well as an upward contribution to the momentum flux as soon as the component in the direction of the mean wind is larger than the mean wind.

We feel that these elements should be used in constructing a dynamic plume model. An important part of the model will be the interaction between plumes. One plume collects the heat that has been released near the surface over an area which is larger than the plume cross section. Therefore over a given area we must have a limited plume population. In this regard it is useful to gain some information concerning the statistics of plume like structures.

2. STATISTICAL CHARACTERISTICS OF PLUMES OVER THE TROPICAL OCEAN

During the GARP Atlantic Tropical Experiment (GATE) we collected wind, temperature, and humidity data in the atmospheric surface layer. The instrumentation was mounted on the boom of the USCGC Dallas. Central to our experimental objectives was the determination of the dissipation rate of turbulent kinetic energy, ε. An estimate of this quantity was obtained with a hot film sensor which provided information about the inertial subrange. The technique has been described in detail by Khalsa and Businger (1977).

The tropical ocean is usually slightly unstable and convective plumes form over the water surface. The vapor flux usually contributes significantly to the buoyancy. The temperature and humidity traces reflect the convective elements that pass by but are not necessarily as distinct and clear as the temperature trace in Fig. 1 indicates. There is a difficulty in determining when the data was taken inside the plume and when outside. By studying the various time series carefully it was concluded that ε was the most promising signal to indicate whether or not the sensor was inside a plume. Inside a plume the turbulence is more intense than outside, therefore ε is larger inside than outside. The next step was to find the level of ε above which we are inside and below which we are outside the plume. This was accomplished by Khalsa and Businger (1977) by generating a binary state intermittency function, I(t), which has the value 1 during the "on" state and zero during the "off" state. This intermittency function I(t) was then used to sample all the signals on the data tape.

The trigger level for the "on" state was based on maximizing the following parameter

$$EX = \frac{(\overline{\varepsilon}_{on} - \overline{\varepsilon}_{run}) \cdot t_{on}}{\overline{\varepsilon}_{run} t_{run}} \qquad (3)$$

where $\overline{\varepsilon}_{on}$ is the average of the dissipation over all "on" states, $\overline{\varepsilon}_{run}$ is the average over the whole run, t_{on} is the time spent in the "on" state and t_{run} is the total period of the run. Therefore EX is the excess of the energy dissipated over the average for the run. This function is maximum when $\overline{\varepsilon}_{run}$ is the trigger level, i.e., $\varepsilon_{on} \geq \overline{\varepsilon}$ and $\varepsilon_{off} < \overline{\varepsilon}$.

Fig. 7 shows the bahavior of I(t) in relation to some actual observations. Some characteristics of the "on" versus "off" state are indicated in Table I, where L is the Obukhov length; $\Delta\varepsilon = \overline{\varepsilon}_{on} - \overline{\varepsilon}_{off}$; $\Delta u = \overline{u}_{on} - \overline{u}_{off}$; $\Delta T_v = \overline{T}_{v\,on} - \overline{T}_{v\,off}$; γ is the intermittency factor given by Townsend (1948)

$$\gamma = \lim_{\tau\to\infty} \frac{1}{\tau} \int_t^{t+\tau} I(t')dt' \tag{4}$$

Table I: Intermittency factor, event period, percent energy dissipated in "on" state and normalized differences of dissipation, wind speed and virtual tempeature between "on" and "off" states. Means of three stability classes are shown.

z/L	γ	T_B	$(\overline{\varepsilon_{on} t_{on}}/\overline{\varepsilon} t_{run})100\%$	$\Delta\varepsilon/\overline{\varepsilon}$	$\Delta U/\sigma_v$	$\Delta T_v/\sigma_{Tv}$
0 to -.1	.462	48.6	76.4	1.25	-.771	1.42
-.1 to -.3	.439	46.8	78.6	1.43	-.436	1.78
< -.3	.412	54.5	81.2	1.65	-.012	1.63

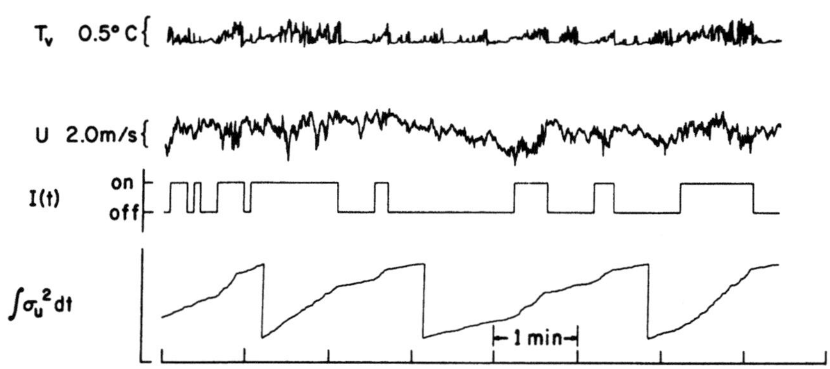

Fig. 7. Traces of virtual temperature, wind speed, the intermittency function and the integrated variance in the 10 - 100 Hz frequency range. The run is 42A. (after Khalsa and Businger, 1977)

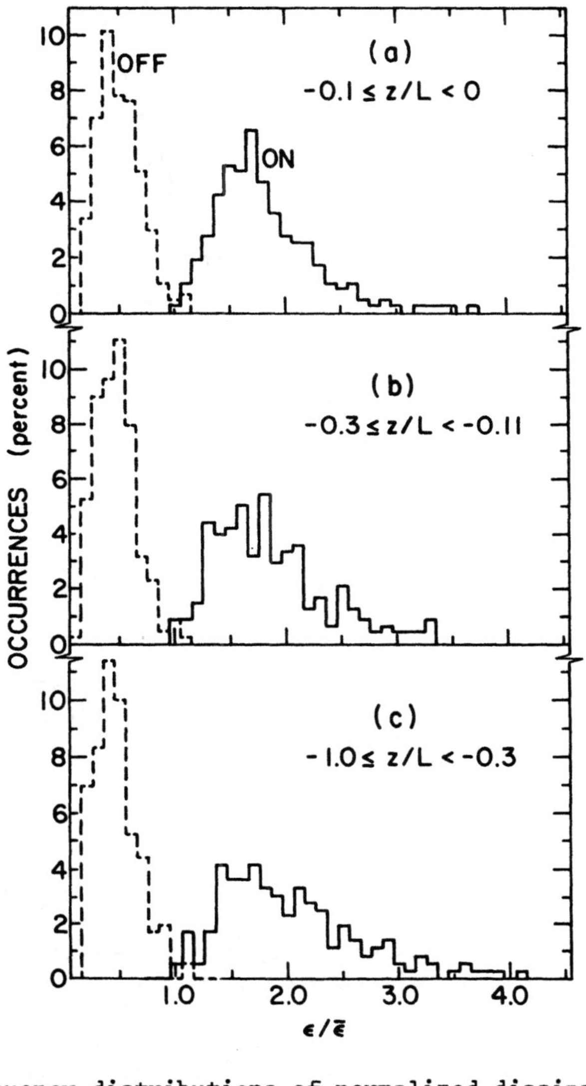

Fig. 8. Frequency distributions of normalized dissipation rate for 3 stability intervals. An occurrence is an average over a single "on" or "off" event. Each of the three figures is a composite of 3 runs. Each occurrence is normalized by the mean of the variable for that run. (after Khalsa and Businger, 1977)

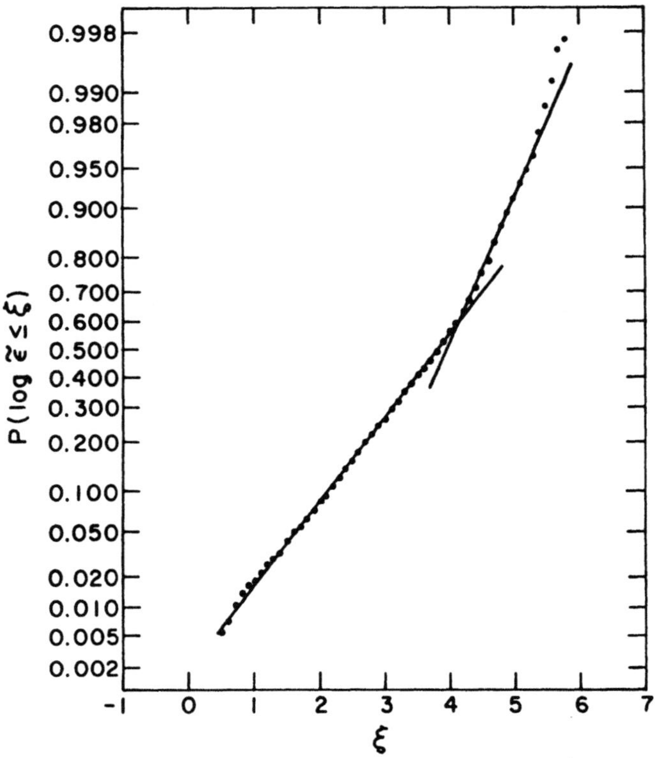

Fig. 9. Cumulative probability function for the natural logarithm of dissipation, $\xi = \log\tilde{\varepsilon}$. The lognormal distribution of $\tilde{\varepsilon}$ would produce a straight line. The run is 44B.

and T_B is the average duration of the "off" period (in seconds). From the table we see that with decreasing stability $\Delta\varepsilon/\overline{\varepsilon}$ increases; $\Delta u/u$ decreases, $\Delta T_v/\overline{T}_v$ does not change; γ decreases and T_B increases. In order to evaluate the distance between plumes T_B should be multiplied by $\overline{u}$. The result is that this product decreases with decreasing stability. This indicates that the plumes are closer together with decreasing stability. If we multiply the duration of the "on" events by the dissipation during those events and divide that by the total energy dissipated, we obtain the fraction of the total energy dissipated during the "on" state. The fourth column in Table I represents this quantity, i.e., $(\overline{\varepsilon_{on}t_{on}}/\overline{\varepsilon}t_{run})100\%$. This percentage increases with decreasing stability and becomes a rather large fraction of the total dissipation.

In order to examine the differences between states more carefully, frequency distributions were generated of ε, u, and T_v. Each

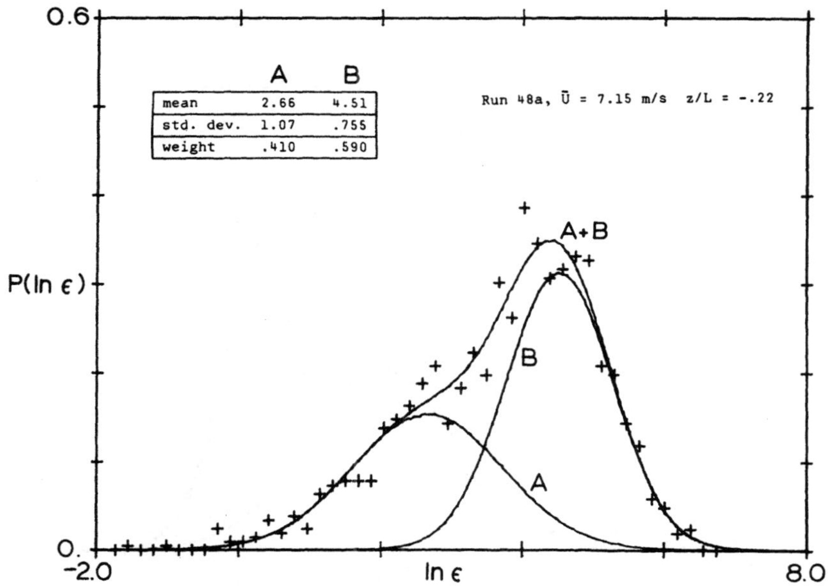

Fig. 10. Empirical probability distribution of the natural log of 2.5 second averages of dissipation along with best-fit combination of two Gaussian distributions.

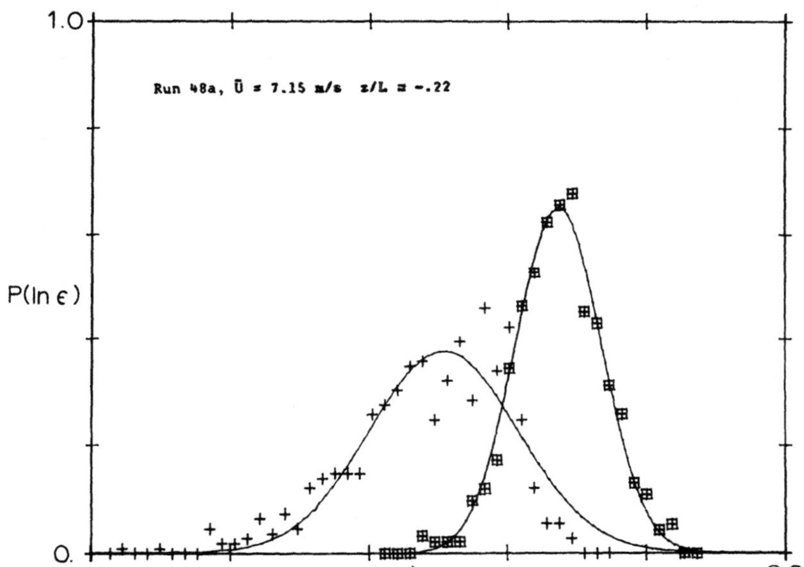

Fig. 11. Empirical probability distributions of the natural log of 2.5 second averages of dissipation rate for states labeled "on" (⊞) and "off" (+). Best-fit Gaussians are shown.

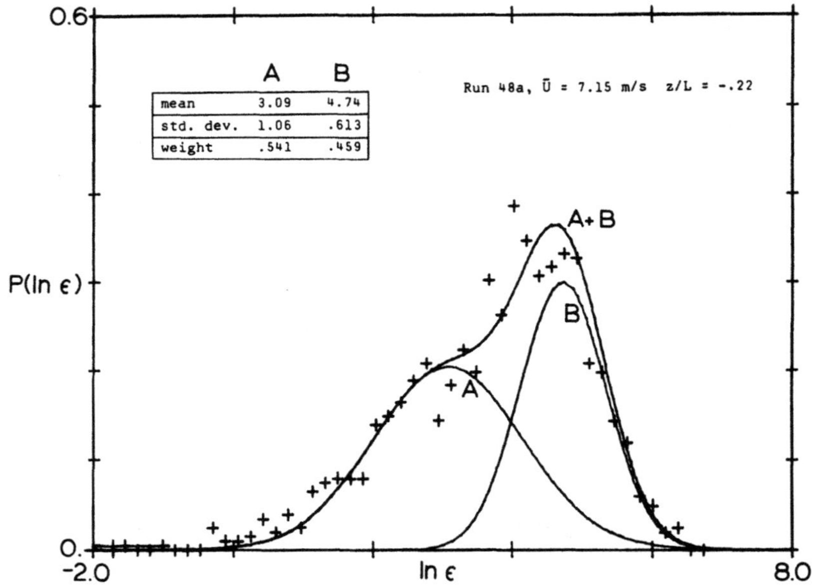

Fig. 12. Empirical probability distribution of the natural log of 2.5 second averages of dissipation rate along with best-fit Gaussians to averages occuring during states labeled "on" (B) and "off" (A). Sum of the two Gaussians, which are weighted using the intermittency factor, is shown in black.

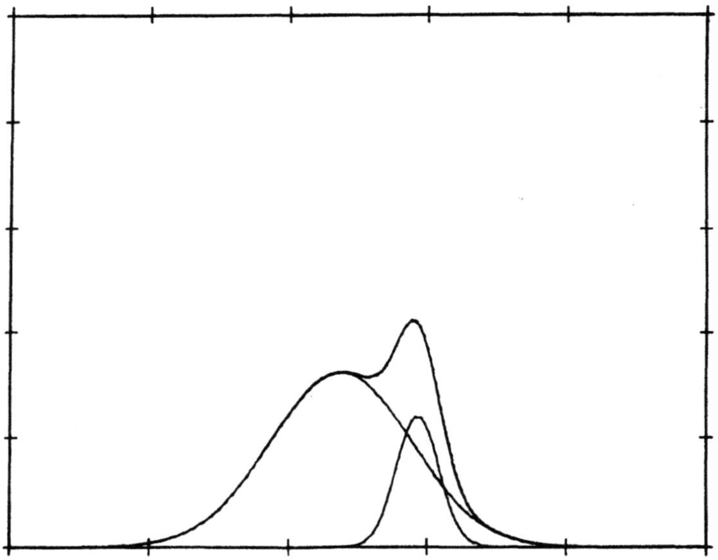

Fig. 13. A best-fit bi-modal Gaussian distribution has been constructed for the "off" distribution of run 48a. The fit is significantly better, than the simple Gaussian, see Fig. 11.

event average was normalized by the mean for the entire run. Fig. 8 contains the results for the dissipation rate, ε. The separation is clear because the states are distinguished on the basis of $\bar{\varepsilon}$. For more detail, see Khalsa and Businger (1977). The figure suggests that ε is bimodally distributed.

To explore this further we calculated the cumulative probability distributions of ε and plotted that versus ℓn ε on probability paper. An example is given in Fig. 9, which presents further evidence of a bimodal distribution. It also suggests that the two distributions may be related to the "off" and "on" states as defined above.

The next step was to analyze the measured probability distribution of ℓn ε into two independent normal distributions. A best fit routine was used and the result shows that two normal distributions do describe the probability distribution quite well. Fig. 10 presents an example of this analysis. However, these distributions do not necessarily correspond to the two states. In order to investigate this the probability distributions for each state were analyzed separately. The result confirms the notion that the bimodal distribution is related to the "on" and "off" states.

Fig. 11 shows the probability distributions of the "on" and "off" states. Fig. 12 shows the same probability distribution as Fig. 10 but now made up of the two normal distributions related to weighted "on" and "off" states. It turns out that by using a χ^2 test that the "on" state normal distribution is always significant but the "off" state is often somewhat skewed and can be made to fit better with two normal distributions, see Fig. 13.

[This research was supported by the National Science Foundation through grants ATM76-00855 and ATM77-02701.]

REFERENCES

Businger, J. A. and A. S. Frisch, 1972: Cold plumes. J. Geophys. Res., 77, 32703271
Kaimal, J. C., 1974: Translation speed of convective plumes in the atmospheric surface layer. Quart. J. R. Met. Soc., 100, 46-52
Kaimal, J. C. and J. A. Businger, 1970: Case studies of a convective plume and a dust devil. J. Appl. Met., 9, 612-620.
Khalsa, S. J. S. and J. A. Businger, 1977: The drag coefficient as determined by the dissipation method and its relation to intermittent convection in the surface layer. (submitted for publication in Boundary-Layer Meteorology).
Wyngaard, J. C. and O. R. Coté, 1971: The budgets of turbulent kinetic energy and temperature variance in the atmospheric surface layer. J. Atmos. Sci., 28, 190-201.

DISCUSSION

C. A. Friehe:

It is interesting to note that temperature ramp structures have been observed in many different turbulent shear flows, their orientation depending upon the mean vorticity and temperature gradient (Mestayer et al 1976; Gibson et al 1977 (both in Phys. Fluids)). These studies suggest that buoyancy may be the cause of the ramp structure.

J. A. Businger:

I agree that convective plume like structures are not the only structures that display the ramp like time series. Kelvin - Helmholz instabilities also appear to generate such time series. It is the vertical coherence of these time series that suggest that in the convective case we are dealing with plume like structures.

O. Phillips:

Can you reliably parametrize the scales and intensities of these plumes in terms of overall parameters? Presumably they occur in the oceanic mixed layer under cooling or evaporative conditions - how big and how intense will they be?

J. A. Businger:

Kaimal's experiment suggest that the surface roughness

length is an important scaling parameter for the plumes. This experiment should be carried out over areas with different roughness lengths in order to verify this assumption. The intensity of the plumes increases with increasing instability but the size does not seem to change much.

K. Hasselmann:

The question of the modulation of the frequency and structure of plumes by larger, mesoscale structures of the atmospheric circulation is clearly extremely important as possible explanation of the observed variability of air-sea fluxes - and also the large variability of wave spectral parameters under apparently ideal homogenous generation conditions. Did you investigate this modulation?

J. A. Businger:

The modulation of the fluxes by mesoscale systems interests us very much. We do not have enough information about mesoscale systems surrounding our GATE observations. However, we hope to obtain information on the mesoscale convergence in the area around the USGG Dallas and correlate that to the intermittency factor that we used in our runs.

F. W. Dobson:

It would be useful, and I believe now it is possible to make measurements of pressure fluctuations in the convective plumes.

J. A. Businger:

I agree that it would be very useful to have such observations. A sensor with a frequency response of 1 to 2 Hz would already give valuable information.

MARINE TURBULENT BOUNDARY LAYER FLUXES OF WATER VAPOR, SENSIBLE HEAT AND MOMENTUM DURING GATE

B. R. Bean and R. F. Reinking

Boundary Layer Dynamics Group, Weather Modification Program Office, ERL/NOAA, U.S. Department of Commerce, Boulder, Colorado, U.S.A.

INTRODUCTION

One of the major objectives of the GARP Atlantic Tropical Experiment (GATE)* was to estimate effects of small-scale tropical weather systems on synoptic-scale circulations (Rasmussen and Murray, 1976). The scales of motion in the 10-5000 m band were measured by aircraft. They complement the coverage of time and space scales by other measurements from buoys, ships and satellite. The gust probe measurements of the means and extremes of temperature, moisture and eddy fluxes in the boundary layer are described. (The entire GATE Program is described by the International and Scientific Management Group for GATE, 1974). A GATE Workshop on data intercomparisons and analysis was recently held in Boulder, Colorado (July-August 1977). One conclusion reached from aircraft intercomparisons was that atmospheric variability was as large as that between aircraft for scales under about 100 m while agreement between aircraft for scales greater than a few hundred meters were well determined by each aircraft in the intercomparison flights. This then means that the results presented below, although based on data most familiar to the authors, would be expected to be the same as those obtained for any one of the aircraft or from a composite of all aircraft data.

THE AIRBORNE GUST PROBE SYSTEM

Given measurements of sufficient duration and/or spatial coverage such that mean vertical motion $\overline{w} \simeq 0$, the mean flux of property

*The <u>G</u>ARP <u>A</u>tlantic <u>T</u>ropical <u>E</u>xperiment (GATE) was the first large-scale experiment of the <u>G</u>lobal <u>A</u>tmospheric <u>R</u>esearch <u>P</u>rogram (GARP).

or constituent ϕ is specified by the average product of the eddy components of ϕ and w; thus, $\overline{F} \simeq \overline{\phi' w'}$. The NOAA airborne gust probe system measures the eddy fluxes of water vapor (or latent heat), $\overline{E} \simeq \overline{\rho_v' w'}$ (or L $\overline{\rho_v' w'}$); sensible heat, $\overline{H} \simeq \rho\ c_p\ \overline{T' w'}$; and momentum, $\overline{M} \simeq -\ \rho\ \overline{u' w'}$. The necessary sensors were mounted on a boom about 8 feet forward of the nose of the aircraft. The system of measurements made to produce the turbulence parameters is shown in Figure 1. The gust probe assembly itself measures the fluctuations (u', v' and w') in the orthogonal components of the wind. The α and β vanes measure the forces of the vertical and cross-axis winds. The force is determined from the distortion of a Wheatstone bridge of sensitive strain gauges set into the rectangular aluminum structure of the vane. The pitot tube differential pressure determines the along-axis wind. The components are corrected for motions of the aircraft and the boom by accelerometers in the probe, and by an inertial navigation system (INS) at the center of gravity of the aircraft. The resolution (rss error) of the gust probe measurements is about ±0.06 m/s for w' and ±0.02 m/s for u' and v'. Air temperature is measured with a small double bead thermister. A correction is made for in-flight dynamic heating. The resolution of T' is about ±0.05°C. Water vapor density is determined with a microwave hygrometer (refractometer) that has very high frequency response. This device provides a measure of rapid fluctuations in atmospheric radio

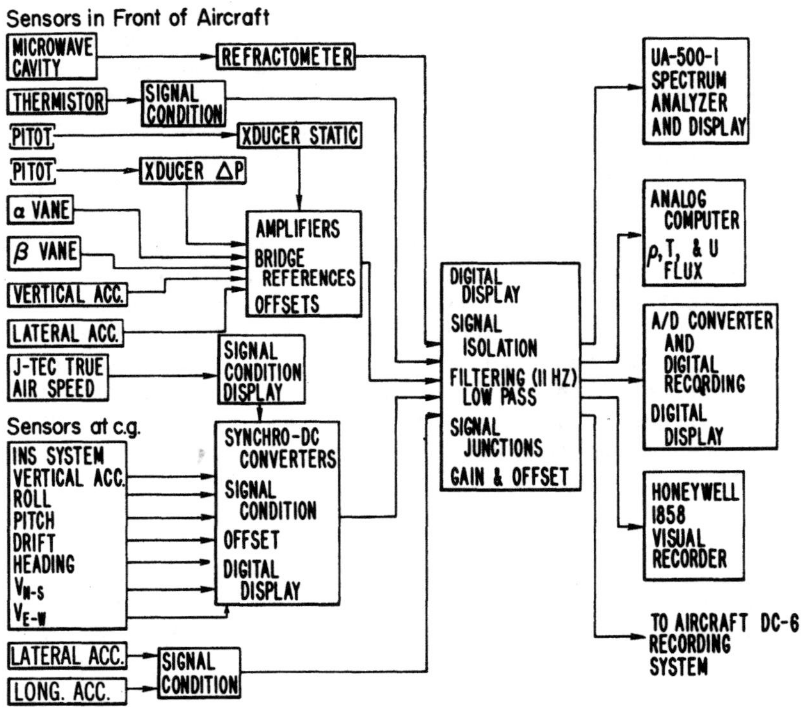

Figure 1. Configuration of the NOAA DC-6 gust probe during GATE.

refractive index (in "N" units). Water vapor density is then calculated from the relationship,

$$\rho_V = 5.805 \times 10^{-4}\ NT - 4.510 \times 10^{-2}\ P.$$

The fluctuations of N, T and P (pitot tube static pressure) can each be measured to a frequency of at least 10 Hz. The resolution, ρ_V', is about ±0.02 g m^{-3}, with an absolute accuracy of about ±0.2 g m^{-3}.

DATA RECORDING AND ANALYSIS PROCEDURES

In all, 14 parameters of atmospheric and aircraft motion and atmospheric properties are measured, electronically filtered to minimize high-frequency noise, and then converted to digital form and recorded on magnetic tape at a rate of 80 samples per second.

The flow chart in Figure 2 illustrates the subsequent procedure for computation and analysis of the five basic variables w', u', v', ρ_V', and T'. The digitized data are again filtered in the computer with a high-pass filter (a linear detrend) and a low-pass filter consisting of a two-stage running mean with the 3 db point at 6 Hz and with a terminal frequency of 15 Hz. Again, identical low-pass filters are used on each parameter. Products of the analyses include time series plots, flight-leg mean values, and spectral analyses of the basic parameters and the corresponding fluxes. Details of the instrument system and the data analyses are presented by Bean et al. (1976). Some 57 hours of turbulence data were gathered on 439 individual flight legs consisting primarily of 3-6 minute durations with extremes of 1 and 41 minutes. Only the high quality data that have been identified by scrutinizing the data logs, raw data, time series plots and power spectra are used in the following analyses.

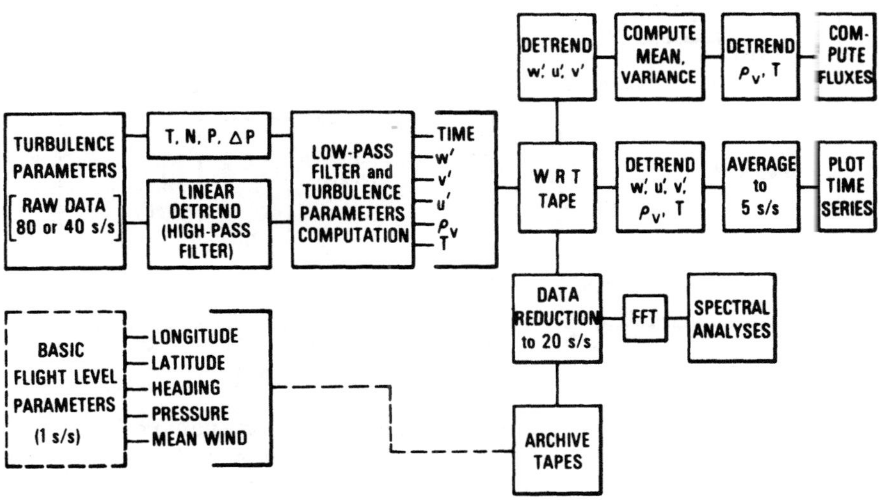

Figure 2. Computer processing procedure for gust probe data.

Phase I (22 June - 12 July 1974), Phase II (28 July - 17 August), and Phase III (30 August - 18 September) of GATE were selected to respectively represent, on a climatological basis, an initial period of relatively constant, fair weather, a period of transition in the weather, and a period of relatively disturbed weather.

AMBIENT TEMPERATURE AND MOISTURE

In Figures 3-5, the extreme values of temperatures, T, and absolute humidities, ρ_v, and the combined averages of the sample mean values, $\overline{T}$ and $\overline{\rho_v}$, are presented as a function of flight altitude for Phases I, II, and III. Samples with rain and with no rain are separated. The "rain" designation does not indicate continuous precipitation which contaminates the ρ_v and T data, but rather that some rain was encountered along the flight path while the sample was being taken. The data closely follow average profiles from radiosonde and tethered balloon data taken within the B-scale ship array (group communications, 1977 GATE Central Program Workshop, Boulder, Colorado).

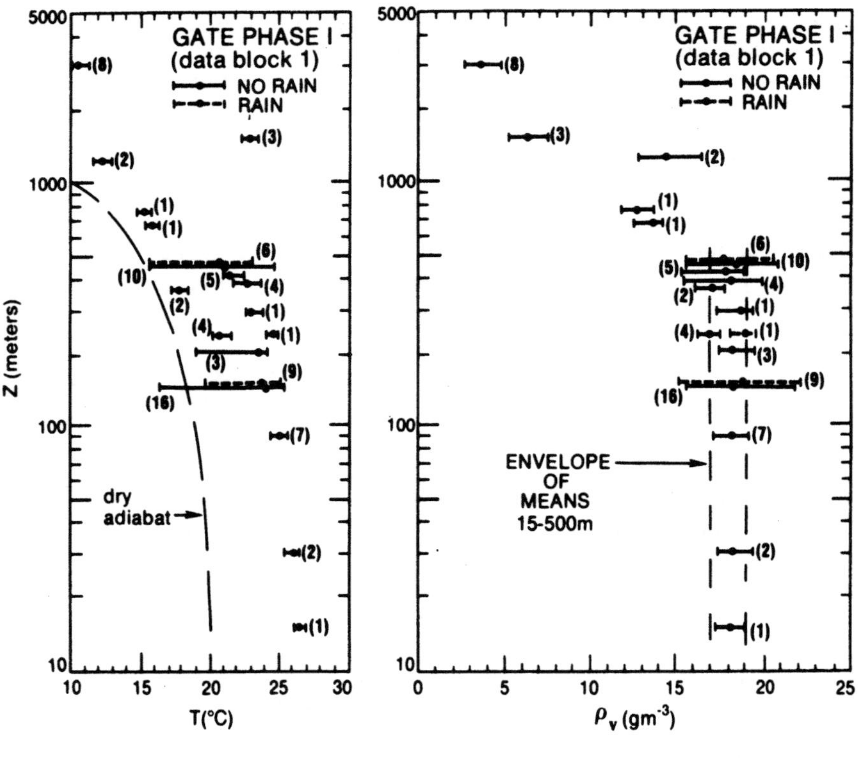

Figure 3. Mean and extreme values of temperature and absolute humidity vs. altitude, measured within or near the B-scale array during Phase I.

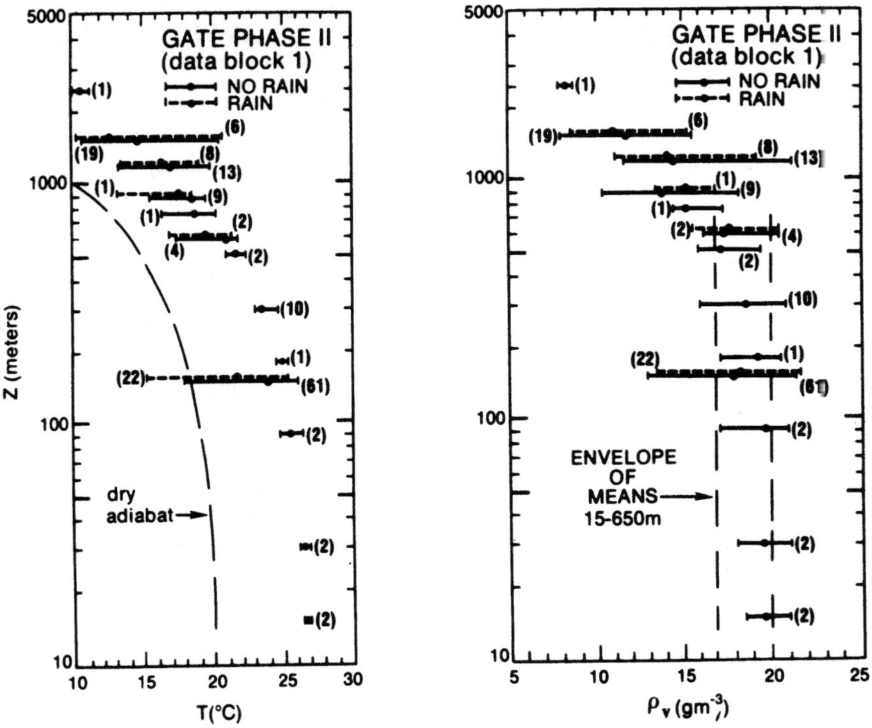

Figure 4. Same as Figure 3 but for Phase II

The mean absolute humidities are nearly uniform from 15 m up to 500 ± 100 m for all three phases (17-19 g m^{-3} during Phase I, 17-20 g m^{-3} during Phase II, and 18-20 g m^{-3} during Phase III). Above this level the humidities begin to drop off significantly with increasing altitude. The low-level uniformity is the result of mixing; the 500 m altitude approximates the mean depth of the well-mixed layer. Analyses of various types of data at the GATE Workshop indicated that this uniformly mixed layer is about 500-600 m deep during undisturbed conditions, is nearly eroded away during convective disturbances such as squall lines, and gradually recovers in depth during post-disturbance, or "wake" conditions when it averages about 300 m. Considerably more samples were taken during undisturbed and wake conditions than in disturbances.

The top of the mixed layer is not well defined by the measured temperatures (Figures 3-5). However, the average temperatures were 26-27°C at 15 m msl cooling to 21-22°C near the 500 m level. The temperature lapses of 5°C/500 m correspond to a well-mixed and therefore adiabatic boundary layer. For Phase II, the lapse of mean

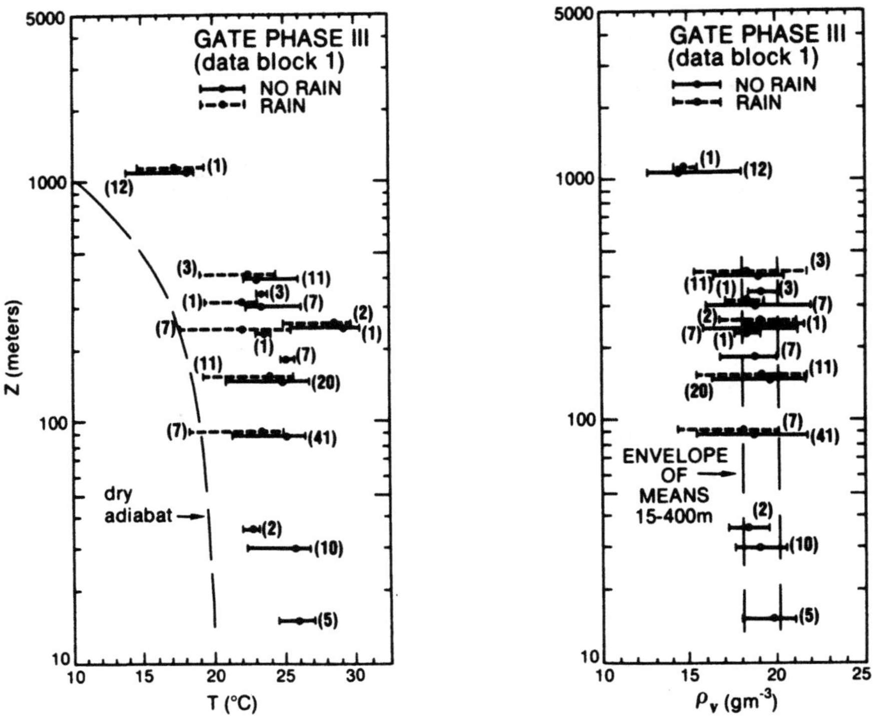

Figure 5. Same as Figure 3 but for Phase III

temperatures shifts toward a moist adiabatic lapse rate above 700 m. Workshop estimates of the lifting condensation level were of the order of 600-700 m for undisturbed conditions and 400-600 m for wake conditions; i.e., convective cloud base was generally found 100-200 m above the mixed layer.

Evaporation in the subcloud and lower cloud layers causes the air to cool by a few tenths to 2°C. Since the aircraft usually flew 17-34 km to collect one sample, the indicated cooling pertains to the difference between samples with and without occasional rain shafts.

Various types of data presented at the Workshop (e.g., radar echo, time series of rainfall, rainfall totals, the various mean temperature and humidity profiles,...) show that the weather was quite similar during the three phases of the experiment, except that squall lines were more frequent during the last phase. The maximum temperature range encountered at any altitude was between 6 and 9°C, depending on the phase, reflecting the moderating effect of the ocean. The measured absolute humidities were always greater than 13 g m^{-3} below

600 m; maximum ranges of 7-9 g m^{-3} about the means were measured. Assuming T = 25°C and p = 1000 mb, an 8 g m^{-3} increase in absolute humidity from 13 g m^{-3} is equivalent to a 4.8 gm^{-3} (or 0.4%) decrease in atmospheric density. Should such a gradient occur between a parcel and ambient air, this density difference would offset hydrostatic balance and induce an acceleration in vertical motion of about 4 cm s^{-2}. Thus, the effect on buoyancy of such differences in absolute humidity would not be inconsequential, although the actual effect would depend on time and space scales of the gradient. By comparison, a temperature increase of 6-9°C would cause a density decrease of about 3%.

The ITCZ region appears to be about 0.5°C warmer and 1.4 g m^{-3} more moist than the main GATE area.

EDDY FLUXES OF SENSIBLE AND LATENT HEAT

Mean fluxes, $\overline{\rho_v'w'}$ and $\overline{T'w'}$, have been computed for each flight leg. Statistics of these fluxes by phase are given in Table 1. The

Table 1. Latent and sensible heat fluxes for the three phases of gate.

	Z ≤ 305 M				Z ≤ 650 M				Z > 650 M			
PHASE	AVERAGE	MEDIAN	STANDARD DEVIATION	N	AVERAGE	MEDIAN	STANDARD DEVIATION	N	AVERAGE	MEDIAN	STANDARD DEVIATION	N
					$\overline{\rho_v'w'}$ (MW CM-2)							
					NO RAIN							
I	4.9	5.2	2.2	(19)	5.3	5.1	4.1	(32)	2.2	0.8	4.6	(8)
II	9.8	8.4	5.9	(33)	9.6	7.6	5.8	(35)	3.8	1.7	11.4	(30)
III	7.9	7.3	4.4	(65)	7.8	7.4	4.4	(75)	-0.1	5.0	11.5	(7)
					RAIN							
II	13.1	9.3	7.9	(7)	(NO SAMPLES, 305-650 M)				5.3	3.2	10.1	(7)
III	9.2	6.8	16.7	(19)	11.4	7.7	17.4	(21)	6.6	---	----	(1)
					$\overline{T'w'}$ (MW CM-2)							
					NO RAIN							
I	-0.4	0.0	1.5	(19)	-0.4	0.0	1.3	(32)	0.0	0.0	0.4	(8)
II	-0.2	-0.2	1.2	(33)	-0.2	-0.2	1.2	(35)	0.3	-0.2	4.5	(30)
III	0.4	0.2	1.3	(65)	0.3	0.2	1.2	(75)	-1.4	-0.4	4.0	(7)
					RAIN							
II	5.9	2.1	18.2	(7)	(NO SAMPLES, 305-650 M)				-3.3	-3.8	4.0	(7)
III	9.0	5.9	11.6	(19)	8.4	5.0	11.3	(21)	-8.8	----	---	(1)

average fluxes are reasonably constant with altitude throughout the well-mixed boundary layer in the absence of rain although the extreme values may vary considerably dependent upon weather conditions.

The sensible heat fluxes, with some notable exceptions, are small at all altitudes. Means values of $\overline{T'w'}$ at $z \leq 305$ m (1000 ft) are small (-0.4 to 0.4 mw cm^{-2}) and statistically not different than zero; the mean latent heat fluxes are more than an order of magnitude larger and all positive, at 4.9 to 9.8 mw cm^{-2}. Bowen ratios of the means are -0.09 (I), -0.02 (II) and +0.06 (III). Most of the radiative heating of the ocean goes to evaporation, very little to sensible heating of the air.

Evaporation rates equivalent to the latent heat fluxes are 0.17 (I), 0.34 (II), and 0.27 cm day^{-1} (III). These rates for the tropical eastern Atlantic are only 30-70% of the 0.5 cm day^{-1} average measured over the western Atlantic during the Barbados Oceanographic and Meteorological Experiment (BOMEX; Bean et al. 1972). Workshop studies showed that the total vertical flux of moist static energy, ρwh, where $h = c_pT + L\ \rho v \rho^{-1} + gz$, is about 5-10 mw cm^{-2} during undisturbed conditions and about 20 mw cm^{-2} in the wake of disturbances. Assuming that the mean vertical motion is small ($w \leq 10$ cm s^{-1}) during these conditions, this total flux is dominated by the eddy flux of latent heat. The mean latent heat fluxes match the estimated total fluxes for the undisturbed conditions; the extreme $\overline{\rho_v'w'}$ values are near those expected during wake conditions.

Means of fluxes measured within the ITCZ are given in Table 2. The $\overline{\rho_v'w'}$ no-rain ITCZ mean is considerably larger than the corresponding Phase I mean outside the ITCZ. The ITCZ mean sensible heat fluxes are near zero even for the rain samples; even the means $\pm 1\ \sigma$ are smaller than those outside the ITCZ.

Composite $\overline{\rho_v'w'}$ and $\overline{T'w'}$ co-spectra for the three phases are given in Figure 6 for $z \leq 305$ m. The aircraft flew at approximately 95 m s^{-1}, so the 0.02-10 Hz band is equivalent to wavelengths of

Table 2. Latent and sensible heat fluxes at 153 M within the intertropical convergence zone (Julian Day 185, July 3, 1974).

	$\overline{\rho_v'w'}$ (MW CM^{-2})			$\overline{T'w'}$ (MW CM^{-2})			
	AVERAGE	MEDIAN	STANDARD DEVIATION	AVERAGE	MEDIAN	STANDARD DEVIATION	N
NO RAIN	8.1	7.7	7.9	0.0	0.0	1.5	(11)
RAIN	7.4	8.0	2.0	-0.6	0.0	3.4	(6)

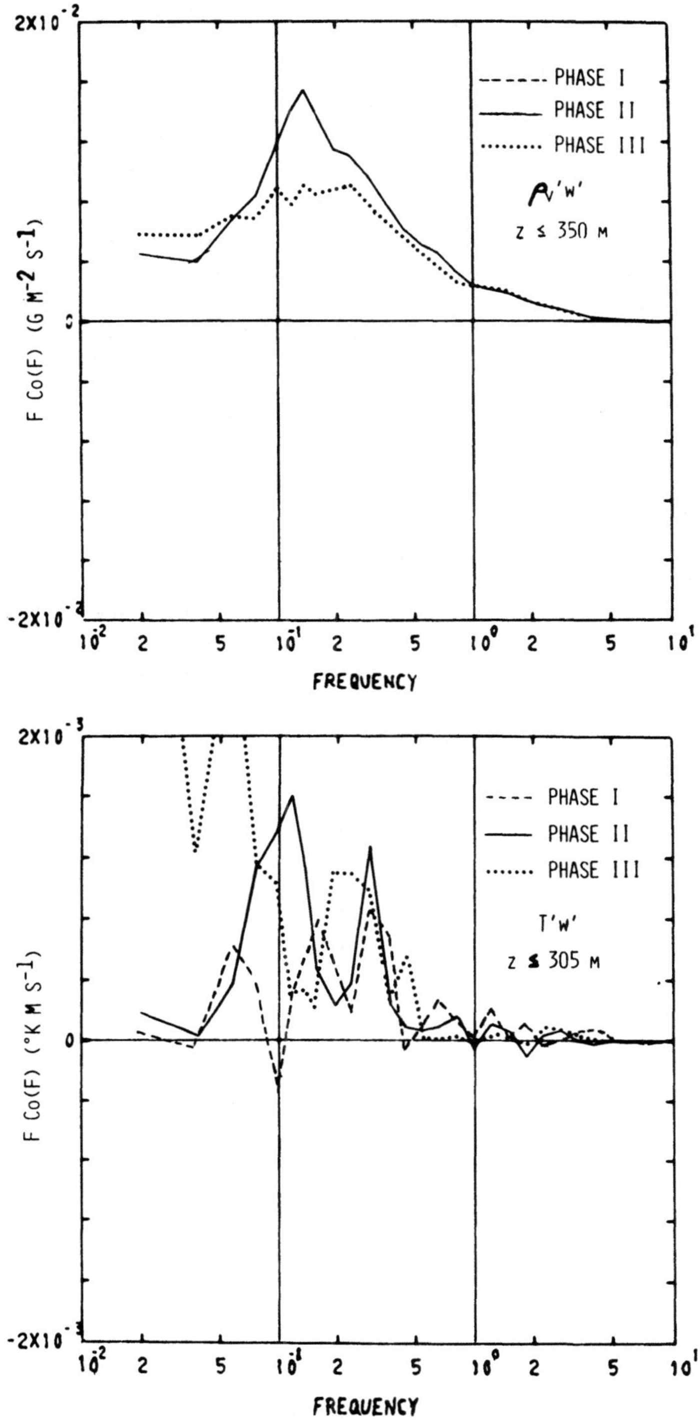

Figure 6. Composite co-spectra of latent heat fluxes (top) and sensible heat fluxes (bottom)

9.5 m - 4.75 km. The composite co-spectra of $\overline{\rho_v'w'}$ are positive at all frequencies between 0.02 and 7.5 Hz, as expected of eddies carrying water vapor upward from the sea surface and bringing dryer air down from aloft. The $\overline{\rho_v'w'}$ spectral peaks occur between 0.14 Hz and 0.24 Hz (693 and 406 m) for all three phases. The scales of motion corresponding to the peaks are greater than the altitudes of measurement, indicating that the eddies are horizontally elongated.

The co-spectra for sensible heat have both positive and negative values; the peaks occur at different frequencies in each phase and have powers about an order of magnitude less than the peaks for latent heat. Rain and no-rain samples were not separated in these composite spectra which may explain why they are positive at most frequencies.

EDDY FLUXES OF MOMENTUM

The vertical eddy fluxes of horizontal momentum, $\overline{u'w'}$, measured at the various altitudes during Phase III are shown in Figure 7. The values and the variability of sign are typical of all three phases. The mean stresses at any level within the mixed layer are small, i.e., usually less than ±1 dyne cm^{-2}; stresses measured nearest the sea surface were usually but not consistently negative. Case-by-case examination of the momentum fluxes and corresponding wind data appears to be necessary for understanding these measurements.

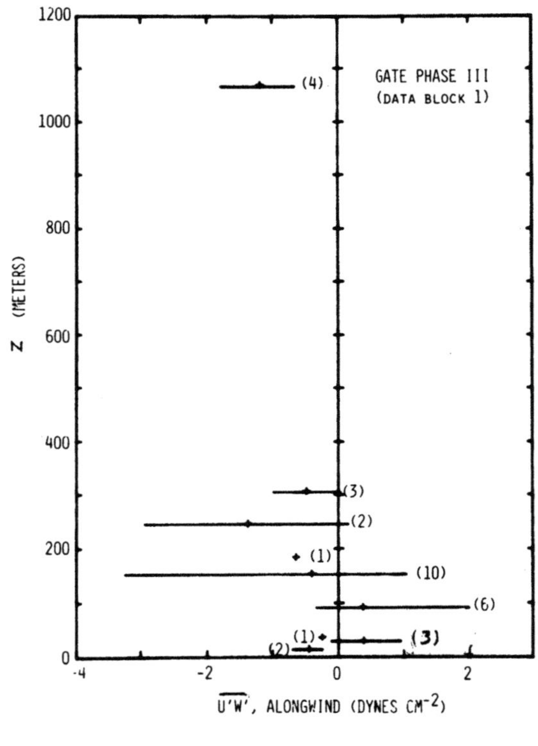

Figure 7. Mean and extreme values of momentum flux vs. altitude during Phase III. Number of samples is given in parentheses.

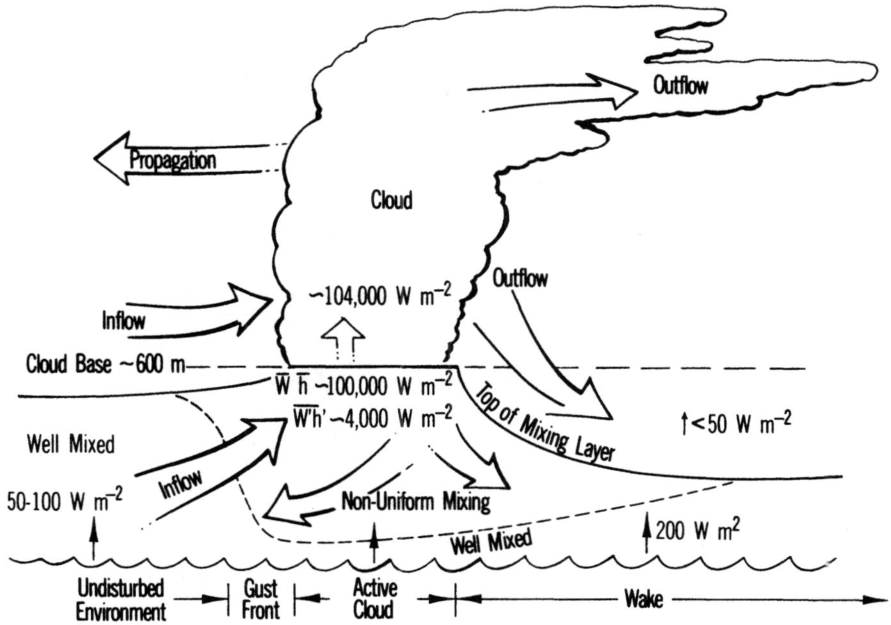

Figure 8. Conceptual model of energy fluxes in the tropical marine boundary layer.

A NOTE OF CAUTION

The GATE Workshop considered the energy balances over the entire array. When tethered balloon data on fluxes for disturbed conditions and aircraft-measured transport of total water within clouds were all considered, a rather startling picture of energy transport in the active areas emerged. Figure 8, compliments of Professor M. Garstang, shows the 50-100 W m^{-2} of the undisturbed environment as reported in Table 1. Consideration of the properties of the mean flow leads to a mean convergence of moist static energy, h, such that $\overline{w}\ \overline{h} \sim 100,000$ W m^{-2} with an eddy component in the active cloud area of $\overline{w'h'} \sim 4,000$ W m^{-2}. The region in the wake of the cloud area is expected to have flux values of about 200 W m^{-2}, in accord with some of the larger flight-leg means reported above. The values for the active region are clearly much greater than the reported 20-30 km flight-leg means, which average many up and down drafts. We will now look at the exchange process in more detail.

A set of time series of w', ρ_v' and their cross product are shown in Figure 9. Periods of the record were selected visually to represent times when the cross product $\overline{\rho_v' w'}$ display marked bursts of vapor transfer. The mean, reference vertical motion, $\overline{w_o}$, was defined as that obtained from the regions <u>outside</u> of the active regions. The value of h is calculated from the equation given earlier. Typical

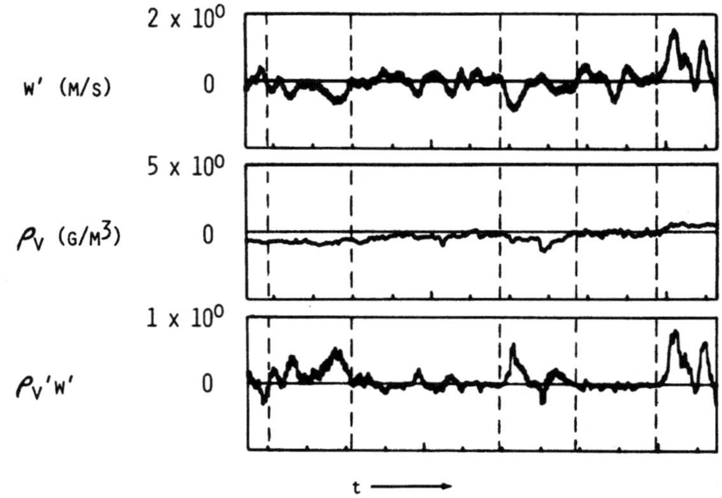

Figure 9. Two-minute time series with selected periods of strong water vapor flux (Julian Day 245, 135515 - 135714 GMT).

values of $\overline{w}\ \overline{h}$ so obtained, where $\overline{w}$ during the event is relative to $\overline{w_o}$, are of the order of 15,000 W m^{-2}. However, consecutive events from a single sample may be as large as plus or minus several hundred thousand watts per square meter. With such large values of energy transfer one must be cautious of the interpretation of the physical processes involved in the transfer process. It is clear, however, that a major portion of the energy exchange takes place in a small portion of the record, perhaps as small as 5 to 10%. This intermittent nature of the energy transfer in the marine boundary layer should be the subject of much research in the coming years.

CONCLUSIONS

A uniformly mixed boundary layer of 500 ± 100 m mean depth is found with average temperatures and absolute humidities nearly the same throughout the experiment. Mean sensible heat fluxes were near zero, whereas, mean latent heat fluxes were 5-10 mw cm^{-2} with the strongest water vapor flux having scales of 400-700 m. Mean stresses, $-\overline{u'w'}$, are less than ±1 dyne cm^{-2}. The stronger vertical motions appear to concentrate and carry fluxes orders of magnitude larger than the spatial means.

REFERENCES

Bean, B. R., R. Gilmer, R. L. Grossman, R. McGavin and C. Travis (1972): An analysis of airborne measurements of vertical water vapor flux during BOMEX. J. Atmos. Sci., 29, 5, 860-869.

GATE: final international scientific plans (1974). Bull. Amer. Meteor. Soc., 55, 711-744.

Rasmussen, J. L. and W. C. Murray (1976): Conference summary on the GATE research program. Bull. Amer. Meteor. Soc., 57, 1005-1011.

Bean, B. R., R. O. Gilmer, R. F. Hartmann, R. E. McGavin, and R. F. Reinking (1976): Airborne measurement of vertical boundary layer fluxes of water vapor, sensible heat and momentum during GATE. NOAA Tech. Memo. ERL WMPO-36, Boulder, Colorado, 83 pp.

DISCUSSION

C. A. Friehe:

Are aircraft suitable for very high wind, low level flux measurements? What are typical noise levels in aircraft flux measurements?

B. R. Bean:

The aircraft turbulence system can certainly measure momentum flux at high wind, 35 mps and larger, and at levels of about 300 m. Below this level sea-spray tends to obscure the surface and make for hazardous flight conditions. The spray tends to render the measurement of sensible and latent heat flux unreliable at best if not impossible. In the absence of spray, however, the noise level of the flux determinations would be of the order of 10 % or less of "average" values.

EDDY FLUXES OF MOMENTUM AND HEAT MEASURED OVER THE ATLANTIC OCEAN IN GALE FORCE WINDS

S. D. Smith

Dept. of Fisheries & Environment
Bedford Institute of Oceanography,
P.O. Box 1006, Dartmouth, N.S., Canada

ABSTRACT

A stable tower has been moored in 58 m of water, 10 km from shore in the Atlantic Ocean near Halifax, Nova Scotia, Canada. This tower is exposed to unlimited fetch for winds from the south and east, and is in a deep-water wave regime for all but the longest waves.

Data from a three-axis thrust anemometer, a micro-thermistor and a resistance wire wave staff are telemetered to a receiver on shore. In addition, a Bendix Aerovane anemometer provides a check on wind speed and direction, and six strain gauges and a three-axis accelerometer monitor tension in the cables supporting the tower and the movement of the tower. The University of British Columbia eddy flux system with a Gill propeller and vane anemometer, thermistors, and humidity sensors was operated simultaneously.

Data were obtained at wind speeds from 8 to 22 m/s, significantly extending the range of available measurements. The sea surface drag coefficient is seen to increase gradually with increasing wind speed.

INTRODUCTION

Wind stress and heat exchange at the sea surface during storms contribute substantially to the wind-driven circulation and the formation of ocean water masses, and their measurement has long been a major goal of air sea interaction studies. Experimental difficulties have limited eddy flux measurements over the sea to winds below 16 m/s (Busch, 1977).

To be representative of the open sea, the chosen site must be open to a long fetch and must be in deep enough water that waves are not steepened and slowed by shoals. The sensors must be mounted in a stable enough way that signals are not contaminated by wave motion, and the supporting structure must be slender enough not to distort the air flow.

THE STABLE PLATFORM AND ITS INSTRUMENTATION

A stable platform was designed to withstand (1) waves of 18 m peak to trough and 10 s period, (2) 45 m/s mean wind with gust factor 1.25, and (3) current of 0.25 m/s. It is a floating tower or spar buoy 47 m long (Fig. 1) with excess buoyancy provided by six naval mooring buoys, rigidly moored by six taut-wire anchor lines near the surface and six more at the bottom, and which protrudes 12.5 m above mean water level (Fig. 2). The structure was installed in June, 1976 at a site 10 km from shore near Halifax, N.S. (Fig. 3). The water depth is 58 m, allowing 10 s waves to travel at 99% of their deep water phase velocity (12 s waves, 96%). The tower is unmanned except during inspection visits, and all equipment used must operate unattended.

Figure 1. The stable platform being lowered into the water. Overall length is 47 m.

Figure 2. The stable platform moored in position. The top of the tower is 12.5 m above the water.

Two sensor systems have been mounted on the tower. The BIO system (Fig. 4) includes a thrust anemometer and a micro-thermistor for eddy flux measurements, a resistance wire wave staff, an Aerovane anemometer, a three-axis accelerometer, and six strain gauges in the six upper anchor cables to monitor tower performance. The data were telemetered to BIO in FM analog form, with a choice of three possible groups of 12 channels of data (Table 1), and were recorded on magnetic tape. These records were subsequently digitized and turbulence levels, fluxes, and spectra were computed from the time series.

A second system was installed by Prof. S. Pond and Mr. W. Large of the Institute of Oceanography, University of British Columbia. Wind turbulence was measured by a modified Gill anemometer (Fig. 5) using a two-bladed, helical styrofoam propeller for the downwind component, a similar propellor with its axis 30° off vertical

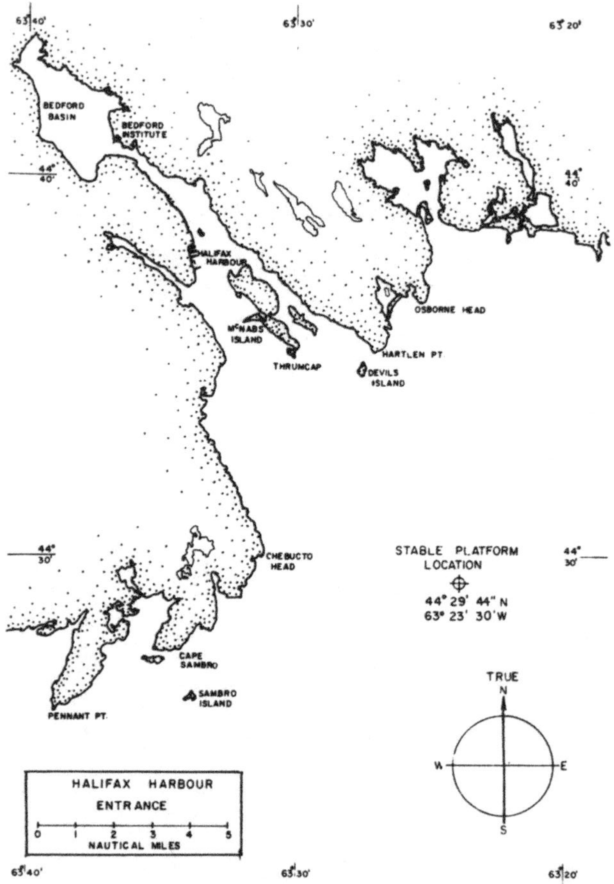

Figure 3. Location of the stable platform.

Figure 4. Instrumentation of the stable platform. At top, from left to right, UBC temperature and humidity sensor shield, UBC Gill anemometer, thrust anemometer, and Aerovane anemometer. At lower left, Aerowatt generator, and at lower right, wave staff. At upper left, two telemetry antennas. In centre, navigation light and radar reflector.

Figure 5. Eddy flux sensors. Left, UBC temperature and humidity shield; centre, UBC Gill anemometer; right, Mk5 thrust anemometer.

TABLE 1. STABLE TOWER SENSORS AND TELEMETRY CHANNELS

Channel	Mode 1 (BIO)	Mode 2 (UBC/BIO)	Mode 3 (Engineering)
1	Aerovane	Air Temp, UBC	Aerovane
2	Anemometer	Humidity, UBC	Anemometer
3	Tension 1	Tension 1	Tension 1
4	Waves	Waves	Waves
5	Humidity, UBC	Humidity, UBC	Tension 2
6		UBC	
7	Accelerometer	Gill	Accelerometer
8		Anemometer	
9	Air Temperature	Air Temperature	Tension 3
10			Tension 4
11	Thrust	Thrust	Tension 5
12	Anemometer	Anemometer	Tension 6

from which the vertical wind was derived by subtracting a fraction of the horizontal wind signal, and a vane to aim the propellers into the wind and sense wind direction. Air temperature and humidity were sensed in a shield. Water temperature was measured by a thermistor at about 5 m depth. The signals were recorded digitally on magnetic tape cassettes, which were changed at the tower at about 1 month intervals. One recorder sampled relatively rapdily in bursts for one hour out of three for eddy flux analysis during periods when winds exceeded 10 m/s. Another recorder ran continuously at a slower rate for estimation of fluxes by the downwind spectrum-dissipation method and to record mean winds and temperatures. The UBC sensor signals were also available through the BIO telemetry system. This will allow us to compare results from both sensor systems and both analysis systems. Power for both systems was supplied by nine automobile batteries connected to give a nominal 36V and charged by an Aerowatt windmill generator.

A preliminary trial was carried out at Sable Island, Nova Scotia, in the fall of 1975 (Smith et.al., 1976) to test the compatibility of the two systems.

The BIO system was operational from October 7 to 27, 1976; November 8 to December 8, 1976; February 5 to 17, 1977; and March 4 to April 7, 1977. A total of 83 data runs were recorded. Of the three periods of down time, the first was to correct design errors and the second and third were due to power regulator failures in the electronics. We also suffered frequent damage to the Aerowatt generator, which has now been relocated 2.4 m higher on the tower.

PRELIMINARY RESULTS

Drag coefficients for 46 thrust anemometer runs at wind speeds up to 22 m/s have been plotted for the period October 1976 to March, 1977 (Fig. 6). The last 19 runs, in April 1977, have not been plotted.

Runs with onshore winds (long fetch) are plotted as solid points, and those in neutral stability conditions ($|z/L|<0.05$) are plotted as larger circles. A regression line fitted to 20 thrust anemometer runs in neutral stability and at long fetch gives

$$10^3 C_{10} = 0.42 + 0.067\ U_{10}\ \text{(m/s)}$$

with correlation coefficient $r = 0.76$. The wind speed dependence is the same as that reported for a compilation of our earlier data (Smith and Banke, 1975), but the present values are lower by 0.2×10^{-3}. Drag coefficients from the UBC Gill anemometer via the BIO telemetry and analysis (crosses) tend to be higher. However, these are preliminary results and corrections for telemetry and recording noise (high frequencies) and for tower motion (wave frequency) await completion of the spectral analysis.

The drag coefficient tends to decrease in stable conditions. In Fig. 7 the April, 1977 data have been added to the earlier runs. The highly stable or unstable runs tend to be at lower wind speeds where we have already noticed lower drag coefficients. In addition, increased wind profile gradients in stable conditions give a higher 13 m wind speed for the same u_*, or a lower u_* for the same wind speed.

The heat fluxes $\overline{tu_3}$ plotted against $U_{10}\Delta T$ (Fig. 8) show less scatter than is commonly observed. Here we have taken the mean air and water temperatures from the slow-response UBC thermistors and the fluctuations from the BIO thermistor and thrust anemometer. The temperature and hence the heat flux are relatively insensitive to platform motion. The slope of the regression line

$$\overline{tu_3} = 0.006°\text{C m/s} + 0.97 \times 10^{-3}\ U_{10}\Delta T,\ r = 0.98$$

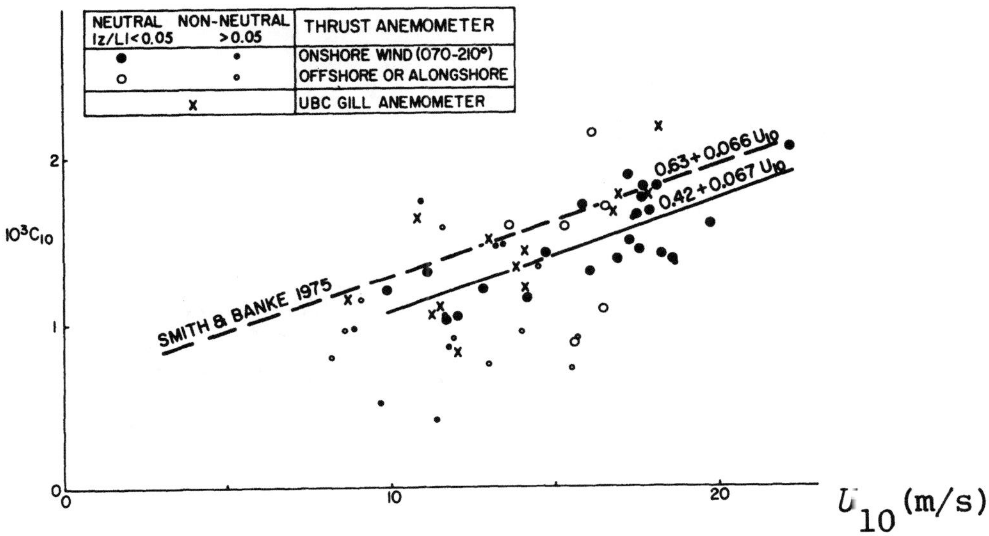

Figure 6. Drag coefficient as function of wind speed at stable platform, October 1976 to March 1977. Solid line is a regression fitted to 21 runs in onshore winds with neutral stability. Dashed line is fitted to a compilation of our earlier data.

indicates a heat flux coefficient $10^3 C_T = 0.97$ which does not appear to be different for stable or unstable cases. This result agrees better with a summary by Friehe and Schmitt (1976) ($10^3 C_T = 0.91$) than with our own data at the beach of Sable Island (Smith and Banke, 1975). The UBC temperature and humidity fluctuation data, which will require corrections for frequency response limitations, will not be reported here.

While spectra have not been completed for all runs, we can examine those for Run 35 in a SSE wind of 18 m/s. These are plotted as log spectrum against log frequency. The data run has been divided into blocks and standard errors of the mean spectral estimates among the blocks are plotted as vertical bars. The downwind spectrum from the thrust anemometer data (Fig. 9a) shows the expected -5/3 slope over two decades, from 0.1 to 10 Hz. The vertical wind spectrum (Fig. 9b) exceeds the downwind spectrum from 1 to 10 Hz by 0.1 on the log scale, which is a factor of 1.3. The downwind spectrum from the UBC Gill anemometer (Fig. 9c) overlaps that from the thrust anemometer within the size of the error bars, tending to have slightly higher values. The vertical wind from the Gill anemometer (Fig. 9d) on the other hand, shows lower spectral levels than the thrust anemometer, typically by 30%, at all

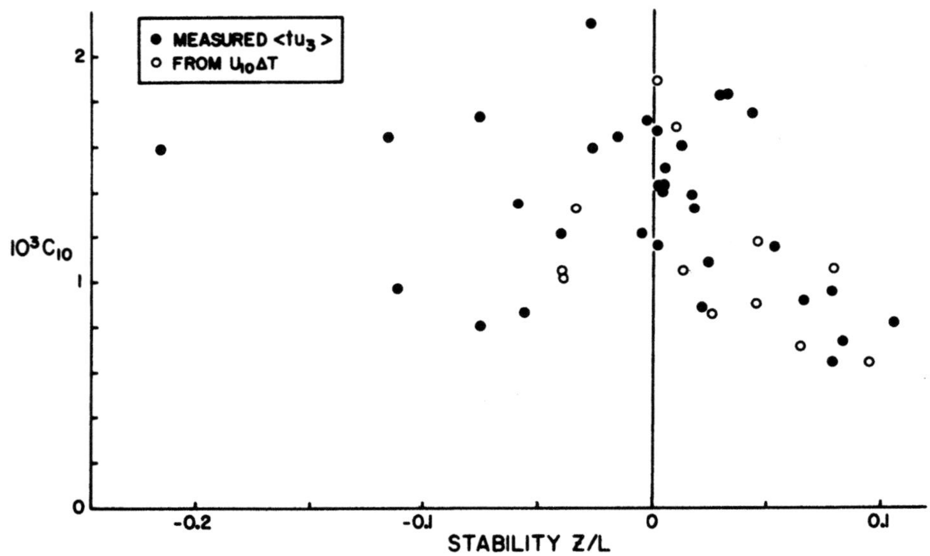

Figure 7. Drag coefficient as function of stability, October 1976 to April 1977.

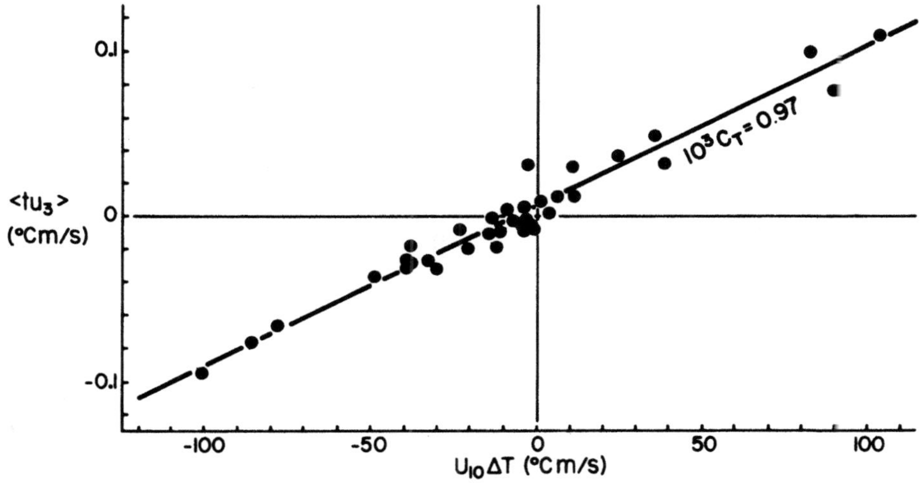

Figure 8. Heat flux $\overline{tu_3}$ as a function of $U_{10}\Delta T$, with regression line.

frequencies, indicating a calibration difference. Based on theory and on past experience (e.g. Smith, 1970) the vertical spectrum should be equal to or greater than the downwind spectrum in the 1 to 10 Hz range, indicating some possible errors in our derivation of vertical wind from the two propellors.

The wave spectrum (Fig. 10a) is sharply peaked at 0.13 Hz and its high-frequency side falls off at a slope of -4.6.

Cospectra of downwind and vertical wind fluctuations give the frequency distribution of the wind stress. The thrust anemometer (Fig. 10b) indicates a peak in stress at the wave frequency and the UBC Gill anemometer (Fig. 10c) shows larger values at frequencies from 2 to 5 Hz, neither of which is believed to be real, and we plan to reduce both these stress estimates by about 8% when spectral corrections are made. For some other runs with a stress minimum at the wave frequency we plan to increase the stress estimates. Large quadrature values (crosses) at the wave frequency are also attributed largely to tower motion. At the wave peak, where the quadrature and cospectra are equal, there is a 45° phase shift from the 180° phase seen elsewhere.

Although the two anemometers were separated horizontally by 1 m, they lay on a crosswind line within ± 10 cm, and so downwind anemometer separation accounts for a relative phase shift of only

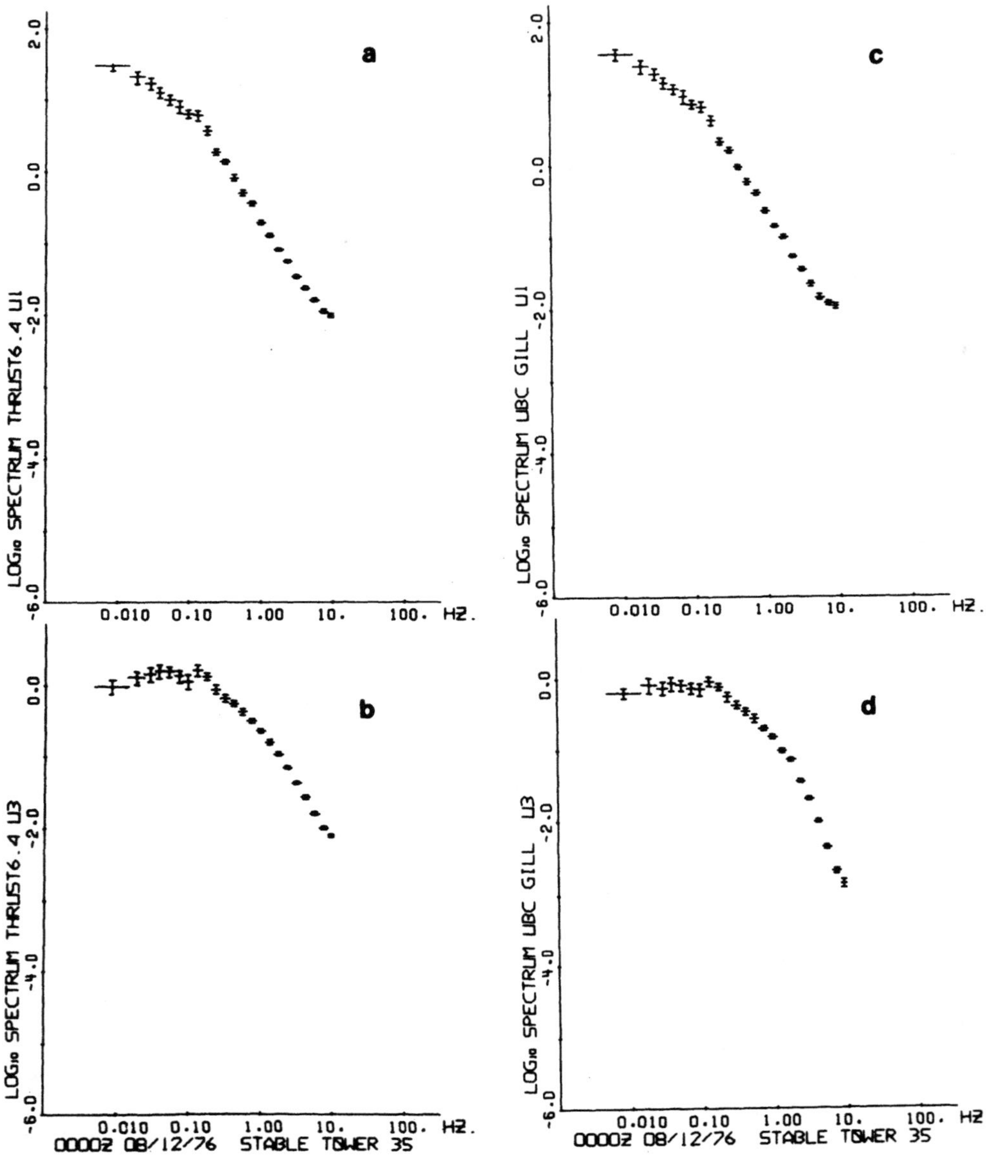

Figure 9. Logarithmic plots of wind spectra for run 35 at 0000Z, Dec. 8, 1976. (a) downwind and (b) vertical, thrust anemometer; (c) downwind and (d) vertical, UBC Gill anemometer.

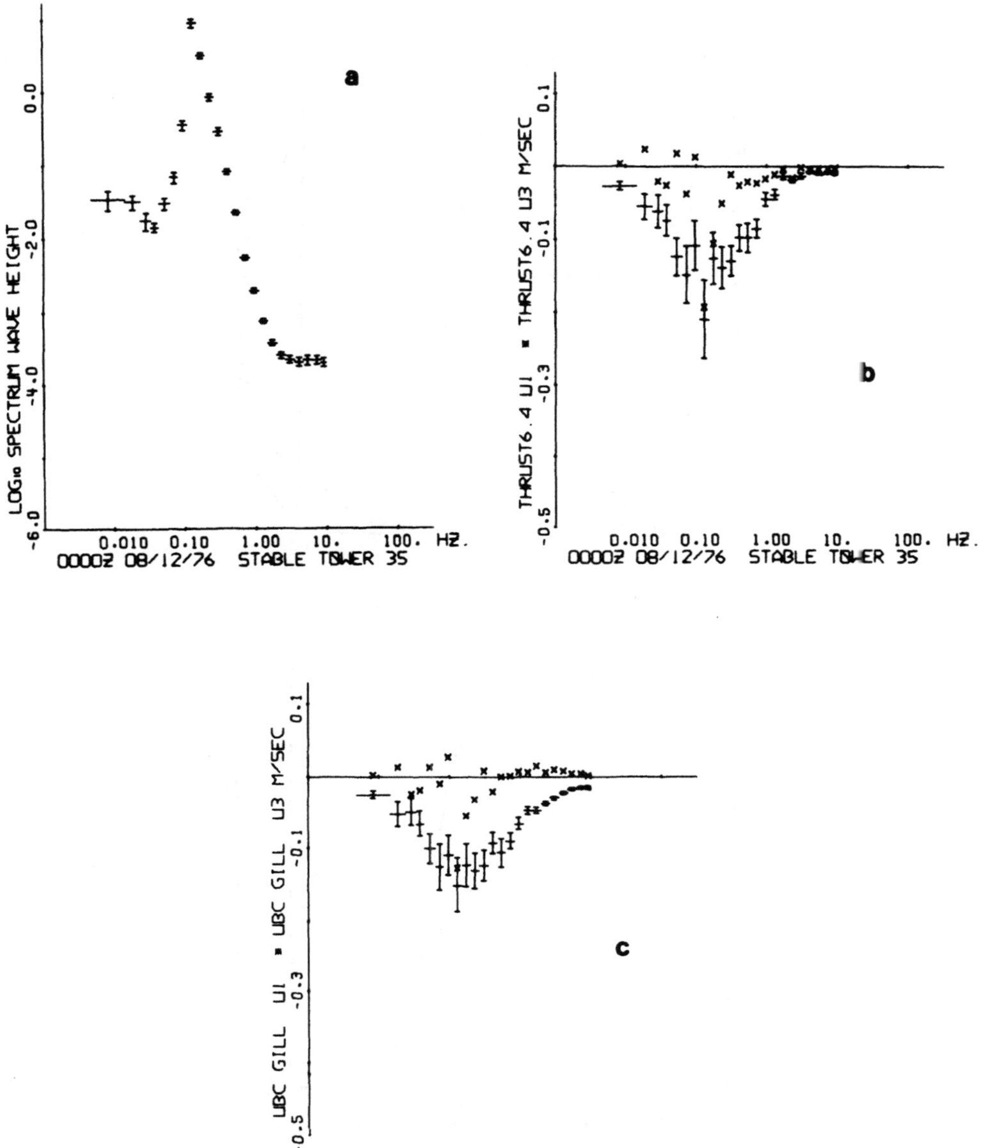

Figure 10. (a) Wave spectrum. (b) Wind stress cospectrum (+) and quadrature spectrum (x) for thrust anemometer. (c) Same for UBC Gill anemometer.

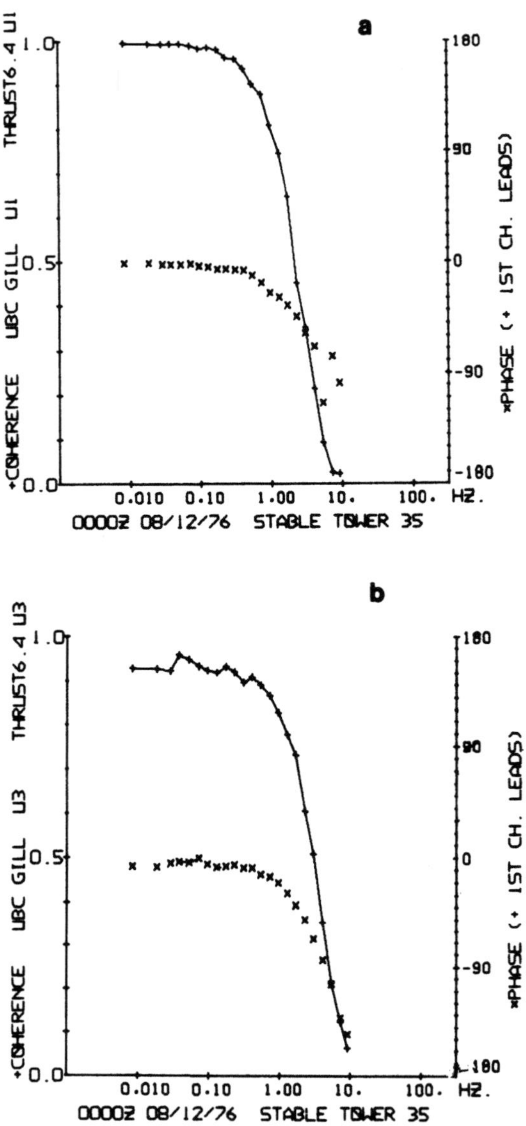

Figure 11. Coherence and phase of Gill and thrust anemometer, (a) downwind and (b) vertical. Negative phase indicates that the Gill anemometer lags.

+ 2° at 10 Hz. The two anemometers show high downwind coherence (>0.8) at frequencies up to 1 Hz, at which point the Gill propeller anemometer lags in phase by 20° (Fig. 11a). The coherence falls and the phase shifts rapidly at higher frequencies. The vertical wind coherence (Fig. 11b) follows a very similar pattern, but with slightly lower coherence at low frequencies where the vertical winds are extremely small and remaining coherent to slightly higher frequencies. Since the phase lags, relative to the thrust anemometer, are virtually identical for both Gill anemometer components, these lags do not introduce significant errors in the measured co- and quadrature spectra.

ONGOING OBSERVATIONS

In the fall of 1976 budget limitations prevented our tightening the lower group of six diagonal anchor cables, and this omission may explain some of our wave-frequency peaks in the wind spectra and cospectra. During the summer of 1977 the slack lower cables have been tightened, and the six upper cables have also been shortened by about 1 m, thereby increasing their tension by 50 to 100%, to approximately 16 MN. This has improved the stability of the platform considerably. Its lowest frequency mode, rotation about the axis of the tower, has been changed from 12 s to 5 s period.

We plan to run a similar experiment, hoping to observe wind speeds as high as 25 to 30 m/s during the coming winter, but the UBC system will be reduced to a Gill anemometer and sea temperature thermistor only. Dr. W. Oost of Royal Netherlands Meteorological Institute plans to operate a new three-component anemometer on the tower during October, 1977 and Mr. Jerry Smith of Dalhousie University proposes to add a two-axis electromagnetic current meter. In the spring of 1978, Dr. E.P. Jones proposes to add a CO_2 sensor to examine gas exchanges. Present plans call for retrieval and dismantling the tower in the summer of 1978.

NOTATION

$C_T = \overline{tu_3}/(U_{10}\Delta T)$	heat flux coefficient
$C_{10} = -\overline{u_1u_3}/U_{10}^2$	sea surface drag coefficient
$g = 9.8$ m/s^2	gravitational acceleration
$K = 0.4$	von Karman constant
$L = -(T+273°C)u_*^3/(gK\overline{tu_3})$	Monin-Obukhov stability length
r	correlation coefficient

u_i	wind fluctuation component, downwind (i=1), crosswind (i=2) and vertical (i=3)
U_{10}	mean wind speed at 10 m level
$u_* = (-\overline{u_1 u_3})^{\frac{1}{2}}$	friction velocity
t	temperature fluctuation
T	mean air temperature at 10 m level
ΔT	difference between sea temperature (5 m depth) and air temperature
z	measurement height, 13 m in this experiment

ACKNOWLEDGEMENTS

The author wishes to thank the members of the air sea interaction group, R.J. Anderson, E.G. Banke, F.W. Dobson, and D.L. Hendsbee, for their dedicated collaboration. Computer analysis of the data was run by M. Spencer and G. Wilson. The UBC data were made available by S. Pond and W.G. Large. The stable platform was designed by R.G. Mills of Whitman, Benn and Associates, Halifax, N.S., and installed under Mr. Mills' supervision by Dominion Diving Ltd., Dartmouth, N.S.

REFERENCES

Busch, N.E., 1977. Fluxes in the surface boundary layer over the sea. In *Modelling and Prediction of the Upper Layers of the Ocean* ed. by E.B. Kraus, Pergamon Press, Oxford, pp. 72-91.

Friehe, C.A. and K.P. Schmitt, 1976. Parameterization of air-sea interface fluxes of sensible heat and moisture by bulk aerodynamic formulas. *J. Phys. Oceanogr. 6,* 801-809.

Smith, S.D., 1970. Thrust-anemometer measurements of wind turbulence, Reynolds stress, and drag coefficient over the sea. *J. Geophys. Res. 75,* 6758-6770.

Smith, S.D. and E.G. Banke, 1975. Variation of the sea surface drag coefficient with wind speed. *Quart. J. Roy. Met. Soc. 101,* 665-673.

Smith, S.D., R.J. Anderson, E.G. Banke, E.P. Jones, S. Pond, and W.G. Large, 1976. A comparison of the air-sea interaction flux measurement systems of the Bedford Institute of Oceanography, & University of British Columbia. *Rep. BI-R-76-17,* Bedford Institute of Oceanography.

DISCUSSION

V. Cardone:

In a recent interwiew, Garrett (1977) has summarized drag coefficient determinations from both profile and eddy flux measurements - including the Smith and Banke (1975) high wind speed data. His amazing techniques suggest strongly that a Charnock relationship is valid with a constant of .012 - .014 and that the differences observed between various investigators in the C_{10} vs U_{10} relationship are due mainly to instrument and measurement errors. However, given a refined single instrument and measurement technique such as yours, there remains a considerable amount of scatter in the data for a given stability class. Could you provide your assessment of the zone of the scatter and particularly how of it might be due to geophysical variability associated with waves, swell, currents, etc.

S. D. Smith:

The scatter in the preliminary results shown here will be somewhat reduced when spectral corrections are applied, perhaps to an rms variability of 0.1×10^{-3} between the Gill and thrust anemometers. I plan to examine the data for wave dependence and proposals for next year include the addition of a current meter. I should have mentioned, as in Smith and Banke (1975) that the Charnock (or Charnock and Ellison) formula describes the observed drag coefficient variation equally well and more elegantly than the regression line shown.

O. Phillips:

Is it possible that the scatter in drag coefficient measurements are associated with larger scale variations in atmospheric structure - temporary inversions, etc.?

S. D. Smith:

Yes, and these can be difficult to sort out from single-point measurements. Our 13 m heigth is the lowest at which the anemometers can avoid wave damage: to minimise effects of accelerations and divergences in the layer below and of larger scale structure. The possible existence of Langmir roll vortices may also effect the results, and Wiebe Oost has indicated that he may take long data runs this fall to study this matter.

K. Hasselmann:

The plots you showed of C_{10} versus wind speed correspond to the traditional representation, but are dimensionally inconsistent. What do you believe is the relevant non-dimensional variable against which the non-dimensional drag coefficient should be plotted? Unless one sorts out one's dimensions, one cannot really be surprised at getting scatter diagrams. It appears that there must be some other dimension besides wind speed determining C_{10}.

S. D. Smith:

I was satisfied with using wind speed in this preliminary analysis, but could make it non-dimensional with parameters which remain constant, such as g.

(If it is g, Charnock's relation results)

G. E. Mattingly

(1) I am interested in hearing a brief description of the calibration procedure for your 3-D thrust anemometer.

(2) Closing comment: Phil Klebenoff's group at the National Bureau of Standards has recently done some work on the characteristics of Gill Anemometers. He finds they follow well accelerating flow but do not follow so well decelerating flow fluctuations.

S. D. Smith:

The anemoneter is calibrated through the entire telemetry and recording system at a number of fixed speeds from 6 to 30 m/s in the B.I.O. windtunnel. A series of tests in 10° horizontal angle increments determines the relative sensitivities of the horizontal components, and a cubic equation relating wind tilt to vertical signal is fitted to a series of tests at 3° tilt increments. Frequency response has been checked by field comparison with a sonic anemometer.

ESTIMATION OF THE FLUXES OVER THE OCEAN BY THE COVARIANCE AND DISSIPATION METHODS

P. G. Mestayer*, F. H. Champagne, C. A. Friehe°,
J. C. La Rue and C. H. Gibson°

Department of Applied Mechanics and Engineering Sciences
University of California at San Diego, La Jolla,
California 92093, U.S.A.

INTRODUCTION

The development of reliable methods for measurement or estimation of the surface fluxes of heat, moisture and momentum over the ocean is of importance in many meteorological and oceanographic problems and programs. The GARP programs AMTEX I and II (Air Mass Transformation Experiment) in February 1974 and 1975 were organized investigations of the atmosphere and ocean responses to the outbreak of cold air from the mainland China across the Kurshio and Ryukyu Islands (Japan). Air-sea exchanges are large during these outbreaks, as wind speeds are of the order of ten m/s, air-sea temperature differences are about 5°C and, following occluded fronts, evaporation rates are also large. A portion of the AMTEX program was concerned with the measurement of surface fluxes by a variety of techniques at several locations under high wind, high flux level conditions.

The purpose of the present work was to compare the estimates of the fluxes of momentum and sensible and latent enthalpies, as obtained by means of the direct or eddy-correlation method (with precision micrometeorological instrumentation), the dissipation technique (with high frequency turbulence anemo- and thermometers), and the bulk aerodynamic formulas.

*Present address: Institut de Mécanique Statistique de la Turbulence
12, Avenue Général Leclerc, 13003 MARSEILLE, France.
°Also, Scripps Institution of Oceanography, U.C.S.D., LA JOLLA.

MEASUREMENT AND DATA ANALYSIS TECHNIQUES

The measurements were made from a tower located on Miyako-jima, a small island in the Ryukyus, at a height of 10.25 m above mean sea level, close to the instrumentation tower of the Meteorological Research Institute of Japan. The wind fetch passed over a reef about 2 km upwind of the tower and over a shallow water bay, before reaching the site.

The micrometeorological instrumentation included a three-dimensional sonic anemometer, a thermistor probe, a Lyman-alpha humidiometer and a cooled-mirror dew-point unit (for calibration and measurement techniques, see Champagne et al., 1977; Mestayer et al., 1975). High frequency fluctuations of longitudinal velocity were measured with a hot film probe and DISA constant temperature anemometer and linearizer. Temperature fluctuations were measured with a 0.625 μm diameter platinum resistance wire operated in a low-noise a.c. Wheatstone bridge. Signals were conditioned and recorded with special low-noise electronic circuits and analog tape recorder to preserve the entire spectral content of the signals.

Considerable attention was given to the total calibration of the hot film anemometer and platinum temperature wire sensors. The probe sensitivities to both velocity and temperature fluctuations were measured at the site in a small heated air jet before and after each data set. During the period selected for this work the sensitivities were

$$e'_{hf} = 0.443\,u' - 0.09\,\theta'$$
$$e'_{cw} = 1.75\,\theta' - 2.66\,10^{-3}\,u' \qquad (1)$$

where e'_{hf} and e'_{cw} are the fluctuating outputs (Volts) respectively from the hot film anemometer and 'cold wire' thermometer; u' and θ' the fluctuating longitudinal velocity (m/s) and temperature (°C). The 3db upper limits of the sensor frequency responses were measured to be 2 KHz for the anemometer and 1.5 KHz for the thermometer (La Rue et al., 1975). The Kolmogorov micro-scale, η, was measured to be $\eta \simeq$ 0.64 mm; the hot film was 0.5 mm long and the cold wire was 0.6 mm long; so, the sensor length averaging effects were not important (Wyngaard, 1968 and 1971). The noise levels of e'_{hf} and e'_{cw} were measured at the site, in the low turbulence level calibration air jet.

Among 17 hours of turbulence measurements recorded on analog tapes, 3 hours were first selected. The signals of all the apparatus were simultaneously recorded on paper charts for closer examination. 67 minutes of data were finaly selected for complete reduction, on technical and meteorological grounds: all apparatus running well, all calibrations good, wind direction and velocity reasonably constant. The data were digitized with a 14-bit analog-to-

TABLE 1. EXPERIMENTAL CONDITIONS

z = 10.25 m	mean	r.m.s.	skewness	kurtosis
Longitudinal velocity	11.20 m/s	1.33 m/s	-0.26	2.89
Vertical velocity	2.12 m/s	0.62 m/s	0.49	4.07
Temperature	16.3 °C	0.45 °C	0.57	3.06
Water vapour density	$8.45\ 10^{-3}$ kg m^{-3}	$0.82\ 10^{-3}$	0.21	2.58

Sea surface: Temperature = 21°C
Saturation water vapour density = $19.2\ 10^{-3}$ kg m^{-3}

$\rho = 1.22$ kg m^{-3} ; $C_p = 248$ cal kg^{-1} °C^{-1}; $L_w = 595\ 10^3$ cal kg^{-1}

$\nu = 1.5\ 10^{-5}$ m^2 s^{-1} ; Pr = 0.63.

digital converter and analyzed on IBM 1130 and TI 980 A digital systems at UCSD (Gibson, 1973); recent analysis was also done at IMST with a HP 2100 A system. The high frequency u' and θ' digital time series were obtained from the digitized voltages, with eq.(1). The fluctuating water vapour density data, q' (kg of water vapour/m^3 of moist air) were obtained from the digitized Lyman-alpha signal calibrated by comparison with that of the dew-point unit. The low frequency velocity components, u' and w' respectively in longitudinal and vertical directions, were obtained from the three digitized voltage outputs from the sonic anemometer. Experimental conditions during the selected period, means and statistics of variables, parameter values, are given in Table 1.

Figure 1-a presents typical time series of u', θ' and q'. One can see the ramp-like structures characteristic of shear flows, in the temperature and humidity signals, emphasized on the expanded graph fig. 1-b (Mestayer et al., 1976; Gibson et al., 1977). However the temperature signal under-shoots at ramp fronts. These 'cold spike' features seem characteristic of over ocean temperature data; here, compared to the high level temperature fluctuations, they are relatively small. Nevertheless, Schmitt et al. (1977) have since shown that these spikes are caused by the humidity sensitivity of temperature sensors polluted by some salt coating; the contaminated data exhibit an anomalously high spectral content in the frequency range just below the dissipation maximum, and a less than -5/3 slope in the inertial sub-range. Because of this contamination, the present temperature results, like most open ocean boundary layer temperature data, have to be considered with caution and much reservation. (See the paper by Friehe and Gibson, this volume.)

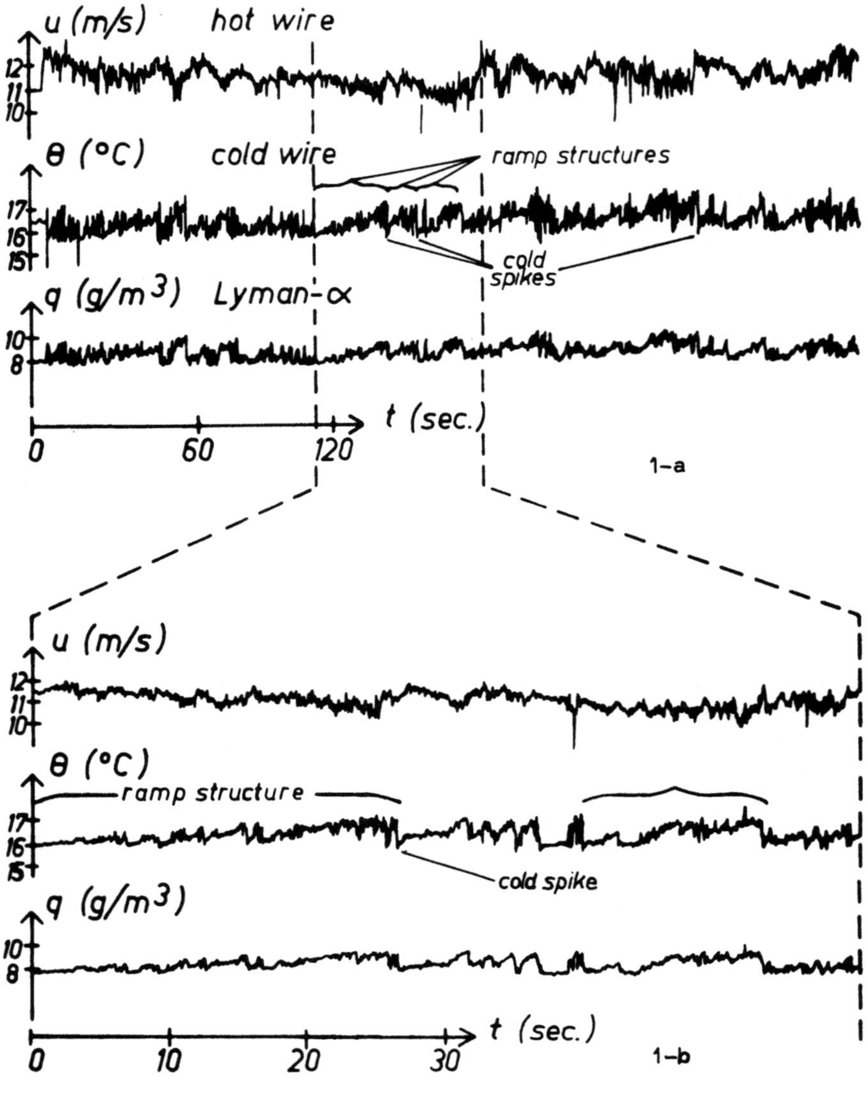

Fig. 1. Time series of longitudinal velocity, temperature and humidity signals.

RESULTS AND DISCUSSION

-a- The dissipation technique is based on theoretical considerations and experimental measurements on the various terms in the budget equations for turbulent kinetic energy, temperature and humidity variances. As proposed by Champagne et al.(1977) it includes semi-empirical relations derived from the Kansas experiment results by Kaimal et al. (1972), Wyngaard & Coté (1971) and (1972), and from the Minnesota experimental results of Champagne et al. In brief, their final relation system is

$$
\begin{aligned}
u_* &= (\overline{-u'w'})^{1/2} = (\varkappa z \epsilon (1+0.5(z/L)^{2/3})^{-3/2})^{1/3} \\
\theta_* &= -\overline{w'\theta'}/u_* = (\varkappa z \chi_\theta \alpha_\theta (1-9z/L)^{1/2}/2u_*)^{1/2} \\
q_* &= -\overline{w'q'}/u_* = (\varkappa z \chi_q \alpha_q (1-9z/L)^{1/2}/2u_*)^{1/2}
\end{aligned}
\qquad (2)
$$

$$
L = \frac{u_*^2 \left(1 + 0.472 \frac{\overline{\theta}}{273} \overline{Q}\right)}{\varkappa g \left(\frac{\theta_*}{\overline{\theta}} + 0.472 \frac{\overline{\theta}}{273} q_*\right)}
$$

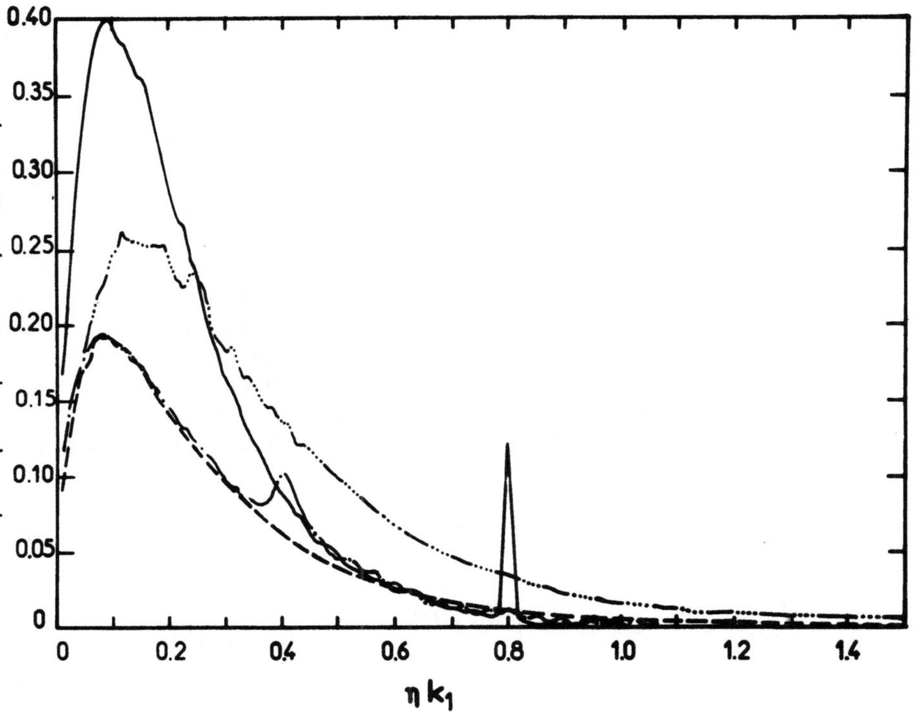

Fig. 2. Kolmogorov normalized velocity and temperature spectra. AMTEX $\partial u/\partial t$(— — —), $\partial \theta/\partial t$(———); Champagne et al. (1977) $\partial u/\partial t$(—·—), $\partial \theta/\partial t$(—····—).

where z/L is the parameter of stability, including humidity fluctuation effects; L is Monin-Obukhov length scale; $\kappa = 0.35$ is Von Karman constant; $\alpha_\theta = \alpha_q = 1.35$. The mean temperature $\overline{\theta}$ and water vapour density $\overline{Q}$, and the dissipation rates for turbulent kinetic energy ϵ, temperature variance, χ_θ, and water vapour density variance, χ_q, are the inputs of the system (2).

The dissipation rates can be computed from the spectral data in the inertial sub-range,

$$E_1(k_1) = \alpha_1 \epsilon^{2/3} k_1^{-5/3} \tag{3}$$

$$E_\theta(k_1) = \beta_\theta \epsilon^{1/3} \chi_\theta k_1^{-5/3} \tag{4}$$

$$E_q(k_1) = \beta_q \epsilon^{1/3} \chi_q k_1^{-5/3} \tag{5}$$

where E_1, E_θ and E_q are the spectra of longitudinal velocity, temperature and water vapour density, $k_1 = 2\pi f / \overline{U}$, assuming Taylor's hypothesis, and α_1, β_θ and β_q are the Kolmogorov universal constants.

Because of sensor resolutions, ϵ and χ_θ were also obtained from the high frequency derivative signals

$$\begin{aligned} \epsilon &= 15\, \nu / \overline{U}^2 \, \langle (\partial u'/\partial t)^2 \rangle = 15\, \nu / \overline{U}^2 \int_0^\infty E_{\partial u'/\partial t}(f)\, df \\ \chi_\theta &= 6\, \mathcal{D} / \overline{U}^2 \, \langle (\partial \theta'/\partial t)^2 \rangle = 6\, \mathcal{D} / \overline{U}^2 \int_0^\infty E_{\partial \theta'/\partial t}(f)\, df \end{aligned} \tag{6}$$

where ν is the kinematic viscosity and $\mathcal{D}$ the thermal diffusivity. The estimates of ϵ and χ_θ from the spectra were corrected for the effects of high frequency noise pick-up by digitizing the derivatives of the noise levels and sustracting the noise spectra from the measured spectra. They were not corrected for Taylor's hypothesis over-estimate the turbulence intensity is only about 10% and these errors are estimated to be less than 3% (Champagne et al., 1977).

Figure 2 shows the derivative spectra in Kolmogorov normalized form

$$\begin{aligned} (\eta k_1)^2 \Phi_1(\eta k_1) &= (2\pi \overline{U})^{-1} \nu^{1/4} \epsilon^{-3/4} E_{\partial u'/\partial t}(f) \\ (\eta k_1)^2 \Gamma_K(\eta k_1) &= (2\pi \overline{U})^{-1} \nu^{1/4} \epsilon^{1/4} \chi_\theta^{-1} E_{\partial \theta'/\partial t}(f) \end{aligned} \tag{7}$$

along with the results of Champagne et al. (1977) obtained over land. The present velocity and temperature spectra have a spike,

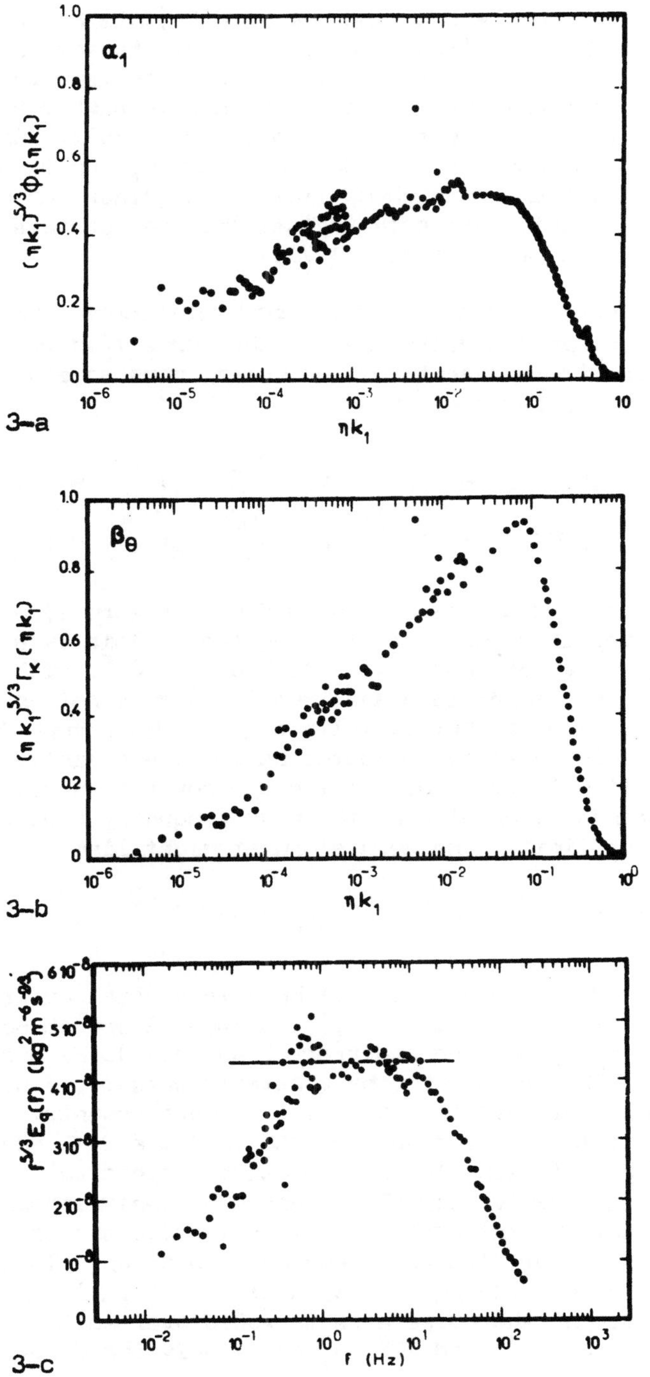

Fig. 3. 'Kolmogorov universal constant plots' for velocity (a) and temperature (b); water vapour density inertial sub-range plot (c).

respectively at $\eta k_1 = 0.4$ and $\eta k_1 = 0.8$ (1125 Hz and 2250 Hz) probably caused by electronic noise pick-up. Except for that spike, the velocity spectrum is very close to the over land result, therefore indicating a universal shape of high wave number surface layer longitudinal velocity spectra (Wyngaard & Pao, 1971; Champagne et al., 1977; Champagne, 1977). On the contrary, the temperature spectrum exhibits a very high maximum and a rather rapid high wave-number cut-off, which are both typical features of data contaminated by cold spikes (Schmitt et al., 1977).

Fig. 3 shows the inertial sub-range behaviours. For u' and θ', the normalized spectra multiplied by the normalized wave-number at power 5/3, should represent the 'universal constants' α_1 and β_θ in the inertial sub-range, if any

$$
\begin{aligned}
(\eta k_1)^{5/3}\Phi_1(\eta k_1) &= \epsilon^{-2/3}\, k_1^{5/3}\, E_1(k_1) &= \alpha_1 \\
(\eta k_1)^{5/3}\Gamma_K(\eta k_1) &= \epsilon^{1/3}\chi_\theta^{-1}\, k_1^{5/3}\, E_\theta(k_1) &= \beta_\theta
\end{aligned}
\qquad (8)
$$

A straight line fit to the data should have a zero slope in the sub-range and a magnitude equal to the constant. This is approximately true for velocity between $\eta k_1 = 10^{-3}$ and $3\ 10^{-2}$, with α_1 about 0.47 (3-a). The temperature spectrum (3-b) does not appear to have an inertial sub-range, but exhibits a slope less than -5/3, characteristic of cold spike contaminated data. The humidity spectrum (3-c) multiplied by frequency to the 5/3 power does exhibit an inertial sub-range, attenuated on the high frequency side by the Lyman-alpha response limit. The zero slope straight line fit gives

$$f^{5/3}E_q(f) = \beta_q\ (\bar{U}/2\pi)^{2/3}\epsilon^{1/3}\chi_q = 4.3\ 10^{-8}\ \mathrm{kg^2\ m^{-6}\ s^{-2/3}} \qquad (9)$$

In Table 2 the results in column 1 were obtained from relation (9), from a similar fit to $f^{5/3}E_1(f)$ between 2 and 15 Hz, from the magnitude of $E_\theta(f)$ for the arbitrary frequency 15 Hz and relations (3) and (4). The values for the constants were $\alpha_1 = 0.52$ and $\beta_\theta = 0.42$ which are averages of previous measurements. β_q was assumed equal to β_θ. For other results in Table 2, ϵ and χ_θ were estimated from relations (6) by computing respectively: the variances of the derivative signals by means of their probability density functions, the integrals of the derivative spectra, uncorrected and corrected for high frequency noise pick-up. In all cases, χ_q was estimated from eq. (9) with $\beta_q = 0.42$. It appears that not taking care of the noise effects would over-estimate the dissipation rates by 11 and 27%, u_* by 3% only and θ_* by 11%.

-b- In the eddy-correlation technique the fluxes are estimated using the directly computed covariances of the signals, with

$$\begin{aligned} \tau &= -\rho \overline{u'w'} \\ H_S &= \rho C_p \overline{w'\theta'} \\ H_L &= L_w \overline{w'q'} \end{aligned} \tag{10}$$

where the bar indicates a time average over the whole data set. u' and w' were obtained from the sonic anemometer, θ' from the cold wire thermometer and q' from the Lyman-alpha unit. A covariance $\overline{w'x'}$, where x' is u', θ' or q', can be computed, either from a time average of time dependent products of fluctuations about some means $\hat{w}$ and $\hat{x}$

$$\overline{w'x'} = \overline{[w - \hat{w}](t) \cdot [x - \hat{x}](t)} \tag{11}$$

or from the integral cospectrum

$$\overline{w'x'} = \int_0^\infty E_{wx}(f)\, df = \int_{f_1}^{f_2} f\, E_{wx}(f)\, d(\text{Ln} f) \tag{12}$$

f_2 is the highest frequency contributing to the cospectrum, or the probe frequency cut-off; f_1 corresponds to the largest scales belonging to turbulence.

The three raw cospectra are shown in fig. 4, multiplied by frequency. The logarithmic abcissas indicate the frequencies and the corresponding lengths of air passed by the probes $l = \overline{U} / f$.

TABLE 2. DISSIPATION TECHNIQUE RESULTS

	Inertial sub-range	Derivative variances	Uncorrected spectra	Corrected spectra
ϵ $(m^2\, s^{-3})$	$1.85\ 10^{-2}$	$3.19\ 10^{-2}$	$2.35\ 10^{-2}$	$2.11\ 10^{-2}$
χ_θ $(°C^2\, s^{-1})$	$1.07\ 10^{-2}$	$1.30\ 10^{-2}$	$9.46\ 10^{-3}$	$7.44\ 10^{-3}$
χ_q $(kg\, m^{-3})^2\, s^{-1}$	$1.84\ 10^{-8}$	$2.21\ 10^{-8}$	$1.99\ 10^{-8}$	$1.93\ 10^{-8}$
$-z/L$	0.475	0.267	0.317	0.313
u_* $(m\, s^{-1})$	0.354	0.441	0.394	0.381
θ_* (°C)	0.410	0.363	0.338	0.304
q_* $(kg\, m^{-3})$	$4.11\ 10^{-4}$	$4.73\ 10^{-4}$	$4.90\ 10^{-4}$	$4.90\ 10^{-4}$

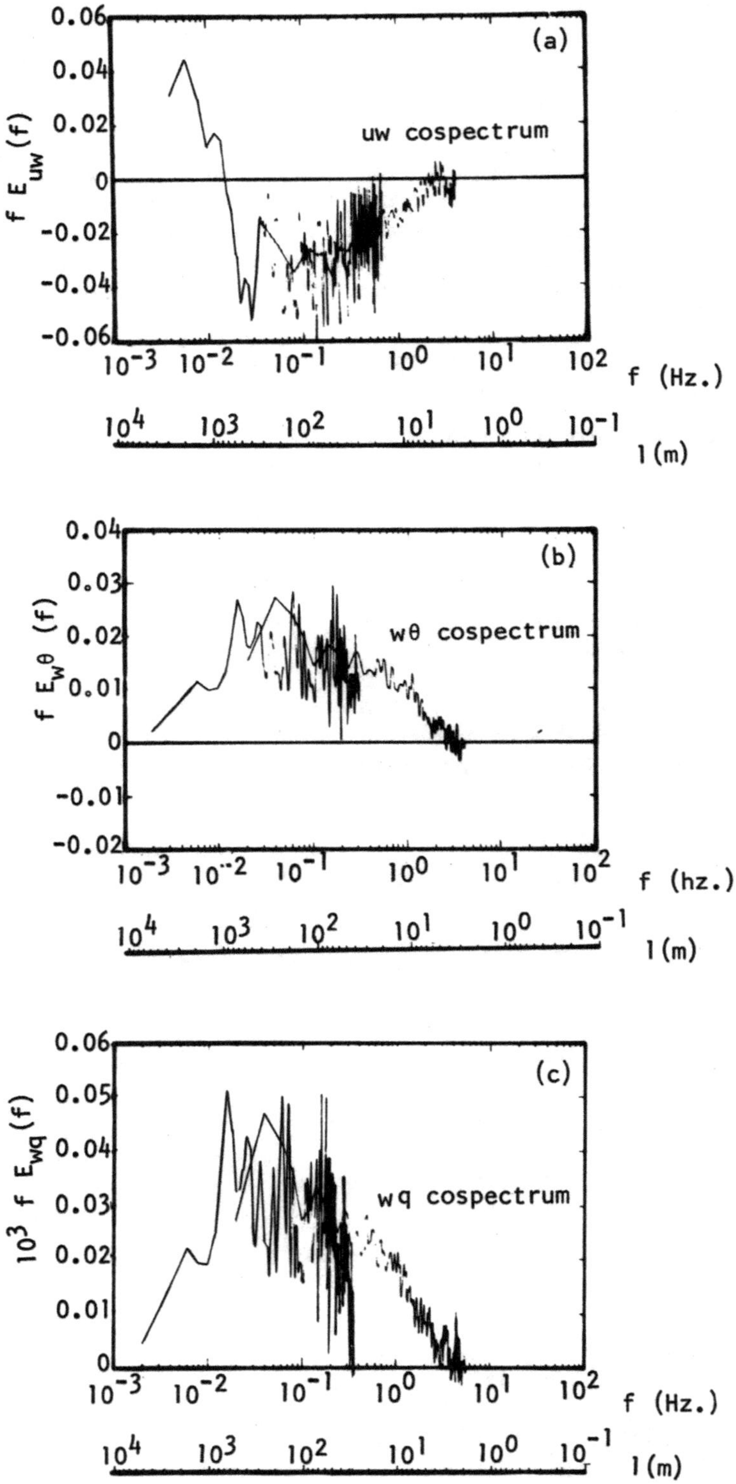

Fig. 4. Cospectra of low frequency signals: u' and w' (a), w' and θ' (b), w' and q' (c).

The three cospectra have similar high frequency roll-offs (the instruments have effective frequency response from d.c. to $\sim$ 3 Hz.) On the low frequency side they largely differ. The scalar fluxes (4-b and -c) receive non-negligible contributions from scales of order of several kilometers. The contributions to a positive stress are limited to scales about 700 m ($1.6\ 10^{-2}$ Hz; 1 minute); at larger scales the flux changes sign.

The choice of the frequency f_1 in eq.(12) is related to the definition of the means $\hat{x}$ and $\hat{w}$ in eq.(11); i.e. the value of f_1 corresponds to the time period over which the signal x is averaged, to form $\hat{x}$..

From examination of fig. 4 plots it is difficult to determine a clear turbulence limit scale. As any choice of f_1 smaller than $1.6\ 10^{-2}$ Hz would be unconsistent with the uw cospectrum plot, this rather high value has been choosen here. Therefore, the three covariances have been computed by time averaging over 67 minutes the products of the fluctuations about means computed over 1 minute centered periods

$$(\overline{w'x'})_{cov} = \int_{t=0}^{67\ \text{min.}} \left(w(t) - \int_{\tau=t-\frac{1}{2}\text{min.}}^{t+\frac{1}{2}\text{min.}} w(\tau)\, d\tau \right)\left(x(t) - \int_{\tau=t-\frac{1}{2}\text{min.}}^{t+\frac{1}{2}\text{min.}} x(\tau)\, d\tau \right) dt \qquad (13)$$

4. CONCLUSIONS

The flux estimates by the various methods are presented in Table 3. The bulk aerodynamic formulas are those proposed by Smith & Banke (1974) and Friehe & Schmitt (1976) from surveys of a large number of results (see for instance Friehe & Gibson, this volume). The covariance results were computed from eq. (13); the 'cospectrum' results, from eq. (12) with f_2 = 3 Hz, $f_1 = 1.6\ 10^{-2}$ Hz for $E_{uw}(f)$ and $f_1 = 2\ 10^{-3}$ Hz for $E_{w\theta}(f)$ and $E_{wq}(f)$. The dissipation method results are issued from the computations presented in Table 2, column 4. The 'error', in per cent, is

$$\text{error}\ \% = 100\ (\text{Dissipation} - \text{Covariance})\ /\ \text{Covariance}$$

The dissipation method appears to estimate the surface fluxes to within about 20%. The sensible heat results have to be considered with reservation: the salt contamination of the temperature sensor has been seen to cause anomalous temperature spectra. For the stress and latent heat flux, the differences between the dissipation technique and direct eddy-correlation estimates are somewhat larger than observed over land. To investigate these discrepancies, the budgets of turbulent kinetic energy, $\overline{e'^2}$, and scalar variances are examined in Table 4. The terms which were not measured are estimated after Businger et al.(1971), Wyngaard & Coté (1971)

TABLE 3. FLUX ESTIMATES

	τ (kg m/s)/(m^2 s)	H_S cal/(m^2 s)	H_L cal/(m^2 s)
Bulk aerodyn. formulas	0.209	28	94
1 minute - covariances	0.153	29	101
Cospectrum integrals	0.135	33	113
Dissipation method	0.177	35	111
'Error' %	+16	+21	+10

TABLE 4. BUDGET EQUATIONS

Turbulent kinetic energy budget (10^3 m^2 s^{-3})

$-\overline{u'w'}\partial\bar{U}/\partial z$	$\frac{g}{T_v}\overline{w'\theta'}$	$\frac{g}{T_v}(.61\ \bar{\theta}\ \overline{w'm'})$	$-\epsilon$	$-\frac{1}{2}\frac{\partial}{\partial z}\overline{w'e'^2}$	I_m	I_{wc}
8.60	3.80	0.90	-21.10	-4.80	-12.60	-13.60

Temperature variance budget (10^3 °C^2 s^{-1})

$-2\ \overline{w'\theta'}\ \partial\bar{\theta}/\partial z$	$-\partial/\partial z(\overline{w'\theta'^2})$	χ_θ
6.15	0.28	7.44

Water vapour density variance budget (10^8 (kg m^{-3})^2 s^{-1})

$-2\ \overline{w'q'}\ \partial\bar{Q}/\partial z$	χ_q
1.76	1.93

and Champagne et al. (1977). I_m is the actual budget imbalance, while I_{wc} is the imbalance predicted from Wyngaard & Coté (1971). They both include the pressure transport term. They do not differ much but seem to have large magnitudes. The scalar budgets also seem more largely imbalanced than in the afore-cited over land results. This may be due to various causes; such as, incorrect choice of turbulence limit time scale in covariance computations; overall effects of contaminated temperature data; inadequacy of over land budget parameterizations and/or simplifications, for turbulence over the wavy ocean surface.

ACKNOWLEDGEMENTS

We would like to thank Pr. Yamamoto, Pr. Mitsuta, Mr. Komatsuzaki and all the AMTEX staff for their assistance and aid in this work. Our thanks also to Dr Di Lullo, from the Division des Affaires scientifiques of NATO for his help to UCSD-IMST cooperation, and to Pr. Favre and Dr. Coantic for their contribution and assistance to carry this work through. We would also like to thank T.K. Deaton, UCSD, who designed the special electronic circuits and took part in the measurements.

The field experiment was performed under grant ATM-74-05459 from the National Science Foundation, Climate Dynamics Office, and AFOSR grant 72-2287. During data analysis, the authors were supported by the following contracts and grants: NATO AS2; CNRS LA 130; NSF DES75-07223; ONR N00014-75-C-0152.

REFERENCES

Businger, J.A., J.C. Wyngaard, Y. Izumi & E.F. Bradley, 1971. J. Atmos. Sci., 28, 181-189.

Champagne, F.H. 1977. J. Fluid Mech., accepted for publication.

_____, C.A. Friehe, J.C. La Rue & J.C. Wyngaard, 1977. J. Atmos. Sci., 34, 515-530.

Friehe, C.A. & K.F. Schmitt, 1976. J. Phys. Oceano., 6, 801-809.

Gibson, C.H., 1973. AGARDograph AG 174.

_____, C.A. Friehe & S.O. McConnel, 1977. Phys. Fluids, accepted for publication.

Kaimal, J.C., J.C. Wyngaard, Y. Izumi & O.R. Coté, 1972. Quart. J. R. Met. Soc., 98, 563-589.

La Rue, J.C., T.K. Deaton & C.H. Gibson, 1975. Rev. Sci. Instrum., 46, 757-764.

Mestayer, P.G., F.H. Champagne, C.A. Friehe, J.C. La Rue & C.H. Gibson, 1975. AMTEX Report 8, 173.

_____, C.H. Gibson, M.F. Coantic & A.S. Patel, 1976. Phys. Fluids, 19, 1279-1287.

Smith, S.D. & E.G. Banke, 1975. Quart. J.R. Met. Soc., 101, 665-673.

Schmitt, K.F., C.A. Friehe & C.H. Gibson, 1977. J. Phys. Ocean. Accepted for publication.

Wyngaard, J.C., 1968. J. Sci. Instrum., 1, 1105-1108.

_____, 1971. Phys. Fluids, 14, 2052-2054.

_____ & Y.H. Pao, 1971. Statistical Models & Turbulence, M. Rosenblatt & C. Van Atta, Eds., Springer-Verlag, 384-401.

_____ & O.R. Coté, 1971. J. Atmos. Sci., 28, 190-201.

_____ & _____, 1972. Quart. J.R. Met. Soc., 98, 590-603.

DISCUSSION

M. Coantic:

I think those data, as well as other recent ones of other investigations, raise two problems:
1) The integral time scales in the lower atmosphere seem to be different for the different second order moments, which means that the classical "spectral gap" characteristics established for the longitudinal velocity could change for other variables.

(2) While the heat and mass flux cospectra seem to go to zero, the stress cospectrum apparently changes sign, which opens the possibility for a countergradient momentum flux by the largest eddies.

K. Katsaros:

Why did you choose a 1 min. averaging time? It would seem a bit short to me.

P. Mestayer:

It did not seem fair to me to take into account in any of the covariance computations the contributions from the scales which contribute to a positive momentum flux, that is from the "contra-gradient" scales, in the sense used by Coantic.

N. B. to Chairman and editor - I would like to add here a small comment from data which were written down on the transpa-r ncies but that I did not clearly express in the talk:

Nevertheless it is interesting to note that if I use instead, what I call the cospectra results, i.e. including the contributions of all frequencies from 10^{-3} Hz, the three budget equations are then very close to being in balance.

H. Charnock:

Referring to Dr. Katsaros' question, it seems to me that the turbulence in the surface layer could have a time scale of order $Z/_{u+}$.

So one minute, or a few minutes, does not seem so short to me as it does to her.

67 minutes would seem to me to include structures of scale larger than those to which the usual surface layer formulation could be expected to apply.

W. Pierson:

Which element in the table is the "right" one and why•

P. Mestayer:

By definitions the "right" estimations of the fluxes are those from the covariance method. But the estimation variability that I have shown exactly revise the problem of the definition; I mean the definition of turbulence, of which perturbations do belong to turbulence and which to mean flow.

L. Hasse:

The observation, that the wθ, wq cospectra go to zero at low frequencies, but the uw cospectrum changes sign, raises suspicion of an instrumental effect. Have you checked the time variation of mean quantities (say $\bar{u}$, $\bar{v}$, $\bar{w}$) for trends or have you removed any trends?

P. Mestayer:

The usual average computations (cumulative average) have been done for all the low frequency signals and no trend can be clearly noticed on any of them. In fact this is one of the reasons why this particular set of data was chosen for complete reduction; along with the stationarity of the wind direction. It can be seen in the body of the paper that $\bar{w}$ is not equal to zero, which is probably due to a wind deflection by shore, but I do not think this can be the cause of the sign change uf $\overline{u'w'}$. In addition I must say that the spectra of $\overline{u'^2}$ and $\overline{w'^2}$ at low frequencies have a behaviour very similar to the unstable results of Kainu et al (1972) (Kansas) but for the lowest frequencies where the $\overline{u'^2}$ spectrum do not exhibit any spectral gap. This could be the cause of a positive $\overline{u'w'}$ estimation, specially if the sonic anemometer had not been set in the right way and the w measurements contaminated by u (as supposed by Dr. Katsaros). Carl Friehe can confirm that the sonic anemometer has been carefully levelled during the experiment and that it has been checked before and after the AMTEX experiment in a large UCSD tunnel: the corrections to apply are in Champagne et al (1977) and they have been applied. On the other hand the $\overline{w'^2}$ spectra do present a gap for the largest turbulent scales. For myself I believe in the reality of large convective perturbations with positive $\overline{u'w'}$.

ESTIMATES OF THE SURFACE FLUXES OVER THE OCEAN

Carl A. Friehe and Carl H. Gibson

Department of AMES and Scripps Institution of Oceanography, University of California, San Diego La Jolla, California 92093 USA

INTRODUCTION

Development of techniques for the measurement and estimation of the vertical fluxes of momentum, heat and moisture over the ocean has been the subject of a number of experiments in the last 10 to 15 years. Interest in these studies was motivated by the importance of the fluxes in wave generation, ocean circulation, large scale atmosphere-ocean interactions and effects of turbulence on propagation of electromagnetic and optical radiation through the atmosphere.

For many of these problems, simultaneous determination of all three fluxes is required. For measurement in the overlying atmospheric surface layer of the heat loss from the ocean, usually only the latent heat flux (proportional to the measured moisture flux) is required, since the sensible heat loss from the ocean is usually small (the ratio of the sensible to latent heat fluxes is the Bowen ratio, B, and is typically taken to be $\sim +0.1$ on average). However, the sensible heat flux is a very important variable to the atmospheric surface layer itself, since the buoyant stability is largely controlled by temperature contributions to density, rather than moisture content (Lumley and Panofsky, 1964). Turbulence levels, energy budgets and drag coefficients are functions of the state of stability in the surface layer. Deardorff (1968), for example, predicted that bulk aerodynamic drag coefficients are increased by ~20% and decreased by ~ 30% from their neutral values for moderately unstable and stable conditions, respectively.

The common measure of stability in the surface layer is the Monin-Obukhov length L which is composed of contributions from the sensible heat flux and moisture flux. L is expressed as $1/L = 1/L_\theta + 1/L_q$ where L_θ is the length due to sensible heat flux (temperature) and L_q is the length due to latent heat flux (moisture). In Figure 1, the ratio L/L_q is shown as a function of B, the Bowen ratio. For B = +0.1, L/L_q is ~ 0.5, which shows that the sensible heat flux is just as important as the latent heat flux in the marine atmospheric surface layer. (A singular point ($L \to \infty$) exists for B = -0.07, where neutral conditions are obtained for finite heat and moisture fluxes.) For B = 0.4, L is dominated by the sensible heat flux. Therefore for complete descriptions of air-sea interaction processes, measurement of the sensible heat flux cannot be neglected.

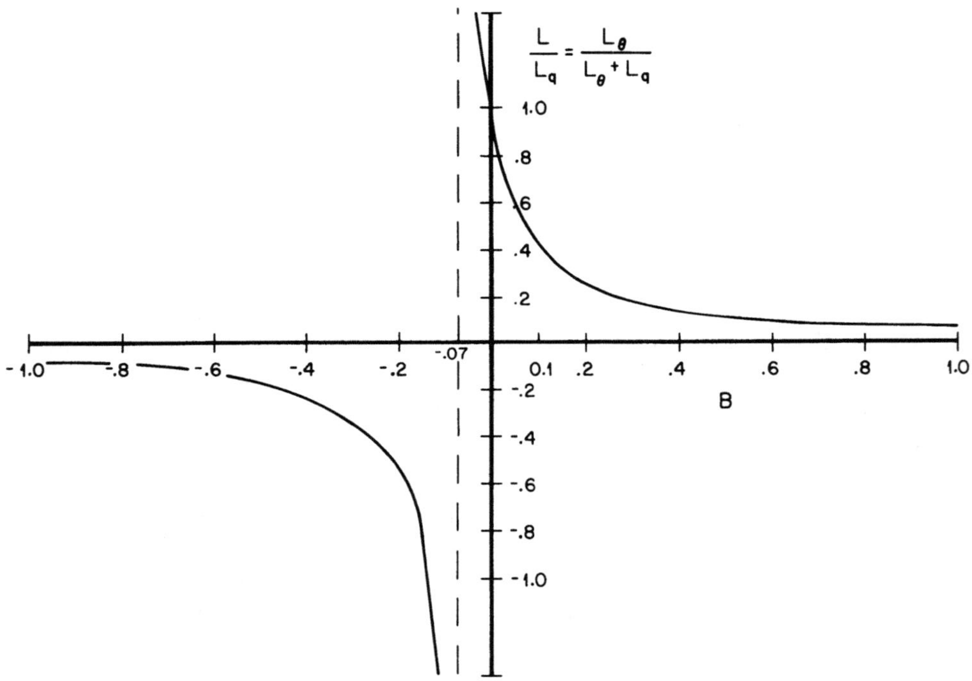

Figure 1. Ratio of L/L_q versus the Bowen Ratio, B.

This paper will attempt to assess the present capabilities that exist for the measurement and estimation of the surface fluxes over the ocean. Emphasis will be placed on recently discovered problems of temperature measurement near the surface of the ocean in the presence of salt spray, which seriously affects our ability to measure the sensible heat flux.

DIRECT FLUX MEASUREMENT TECHNIQUES

The fluxes of momentum (τ), sensible heat (H_S), moisture (E) and latent heat (H_L) for a constant density fluid are defined by

$$\tau = -\rho\, \overline{uw}$$

$$H_S = \rho C_p \overline{w\theta}$$

$$E = \overline{wq} \quad ; \quad H_L = VE$$

where the nomenclature and units are those commonly used (see for example, Champagne et al., 1977; q is water vapor density, $g \cdot m^{-3}$; V is heat of vaporization, cal/g).

The fluxes are difficult to measure as they are the average covariances of the appropriate fluctuating turbulent velocity component and scalar quantities. Correlation coefficients can be quite low (absolute value < 0.5), and frequency response and signal-to-noise requirements reasonably high. The averaging times (noted by the overbars in the above formulas) are typically of the order of 15-60 min for measurements made at fixed points near the surface. At larger elevations, the necessary averaging times may become prohibitively long, encompassing diurnal and weather system changes (Wyngaard, 1973). Measurements from aircraft have been suggested by Wyngaard as a way around the problem.

Instrumentation for the direct covariance measurements has been developed for over land and over ocean use. Perhaps the most important instrument developed for velocity component measurements has been the sonic anemometer. (See Mitsuta, ed., 1971, for a compilation of articles on the development of the sonic anemometer.) The device is particularly useful because it gives an unambiguous measurement of the vertical velocity component w with adequate spatial resolution and frequency response for most applications. Propellor anemometers and other mechanical devices have been used, but limitations in frequency response and threshold starting values are usually not precisely known. Hot-wire anemometers are fragile and change calibration with the build-up of spray (Schacher and Fairall, 1976). At sea, motion of the instrument platform due to wind and wave forces is a severe problem since the measurement of w is easily contaminated by response to u once the anemometer is tilted (Kaimal and Haugen, 1969). Approximate corrections for motion have been made by correcting for the instantaneous pitch angle of the anemometer (Mitsuta and Fujitani, 1974 and Friehe et al., 1975) or by mounting instrumentation on a gyro-stabilized buoy (Dunckel et al., 1974). Total sensor motion correction has been achieved by using inertial navigation platforms on research aircraft (Lenschow, 1972).

Fine scale humidity measurements have been made with thermocouple psychrometers (Leavitt, 1975), Lyman-alpha humidiometers (Pond et al., 1971) and microwave cavities on aircraft (Bean et al., 1972). Salt spray contamination and slow frequency response of the psychrometers is a potential problem, as are uncertain calibration and drift of the Lyman-alpha device. The infra-red hygometer may have a high noise level for flux measurements (Hyson and Hicks, 1974).

Temperature fluctuation measurement at frequencies sufficient for direct covariance data is in principle simple, as fine wire resistance elements, thermistors and thermocouples are adequate. The sonic anemometer can be used as a virtual temperature fluctuation device, but suffers from velocity sensitivity when used in the usual mode of operation (Kaimal, 1968).

Over land, successful temperature fluctuation measurements appear to have been made (Kaimal et al., 1971; Champagne et al., 1977); however, over the ocean measurements using sensors with exposed surfaces have been anomalous and difficult to interpret. The difference has been recently shown to be due to salt spray contamination on the sensor surfaces (Schmitt et al., 1977a). Because of the salt, spray drops can exist when the water vapor concentration is considerably less than saturated. Spray accumulates on the sensor surface soon after exposure to the marine air (~ 5 to 30 min). In the marine surface layer there are humidity fluctuations on all scales, positively correlated with the temperature fluctuations when the sensible heat and moisture fluxes are upward (from the ocean to the atmosphere). The salt spray on the sensor surfaces is therefore in a changing humidity environment, and water vapor evaporates from, and condenses on, the spray. The latent heat of vaporization is large, and the heat exchanges associated with these processes are sensed by the probe. Figure 2 illustrates typically observed signals in the marine boundary layer, including extraneous "cold spikes", when water evaporates from the spray due to the drop in humidity and the sensor supplies the heat of vaporization, and also "hot spikes" when heat is liberated by condensing water vapor when the humidity suddenly increases. These temperature contamination effects have not been found over land; the temperature signals there are much like the humidity signals of Figure 2; with no hot or cold spikes. The contamination of the measured temperature signal by the (unknown) response to the humidity field has severely affected most open ocean boundary layer experiments to date.

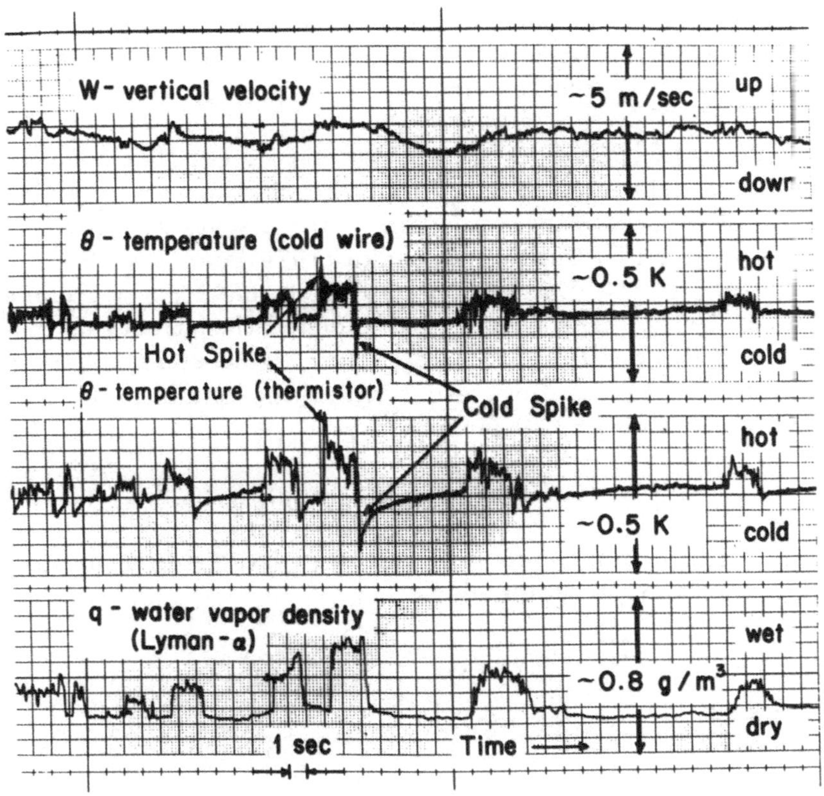

Figure 2. Time series of temperature, humidity and vertical velocity obtained over the ocean (Schmitt et al., 1977).

A suitable remote-sensing technique or spray protection device for the surface sensors has not been developed. The effects on the contamination will be discussed below for the different flux estimation techniques. For the direct covariance method, it appears that increased values of $\overline{w\theta}$ are measured with a contaminated θ-sensor under unstable conditions (Friehe and Schmitt, 1976). It also should be noted that the salt spray effect may cause errors in the measurement of mean temperature (Holmes and O'Brien, 1975; Schmitt et al., 1977a).

FLUX ESTIMATES FROM THE BULK AERODYNAMIC FORMULAS

The bulk aerodynamic formulas, which relate the turbulent covariances to 10 m-mean wind speed, air temperature, dew point and sea temperatures by empirical drag, Stanton and Dalton number coefficients, are appealing because of the existence of a large body of data of the mean variables obtained from ship reports over the oceans. Mean data from buoys can also be readily used for flux estimations.

Considerable scatter is shown by most direct flux covariance and mean data used to determine the coefficients; standard deviations of ~ 0.25 dynes/cm^2 for τ for a wind speed of 10 ms^{-1} are given by Smith and Banke (1974) and 2×10^{-4} cal/cm^{2}s and 4×10^{-4} cal/cm^{2}s for H_S and H_L by Friehe and Schmitt (1976). The scatter has clouded the effects of the dependence of the coefficients on stability, wind speed, wave effects, surface skin cooling, etc. Smith and Banke propose for the drag coefficient C_D ($-\overline{uw} = C_D U_{10}^2$):

$$C_D = [0.63 + 0.066\, U_{10}(\text{m/s})] \times 10^{-3}$$

Friehe and Schmitt surveyed $\overline{w\theta}$ and $\overline{wq}$ data from a number of experiments and recommended

$$\overline{w\theta} = A + C_H U_{10}(T_S - \theta_{10})$$

$0 < U_{10}(T_S - \theta_{10}) < 25$ mKs^{-1}	$U_{10}(T_S - \theta_{10}) > 25$ mKs^{-1}
$A = +0.002$ mKs^{-1}	$A = 0$
$C_H = 0.97 \times 10^{-3}$	$C_H = 1.46 \times 10^{-3}$

and

$$\overline{wq} = C_E U_{10}(Q_S - Q_{10}) \; ; \; C_E = 1.32 \times 10^{-3}$$

In surveying the $\overline{w\theta}$ data, Friehe and Schmitt rejected those values for which the original data exhibited cold spikes in the θ time series; these gave anomalously large values of $\overline{w\theta}$ and suggested $C_H \doteq 5 \times 10^{-3}$. The spikes are usually largest at low heat flux levels because the humidity contamination is greatest. At higher heat flux levels, the temperature fluctuations are greater and the humidity contamination effect decreases and reasonable values of $\overline{w\theta}$ were obtained. Results over fresh water (Smith, 1974; Hicks, 1972) where salt spray is not a problem also give reasonable values of $\sim 1.3 \times 10^{-3}$.

As mentioned above, Deardorff (1968) predicted the stability dependence of the coefficients, and found supporting data from earlier over ocean profile experiments in agreement with the theory. Since stability is largely determined by the sensible heat flux, and since there is reason to suspect $\overline{w\theta}$ values, the verification of the stability dependence still appears to be open.

FLUX ESTIMATES FROM DISSIPATION RATE MEASUREMENTS

It is possible to estimate the fluxes from appropriate fine scale measurements of the turbulent velocity and scalar fields. An advantage of this technique is that measurement of the w component is not required. Also, measurement and interpretation of small mean quantities such as the air-sea temperature is avoided. The turbulent kinetic energy and scalar variance budget equations are employed, and the production terms - the product of the turbulent covariances and the mean vertical gradients - are estimated from prior knowledge of the partitioning of energy among the various terms in the equations and measurement of the dissipation rates of turbulent kinetic energy, temperature variance and humidity variance. (The methods have been reviewed recently by Champagne et al., 1977.) The dissipation rates can also be estimated from inertial subrange spectral measurements. The inertial subrange method is required for the humidity budget, since no sensor presently exists for the necessarily small scale dissipation rate measurement of water vapor fluctuations.

For the surface layer over land, the terms in the budget equations are reasonably well known for a wide range of stability conditions, and good flux estimates have been obtained (Wyngaard and Coté, 1971; Champagne et al., 1977). Over water, the terms in the budgets are less well known, due partly to poor measurements of $\overline{w\theta}$ and partly since the divergence of the pressure velocity covariance may be important over a wavy, moving surface. Stress estimates over the ocean using the dissipation technique have given various results, usually equal to or greater than the directly measured values. Results from some recent experiments are given in Figure 3. The scatter is large, indicating that under some conditions the wave field may affect the kinetic energy budget equation through the pressure term and consequently give high stress estimates using the over-land budget equation. Also shown in Figure 3 are the over land results of Champagne et al.; the average value agrees well and the percentage error in the estimates is acceptable because the stress was high, but the scatter is still rather large.

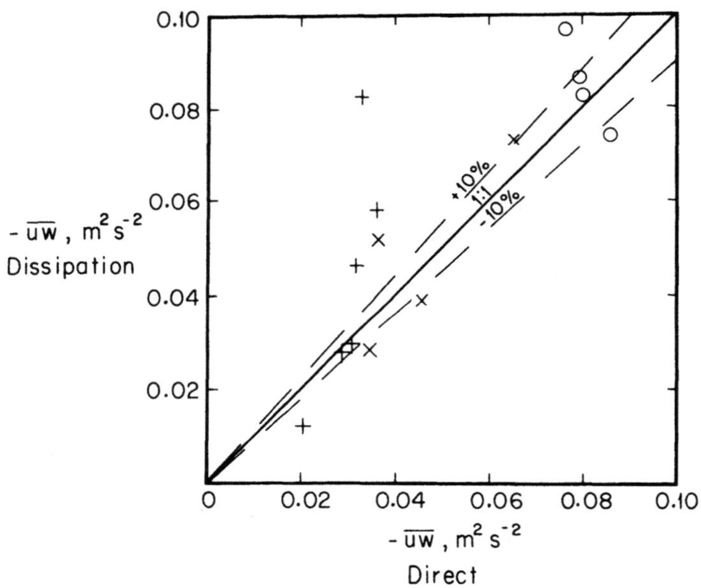

Figure 3. Directly-measured stress values versus dissipation rate estimates from over water experiments of Schmitt et al. (+) (1977b) and Dreyer (1974) (x). Over land dissipation estimates of Champagne et al. (1977) are also shown (o).

Champagne et al. also estimated the sensible heat and moisture fluxes from the dissipation technique, and found reasonably good agreement with the directly measured values. Over the ocean, Pond et al. (1971) and Leavitt (1975) found reasonable inertial dissipation moisture flux estimations but agreement was poor for the sensible heat flux estimates. Stegen et al. (1971) however found reasonable agreement using the temperature dissipation rate obtained from fine scale derivative measurements rather than estimates from inertial subrange. Schmitt et al. (1977a) show that extremely large estimates of the sensible heat flux are obtained from temperature data contaminated by the salt-spray-induced humidity sensitivity. The spectral shape is also altered so that inertial subranges are not always found for contaminated temperature signals (Friehe et al., 1975). Large, widely scattered values for the temperature inertial subrange Kolmogoroff constants found by Gibson et al. (1970) and Leavitt (1975) are probably associated with this problem, although the reasons are more complicated than simple loss of frequency response due to the added mass of salt spray on the sensors. The value obtained by Boston and Burling (1972), also about twice that found in laboratory and over land studies, may not be due to the salt spray contamination since the data did not exhibit "cold spikes" and an extensive inertial subrange was observed.

DISCUSSION AND CONCLUSIONS

In view of the present inability to make reliable temperature fluctuation measurements over the ocean, surface fluxes of heat, moisture and momentum appear to be best estimated by the bulk aerodynamic formulas. The errors in the bulk coefficients and mean value difference data become less significant under high flux conditions, which are generally of most interest in air-sea interaction studies. However, continued verification of the formulas is required, especially for high wind conditions and with temperature sensors insensitive to humidity fluctuations.

Further research on surface layer turbulence is required to develop flux estimation techniques which are better than the bulk formulas. Our basic knowledge of the turbulent budget equations and effects of waves and stability is limited over the ocean. Until proper sensors are developed which are not affected by salt spray, some research may best be performed over fresh water lake surfaces and in simulation wind tunnels, such as the facility at IMST, Luminy, France (Coantic and Bonmarin, 1975).

ACKNOWLEDGEMENTS

Our work in air-sea interaction has been supported by the National Science Foundation (Grants DES75-07223 and ATM-74-05459), the Office of Naval Research (Contracts N00014-69-A-0200-6043 and N00014-75-C-0152), NATO and the National Oceanic and Atmospheric Administration.

Much of the work presented here has been the result of several field experiments performed on the Scripps Institution of Oceanography research platform FLIP; we would like to acknowledge the contributions made by S.O. McConnell, G.F. Dreyer, J.P. Clay, F.H. Champagne, T.K. Deaton and K.F. Schmitt.

REFERENCES

Bean, B.R., R. Gilmer, R.L. Grossman and R. McGavin (1972) J. Atmos. Sci. 29, 860-869.

Boston, N.E.J. and R.W. Burling (1972) J. Fluid Mech. 55, 473-492.

Champagne, F.H., C.A. Friehe, J.C. LaRue and J.C. Wyngaard (1977) J. Atmos. Sci. 34, 515-530.

Coantic, M. and P. Bonmarin (1975) Atmos. Technol. 7, 72-79.

Deardorff, J.W. (1968) J. Geophys. Res. 73, 2549-2557.

Dreyer, G. (1974) Ph.D. Dissertation, University of California at San Diego.

Dunckel, M., L. Hasse, L. Krügermeyer, D. Schriever and J. Wucknitz (1974) Boundary-Layer Meteor. 6, 81-106.

Friehe, C.A., C.H. Gibson, F.H. Champagne and J.C. LaRue (1975) Atmos. Technol. 7, 15-23.

Friehe, C.A. and K.F. Schmitt (1976) J. Phys. Oceanogr. 6, 801-809.

Gibson, C.H., G.R. Stegen and R.B. Williams (1970) J. Fluid Mech. 41, 153-167.

Hicks, B.B. (1972) Boundary-Layer Meteor. 3, 201-213.

Holmes, J.F. and J.F. O'Brien (1975) ISA Transactions 15, 372-377.

Hyson, P. and B.B. Hicks (1975) J. Appl. Meteor. 14, 301-307.
Kaimal, J.C. (1969) Radio Sci. 4, 1147-1153.
Kaimal, J.C. and D.A. Haugen (1969) J. Appl. Meteor. 8, 460-462.
Kaimal, J.C., J.C. Wyngaard, T. Izumi and O.R. Coté (1972) Quart. J. Roy. Meteor. Soc. 98, 563-589.
Leavitt, E. (1975) J. Phys. Oceanogr. 5, 157-163.
Lenschow, D.H. (1972) NCAR Technical Note, NCAR-TN/EDD-74, 39 pp.
Lumley, J.L. and H.A. Panofsky (1964) The Structure of Atmospheric Turbulence. New York, Interscience.
Mitsuta, Y., ed. (1971) Development of Sonic Anemometer-Thermometer and its Applications to the Study of Atmospheric Surface Layer (Final Report), Japan-U.S. Joint Study Group, WDD Technical Note No. 6, Kyoto, 250 pp.
Mitsuta, Y. and T. Fujitani (1974) Boundary-Layer Meteor. 6, 203-217.
Pond, S., G.T. Phelps, J.E. Paquin, G. McBean and R.W. Stewart (1971) J. Atmos. Sci. 28, 901-917.
Schacher, G. and C.W. Fairall (1976) Rev. Sci. Instrum. 47, 703-707.
Schmitt, K.F., C.A. Friehe and C.H. Gibson (1977a) Humidity sensitivity of atmospheric temperature sensors by salt contamination (to be published in Journal of Physical Oceanography).
Schmitt, K.F., C.A. Friehe and C.H. Gibson (1977b) Sea surface stress measurements (to be published).
Smith, S.D. (1974) Boundary-Layer Meteor. 3, 235-255.
Smith, S.D. and E.G. Banke (1975) Quart. J. Roy. Meteor. Soc. 101, 665-673.
Stegen, G.R., C.H. Gibson and C.A. Friehe (1973) J. Phys. Oceanogr. 3, 86-92.
Wyngaard, J.C. (1973) Chapter 3 in Workshop on Micrometeorology, D.A. Haugen, ed., Boston, American Meteorological Society.
Wyngaard, J.C. and O.R. Coté (1971) J. Atmos. Sci. 28, 190-201.

DISCUSSION

M. S. Longuet-Higgins:

In future laboratory measurements of the momentum flux and drag coefficient, would it not be a good idea to try varying the surface tension? The small-scale roughness in the sea surface is very probably involved with the drag; and the minimum velocity of capillary-gravity waves might be tried as a quantity with which to normalize the wind-speed or the friction velocity.

C. A. Friehe:

A laboratory study would be useful to determine the effects of varying small waves on the air turbulence structure. Problems of extrapolation of the results to the open ocean case should also be recognized. La Rue et al. of U.C.S.D. have begun some experiments to investigate the turbulence structure, heat and momentum fluxes following the wave motion at the IMST Annex de Luminy facility.

W. Pierson:

Were the Navy winds and the Pata Buoy winds referred to the same elevation above the sea surface? Was there a net difference between the two wind measurements?

C. A. Friehe:

Pierson's question refers to material presented in the talk at the symposium, but due to space limitations was not included in the written version. Results of a study by Friehe and Pazan (to be published) of 30 days of 10-m surface meteorological data from a NOAA buoy were shown and compared to the independent surface analysis by US Navy, Fleet Numerical Weather Central, based on geostrophic calculations. Overall, good agreement was shown by the buoy data and the Navy analysis. The answers to the questions are:

(1) The FNWC Navy winds were given at 19.5 m, and were referred to 10 m (the height of the actual buoy wind measurements) by using an iterative technique based on the Businger et al. (1971) J. Atmos. Sc.) flux-profile relationships.

(2) For the 768-hour average, the actual buoy winds were 7.5 ms^{-1}, whereas the FNWC analysis gave 8.2 ms^{-1}.

O. H. Shemdin:

In previous laboratory measurements of the drag coefficient as a function of wave height and frequency in the Stanford University wave tank the results indicated systematic dependences of the drag coefficient on wave height and wave frequency as generated by a mechanical wave piston. The variation is not attributed to instrument errors. It would be useful to isolate processes under laboratory conditions to explore the processes which affect the drag coefficient scatter. I would be interested to know if others feel that waves play an important role in affecting the drag coefficient.

C. A. Friehe:

Our results indicate that the turbulent kinetic energy budget and structure of the velocity field (e.g., local isotropy in the inertial subrange) are measurably different close to ocean waves (within about 5 wave heights) from surface layer turbulence over land and at high elevations over the sea. The wave field must be the source of the modification of the turbulence field, probably through the pressure field. The mechanism whereby the drag coefficient is altered is not yet known.

EXPERIMENTAL STUDY OF MOMENTUM TRANSFER, HEAT AND WATER VAPOR FLUXES UNDER DIFFERENT STABILITY CONDITIONS

F. Resch, J. P. Selva

Institut De Mecanique Statistique de la Turbulence
12 Avenue General Leclerc 13003 Marseille, France

INTRODUCTION

It is well known that the presence of a significant buoyancy flux associated with sensible and latent heat fluxes affects the turburlent structure of the surface layer of the marine atmosphere, as is the case when air and water temperatures get different values: under such conditions the classical "neutral" description cannot be taken into account any more. Stability parameters have then to be introduced in the formulation of the various fluxes. This has already been studied by many authors and among them Monin and Obukov (1954), Businger (1973) and Kondo (1975) can be cited. The I.M.S.T. air-sea interaction facility has been specially designed to simulate -at least partially- the atmospheric stability effects. After preliminary studies on a one fifth scale model by Marmottant (1974) as well as on the large facility by Peri (1975) the authors have undertaken in 1975 a rather extensive laboratory study of the influence of stratification upon the momentum, heat and water vapor fluxes. Although the full program is not yet completed, the first experimental results are to be presented here.

STRATIFICATION PARAMETERS

To take into account the stratification effects, it is necessary to introduce an additional parameter characterizing the production, or destruction, of turbulent kinetic energy by buoyancy forces. This is usually achieved by the use of the Richardson number under different formulations such as the "flux Richardson number" or the "gradient Richardson number". Another approach using the dimensional analysis, leads to the Monin Obukhov (1954) similarity theory (see also Monin and Yaglom 1971) and the dimensionless parameter $\zeta = z / L$ where L is the well known Obukhov (1946) length :

$$L = -u_*^3 / \kappa \left[g/\theta \; H/\rho C_p + g(r_e - r_a)/ r \; J_0/\rho \right]$$

The second term within the brackets takes into account the influence of the water vapor flux upon buoyancy, which has been shown to be of sometimes dominant importance in the lower marine atmosphere. It is to be noted that both ζ and Richardson numbers are functions of the elevation, z , and consequently do not easily characterize a given situation. Therefore a "bulk Richardson number" was selected as a global dimensionless stability parameter:

$$Ri_b = \frac{gh}{\theta_a} \; \frac{\theta_h - \theta_0}{U_h^2}$$

or more precisely, including the influence of humidity upon density variations:

$$Ri_b = 3.4 \; 10^{-3} \; (\Delta\theta + .18 \; \Delta C) \; \frac{gh}{\rho U_\infty^2} \; ,$$

where h is taken as the thickness of the surface layer where the respective fluxes are approximately constant. For laboratory situations this thickness can be taken as one quarter of the conventional boundary layer thickness. As a reference in the present study the last test section, where the boundary layer is approximately one meter thick, has been selected; therefore a value of h equal to .25 meter has been adopted. In a first approximation, Ri_b can be expressed here in terms of $\Delta\theta$, the difference between air and water temperatures, and U_∞^2 .

$$Ri_b = 7 \; 10^{-2} \; \frac{\Delta\theta}{U_\infty^2} \qquad \text{or} \qquad L = 3.7 \; \frac{U_\infty^2}{\Delta\theta}$$

Then, judicious combinations of both parameters allow to generate various stratification levels. In our study, the following values have been selected : $Ri_b = -.009$ ($L = -2.2$ m), and $Ri_b = -.003$ ($L = -6.7$ m) for the unstable cases, $Ri_b \simeq 0$ ($L \rightarrow \infty$) for the neutral case, and $Ri_b = + .006$ ($L = +3.3$ m) for the stable case.

EXPERIMENTAL FACILITY AND PROCEDURE

The flexibility of the I.M.S.T. air-sea interaction facility allows the above mentioned values to be reached easily. This facility can be briefly described as a combination of a micrometeorological wind tunnel with a wind wave tank, see figure 1. The wind tunnel test section is 40 meters long, 3.2 meters wide and 1.5 meters high; the water tank is 40 meters long, 2.6 meters wide and .75 to 1 meter deep. (See Coantic, Bonmarin, Pouchain, Favre, 1969). The main characteristics concerning the turbulent transfers are as follows : the air velocity can be accurately varied between .3 and 14 m/s, the air and water temperatures can be varied independently between 5 and 35°C -with a precision of around .1°C in time and space-. The wet bulb temperature can be varied between 1 and 20°C. These variations are monitored by heaters, coolers and vapor injectors if necessary, see figure 2. Turbulent momentum, heat and mass transfer processes are thus effectively produced within 40 meters length of interfacial boundary layers. Turbulence is generated at the entrance section by means of large triggering elements.

Velocity, temperature and humidity profiles have been determined at various sections along the tunnel using Pitot static tubes, chromel-constantan thermocouples and Peltier effect dew point hygrometer - with a precision of .2°C on the dew point temperature - . Special attention was paid to the time and space stability of the flow to ensure repeatable and comparative experiments. Surface values were obtained using floating particles for the surface velocity, an infrared radiometer for the surface temperature, and the value of the saturation water vapor pressure at the surface temperature for humidity.

To obtain meaningful mean values in spite of the presence of low frequency turbulent fluctuations, data were digitized and recorded through a numerical system, see Coantic and Favre, 1973). The sampling frequency used was .4 H_z and a total averaging time of the order of 1 to 3 minutes per point was adapted. Around 40 to 50 data points were recorded for each profile, providing between 10 and 40 points in the linear part of the logarithmic profiles : this is essential for an accurate evaluation of the fluxes by the "profile" method.

During the data processing, a correction was introduced to compensate for the small time fluctuations due to the regulating system and the small discrepancies in the overall flow conditions, which cannot be avoided when experiments are not continuously performed.

FLUX DETERMINATION

From these accurately determined velocity, temperature and humidity profiles, the corresponding momentum, heat and water vapor fluxes can be evaluated. In the present experiment, this was done by the so-called "profile method" and "integral method". As a first step, profile relationships were taken as :

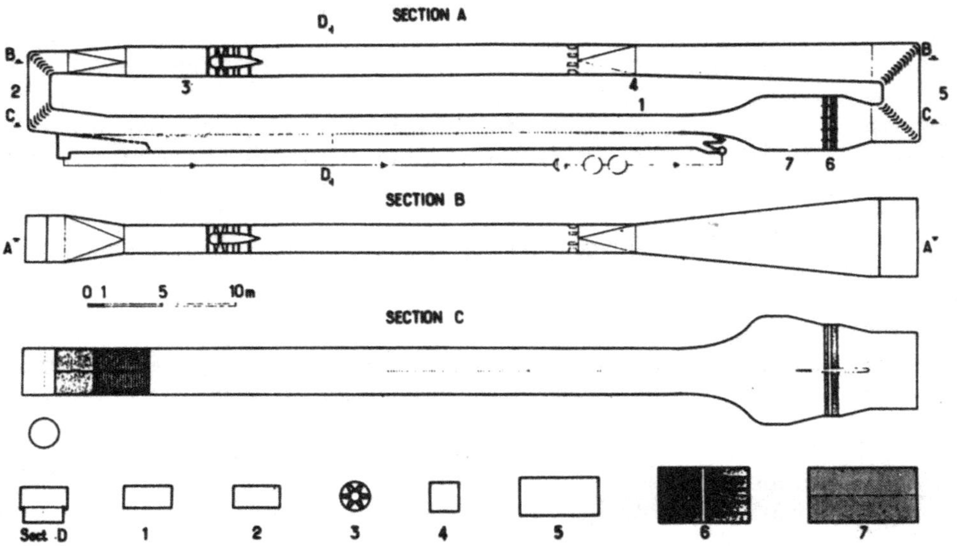

Fig. 1. General Scheme of the I.M.S.T. air-sea interaction facility.

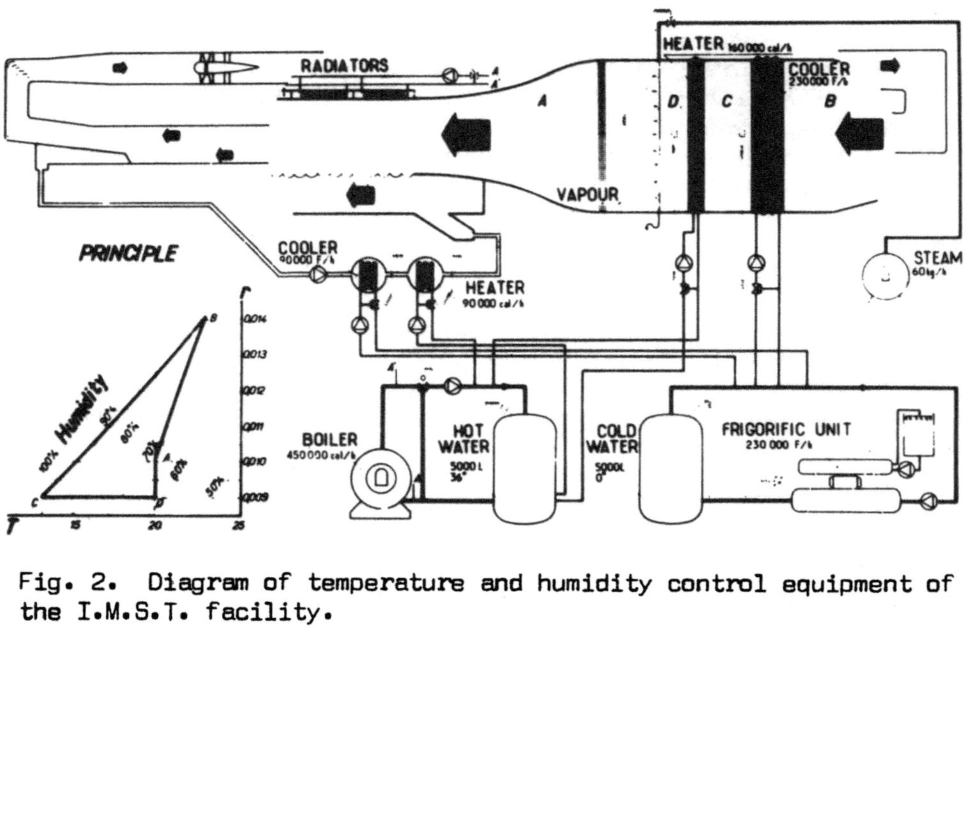

Fig. 2. Diagram of temperature and humidity control equipment of the I.M.S.T. facility.

$$u(z) - u_0 = \frac{u_*}{\kappa} \ln \frac{z}{z_0}$$
$$\theta(z) - \theta_0 = \frac{\theta_*}{\kappa} \ln \frac{z}{z_{0\theta}}$$
$$c(z) - c_0 = \frac{c_*}{\kappa} \ln \frac{z}{z_{0c}}$$

The friction velocity u_*, the scaling temperature θ_* and the scaling humidity c_* are obtained by measuring the slope of the linear portion of the profile plotted in log-linear coordinates : special attention should be paid when evaluating the slope values under stable or unstable conditions. The deviations from the neutral profile have been analytically introduced, by means of the semi-empirical functions of the dimensionless parameter $\zeta = z/L$, given in Businger (1973). By least-squares fitting the corrected data, new estimates for u_* , θ_* and c_* are obtained as well as a new value of L. This has to be repeated by iteration once or twice to get the appropriate scaling parameters and consequently the corresponding fluxes. Turbulent Prandtl and Schmidt numbers have been chosen equal to .74 which is the commonly accepted value for low speed turbulent boundary layers.

The formulation leading to the evaluation of flows by the integral method is as follows :

$$\tau_0(x) = \frac{d}{dx} \rho U_\infty^2 \delta^{**} + \rho U_\infty \delta^* \frac{dU_\infty}{dx}$$
$$Q_0(x) = \frac{d}{dx} \rho C_p U_\infty (\theta_0 - \theta_\infty) \delta_i$$
$$E_0(x) = \frac{d}{dx} \rho U_\infty (C_0 - C_\infty) \delta_e$$

The various conventional thicknesses δ^* (displacement), δ^{**} (momentum), δ_i (enthalpy) and δ_e (concentration) have been determined by integration of the relevant functions. For this purpose, it was necessary to extrapolate the linear parts of the log profiles down to $z = 0$ as already noted by Chambers and al. (1970).

It should be mentioned at this stage that another method for evaluating the water vapor fluxes has also been used: it is a global technique consisting in weighing the amount of water condensed under steady conditions on the drying system of the closed loop facility.

EXPERIMENTAL RESULTS

Three series of experiments have been completed for different values of the Richardson number. The data have been only partially processed, leading to the first results, presented here. Velocity,

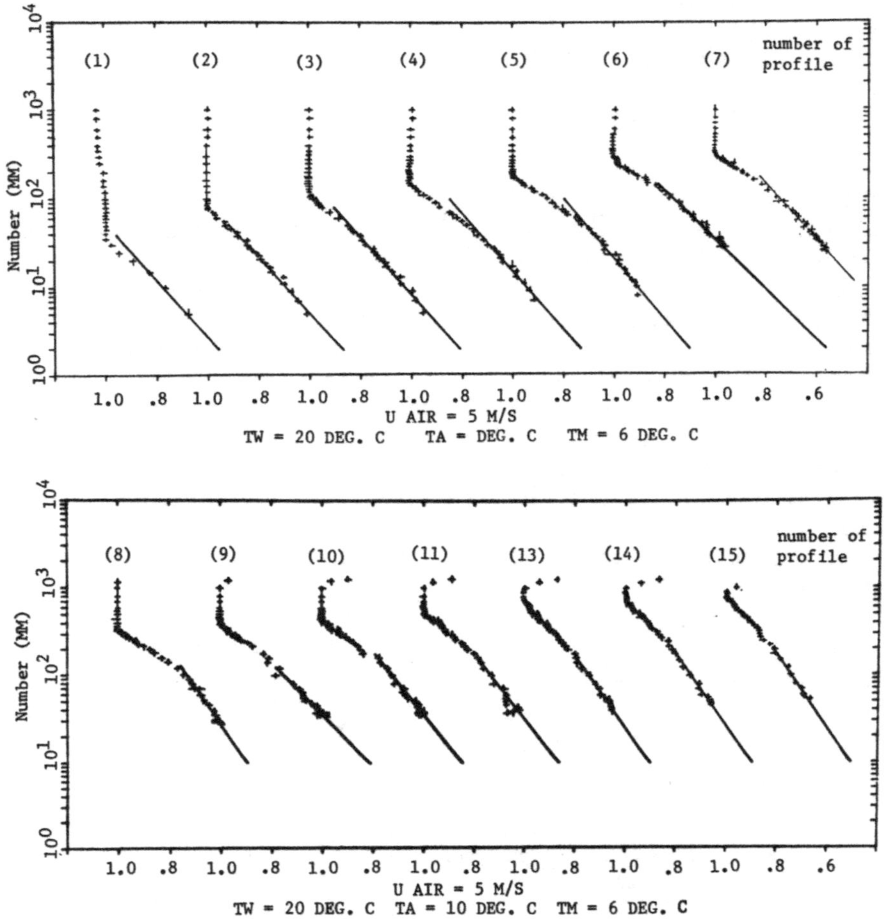

Fig. 3. Normalized velocity profiles versus height at different sections along the wind tunnel center line.

temperature and humidity profiles are shown on figures 3, 4 and 5 in log-linear coordinates, normalized according to : $G-G_o/G_\infty-G_o$, where G represents alternatively u, θ or c; the indexes, "o", and, "∞", identify surface and free stream values. The evolution of the profiles as a function of the distance from the leading edge is clearly seen as well as the progressive increase of the boundary layer thickness. Results shown on figures 3,4 and 5 correspond to a Richardson number $Ri_b = -.003$. The experimental conditions are : θ water = 20°C, θ air = 10°C , θ dew point = 6°C , U_∞ = 5 m/s . The two-dimensionality of the flow was carefully checked by recording vertical profiles at three spanwise positions, in several sections as shown on figure 6.

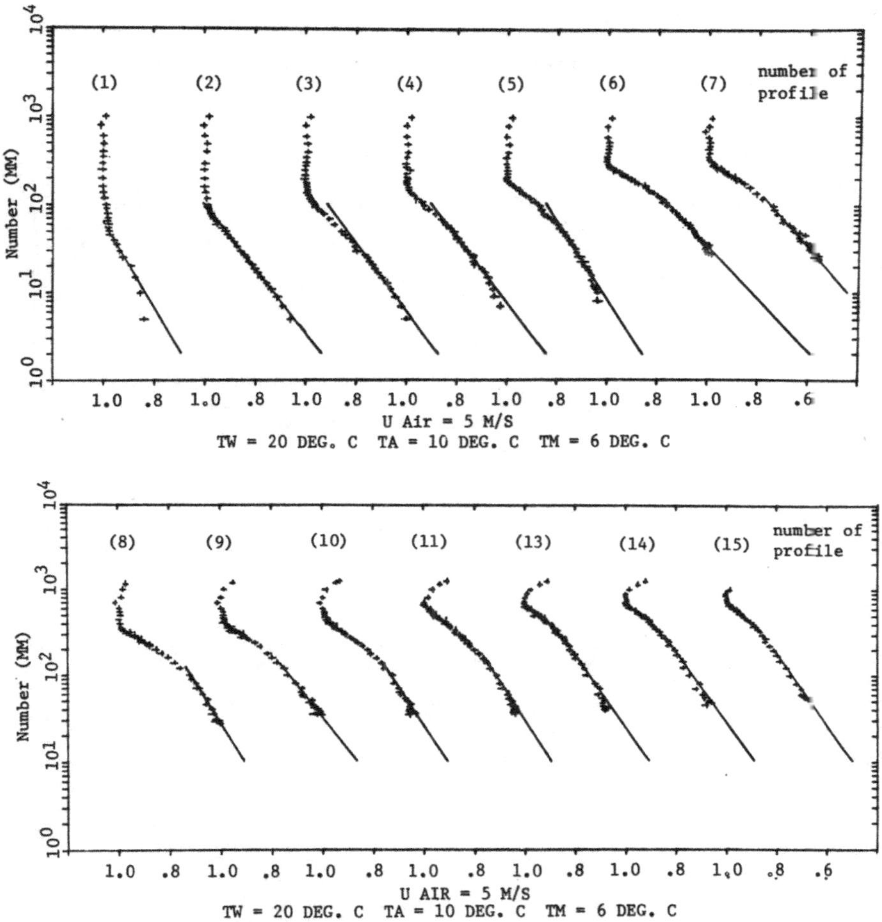

Fig. 4. Normalized temperature profiles versus height.

From these profiles, fluxes were determined by the two above mentioned methods. They are presented in terms of dimensionaless exchange parameters such as the friction coefficient, the Stanton number and the Dalton (or mass Stanton) number :

$$C_f = \tau_0 / \tfrac{1}{2} \rho U_\infty^2$$

$$St = Q_0 / \rho C_p U_\infty (\theta_0 - \theta_\infty)$$

$$Da = St_m = E_0 / \rho U_\infty (C_0 - C_\infty)$$

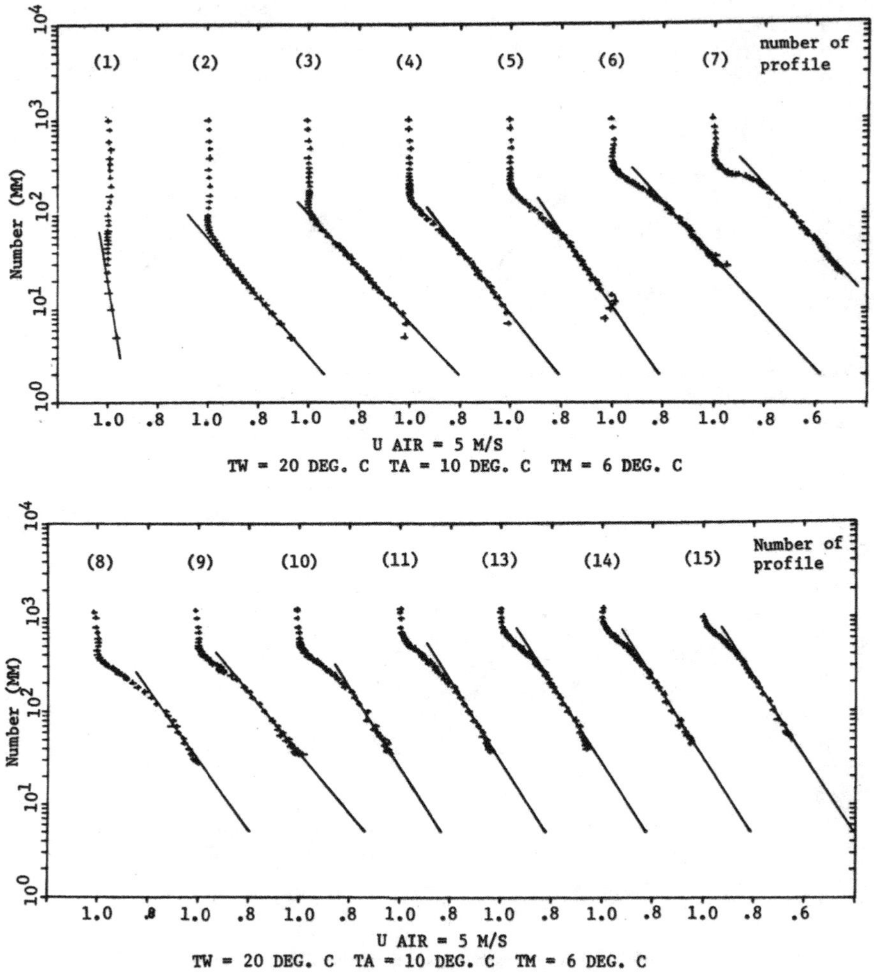

Fig. 5. Normalized humidity profiles versus height.

Results obtained in the two unstable cases by both methods are given in figure 7. The crosses and squares correspond to the profile method, the lines to the integral method. As already shown by Chambers and al.(1970), results obtained from the profile technique are, on the whole, above those from the integral technique. In addition, a slight decrease in the exchange parameters with fetch is observed, particularly with the integral method.

The scatter apparent in profile estimates, although noticeable as usual, appears however to be of the same relative importance as in similar experiments. The following average values can be evalu-

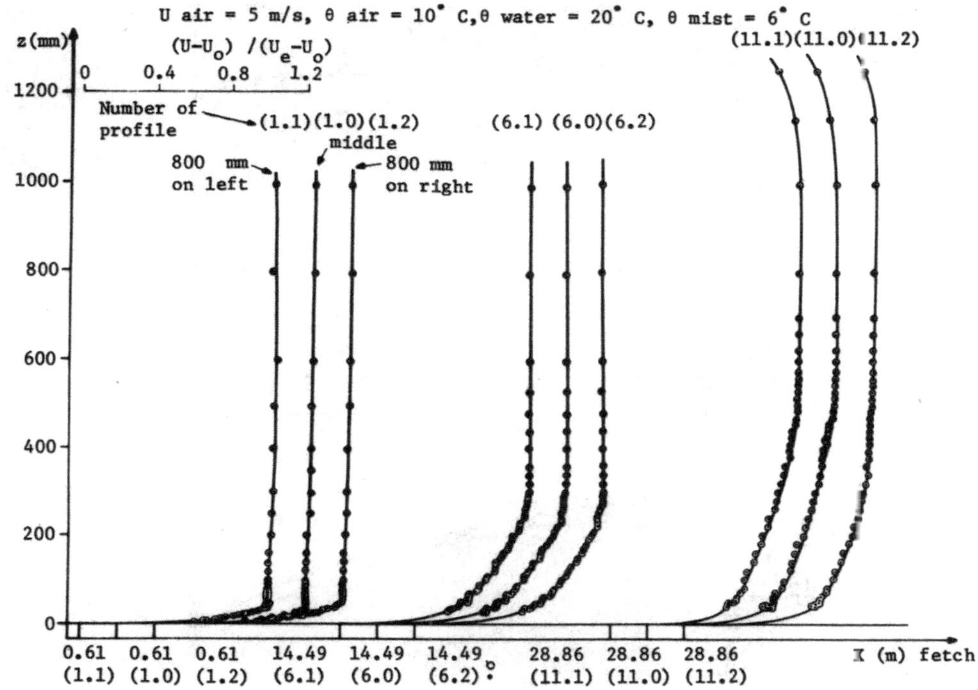

Fig. 6. Spanwise velocity profiles.

ated from those profile estimates:

$$Cf = 4.75\ 10^{-3}, \quad St = 3.01\ 10^{-3}, \quad Da = 3.45\ 10^{-3}$$

The water vapor flux can also be presented in terms of a Sherwood number :

$$Sh = E_o L / \rho \mathcal{D} \ (C_o - C\ )$$

versus the Reynolds number based on fetch : $Re = U\Lambda/\nu$ where Λ is the length of the wind tunnel test section (=40 meters). Results for the unstable cases compare rather favourably with results obtained by other authors and with the semi-empirical relationship proposed by Coantic and Favre (1973) $Sh = .08\ Re^{.75}$, see figure 8.

Results obtained by the global method are displayed on figure 9 for different stratification levels. As pointed out here above,

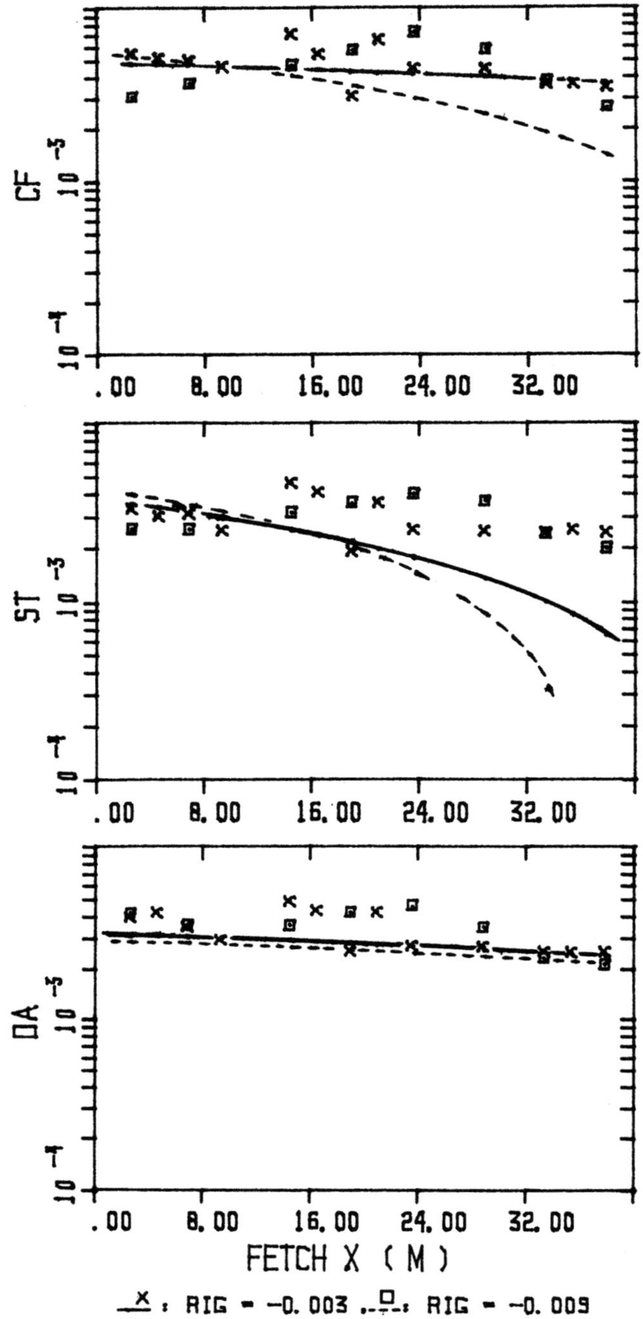

Fig. 7. Transfer coefficients versus fetch.

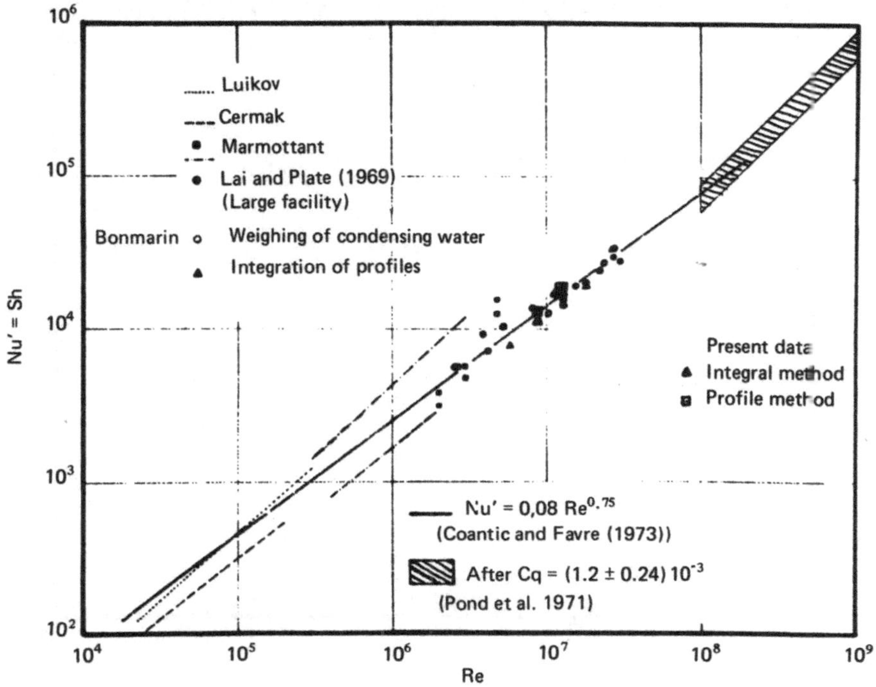

Fig. 8. Sherwood number versus Reynolds number based on fetch.

data from the two unstable cases seem to be in accordance with the semi empirical relationship $Sh = .08\ Re^{.75}$. But, a very clear discrepancy appears for the neutral case, and even more for the stable case. This would tend to show that the proposed relationship is suitable only for an unstable atmosphere. A stability parameter has to be taken into account in the correlation, for a proper description of stratification effects. As an example, one could investigate the use of the dimensionless heat flux $B^* = g/\theta\ H/\rho C_p\ (U_* g)^{-1}$ recently proposed by Laï (1977).

SPRAY EFFECT RESULTS

Data concerning the influence of the spray on water vapor fluxes are now to be presented, as preliminary results of an international cooperation with Drs. R. Laï and J. Wu from the College of Marine Sciences of the University of Delaware. The experiments have been performed by R. Laï and F. Resch in 1976 in the I.M.S.T.

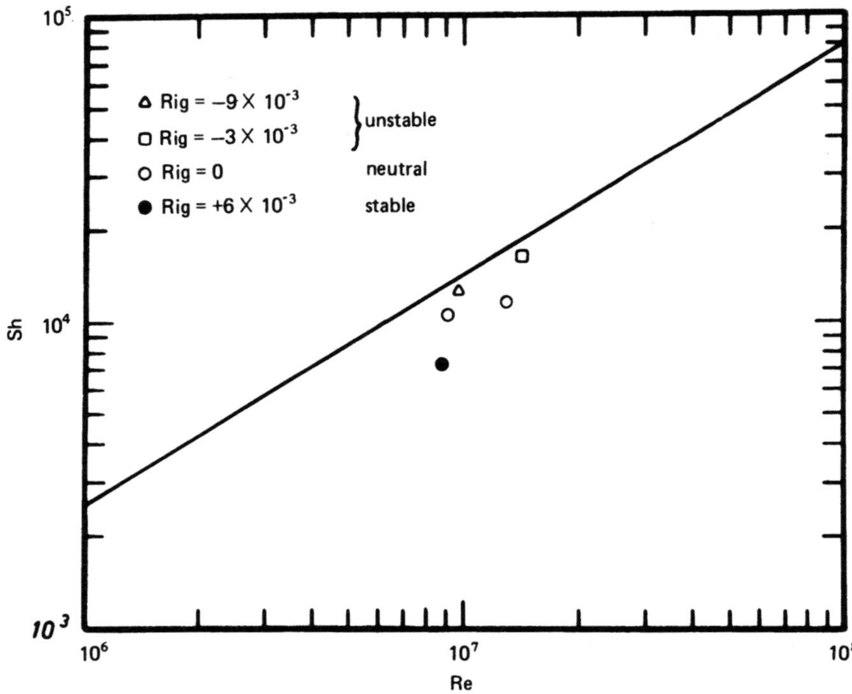

Fig. 9. Results of the global method for different stratification levels.

facility. The water vapor fluxes have been measured by the profile method for wind velocities higher than 8 m/s, while the spray characteristics (namely, number of droplets, droplets size spectrum and droplets velocities) have been determined with a dual laser beam technique. Water vapor estimates are presented on figure 10 as Sherwood numbers, versus Reynolds numbers. One can see a very strong deviation from the previous relationship. One can conclude that the Reynolds number alone does not take into account spray effects, so that a new relevant parameter should be again introduced. This will be done in the near future when results concerning the spray characteristics will be available.

CONCLUSIONS

Preliminary results of the present laboratory experiment indicates that the influence of density stratification significantly affects the evaporation rate.

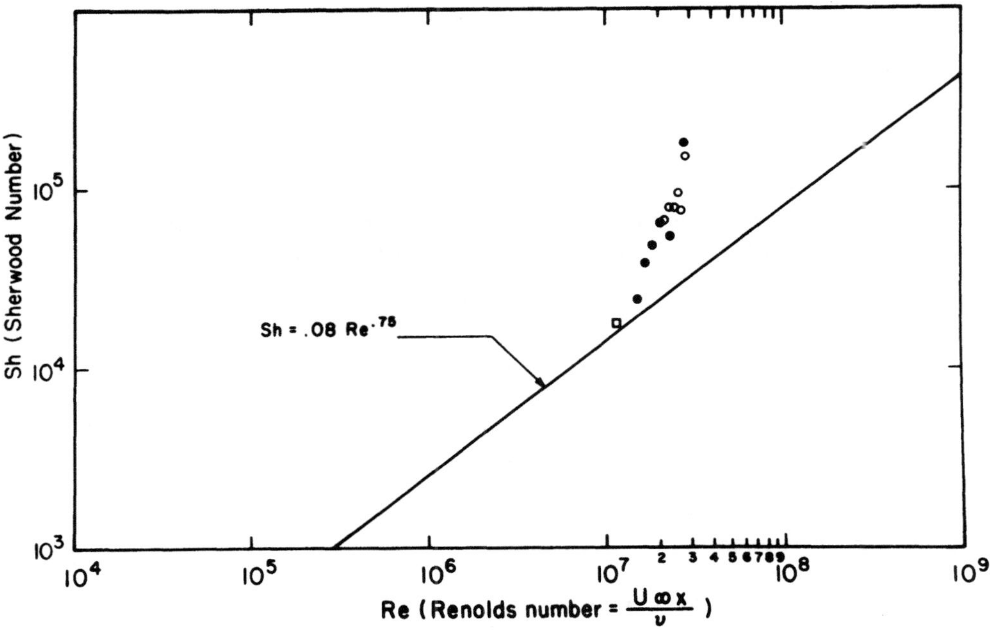

Fig. 10. Influence of spray on evaporation.

Friction coefficients, Stanton numbers and Dalton numbers obtained by the two indirect methods are presented as a function of fetch. As already noted by other authors, those dimensionless exchange coefficients seem to be slightly decreasing with the fetch.

The classical semi-empirical relationship concerning the water vapor fluxes have to be improved to take into account the stratification effects as well as the spray effects at high velocities.

ACKNOWLEDGMENTS

The authors would like to acknowledge herewithin the assistance of organisations and individuals, in particular Dr. M. Coantic, directeur de Recherches au C.N.R.S., for his valuable comments on the flux program and Pr. J. Wu for his continous interest in the spray program. The Electricité de France (Contract N° 16175), and the Centre National Pour L'Exploitation des Océans (Convention N° 771673) have provided their financial and material support. The Centre National de la Recherche Scientifique has also contributed to the present study. Finally, thanks are due to the NATO Scientific Affairs Division (Air-Sea Interaction Panel) which has sponsored the international cooperation between I.M.S.T. and the University of Delaware for the spray program.

NOTATION

B^+	Normalized heat flux
c	Specific humidity
c_*	Scaling humidity
C_f	Friction coefficient
C_p	Specific heat at constant pressure
	Molecular mass diffusion coefficient of water vapor
Da	Dalton number (or mass Stanton number = Stm)
E_o	Water vapor flux
g	Acceleration due to gravity
h	Surface layer thi kness
H	Heat flux across the boundary
L	Obukhov length
r	Perfect-gaz constant (r_a for dry air, r_e for water vapor)
Re	Reynolds number based on fetch
Ri_b	Bulk Richardson number
Sh	Sherwood number
St	Stanton number
u	Mean velocity
U_h	Mean velocity at elevation h
U_∞	Free stream velocity
u_*	Friction velocity
z	Elevation
z_0	Roughness parameters ($z_0, z_{0_\theta}, z_{0_c}$)
δ	Conventional thicknesses ($\delta^*, \delta^{**}, \delta_i, \delta_e$)
κ	Von Karman constant
Λ	Test section length (=40 meters)
ν	Kinematic viscosity
τ_o	Shear stress
θ_a	Absolute temperature $\left(=\frac{\theta h + \theta o}{2}\right)$
θ_h	Temperature at elevation h
θ_*	Scaling temperature
ζ	Dimensionless height $\left(=\frac{z}{L}\right)$

REFERENCES

Businger, J.A., 1973: Turbulent Transfer in the Atmospheric Surface Layer. Workshop on Micrometeorology, D.A. Haugen Ed., American Meteorological Society, 67–100.

Chambers, A.J., P.A. Mangarella, R.C. Street and E.Y. Hsu, 1970: An experimental Investigation of Transfer of Momentum at an Air-Water Interface, Dept. Civil Engr., Stanford Univ., Tech. Dep. N° 133.

Coantic,M.,P. Bonmarin, B. Rouchain and A. Favre, 1969: Etude d'une Soufflerie pour Recherches sur les Echanges Atmosphère-Océans. Aerodynamics of Atmospheric Shear Flow, AGARD Conference Proceedings N° 48, 17–0 to 17–14.

Coantic, M. and A. Favre, 1973: Activities in, and preliminary results of, air-sea interactions research at I.M.S.T. Advances in Geophysics, Vol. 18, 391–405.

Deardorff, J.W., 1968: Dependence of Air-Sea Transfer Coefficients on Bulk Stability. J. Geophys. Res., 73, 2549–2557.

Kondo, J., 1975: Air-Sea Bulk Transfer Coefficients in Diabatic Conditions. Boundary-Layer Meteorology, 9, 91–112.

Laï, R.J., 1977: Evaporation from a Warm, Wavy Surface: A Laboratory Study. J. Phys. Oceanography, Vol.7, N°3, 431–435.

Mangarella, P.A., A.J. Chambers, R.L. Street and E.Y. Hsu, 1971: Energy and Mass Transfers through an Air-Water Interface. Dept. Civ. Engr., Stanford University, Tech. Dep. N°134.

Marmottant, B., 1974: Contribution à l'Etude de l'Evaporation dans une Couche Limite Turbulente d'Interface Air-Eau. Doctoral Thesis, Univ. of Provence, Marseille, France.

Monin, A.S., 1970: The Atmospheric Boundary Layer. Annual Review Fluid Mechanics, Vol.2, 225–250.

_____ and A.M. Obukhov, 1954: Basic Laws of Turbulent Mixing in the Atmosphere near the Ground. Tr. Akad. Nauk SSSR Geofiz. Inst., N°24 (151), 163–187.

_____ and A.M. Yaglom, 1971: Statistical Fluid Mechanics, Vol.1, Tr. M.I.T. Press, Cambridge, Mass. .

Obukhov, A.M., 1946: Turbulence in an Atmosphere with a Non-Uniform Temperature. Tr. U.S.S.R. Inst. Teoret. Geofiz., N°1 (Transl. in Boundary Layer Meteor., 2, 7 –29).

Peri, G., 1975: Application de Méthodes Hygrométriques à l'Etude du Mécanisme de l'Evaporation lors de la simulation des Echanges Océan-Atmosphère. Doctoral Thesis, Univ. Aix-Marseille, Marseille, France.

DISCUSSION

J. A. Businger:

The number $B^{+} = \frac{g}{\theta} \frac{H}{\rho C_{\rho}} (u_{*}g)^{-1}$ can be simplified into $\theta_{*}/$ θ_{*}/θ . Is this what was intended? (θ_{*} is the scaling temperature: $\theta_{*} = \frac{\overline{w'\theta'}}{u_{*}}$).

F. Resch:

As I said in my presentation this number B^{+} is proposed by Dr. R. Lai in his publication in the Journal of Physical Oceanography. It represents the buoyancy flux $g/\theta\ H/\rho C_{\rho}$ non dimensionalised by $(u_{*}g)^{-1}$. Then it represents a dimensionless heat flux.

E. J. Plate:

The remark by J. A. Businger that the coefficient

$$B = \frac{g\,H}{\theta\rho C_{\rho}} \frac{1}{u_{*}g} = \theta_{*}/\theta$$

as used by R. Lai ist the ratio of the shear temperature to the mean temperature should not be understood to imply that such a parameter has no meaning. Our own experiments which have recently been published in the Proceedings, 17 IAHR Congress, Baden-Baden, August 1977 (by Zengefeld and Plate) show that a relationship of Sherwood vs. Reynolds number as used by you is not satisfactory in explaning the observed curves. Other factors enter, and dimensional analysis yields a temperature ratio as an additional factor. The Russians have called such a ratio a "Gukhman" number. We are further investigating this problem, and I hope that we shall have an opportunity soon to compare notes and results with Drs. Resch and Selva.

F. Resch:

This will be very fruitful and I do hope it will be possible very soon.

W. Pierson:

(1) In this magnificent facility, why do the data still

scatter?

(2) What are the standard deviations on the points in your various scatter plots?

F. Resch:

(1) First of all the scatter is comparable to the scatter of similar measurements in laboratory by other authors. Then results from the two indirect methods are gathered and it is known that results obtained by the profile method are usually greater than those obtained by integral method.

(2) Standard deviations obtained are as follows:

$$C_f = 4.75.10^{-3} \pm 1.4\ 10^{-3}$$

$$St = 3.01.10^{-3} \pm 7.5.10^{-4}$$

$$Do = 3.45.10^{-3} \pm 9.10^{-4}$$

M. Coantic:

The only explanation I can offer is that such measurements are very difficult, so that estimates of heat and mass transfer rates by the profile technique exhibit in general a fairly large scatter.

S. C. Ling:

(1) Will F. Resch care to comment on the same question Prof. Street of Stanford University had put to me yesterday concerning a sweeping statement that he had made in his presentation that the effect of water droplets are unimportant to the heat transfer process? Since your research is very similar to that of R. L. Street, I wonder if you have found the same conclusion? I had no wish to answer his question, because my work is intended for use over an open ocean, while his was for a very small confined tank with very low Reynolds number.

(2) If indeed R. L. Street's statement was valid only for a very small tank and under a specific artificial set up of his test system as Prof. Wu had just pointed out very clearly, I wonder if R. L. Street would like to have a final word? I must apologize for reopening this public debate at the insistence of Prof. Street - a consequence I had tried very hard to avoid.

F. Resch:

I do not want to comment on this point at that stage of our work. We should first, as I have said, combine evaporation results with droplet characteristics measured by Dr. Lai. Unless Dr. Wu wishes to comment on it.

J. Wu:

In Figure 6 of the paper "Fluxes through the boundary layers at an Air-Water Interface: Laboratory Studies" by R.L. Street et al. the results were not compared on solid physical grounds. The wind velocity instead of the wind-friction velocity was used, the elevation was not nondimensionalized with respect to the boundary-layer thickness, and no parameter describing the wave conditions was used.

There are two length scales involved here: one is the ejection height of jet drops and the other is the boundary-layer thickness. The vertical distribution of large droplets is likely to be governed by the ejection height, while that of small droplets is likely to be governed by the boundary-layer thickness. In as much as we are dealing with rather small droplets here, the boundary-layer thickness should be used.

It appears that we all agree that the production of spray is related to breaking waves. Therefore, a parameter describing the wave conditions must also be used, if the data obtained from various wind-wave tanks are compared. In any event, effort should be made to compare these sets of data properly and to resolve the differences among them. As for the importance of spray to the evaporation from the sea surface, we can perhaps postpone our discussion to a later date as both myself and Prof. Ling have just started our field experiments.

R. L. Street:

In response to the questions and the subsequent comment by Wu, I can add:

(1) The measurements under spray conditions by Dr. Resch et al. and reported as Sherwood numbers are entirely consistent with data under spray conditions obtained at Stanford and shown in the paper by Street et al. at this meeting.

(2) I agree with Dr. Wu that facility conditions do affect spray generation process, in the Stanford facility we find that the evaporation from the spray, while it increases rapidly with wind speed, does <u>not</u> make a significant contribution to the total evaporation.

FLUXES THROUGH THE BOUNDARY LAYERS AT AN AIR-WATER INTERFACE: LABORATORY STUDIES

R. L. Street and C. S. Wang*, D. A. Mc Intosh**, and
A. W. Miller Jr.***

*Stanford University, Stanford, California, U.S.A.
**Michigan State University, East Lansing, Michigan, U.S.A.
***Brigham Young University, Provo, Utah, U.S.A.

The energy and mass exchange processes and temperature structure within the surface layers at an air-water interface have been examined in a large laboratory facility. Measurements in the air surface layer at low and moderate windspeeds ($\leq$10 m/s) establihed that the air flow is fully rough for $U_\infty > 4$ m/s and that, while the magnitude of heat transfer depends on surface conditions, the mechanisms of transfer apparently do not. Measurements in the water surface layers plus transport data from the air surface layer led first to determination of the aqueous sublayer thickness and subsequently to a coupled surface layer theory (based on rough boundary theory) which is in quantitative agreement with experiment. Measurements at high wind-speed (>10 m/s) and under conditions of wave breaking and spray formation are used to show (1) that spray evaporation does not make a significant contribution to the total evaporation and latent heat transport and (2) that, when experiment and theory are compared, wave breaking and spray formation apparently act to decrease the transport of sensible and latent heat compared to that expected for the given wave height and surface drag.

INTRODUCTION

The energy and mass exchange processes and the temperature structure within the boundary layers at an air-water interface have been examined in experiments in the Stanford Wind, Water-Wave Research Facility. The data, which are for neutrally stable layers where forced convection dominates, created the opportunity to draw some conclusions about the behavior of the interface fluxes and our ability to predict them. McIntosh, et al. (1975b) made measurements in the air boundary layer above the interface with a three-wire probe, consisting of two hot wires and a resistance thermometer. In a second experiment, Wang (1977) made measurements with an electrostatic capacitance-wire probe system to detect spray at high windspeeds. Miller, et al. (1975) made measurements of the conditions in the aqueous layer beneath the interface with an infrared radiometer and thermister system.

Based on the aqueous layer measurements, Street and Miller (1977) derived an estimate of the viscous and thermal sublayer thicknesses in the water when the interface was fully rough. The theory for rough boundary flows has been extended to both air and water surface layers by Street (1977).

In this paper we first summarize key results from the above and, then, draw them together with the work of Banner and Melville (1976) and Easterbrook (1968) to generate a hypothesis regarding the changes that occur in the fluxes of heat and mass across the interface as a consequence of high windspeed and spray generation.

THE FACILITY AND INSTRUMENTATION

The Stanford Wind, Water-Wave Research Facility (Fig. 1) is about 35 m long; the test section is approximately 20 m long, 0.9 m wide and 1.93 m high. The channel is filled with water to a depth of about 1 m, leaving a 1 m deep air flow section. Air flow is produced by drawing air through the test section with a fan at the downstream end of the channel. Honeycombs, a series of turning

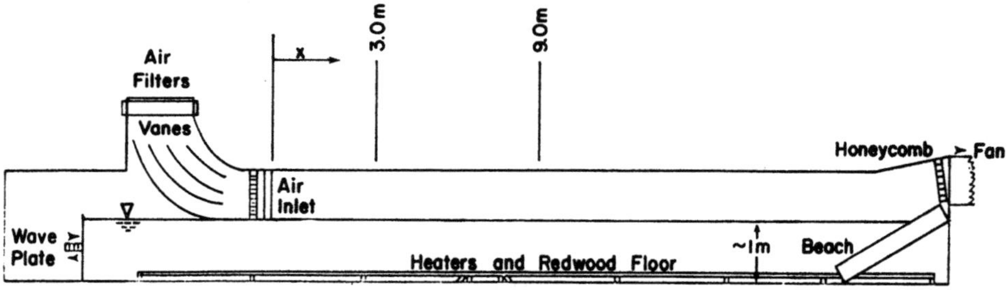

Figure 1. Schematic of the Stanford Wind, Water-Wave Research Facility

vanes in the inlet, a set of filters, and several small mesh screens straighten and condition the air flow. In the water a "beach" is used to minimize wave reflections into the text section. The water is heated from below by electric heating cables giving relatively constant water-air temperature differences (±0.1°C) over an experimental run. In all cases free-stream air velocity (U_∞) was monitored with a standard pitot-static tube and the water surface elevation (η) was measured with a capacitance-type wave gauge. Mean temperatures were measured with bead-in-glass thermisters.

In the McIntosh, et al. (1975b) studies of the air surface layer, fluctuating horizontal and vertical velocity components (u',v') and temperature (T) were measured with a triple wire sensor which consisted of two platinum hot-wires (5.0 μm d) in an X-array and a platinum cold-wire (2.5 μm d). The cold wire, which served as a resistance thermometer, was located 0.6 mm upstream of the imaginary intersection of the X-array hot-wires. They were operated at a sufficiently large overheat ration (1.80) to limit contamination of the velocity signal by temperature fluctuations. Tests confirmed that the cold-wire signal was not influenced by the hot wires.

In the Wang (1977) study of spray generated at high wind speeds the wet-bulb temperature, necessary for evaluating mean humidity, was obtained by attaching a wet wick to a thermister. Special precautions were taken to prevent the dry-bulb temperature thermister from being wetted by sprays. Spray data was obtained by use of an electrostatic capacitance wire probe 63 μm in diameter and 60 mm long (Wang, et al., 1974). This probe detects water drops in a range from 25 to 1250 μm diameter. Drops impacting the detector cause a change in the probe's capacitance; this change is recorded and processed by a digital data acquisition system.

In the Miller, et al. (1975) experimental determination of temperature and flux through the aqueous surface layer the near surface water temperature (at a depth z_{rad} of 140 μm) was measured with an infrared radiometer with an indium antimonide detector. Bulk water temperature was measured at a depth z_b of 100 mm. Details of the process of extracting the near-surface/bulk-temperature difference ($T_{rad} - T_b$) are given in Miller, et al. (1975). Calibrations showed the temperature difference to be accurate within ±0.01°C. Other parameters, such as the mean free-stream windspeed, temperature and humidity and the water surface elevation, were obtained as well, as described above. They formed the basis for calculating the total heat transfer Q_T in the aqueous surface layer and the surface roughness through use of data collected by previous investigators in the Stanford facility (see, e.g., Mangarella, et al., 1973; Bole and Hsu, 1969). The latent and sensible transfers from the interface, Q_H and Q_S, were computed from previous data; radiative transfer was estimated from the measured near-surface temperature T_{rad}.

BASIC RESULTS

Characteristics of the Air Flow

McIntosh, et al. (1975b) measured the effect of interfacial wave conditions on the transport of sensible heat and momentum. For our purposes here their data subset of mean flow characterizations and turbulent fluxes for wind-generated waves provide the necessary information.

To avoid the most rapid fetchwise gradients measurements were taken at a location 12 m downstream from the air inlet. Data were taken at windspeeds ranging from 2 to 10 m/s with a water-air temperature difference of 10°C. Turbulent fluxes were measured at five points in the boundary layer, ranging in elevation from 0.6 to 4 cm at U_∞ = 2 m/s to 2.6 to 6 cm at U_∞ = 8 m/s, for example. The eight channels of data recorded for these runs included freestream velocity, the outputs of the two hot-wires and the resistance thermometer, wave height, and the bulk-water temperature. Mean measurements of temperature and velocity were made at approximately 25 points across the boundary layer profile. The data was acquired and subsequently analyzed on the laboratory's data-acquisition and reduction system.

Direct measurements of the fluxes of heat and momentum at various elevations (z) above the mean water level were determined from the correlations of the velocity and temperature fluctuations:

$$Q_S(z) = \rho_a c_{p_a} \overline{v'T'} \quad (1)$$

$$\tau(z) = -\rho_a \overline{u'v'} \quad (2)$$

where the overbar denotes a time average, Q_S is the sensible heat flux, u', v', T' denote the deviation of the horizontal and vertical velocity components and temperature from their respective mean values, τ is the shear stress, ρ_a is the air density and c_{p_a} is the specific heat of air at constant pressure. The procedure used to obtain the interfacial fluxes was a simple extrapolation of the $\overline{u'v'}$ and $\overline{v'T'}$ vertical distributions to the mean water level (z = 0). It is the values $Q_S(0)$ and $\tau(0)$ which are used to normalize the profiles defined below.

Variations in the mean velocity and temperature profiles can be used to assess the degree of roughness of the air-water interface. Typically,

$$\frac{u}{u_{*_a}} = \frac{1}{k'} \ln \frac{z}{K_s} + B_v \qquad (3)$$

where $u_{*_a} = [\tau(0)/\rho_a]^{1/2}$ is the friction velocity, $k' = 0.40$ is the von Karman constant, K_s is the equivalent sandgrain roughness and B_v is a function of the surface roughness; $B_v = 8.5$ for fully rough flow, i.e., for $K_s u_{*_a}/\nu_a > 55$ (Hinze, 1975).

The present velocity data has been corrected (according to Wu, 1975) to exclude the effect of the surface drift current, i.e., the drift velocity $u_d = 0.55\ u_{*_a}$ was subtracted from the measured windspeed u_w. The data is plotted in law of the wall coordinates, viz., $u^+ = u/u_{*_a}$, $z^+ = u_{*_a}z/\nu_a$ and $u = u_w - u_d$. For the demarcation of the fully rough regime we introduce $B_v = 8.5$ and $K_s = 55\ \nu_a/u_{*_a}$ plus law of the wall coordinates into Eq. (3). The result is

$$u^+ = 2.5 \ln z^+ - 1.52 \qquad (4)$$

This equation plus the measured profiles are shown in Fig. 2. At the lowest windspeed the velocity profile is almost identical to the accepted smooth flat plate correlation as expected because at U_∞ = 2 m/s there are no visible waves generated at the interface. However, as the windspeed is increased and wind-generated waves appear at the interface a progressive shift in the velocity is observed. For $U_\infty \approx 4.0$ m/s the wind-generated waves just begin to form a fully rough surface. The interface is fully rough for $U_\infty \geq 5$ m/s at this fetch (12 m).

The temperature profiles are also presented in wall coordinates, viz.,

$$t^+ = \frac{1}{K_T} \ln z^+ + A_T \qquad (5)$$

where $t^+ = [T_b - T_a(z)]/t_*$, T_b is the bulk water temperature, $T_a(z)$ is the air temperature, and $t_* = Q_S(0)/\rho_a c_{p_a} u_{*_a}$ is the friction temperature. At present the values of K_T and A_T are not firmly established. Yaglom and Kader (1974) suggest a value of $K_T \approx 0.47$, independent of surface conditions and close to $k' = 0.40$. The data of Kays (1969) in a zero pressure gradient flow over a smooth flat plate yield $K_T = 0.42$ and $A_T = 1.80$. In Fig. 3 the close agreement between the $U_\infty = 2$ m/s and the flat plate profiles suggests again that the surface is smooth at this windspeed. As the windspeed is increased and the wind-generated waves grow, the profiles tend to congregate about a single line despite the progressive increase in surface roughness previously observed in reference to the velocity profiles. The K_T value for the individual profiles ranges from

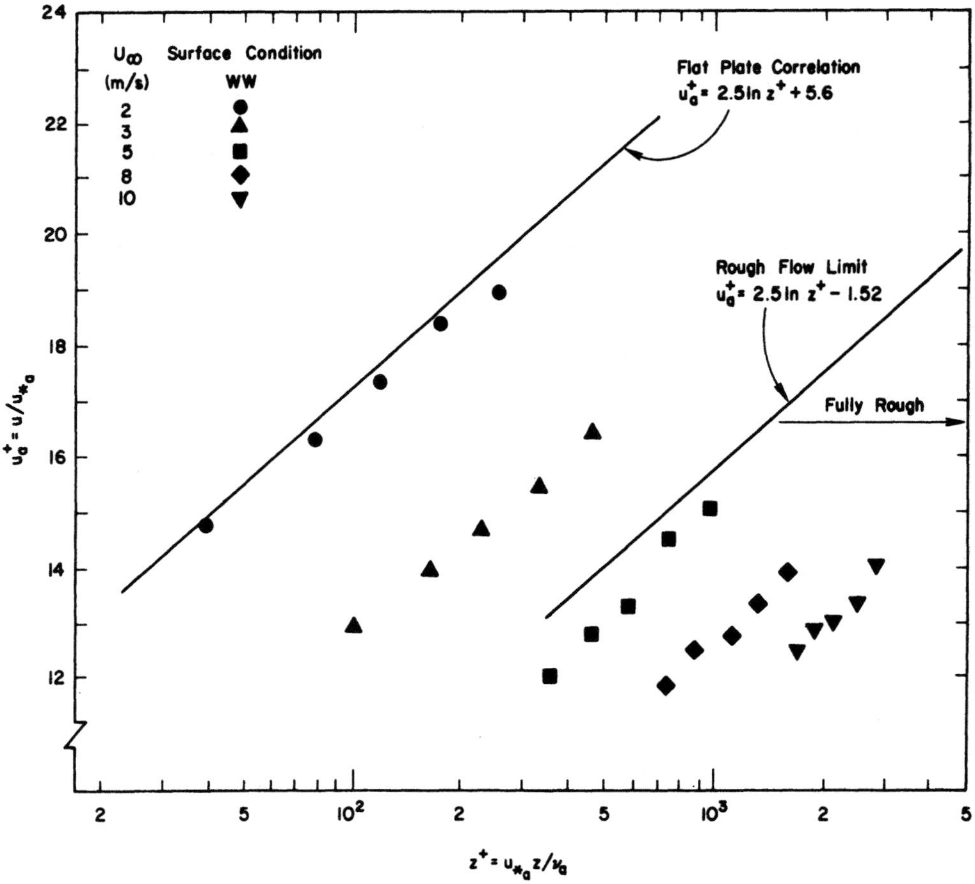

Figure 2. Mean Velocity Profiles in Law of Wall Coordinates

about 0.46 at U_∞ = 2 m/s to 0.56 at U_∞ = 3 m/s to 0.42 for U_∞ > 3 m/s. On average these results are consistent with Yaglom and Kader'.

In summary, using these and other results McIntosh, et al. (1975b) demonstrate that the transfers of heat and momentum are fundamentally different at an air-water interface. The basic air flow is fully rough for windspeeds greater than 4 m/s, and the normalized temperature profiles are similar, suggesting that a similar mechanism of heat transfer is operating in all cases. However, while the mechanism of transfer of heat away from the boundary is not affected by the surface roughness changes, the magnitude of the actual transfers is. We demonstrate this dependence below.

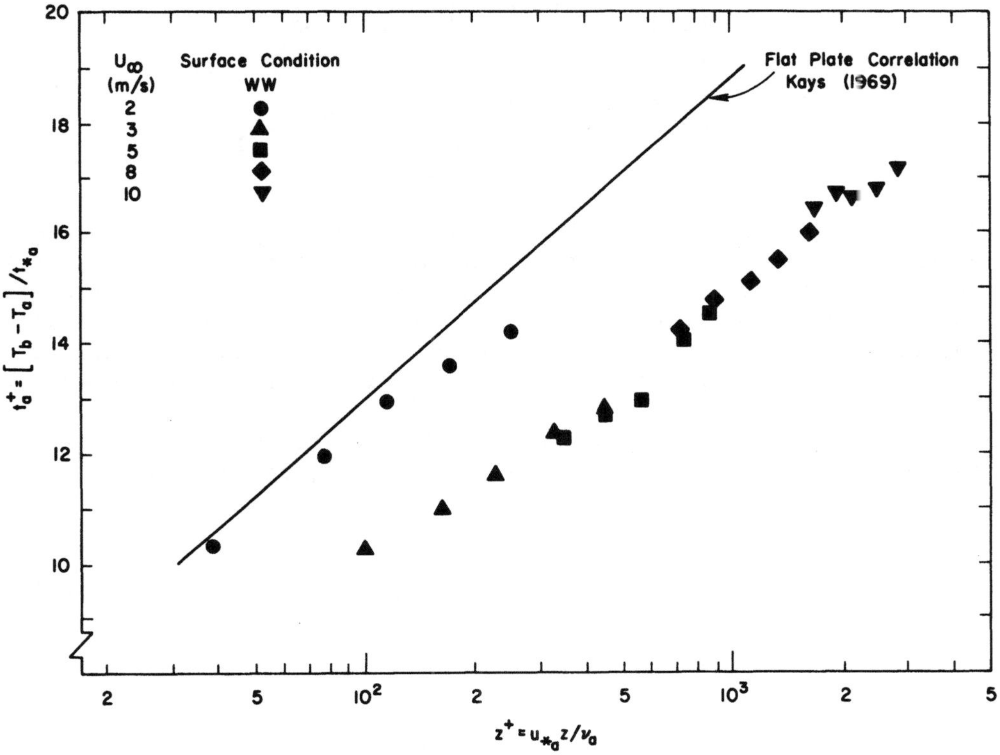

Figure 3. Mean Temperature Profiles in Wall Coordinates

Theoretical Considerations

In Street and Miller (1977) the thicknesses of the viscous and thermal sublayers in the water beneath the air-water interface were obtained by an application of the theory of rough-wall flows to a subset of the data obtained by Miller, et al. (1975) for neutrally stable air boundary layers. For fully rough flow the dimensionless viscous sublayer thickness $\delta_{v_w}^+$ is proportional to the square root of the roughness Reynolds number $h_w^+ = u_{*_w}h/\nu_w$ based on mean roughness height h. Similarly the dimensionless thermal sublayer thickness $\delta_{t_w}^+$ is proportional to $(h_w^+)^{1/2}$ and to $(Pr_w)^{-1/3}$ where $Pr_w = \nu_w/\kappa_w$ is the molecular Prandtl number, ν_w is the kinemetic viscosity of water and κ_w is the molecular diffusivity of water.

Rough wall theory is valid for $h_w^+ \geq 100$. Under this condition one can write the effective thermal diffusivity K_w for water as follows (Street and Miller, 1977; Street, 1977):

$$K_w(z) = \begin{cases} \kappa_w\left[1 - \left(z^+/\delta^+_{t_w}\right)^3\right], & z^+ > -\delta^+_{T_w} \\ \kappa_w\left[1 - k'P_{r_w} z^+ / P_{r_{t_w}}\right], & z^+ < -\delta_{T_w} \end{cases} \tag{6}$$

Here, $P_{r_{t_w}}$ is the turbulent Prandtl number, and δ_{T_w} is the thickness of the matched layer, viz., the point at which the cubically-varying estimate of $K_w(z)$ is equal to the $K_w(z)$ appropriate to a logarithmic variation of temperature in the fully turbulent part of the boundary layer; $\delta^+_{T_w} = u_{*_w}\delta_{T_w}/\nu_w$. A schematic of the thickness structure in the interface surface layers is given in Fig. 4; $z \leq 0$ in the water.

For a steady mean flow with negligible fetchwise gradients the equation for constant vertical heat flux Q_T in the aqueous surface layer is

$$\frac{dT_w}{dz} = \left(\frac{-Q_T}{\rho_w c_{p_w}}\right) \frac{1}{K_w(z)} \tag{7}$$

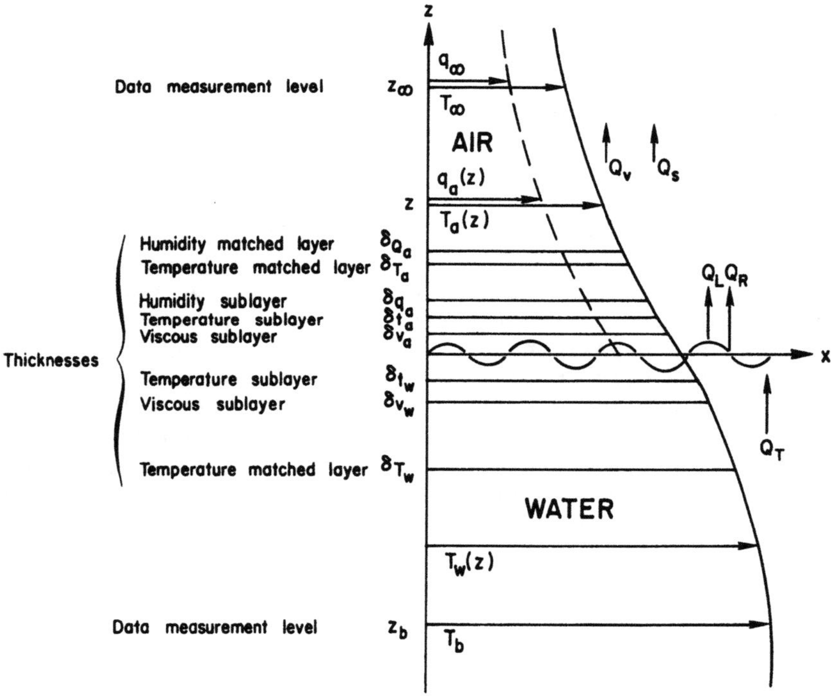

Figure 4. Schematic of the Surface Layers at an Air-Water Interface

With Q_T, ρ_w and c_{p_w} taken as constant and $K_w(z)$ as a known function of z, κ_w, h', Pr_w, Pr_{t_w}, δ^+_{tw} and $\delta^+_{T_w}$, Eq. (7) can be integrated.

Miller, et al. (1975) obtained the bulk water temperature $T_b = T_w(-100\ \text{mm})$, the near surface temperature $T_{rad} = T_w(-140\ \mu\text{m})$ with the radiometer, and the fluxes of radiation Q_R, sensible heat Q_S and latent heat Q_L plus the surface conditions h and u_{*_w}. The shear velocity u_{*_w} was related to the shear velocity u_{*_a} according to

$$u_{*_w} = (\rho_a/\rho_w)^{1/2} u_{*_a} \tag{8}$$

which implies stress continuity at the interface. Using this data, the fact that

$$Q_T = Q_S + Q_L + Q_R \tag{9}$$

and the analytic expression for Q_T obtained from integration of Eq. (7) between $z_{rad} = -140\ \mu\text{m}$ and $z_b = -100$ mm, Street and Miller (1977) showed that

$$\delta^+_{v_w} = 0.37(h^+_w)^{1/2} \ ; \quad \delta^+_{t_w} = 0.37(h^+_w)^{1/2}(Pr_w)^{-1/3} \ ;$$

$$\delta^+_{T_w} = (k'/Pr_{t_w})^{1/2}(\delta^+_{v_w})^{3/2} \tag{10}$$

For cases in which the interface is rough for both air and water surface layers, it seemed logical to extend this theory to the air boundary layer and to model the sensible and latent heat fluxes there in the same manner. Figure 4 gives a schematic of the thickness and variable layout as used in Street (1977). The essence of the technique is summarized briefly. The basic concept is the expression of the fluxes Q_T, Q_S and Q_L in terms of gradients under conditions of steady, constant-flux-layer flow. Thus,

$$Q_T = -\rho_w c_{p_w} K_w \frac{dT_w}{dz} \ , \quad 0 \geq z \geq z_b \tag{7}$$

$$Q_S = -\rho_a c_{p_a} K_a \frac{dT_a}{dz} \ , \quad 0 \leq z \leq z_\infty \tag{11}$$

$$Q_L = -\rho_a L K_q \frac{dq_a}{dz} \ , \quad 0 \leq z \leq z_\infty \tag{12}$$

At the interface the net radiation flux Q_R was given in Miller, et al. (1975) by

$$Q_R = E_w \sigma (T^4_{s_{ABS}} - E_a T^4_{\infty_{abs}}) \tag{13}$$

Here, q is the specific humidity; K_w, K_a, K_q are the effective diffusivities of heat in water [see Eq. (6)], of heat in air, and of water vapor in air, respectively. L is the latent heat of vaporization of water, σ is the Stefan-Boltzmann constant, and E is the emissivity ($E_w = 0.97$; $E_a = 0.80$).

Equations (7), (11) and (12) are integrated from the interface, where values are T_s and q_s, to the appropriate measuring level, where the values are T_∞, q_∞ and z_∞, or T_b and z_b, in the same manner illustrated above for the water surface layer. Table 1 shows the key theoretical expressions. For a given windspeed we assume that u_{*a} and u_{*w}, h and the medium properties κ_w, κ_a, κ_q, c_{pa}, ρ_a, etc. are known. The values T_b, z_b, and T_∞, q_∞, z_∞ are measured. The only unknown is T_s which determines the fluxes because q_s and L are functions of T_s alone. We find T_s under the constraint (see Eq. (9))

$$Q_T = Q_S + Q_L + Q_R \tag{14}$$

To test the theory, we used 26 cases of wind-generated water waves (Miller, et al., 1975) with water-air temperature differences ($T_b - T_\infty$) between 0 and 15°C, fetches (x) of 9.5 and 14.5 m, and water roughness Reynolds numbers $h^+_w > 100$ (actually $U_\infty \approx 7.5$ and 10 m/s; while $U_\infty = 5$ m/s data agree with the theory as well, $h^+_w < 100$ for them). These data are from the same data set used by Street and Miller (1977) and described above. From the discussion of the McIntosh, et al. (1975b) work above we see that the theory should be applicable over these ranges.

Figure 5 shows the comparison between the experimental and theoretically-predicted Q_T values. These results are typical of Q_S and Q_L as well. In the calculation $Sc_{t_a} = 1.19$ was used, while $Pr_{tw} = 1.0$, $Sc_a = 0.6$, $Pr_a = 0.71$ and $Pr_w \geq 1$ was computed for each T_b. Independent experiments by McIntosh, et al. (1975a) showed $Sc_{t_a} = 1.19 \pm 0.08$ for $U_\infty = 3.5$ to 11.0 m/s in the Stanford facility; $Sc_{t_a} = \phi_E = \varepsilon_M/\varepsilon_E$ is the dimensionless humidity gradient for neutrally stable cases (Monin-Obukov length $\rightarrow \infty$) and ε_M and ε_E are the eddy diffusivities of momentum and water vapor.

These results suggest that the coupled surface layer theory based on rough wall concepts is valid under the assumptions of steady-state, negligible fetchwise gradients and constant flux surface layers. We will use this theory below to develop a hypothesis for high windspeed cases in which spray occurs.

Table 1 Summary of Theoretical Results

WATER:

$$\delta^+_{v_w} = 0.37\left(h^+_w\right)^{1/2} \; ; \; \delta^+_{t_w} = Pr_w^{-1/3}\delta^+_{v_w} \; ; \; \delta^+_{T_w} = \left(\frac{k'}{Pr_{t_w}}\right)^{1/2}\delta^{+\,3/2}_{v_w}$$

$$K_w(z^+) = \kappa_w\left[1 - \left(z^+/\delta^+_{t_w}\right)^3\right], \; z^+ > -\delta^+_{T_w} \; ; \; K_w(z^+) = \kappa_w\left[1 - k'Pr_w z^+/Pr_{t_w}\right], \; z^+ < -\delta^+_{T_w}$$

$$Q_T = \rho_w c_{p_w} u_{*_w}(T_b - T_S)/r_{H_w}$$

$$r_{H_w} = \frac{Pr_w\,\delta^+_{t_w}}{3}\left\{\frac{1}{2}\ln\left[\frac{\left(\delta^+_{t_w} + \delta^+_{T_w}\right)^2}{\delta^{+\,2}_{t_w} - \delta^+_{t_w}\delta^+_{T_w} + \delta^{+\,2}_{T_w}}\right] + \sqrt{3}\left[\tan^{-1}\frac{2\delta^+_{T_w} - \delta^+_{t_w}}{\sqrt{3}\,\delta^+_{t_w}} - \tan^{-1}\frac{1}{\sqrt{3}}\right]\right\} + \frac{Pr_{t_w}}{k'}\ln\left\{\frac{1 - k'Pr_w z^+_b/Pr_{t_w}}{1 + k'Pr_w\delta^+_{T_w}/Pr_{t_w}}\right\}$$

AIR: HEAT

$$\delta^+_{v_a} = 0.37\left(h^+_a\right)^{1/2} \; ; \; \delta^+_{t_a} = Pr_a^{-1/3}\delta^+_{v_a} \; ; \; \delta^+_{T_a} = \left(\frac{k'}{Pr_{t_a}}\right)^{1/2}\delta^{+\,3/2}_{v_a}$$

$$K_a(z^+) = \kappa_a\left[1 + \left(z^+/\delta^+_{t_a}\right)^3\right], \; z^+ < \delta^+_{T_a} \; ; \; K_a(z^+) = \kappa_a\left[1 + k'Pr_a z^+/Pr_{t_a}\right], \; z^+ > \delta^+_{T_a}$$

$$Q_S = \rho_a c_{p_a} u_{*_a}(T_s - T_\infty)/r_{H_a}$$

$$r_{H_a} = \frac{Pr_a\delta^+_{t_a}}{3}\left\{\frac{1}{2}\ln\left[\frac{\left(\delta^+_{t_a} + \delta^+_{T_a}\right)^2}{\delta^{+\,2}_{t_a} - \delta^+_{t_a}\delta^+_{T_a} + \delta^{+\,2}_{T_a}}\right] + \sqrt{3}\left[\tan^{-1}\frac{2\delta^+_{T_a} - \delta^+_{t_a}}{\sqrt{3}\,\delta^+_{t_a}} - \tan^{-1}\frac{1}{\sqrt{3}}\right]\right\} + \frac{Pr_{t_a}}{k'}\ln\left\{\frac{1 + k'Pr_a z^+_\infty/Pr_{t_a}}{1 + k'Pr_a\delta^+_{T_a}/Pr_{t_a}}\right\}$$

WATER VAPOR

$$\delta^+_{v_a} = 0.37\left(h^+_a\right)^{1/2} \; ; \; \delta^+_{q_a} = Sc_a^{-1/3}\delta^+_{v_a} \; ; \; \delta^+_{Q_a} = \left(\frac{k'}{Sc_{t_a}}\right)^{1/2}\delta^{+\,3/2}_{v_a}$$

$$K_q(z^+) = \kappa_q\left[1 + \left(z^+/\delta^+_{q_a}\right)^3\right], \; z^+ < \delta^+_{Q_a} \; ; \; K_q(z^+) = \kappa_q\left[1 + k'Sc_a z^+/Sc_{t_a}\right], \; z^+ > \delta^+_{Q_a}$$

$$Q_L = \rho_a L u_{*_a}(q_s - q_\infty)/r_{E_a}$$

$$r_{E_a} = \frac{Sc_a\delta^+_{q_a}}{3}\left\{\frac{1}{2}\ln\left[\frac{\left(\delta^+_{q_a} + \delta^+_{Q_a}\right)^2}{\delta^{+\,2}_{q_a} - \delta^+_{q_a}\delta^+_{Q_a} + \delta^{+\,2}_{Q_a}}\right] + \sqrt{3}\left[\tan^{-1}\frac{2\delta^+_{Q_a} - \delta^+_{q_a}}{\sqrt{3}\,\delta^+_{q_a}} - \tan^{-1}\frac{1}{\sqrt{3}}\right]\right\} + \frac{Sc_{t_a}}{k'}\ln\left\{\frac{1 + k'Sc_a z^+_\infty/Sc_{t_a}}{1 + k'Sc_a\delta^+_{Q_a}/Sc_{t_a}}\right\}$$

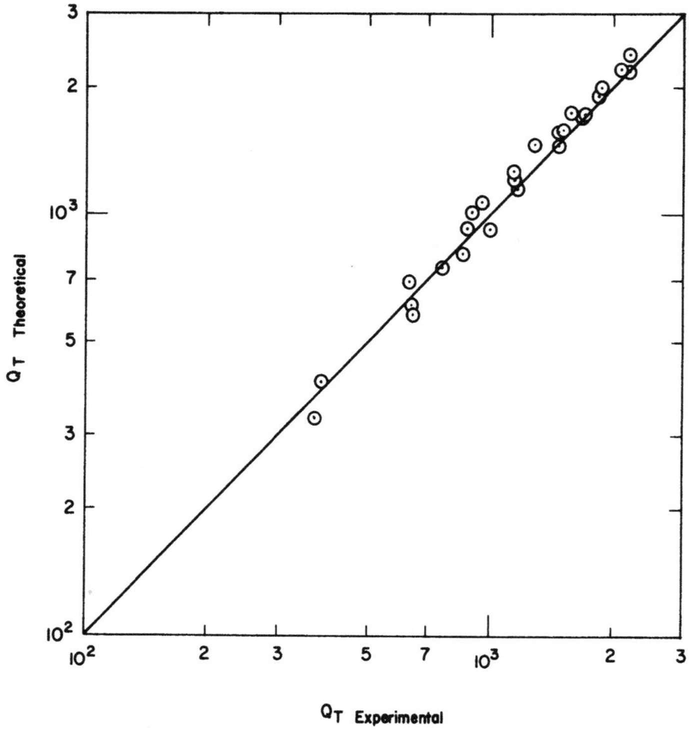

Figure 5. Comparison of Flux Prediction with Laboratory Experimental Results: Total Flux $Q_T = Q_S + Q_R + Q_L$

Conditions Under Which Spray Occurs

Wang (1977) used his data, which was taken at windspeeds between 12.5 and 14.5 m/s and at fetches between 3 and 12.3 m, to determine the effect of spray on the energy transfer across an air-water interface. Mean transfers of momentum, heat and mass were determined by a standard boundary-layer integral technique using the instrumentation described above. Droplet distributions were measured with the electrostatic-capacitance probe.

Profile data for the integral technique were taken from the free stream down to heights very near the wave crests at five locations along the channel. Output signals from the pitot-static tube (for velocity) and five temperature sensors (wet- and dry-bulb reference probes in the free stream, two probes in the boundary layer, and one in the water) were simultaneously digitized, stored and then coverted to velocity, temperature and specific humidity by the data acquisition system. Spray data was taken at the same fetches with data being taken in profiles from the highest level at which spray was detected downward in a series of small intervals to near the wave crests.

The droplet data was used to evaluate the evaporation from droplets moving in air. The evaporation rate equation for each droplet is

$$\frac{dm}{dt} = 2\pi d \kappa_q (\rho_s - \rho_a)(1 + 0.276\ Re^{1/2} Sc_a^{1/3}) \tag{15}$$

where d is droplet diameter, $(\rho_s - \rho_a)$ is the water vapor density difference between the drop surface and the ambient fluid, Re is the Reynolds number based on the relative velocity (here taken as 0.25 u(z)) and droplet diameter. (Wu, 1973, shows that the average droplet speed is 0.7 to 0.8 u(z).) If we define $\gamma(d, z)$ as the number of droplets in the size range $(d \pm \delta d/2)$ counted by the detector in the sampling period T (in seconds) with a sampling interval δd (= 50 μm here), then the spectral drop size distribution per unit volume can be defined as

$$\alpha(d) = \frac{\gamma(d,z)}{f A_s T U_d \delta d} \qquad \left\{\frac{\text{no./cm}}{\text{cm}^3}\right\} \tag{16}$$

where f is the detector collection coefficient, A_s is the cross-section area of the detector, and $u_d = 0.75\ u(z)$ is the droplet velocity.

The droplet evaporation rate is estimated, first, by summation over all drop diameters, viz.,

$$E(z) = \sum_d \frac{dm}{dt}\,\alpha(d)\,\delta d \qquad \{\text{g/cm}^3 \cdot \text{sec}\} \tag{17}$$

and, second, by integration across the boundary layer

$$E_d = \int_0^{z_{od}} E(z)\,dz \qquad \left\{\frac{\text{g}}{\text{cm}^2 \cdot \text{sec}}\right\} \tag{18}$$

It turns out that E(z) is a logarithmic function of z/z_{od} for any windspeed and fetch, where z_{od} is the thickness of the droplet boundary layer. Accordingly the integral is evaluated analytically. This allows us to extrapolate the curve below the lowest measurement point; in fact $\lim_{z\to 0} E(z) = \infty$ under this approximation so we expect E_d to be an overestimate.

Wu (1974) estimated, on the basis of results from his facility, that the evaporation from spray would be almost 13 percent of the total evaporation at a windspeed of 13.4 m/s and would increase rapidly with U_∞ thereafter. Figure 6 shows a comparison of droplet spectral distributions for several sets of data. Wu's data lie well above all other data and have a much different shape; Wu's droplet distribution has many more and larger droplets, leading to

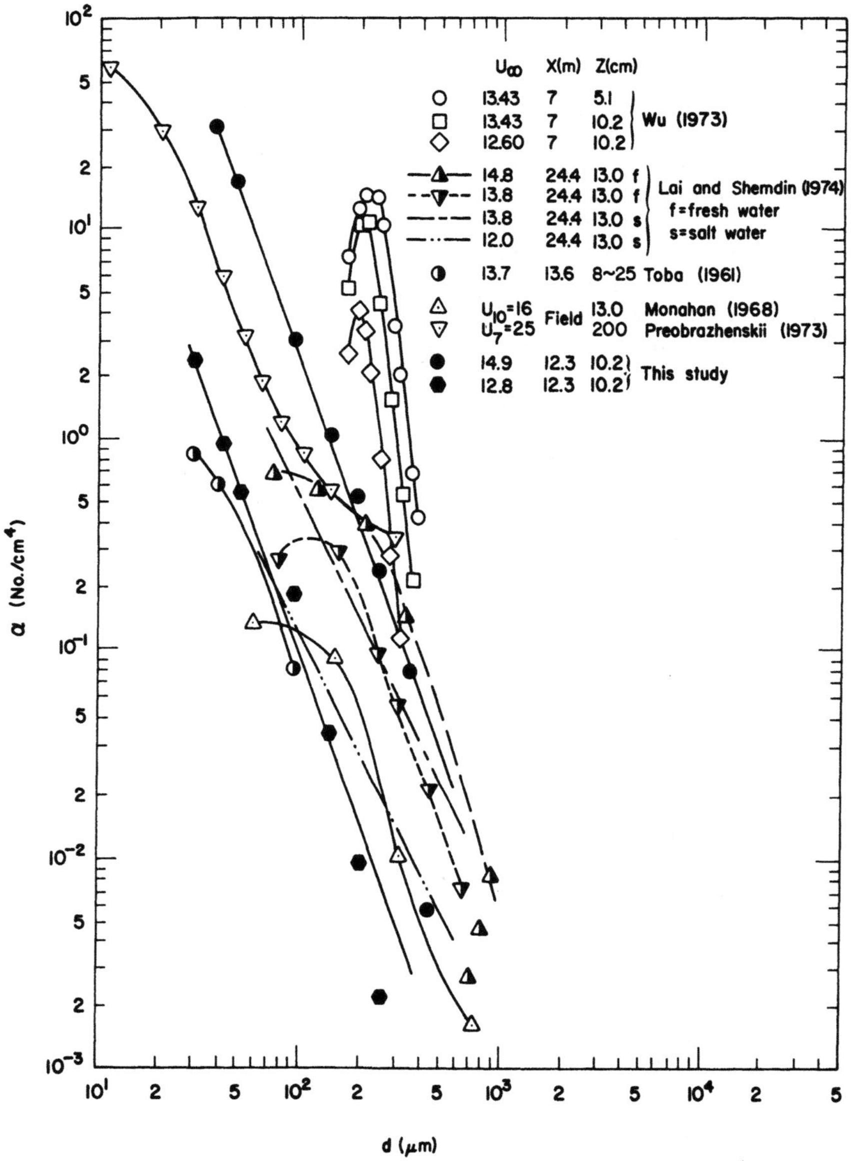

Figure 6. Comparison of Droplet Spectral Distributions

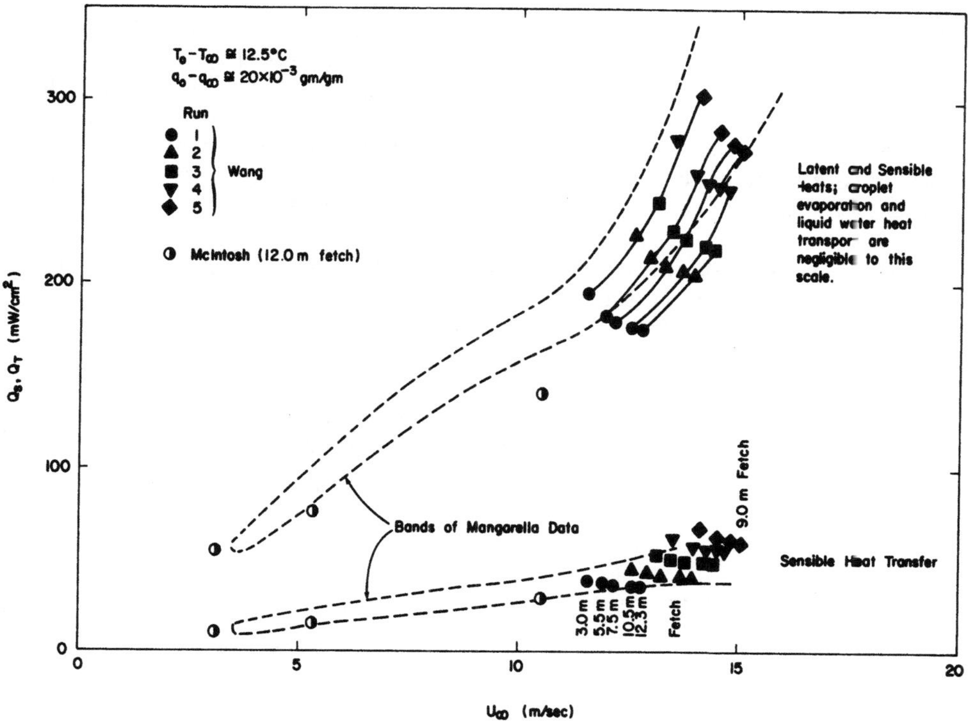

Figure 7. Latent Plus Sensible Heat Transfer Rates Versus Windspeed

high evaporation rates. Wu's data are not typical of most laboratory and field data; in Wu's facility the droplet production mechanism is probably different, being tearing of wave crests contrasted to bubble bursting for all other reported data.

Our estimates of evaporation from droplets for our data show that E_d is less than 2 percent of the total evaporation at all windspeeds and fetches (Wang, 1977). However, as Figure 7 shows, there is a marked increase in measured latent and, hence, total heat transport under spray conditions compared to what would be expected from an extrapolation of the low windspeed curves. Wang (1977) hypothesized that the increase is due to increased transfer from the roughened surface caused by wave breaking and bubble bursting; however, this hypothesis may be wrong as shown below.

A HYPOTHESIS FOR HIGH WINDSPEED CASES

First, McIntosh, et al. (1975b) established that the air flow layer does indeed behave as a fully rough flow for windspeeds $U_\infty >$ 4 m/s and that the mechanism for sensible heat transfer does not change with windspeed. Second, Miller, et al. (1975) and Street and Miller (1977) confirmed that the water surface layer flow is rough for $h_w^+ > 100$. Third, the theory of Street (1977) adequately predicts Q_T, $\overline{Q}_L$ and Q_S in the windspeed range U_∞ = 5 to 10 m/s. (The results at 5 m/s are good because $h_a^+ > 100$ and aqueous surface layer effects where $h_w^+ < 100$ are not dominant.) Following on these results, we have generated a speculative hypothesis.

To begin we constructed a composite data set based on the data of Bole and Hsu (1969), McIntosh, et al. (1975b), Wang (1977) and Mangarella, et al. (1971). These data span the windspeed range U_∞ = 5 to 15 m/s. Figure 8 shows the root-mean-square wave amplitude $(\overline{\eta^2})^{1/2}$ and u_{*a} for this data set at a fetch of 10.5 m. [$h = 2.5\ (\overline{\eta^2})^{1/2}$ in the Stanford facility (Street and Miller, 1977).] Next we selected the water-air temperature difference $T_b - T_\infty$ =

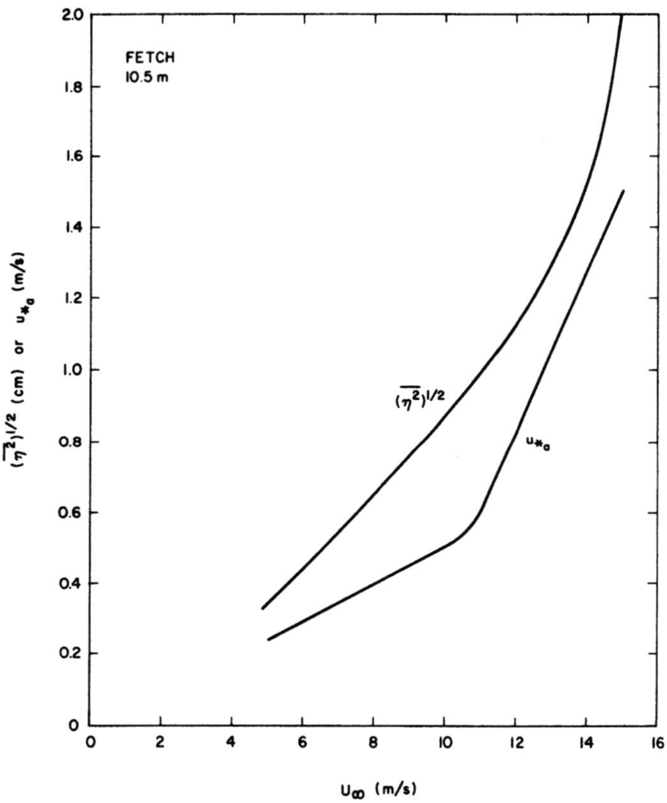

Figure 8. Root-Mean-Square Wave Amplitude and Air Friction Velocity Versus Windspeed for the Composite Data Set

12.5°C and the specific humidity difference $q_s - q_\infty$ = 20 mg/g used by Wang (1977) in his experiments; T_b was set at 20°C. Using this data set we applied the coupled surface layer theory (Street, 1977) described above over the total range of experimental conditions U_∞ = 5 to 15 m/s, computing Q_T, Q_L, Q_S, Q_R, T_s and the parameters such as h_w^+, h_a^+, etc., for each windspeed. Results for the 10.5 m fetch are shown in Figure 9 along with Wang's experimental data at the same fetch.

The results shown in Figure 9 are surprising. In the 5 to 10 m/s range the experiments and theory are known to be in agreement. An extrapolation of the theory curve from this range to larger U_∞ intersects the first data point in Wang's set. However, due to the rapid increase in u_{*a} which occurs for U_∞ > 10 m/s, the theory actually rises well above the experimental data. The theory curve only levels off near 15 m/s where the very rapid increase in $h[\overline{(\eta^2)}^{1/2}]$ seen in Figure 8 causes increased sublayer thicknesses

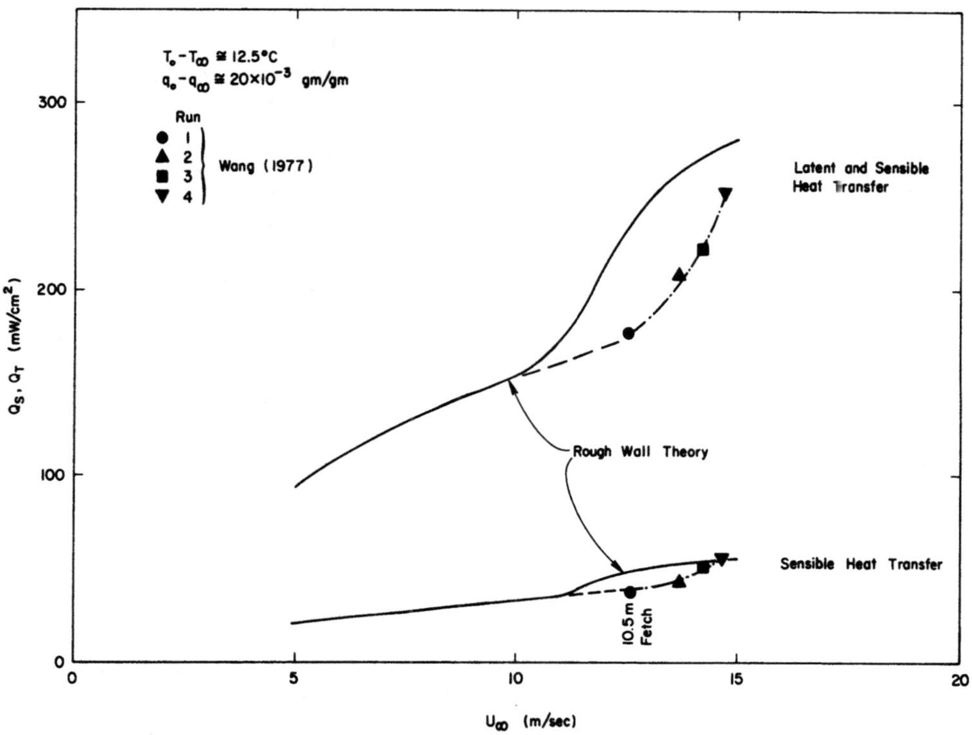

Figure 9. Comparison of Rough Boundary Theory and Experimentally-Measured Fluxes at High Windspeed

and, hence, increased resistance to energy transfer. Thus, according to the rough wall theory, the transfers should be greatly augmented when $U_\infty \geq 11$ m/s, but experiments show that they are not. Wang's conclusion, cited above, that droplet evaporation is negligible is entirely consistent with these results. Indeed, there is a drop in the actual transfer compared to that predicted by theory for $U_\infty > 10$ m/s. Wang's hypothesis that the transfer is enhanced by increased surface roughness is correct for the theory, but not for the experimental data.

An observation regarding the water surface states for $U_\infty >$ 10 m/s plus reference to Easterbrook (1968) and Banner and Melville (1976) lead us to a plausible hypothesis. Wave breaking leading to air bubble entrainment, breaking, and spray formation is well established for $U_\infty = 12.5$ m/s in our facility. According to Banner and Melville (1976), when wind blows over surface gravity waves, the air flow will not separate unless wave breaking is occurring, and there is a large increment of drag over a breaking wave compared to that over a finite amplitude unbreaking wave. We infer from this and Figure 8 that wave breaking begins to occur in significant amounts for $U_\infty \geq 11$ m/s in our facility. This suggests that flow separation on a significant scale is also possible.

In air flow over well developed waves, Esterbrook (1968) observed flow separation in the form of vortices and dead air regions where the build-up of high humidity tends to reduce water vapor transfer across the air-water interface. He suggested that the reduced turbulent mixing between the dead air spaces and the outer flow inhibits vertical transport of moisture. The difference between the rough wall theory and our experiments suggests that the latent heat and sensible heat transports are being inhibited in the region where wave breaking and spray formation are occurring.

Apparently, the coupled surface layer, rough boundary theory is a valid tool for predicting surface fluxes, provided there is not significant wave breaking and spray formation. We tentatively conclude then that the effect of wave breaking and spray formation is to increase the surface drag (i.e., u_{*a}) and to decrease both sensible and latent heat transport compared to the theoretical prediction under the conditions obtained in our facility.

ACKNOWLEDGMENT

This work was supported by the Atmospheric Research Section (Grant ATM 73-06538) and the Engineering Chemistry and Energetics Section (Grant ENG 76-80127) of the National Science Foundation, U.S.A.

REFERENCES

M. L. Banner and W. K. Melville (1976). On the separation of air flow over water waves. J. of Fluid Mech., 77, 4, pp. 825-842.

J. B. Bole and E. Y. Hsu (1969). Response of gravity water waves to wind excitation. J. Fluid Mech., 35, pp. 657-676.

C. C. Easterbrook (1968). A study of the effects of waves on evaporation from free water surfaces. Cornell Aero. Lab. Rept. No. RM-2151-P-1.

J. O. Hinze (1975). Turbulence, 2nd Ed., McGraw-Hill.

W. M. Kays, R. J. Moffat, and W. H. Thielbahr (1969). Heat transfer to the highly accelerated turbulent boundary-layer with and without mass addition. Dept. M.E. Tech. Rep. HMT-6, Stanford University.

J. R. Lai and O. H. Shemdin (1974). Laboratory study of the generation of spray over water. J. Geophys. Res., 79, pp. 3055-3063.

P. A. Mangarella, A. J. Chambers, R. L. Street, and E. Y. Hsu (1971). Energy and mass transfer through an air-water interface. Dept. C.E. Tech. Rep. 134, Stanford University.

P. A. Mangarella, A. J. Chambers, R. L. Street and E. Y. Hsu (1973). Laboratory studies of evaporation and energy transfer through a wavy air-water interface. J. Phys. Ocean., 3, 1, pp. 93-101.

D. A. McIntosh, R. L. Street, and E. Y. Hsu (1975a). The influence of air-water interfacial conditions on turbulent transfer of latent and sensible heat. Geophys. Res. Ltrs., 2, 1, pp. 12-14.

D. A. McIntosh, R. L. Street, and E. Y. Hsu (1975b). Turbulent heat and momentum transfer at an air-water interface: the influence of surface conditions. Dept. C.E. Tech. Rep. No. 197, Stanford University.

A. W. Miller, Jr., R. L. Street, and E. Y. Hsu (1975). The structure of the aqueous thermal-sublayer at an air-water interface. Dept. C.E. Tech. Rep. No. 195, Stanford University.

E. C. Monahan (1968). Sea surface as a function of low elevation windspeed. J. Geophys. Res., 73, 4, pp. 1127-1137.

L. Y. Preobrazhenskii (1973). Estimation of the content of spray drops in the near water layer of the atmosphere. Fluid Mech.--Soviet Res., 2, 2, pp. 95-100.

R. L. Street (1977). Turbulent heat and mass transfers across a rough, air-water interface: a simple theory. Submitted for publication.

Y. Toba (1961). Drop production by bursting of air bubbles on the sea surface (III): study by use of a wind flume. Mem. of Coll. Sci., Kyoto U., Ser. A., 29, 3, Art. 4.

C. S. Wang (1977). The study of energy, mass, momentum transfer and spray measurement at an air-water interface. Ph.D. dissertation, Dept. C.E., Stanford University.

C. S. Wang, T. R. Mogel, D. F. Hollingshead, R. L. Street, and E. Y. Hsy (1974). A sensor for liquid atmospheric aerosols. AMS Symp. on Atmos. Diff. and Air Pollu., Santa Barbara, California.

J. Wu (1973). Spray in the atmospheric surface layer: laboratory study. J. Geophys. Res., 78, pp. 511-519.

J. Wu (1974). Evaporation due to spray. J. Geophys. Res., 79, pp. 4107-4111.

A. M. Yaglom and B. A. Kader (1974). Heat and mass transfer between a rough wall and turbulent fluid flow at high Reynolds and Peclet numbers. J. Fluid Mech., 62, pp. 601-623.

DISCUSSION

M. F. Coantic:

Until quite recently, I should have agreed with your transposition of the solid rough wall models to the aqueous surface sublayer. There are, however, some intuitive physical reasons for not doing so. In effect, considering the air-water density ratio (1/1 000), it appears reasonable to think that the small scale air motions very close to the air-water interface are restrained by the water almost as though it is a solid boundary. On the other hand, the water motions should not be restrained by air to the same extent, and should therefore be free to converge and diverge at the interface itself. From the continuity relations: the normal velocity fluctuation component could therefore vary as the first power of the distance from the interface, instead of the second power. As a consequence, the turbulent diffusivity should increase more rapidly close to the interface, with a subsequent increase in the transfer efficiency of any scalar contaminant of relatively large Prandtl or Schmidt number. Two different pieces of evidence seem to favour such a tentative suggestion:

(i) One recent set of radiometric data by DESCHAMPS (1977) seems to indicate a very abrupt decrease in the temperature drop across the aqueous sublayer with increasing wind velocity, which could be explained on the preceding basis.

(ii) One can find from the chemical engineering literature that the difference between the structure of turbulence on the gas and on the liquid sides of a clean gas-liquid interface has been accounted for in models, and verified by experiments (see e.g. Levich (1962), Davies (1972)).

I would propose that further investigations of this concept could help in understanding not only energy but also mass exchanges across the air-sea interface.

R. L. Street:

The argument regarding the behaviour of the aqueous surface layer at the interface is physically appealing. It will be interesting to investigate the effect of using the resulting z^2 dependence of diffusivity in the water layer. While our flux prediction results are already in good agreement with the experimental data, the z^2 dependence should alter the sublayer thickness coefficient (now 0.37) and may yield a more accurate estimate of the effective sublayer thicknesses in the water.

K. Katsaros:

Since the functional form of the Prandtl number dependence of S will be to the 1/2 or 1/3 power depending on the assumed profile shapes, one should be able to check which is the valid assumption by looking at the Prandtl number dependence.

R. L. Street:

I agree; however, the relatively small range of Pr in the water for our experiments and the scatter in the flux measurement data tend to mask the Pr-power effect. We are currently conducting LDA/thermister measurements in the water surface layer as mentioned in my presentation. These measurements will provide both temperature profiles and direct flux measurements which will make evaluation of the Pr dependence feasible.

J. Wu:

The roughness length was used in the formula for estimating the thickness of the thermal sublayer over the solid surface. However you have used the height of dominant waves in your formula.

The dominant waves are not the roughness elements. Will you please comment on this?

R. L. Street:

The usual formulation of rough wall theory uses the mean height of the roughness elements, which typically have been of uniform shape. The key is that the roughness should be relatively closely packed. In our case we have used the mean wave height which is 2.5 times the RMS wave amplitude in our facility because the heights are Rayleigh-distributed. Kondo has shown (as have our experiments with mechanically-generated waves) that the wind generated waves act as roughness as long as they are relatively short. The cases reported herein fall in this category so the entire spectrum does act as roughness elements. That we get the proper flow behaviour, i.e. sublayer thickness proportional only the $h^{1/2}$, was demonstrated in our Journal of Physical Oceanography paper cited above.

M. Donelan:

Rather than use the rms height of the waves as the roughness parameter, why not adopt a suggestion by Kitaigorodskii and Volkov (IZV, Atmos and Oceanic Physics Series 1, 9, 978-88) and include the phase velocity difference of the waves and the wind to bias the roughness length in favour of the short waves. In this way the roughness length h^t would be facility independent.

R. L. Street:

This is a useful suggestion which we will investigate. It has the advantage of not requiring us to distinguish between ripples which act as roughness and swell which do not.

F. W. Dobson:

Why are the velocities in the work not scaled by the waves phase velocity?

R. L. Street:

In this first and rather simple approach we have excluded from the analysis all wave characteristics except their mean height. Now that we have good basic agreement with experiment, we will look at ways to include the mobility-character of the interface.

ON AN ISOTOPIC METHOD TO INVESTIGATE MASS TRANSFER AT AN AIR-WATER INTERFACE: THE DEPENDENCE OF BULK EVAPORATION COEFFICIENTS ON WATER SURFACE CONDITIONS

Liliane Merlivat

Department de Recherche et Analyse - CEN/SACLAY
B.P. No. 2-91190 Gif sur Yvette - France

We propose in this paper, first, to show what specific contribution to an improved understanding of transfer between air and water can be inferred from the analysis of the distribution of the isotopes of hydrogen and oxygen in the water vapor above a liquid surface and then, summarize the main results so far obtained from laboratory experiments run in the I.M.S.T. air-sea interaction simulation facility at Marseille.

PARAMETERIZATION OF AIR-WATER INTERFACE FLUXES BY THE BULK AERODYNAMIC FORMULAE

In the turbulent constant flux layer developed above a water surface, the transfer of momentum and water vapor are given by the bulk aerodynamic formulae :

$$\frac{\tau}{\rho} = u_*^2 = C_D\, u^2 \qquad (1)$$

$$\frac{E}{\rho} = C_q\, u\, (q_s - q) \qquad (2)$$

or

$$\frac{E}{\rho} = \Gamma\, (q_s - q) \qquad (3)$$

where τ is the shearing stress, E the evaporation rate, ρ the air density, u_* the friction velocity, aud q_s the saturated specific humidity at the water surface temperature. u, q and Γ are the wind speed, specific humidity and profile coefficient respectively at height z. C_D and C_q are the bulk transfer coefficients for momentum (drag coefficient) and water vapor respectively. Their value is a function of the height z.

A central question in the field of air-sea interaction has been the determination of the applicability of the bulk aerodynamic formulae and mainly the evaluation of the associated coefficients i.e, C_D and C_q for momentum and mass respectively. A first question is to examine if C_D and Cq are constant quantities, as usually postulated, then to know the relation between these two coefficients.

Recently a relatively large number of investigations of direct measurements of momentum, heat and moisture fluxes over the sea have been reported. The water vapor flux is obtained by the direct measurement of the mean vertical velocity-water vapor density covariance, $E = \overline{wq}$. A compilation of existing data on direct moisture fluxes measured at sea together with the mean values of u, q and q_s has recently been reported [1]. From the plot of E versus u. $(q_s - q)$ a value of $C_q = 1.32\ 10^{-3}$, for z = 10 meters, is derived with extreme values ranging from 1.10^{-3} to $1.8\ 10^{-3}$. It appears doubtful at the present time that such a large scatter has to be imputed solely to uncertainties in experimental techniques.

The unique dependence of E on u $(q_s - q)$ appears uncorrect. The dependence of other physical quantities involved in the transport processes has to be investigated.

Several theoretical models have been proposed to describe evaporation at an air-water interface, leading to different expressions of the profile and evaporation coefficients. The relative degree of physical meaning of each of these expressions is difficult to assess from the only comparison between experimental and theoretical values of E as all the expressions of the coefficients C_q and Γ contain numerical constants just adjusted in that way.

It has been shown [2] that the profile coefficients put forward in the different evaporation theories can be written as :

$$\Gamma = \frac{u_*}{\rho_M + \rho_T} \qquad (4)$$

where ρ_M and ρ_T stand for molecular and turbulent resistance.

TABLE 1

EXPRESSIONS OF THE PROFILE COEFFICIENT

MODEL	$\Gamma = \dfrac{u_*}{\rho_M + \rho_T}$
1) Reynold's analogy	$\Gamma = u_*\left(\frac{1}{\chi} \ln \frac{z}{z_0}\right)^{-1}$
2) Sheppard's theory	$\Gamma = u_*\left(-\frac{1}{\chi} \ln D + \frac{1}{\chi} \ln \chi u_* z\right)^{-1}$
3) Sverdrup's theory	$\Gamma = u_*\left(\frac{u_* d}{D} + \frac{1}{\chi} \ln \frac{z + z_0}{d + z_0}\right)^{-1}$
4) Kitaigorodskiy and Volkov'theory	$\Gamma = u_*\left(\frac{u_* d}{D} + \frac{1}{\chi} \ln \frac{z}{d}\right)^{-1}$
5) a) Brutsaert's theory (1975) smooth surface (Re_s < 0.13)	$\Gamma = u_*\left[13.6\left(\frac{\nu}{D}\right)^{2/3} + \frac{1}{\chi} \ln \frac{u_* z}{30\nu}\right]^{-1}$
b) rough surface (Re_s > 2)	$\Gamma = u_*\left[7.3 Re_s^{1/4}\left(\frac{\nu}{D}\right)^{1/2} + \frac{1}{\chi} \ln \frac{z}{z_0} - 5\right]^{-1}$
6) Owen and Thomson' theory (1963) rough regime	$\Gamma = u_*\left[2.4 Re_s^{0.45}\left(\frac{\nu}{D}\right)^{0.8} + \frac{1}{\chi} \ln \frac{z}{z_0}\right]^{-1}$

Combining the equations (1) to (4) we get for the bulk evaporation coefficient :

$$C_q = \frac{\sqrt{C_D}}{\rho_M + \rho_T} \qquad (5)$$

The table 1 summarizes the expressions of Γ in six different models. In the table, ν is the kinematic viscosity of air, χ the van Karman constant, z_o the surface roughness parameter of the liquid and Re_s the surface roughness Reynolds number. d represents the thickness of the laminar sublayer in the models 3 and 4.

In a very recent work Deacon [3] has derived an equation (his equation 5) similar to (4) to compute the water vapor and gas fluxes at an air-water interface. It is worthwhile to mention also that very close expressions to those proposed by Brutsaert [4, 5] have been derived by Yaglom and Kader [6]. These last expressions are discussed in [3] and comparison between the two evaporation models in found in [5].

A quite different dependency of the transfer coefficients listted in table 1 on the molecular diffusion coefficient of water vapor in air, or on the Schmidt number, $Sc = \frac{\nu}{D}$, of water vapor in air, is noticed.

We have shown in previous works [2, 7] that the relationship between the transfer coefficients, C_q or Γ, and the Schmidt number can be inferred from the study of the distribution of the isotopes of hydrogen and oxygen in water vapor.

THE BASIC POINTS OF THE ISOTOPIC METHOD

The main features can be summarized as follows :

- the deuterium and oxygen 18 are present in natural water at the trace level as molecules $HD^{16}O$ and $H_2{}^{18}O$. The main component is $H_2{}^{16}O$. The relative proportions of $H_2{}^{16}O$: $H_2{}^{18}O$: $HD^{16}O$ are $1 : 2.10^{-3} : 3.10^{-4}$.

- the three molecules have

  a/ different saturation vapor pressure at a same temperature, and therefore, different specific humidity.

  b/ different molecular diffusion coefficients in air, and therefore distinct Schmidt numbers.

- $H_2{}^{18}O$ and $HD^{16}O$ are ideal tracers to study the transfer of water vapor in air.

- in the constant flux layer developed above an air-water interface, an equation analog to 2 and 3 can be written to compute the evaporating flux of each isotopic species. From these relations, it follows (details are given in [2] and [7]) that the deuterium and oxygen 18 concentration in the water vapor within the turbulent boundary layer at different levels above the water surface are linearly related. It is shown that the coefficients of this relationship are dependent on the relative values of ρ_M and ρ_T

and the exponent, n, of the Schmidt number in the expressions of Γ . From the experimental determination of the isotopic profiles, we compute the quantity :

$$B = \frac{n\rho_M}{\rho_M + \rho_T} \qquad (6)$$

In summary, the comparison between experimental results and theoretical expressions as listed in table 1, is tested through the isotopic method, on the values of ρ_M, ρ_T and n as given in B (equation 6), whereas the measurement of evaporation rate and mean values of wind speed, surface temperature and specific humidity allows to establish a comparison between experimental and theoretical values of the evaporation coefficient C_q, i.e (from equation 5) the sum $(\rho_M + \rho_T)$.

EXPERIMENTAL PROCEDURE

The experiments were conducted in the large air-sea interaction simulating facility at the Institut de Mecanique Statistique de la Turbulence (I.M.S.T.) at Marseille [8]. In steady state working conditions, a constant flux layer was developed above the water surface. Sampling of water vapor for isotopic analysis was made within this layer, which was about 15 to 20 cm thick at a fetch of 34 meters. Details of the whole experimental procedure were given in [2]. Experiments were run either with wind waves or artificially generated ones. A list of experimental parameters is given in table 2. The frequency of the generated waves was 1 hertz.

TABLE 2

MEAN AIR VELOCITY m/sec	SURFACE CONDITIONS H, waves height in cm	NUMBER OF EXPERIMENTS
0.7	Wind waves	1
2.0	Wind waves	3
"	$1.2 \leq H \leq 9.6$	5
3.5	Wind waves	2
"	$4.2 \leq H \leq 9.8$	4
5.0	Wind waves	3
"	$H = 10.1$	1
6.0	$7.0 \leq H \leq 11.8$	2

Near neutral stability conditions have been maintained in all experiments.

RESULTS

1 - Momentum transfer.

From the mean wind profiles measured for each experiment, we have computed the values of u_* and z_o through the usual relation :

$$u = \frac{u_*}{\chi} \ln \frac{z}{z_o}$$

Figure 1 represents the variation of the drag coefficient, $C_D = \left(\frac{u_*}{u}\right)^2$, calculated for z = 10 meters as a function of the surface roughness number $Re_s = \frac{u_* z_o}{\nu}$. Also are plotted the empirical relation observed by KITAIGORODSKIY et al (1973) [8] in nature, $C_{D\ (z = 10)} = 1.2\ 10^{-3}\ Re_s^{\ 0.15}$ and the relation between C_D and Re_s as calculated with the Charnock's relation,

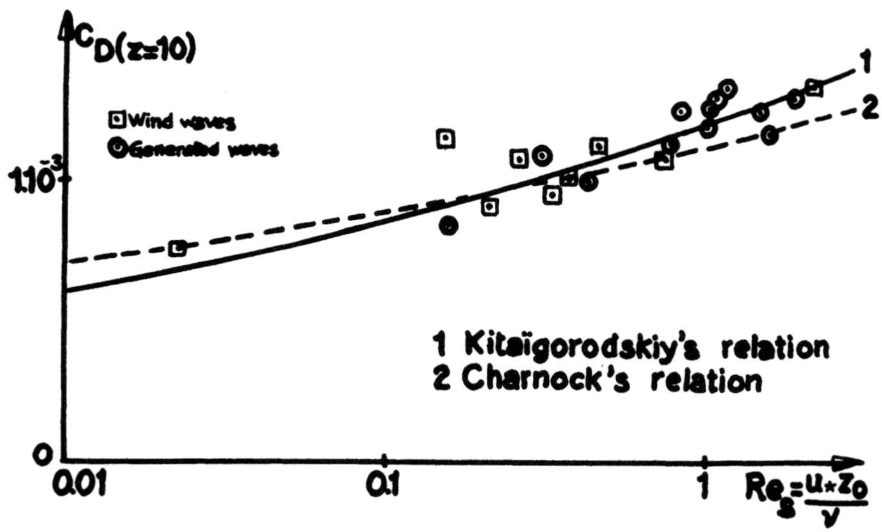

Fig. 1. Drag coefficient variation with the Reynolds surface roughness number Re_s.

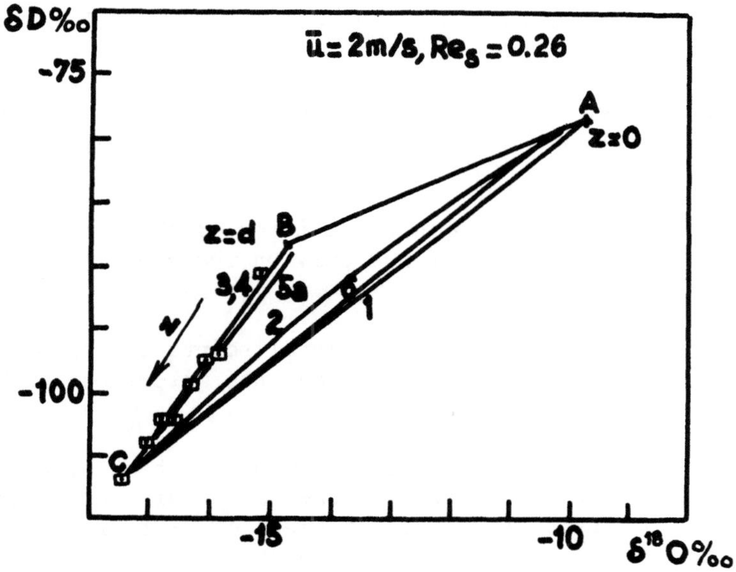

Fig. 2 - Experimental and theoretical distribution of the deuterium and oxygen 18 in the water vapor at differents heights above the water surface in the boundary constant flux layer.
- Curve 1 : Reynold's analogy
- Curve 2 : Sheppard's model
- Curve 3, 4 : Sverdrup's and Kitaigorodskiy and Volkov' model
- Curve 5, a : Brutsaert's model for a smooth surface
- Curve 5, b : Brutsaert's model for a rough surface
- Curve 6 : Owen and Thomson'model

$z_o = \frac{u_*^2}{ag}$ with a = 81.1. This latter relation has been quite well confirmed by recent measurements [10].

Two conclusions can be drawn :

a) there is a reasonnably good agreement between measurements made in the laboratory and those made in nature. The value of the drag coefficient is not constant; it is an increasing function of Re_s, or also of the mean wind speed.

b) the existence of wind waves or generated waves with the various heights which have been observed in the course of our experiments, does not affect significantly the relation between C_D and Re_s. However, for a same mean wind speed at the entrance of the wind tunnel, the value of Re_s is changed by generating waves.

2 - Isotopic results.

Figure 2 shows, for one experiment, the variations of deuterium and oxygen 18 measured in the water vapor at different levels above the liquid surface. A qualitative description of the figure is as follows :

- A represents the isotopic composition of the water vapor right at the interface. It depends only on the surface temperature of the liquid.

- Going from B to C, the experimental points correspond to increasing values of the height z above the interface, within the constant flux layer.

- The segment AB represents the isotopic content of the water vapor within the viscous sublayer close to the air-water interface. The influence of the molecular coefficient of diffusion of water vapor in air, D, has been introduced in various ways in this interfacial sublayer in the models listed in the table 1. The slope of AB is fixed by the numerical values of $D_{H_2^{16}O}$, $D_{HD^{16}O}$ and $D_{H_2^{18}O}$. Its length depends on the importance of the viscous regime in the whole evaporation process. The curves 1 to 6, are computed from the theoretical models listed in the table 1. (details of the calculations are given in [2]). We observe on the figure a good sensitivity of the isotopic method to check the agreement between experimental data and theoretical models.

Quantitatively, this check is made by comparing theoretical B values (equation 6) and experimental ones, which are given by a numerical function of the slope, a, and the intercept, b, of the experimental line (Cf. [2] and [7]). The figure 3 shows experimental and theoretical variations of B versus the surface roughness Reynolds number, Re_s.

Two main experimental difficulties account for a large change in the error bars calculated for each experiment. When the wave height increases, the sampling must be done in the upper part of the boundary layer, to avoid sampling of the liquid. The humidity as well as the δD and $\delta^{18}O$ variations are small in that region. The accuracy of the isotopic measurements is constant (0.3 ‰ for deuterium and 0.1 ‰ for oxygen 18), and thus when the range of variation of the isotopic data becomes smaller, the precision on a and b in the equation of the experimental line becomes poorer. The same trend is observed for the higher wind speed experiments because no large gradients in the humidity and in the isotopic composition can be maintained as they are produced by the refrigeration unit of the system which has a limited efficiency.

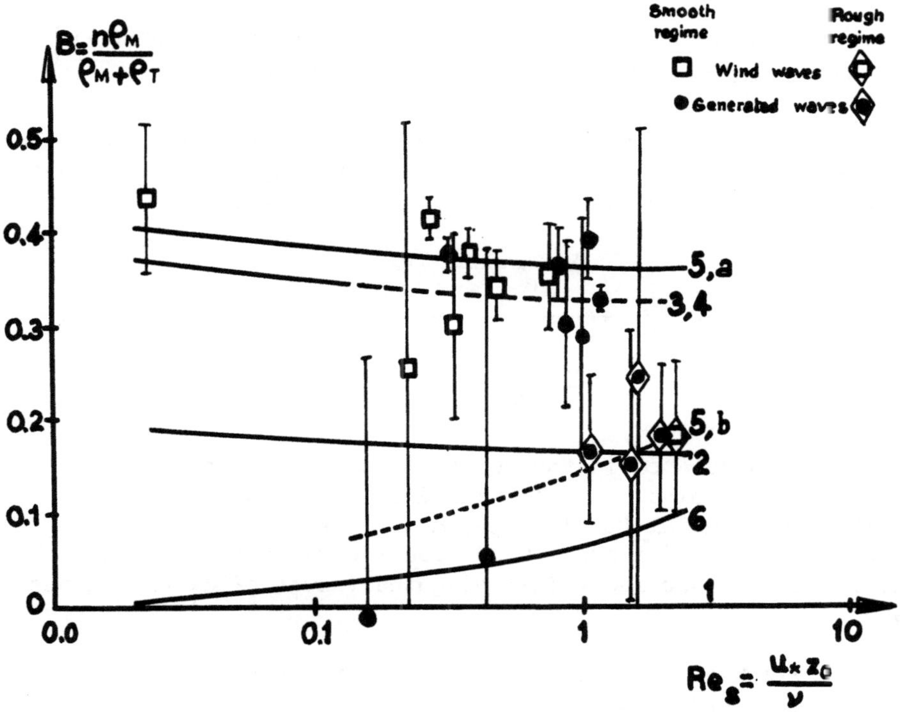

Fig. 3 - Variation of $B = \dfrac{n\rho_M}{\rho_M + \rho_T}$ as a function of Re_s.

(For the curve numbers see caption of the figure 2).

A qualitative examination of the figure shows that :

a) The Reynolds analogy and the Owen and Thomson'models (curves 1 and 6) do not reproduce the isotopic fractionations observed in our experimental range of values of Re_s.

b) The agreement with the Sheppards's theory (curve 2) appears correct only for the larger values of Re_s ($Re_s > 1$).

c) A relatively good agreement is found between the experimental and theoretical values of $\frac{n\rho_M}{\rho_M + \rho_T}$ calculated with Brutsaert's theory for a smooth surface if $Re_s < 1$ (curve 5 a) and for a rough surface if $Re_s > 1$ (curve 5, b).

To check quantitatively this agreement we have computed the difference (d_i) between the theoretical and experimental values of B. σ_i being the standard deviation of each value for B_i, the mean value of d_i is :

$$\bar{d} = \frac{\Sigma \frac{d_i}{\sigma_i^2}}{\Sigma \frac{1}{\sigma_i^2}}$$

$\bar{d}$ has been calculated in the rough surface model for 5 experiments when $Re_s > 1$ (points with a diamond on the figure) and in the smooth surface model for 15 experiments when $Re_s < 1$. The respective value of $\bar{d}$ are -0.001 and -0.027 with a gaussian distribution of the individual weighted values of d_i around the mean.

We have then 20 experimental observations which allow to draw the following conclusions :

a) The Brutsaert's model describes quantitatively the distribution of the isotopes in the transfer of the water vapor in the atmosphere above a water surface.

b) The characteristics of the air flow above the sea surface are those of a smooth regime for values of Re_s up to 1, larger than the theoretical number, $Re_s = 0.13$, established for a solid wall. This is of importance because it is difficult to establish when the sea surface behaves as a smooth or a rough boundary.

c) No systematic difference between the values of $\frac{n\rho_M}{\rho_M + \rho_T}$ is observed for the experiments, run either with wind waves only or with superposed generated waves. This suggests that the presence of waves (with a frequency of 1 Hz and a maximum wave height equal to 12 cm) does not modify the evaporation process, which is in disagreement with the measurements of EASTERBROOK (1968) [11] for similar conditions.

CALCULATION OF THE EVAPORATION COEFFICIENT

The property of simulating field conditions in our experiments in the wind water channel, seems well demonstrated by the wind profile determinations and the deduced drag coefficients as shown in figure 1. A good agreement with relations established from field data is observed. Furthermore from the study of the distribution of the isotopes of hydrogen and oxygen in the water vapor, it is shown that the mass transfer model proposed by Brutsaert describes correctly the evaporation processes.

We will then compute the values of the evaporation coefficient, C_q, from the equation (5), in which C_D is obtained assuming that u_* and z_o obey the Charmock's relation and $(\rho_M + \rho_T)$ takes the values given by the Brutsaert's expressions (cf table 1), for the smooth regime if $Re_s < 1$, for the rough regime if $Re_s > 1$.

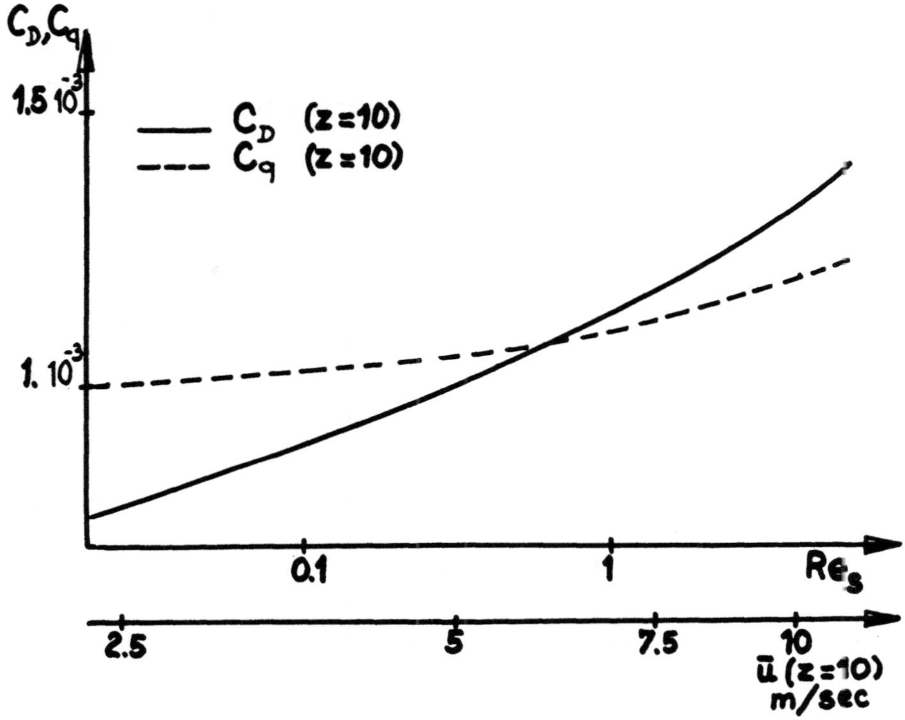

Fig. 4 - Variation of the drag coefficient and the evaporation coefficient for z = 10 meters, versus the surface Reynolds number and the mean wind speed.

We show in the table 3 and the figure 4 the values of C_q, C_D, $\frac{C_q}{C_D}$ for near neutral stability conditions, at z = 10 meters, when Re_s varies from 0.02 to 10, or, for a mean wind speed range from 2 to 13 m/sec, at z = 10 meters.

A variation of about 25 % of the evaporation coefficient is predicted for a commonly occuring range of variation of the wind speed at sea. This effect can readily be accounted for a part of the apparent large scatter of the C_q values as computed from moisture fluxes and u. $(q_s - q)$ field data by the equation (2), as discussed in the first part of this report.

The values of C_q in the table 3 are rather on the lower side of the range of data measured in the field (from about 1.10^{-3} to $1.8\ 10^{-3}$ in the compilation given in [1]). This is interpreted as the result of the influence of the stability conditions on the transfer coefficients. In most cases, unstable stratification conditions are prevailing over the ocean and then the evaporation coefficients are larger. Variations of about 15 % of C_q have been computed [12] when the bulk Richardson number varies from -0.02 to 0.

TABLE 3

Re_s	$\bar{u}$ m/sec (z = 10)	C_q (z = 10)	$\frac{C_q}{C_D}$ (z = 10)
0.02	2.2	$0.99\ 10^{-3}$	1.31
0.1	3.6	1.02	1.16
1.	6.8	1.09 "	0.97
10	12.7	1.24 "	0.84

Finally, in this work, we have tried to point out that, in addition to taking into consideration the influence of the stability conditions on moisture fluxes, the dependence of the values of the evaporation coefficient on wind speed must be taken into account. Values of C_q are calculated for neutral stability conditions. Effect of spray and breaking waves has not been included in the above study.

REFERENCES

[1] FRIEHE, C.A., and SCHMITT. K.F. (1976). Parameterization of air-sea interface fluxes of sensible heat and moisture by the bulk aerodynamic formulas. J. Phys. Oceanogr. 6, 801-809.

[2] MERLIVAT L., COANTIC M. (1975) Study of mass transfer at the air-water interface by an isotopic method. J. Geophys. Res., 80, 3455-3464.

[3] DEACON, E.L. (1977). Gas transfer to and across on air-water interface. Tellus, 29, 363-374

[4] BRUTSAERT, W. (1975). Local evaporation (or heat transfer) from rough and smooth surfaces at ground level. Water resources research., 11, 543-550.

[5] BRUTSAERT, W. (1975). The roughness length for water vapor, sensible heat, and other scalars. J. Atmos. Sci, 32, 2028-2031

[6] YAGLOM, A.M., and B.A. KADER, 1974 : Heat and mass transfer between a rough wall and turbulent fluid flow at high Reynolds and Peclet numbers. J. Fluid. Mech., 62, 601-623

[7] MERLIVAT, L. (1977).The dependence of bulk evaporation coefficients on air-water interfacial conditions as determined by the isotopic method - to be published in J. Geophys. Res.

[8] COANTIC, M., and FAVRE, A., (1974) Activities in and preliminary results of, air-sea interactions research at I.M.S.T. Proceedings of the Second UITAM-IUFF Sympcsium on Turbulent diffusion in Environmental Pollution, Charlottesville, Va., April 8-14.

[9] KITAIGORODSKIY, S.A., O.A. KUZNETSOV, and G.N. PANIN Coefficients of drag, sensible heat, and evaporation in the atmosphere over the surface of a sea. Izv. Akad.Sci. USSR, Atmos. Oceanic Phys., 9 (11), 1135-1141.

[10] SMITH, S.D., and BANKE, E.G. (1975). Variation of the sea surface drag coefficient with wind speed. Quart. J. Roy Meteor. Soc., 101, 665-673.

[11] EASTERBROOK, C.C. (1968), A study of the effects of waves on evaporation from free water surfaces. Research report n° 18; U.S. Department of the Interior.

[12] DEARDOFF, J.W. (1968). Dependence of air-sea transfer coefficients on bulk stability. J. Geophys. Res., 73, 2549-2557.

DISCUSSION

L. Hasse:

I am puzzled that the ratio of the transfer coefficients between momentum and water vapor varies so much with wind speed. In the open sea, they go much more parallel. There seems to be a systematic difference between sea and laboratory, which should be clarified in future.

L. Merlivat

To my knowledge, variations of the $\frac{C_q}{C_D}$ ratio have been reported previously in 1973 by Kitaigorodskii et al. from observations made at sea. The values of the ratios Cq and C_D we have derived come from the straightforward application of the Charnock's relation and the Brutsaert's model to calculate C_D and C_q respectively. We have checked their validity in the laboratory but I do not think that their application is restricted to laboratory measurements but may be extended to field situations.

F. Resch:

How do you define point B on the diagram?

L. Merlivat:

The point B represents the isotopic composition of the water vapor at the top of the diffusive sublayer. If samples could be taken within this layer their isotopic composition would be represented on A B . As obviously the sampling is only possible in the turbulent part, the experimental points lie on the portion B C but the length of A B represents a signature of the molecular effects right at the interface.

E. J. Plate:

You assumed a constant flux layer.
Did you verify if this assumption is valid in the laboratory

facility?

L. Merlivat:

Yes. We have been working in the semi-logarithmic profile zone which had a height of 15 to 20 cm at our working value of the fetch at 34 meters.

EDDY CORRELATION MEASUREMENT OF SEA-AIR CO_2 FLUX

E. P. Jones and S. D. Smith

Dept. of Fisheries & Environment,
Bedford Institute of Oceanography
Dartmouth, N.S., Canada

ABSTRACT

CO_2 fluxes between the ocean and atmosphere were measured off a beach on Sable Island using the eddy correlation method. These gas fluxes are measured directly over a short enough time period and a local enough region that weather and other conditions which determine the fluxes are nearly constant. Values for the fluxes measured were about 0.8 $\mu mol/m^2s$ out of the ocean.

INTRODUCTION

A major objective of direct field measurements of air-sea gas fluxes is to find out what environmental quantities control the gas exchange process. Although we can measure concentrations of many natural and anthropogenic gases in the air and sea, we do not know how wind speed, temperature, humidity, and other physical and chemical quantities may affect exchange processes. Only two field measurements of air-sea gas fluxes, CO_2 and radon, have been previously reported (Broecker and Peng, 1974; Peng *et al.*, 1974; Peng and Broecker, 1976). Both involve radioactive nuclei. CO_2 fluxes measured using relative carbon-14 ($T_{1/2}$ = 5700 yr) concentration in the air and water are long term, global averages. In

Adapted from Journal of Geophysical Research (In press, 1977).

contrast, radon fluxes measured using radon-222 ($T_{1/2}$ = 3.8 days) concentration profiles in the water column are local rather than global and respond to shorter duration events. Because of possible advection in the mixed layer and because weather and mixed layer conditions can change in very short times (hours) a particular radon-222 concentration profile could be far from a steady-state condition.

We report the first measurement of a gas flux between the air and sea made over a short enough time period and a local enough region to preclude major changes in the quantities which may affect gas exchange. Carbon dioxide was chosen becuase it is the most abundant of atmospheric gases after nitrogen, oxygen, argon and water vapour, and is readily detected by strong infrared absorption at a wavelength of 0.426×10^{-3} cm. The eddy correlation technique has been used to determine water vapor, heat and momentum fluxes over a wide variety of conditions (Pond, 1971, 1972; Businger, 1975). Applied to gas fluxes, it correlates fluctuations in the gas concentration, q, at a fixed point with vertical wind fluctuations, u_3. Above a very thin laminar boundary layer (< 1 mm) there exists a layer several tens of metres thick in which mixing is effected almost entirely by wind turbulence and the average flux

$$F = \overline{qu_3} \tag{1}$$

is constant with height and equal to the average flux at the surface. The overbar denotes a time average which, for measurements at 10 m height, is taken over 15- to 60-minute intervals. Shorter intervals do not yield the correct average flux since the effects of individual gusts begin to be important, while for longer intervals environmental quantities may not be constant.

MEASUREMENT TECHNIQUE

The CO_2 sensor (Jones *et al.*, 1977) is the key to the direct measurement of a CO_2 flux since previously available devices were either not fast enough or not sensitive enough. It is a variant of a non-dispersive gas cell analyser developed by Barringer Research Co. Ltd., Rexdale, Ontario, Canada, to detect infrared emissions from trace quantities of tropospheric gases such as CO_2, CO, NH_3, CH_4, HCl, NO and C_2H_6. A radiant source black body is housed at the top of the sensor (Fig. 1), connected electrically and pneumatically to the main sensor volume through supporting tubes. The source radiant energy passes via a germanium lens and window through the sampling path, which is 10 cm long and 1 cm in diameter, is partially absorbed, and passes down a snout to the main sensor body. The source signal and a signal from a reference black body are alternated at 1 KHz by a chopper disk. A beam

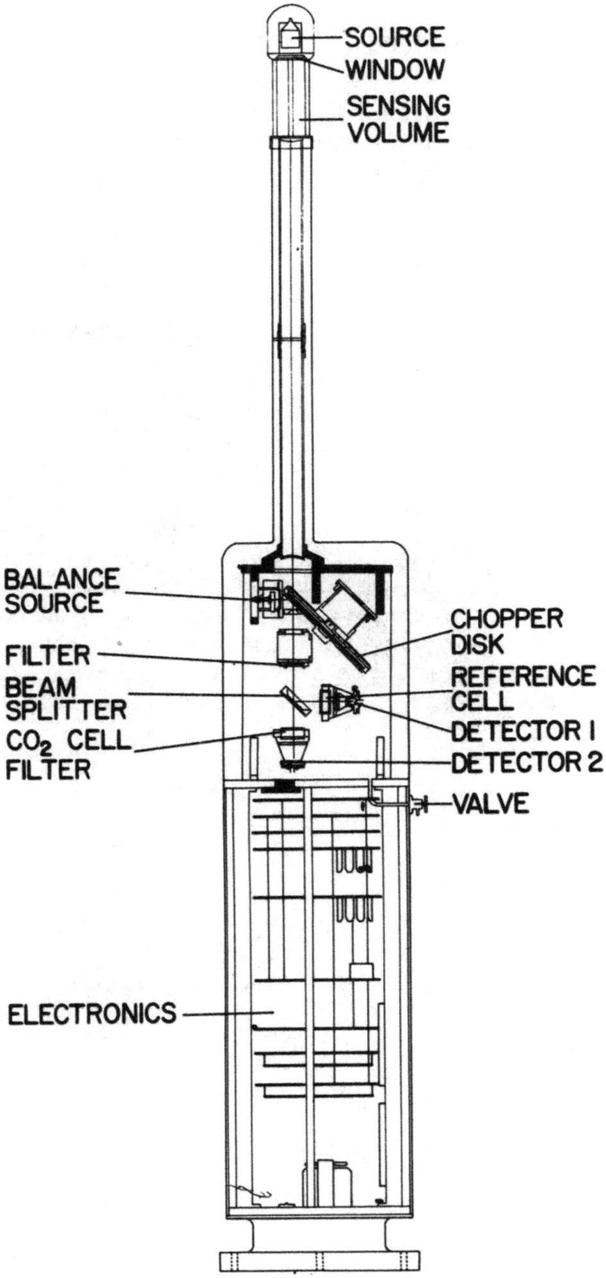

Figure 1. The Barringer CO_2 sensor.
The sensing path near the top is 10 cm long.

splitter directs the chopped signal through gas cells to two thermoelectrically cooled PbSe detectors. The whole measuring path except for one gas cell is evacuated to remove CO_2. The gas cell is filled with CO_2 to serve as a filter opaque to CO_2 absorbtion lines and its signal is insensitive to CO_2 fluctuations. Circuits which take the difference of the two detector outputs are balanced using the reference part of the chopped signal. Over a 10 Hz bandwidth, determined by electrical filtering, the rms noise is 0.3 ppm or 0.1% of the ambient concentration. The sensing path, which is 10 cm long and 1 cm in diameter, is open to the wind, and air passes through it in less than 10 ms. To measure wind and temperature fluctuations, we used a Kaijo Denki PAT 311-1 three axis sonic anemometer, which averages velocity fluctuations over 20 cm sound paths, and a VECO microbead thermistor, which responds with a time constant of approximately 0.1 s. The characteristics and performance of these instruments are discussed in detail elsewhere (e.g. Smith *et al.*, 1976).

Our flux measurements were made in July 1976 on Sable Island, Nova Scotia, a low sand island in the Atlantic Ocean 300 km east-southeast from Halifax (Fig. 2). The weather at this time of year is characterized by high humidity, fog, relatively gentle winds, and rising sea surface temperature. The season was chosen because the rising water temperature reduces the solubility of CO_2 and should result in a relatively large, and hence more easily measured, upward flux. Beach sites on Sable Island have been found during

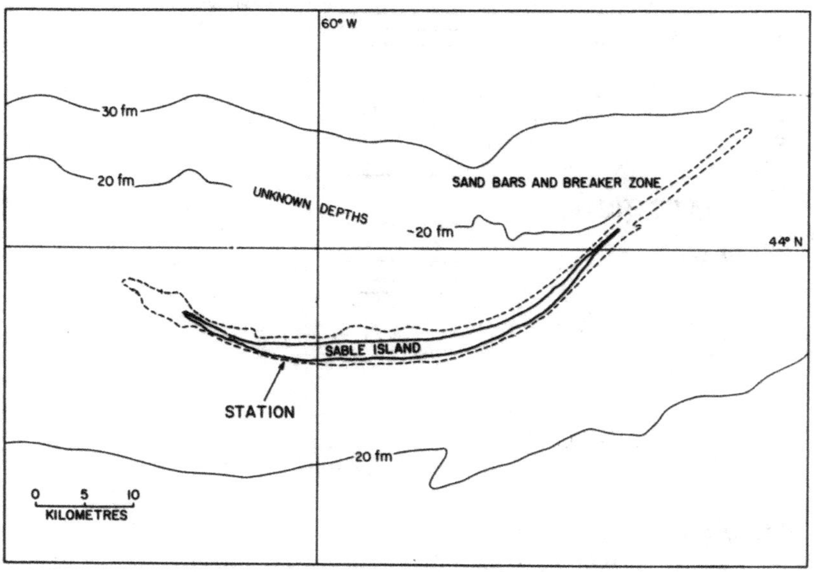

Figure 2. Sable Island experimental site.

periods of onshore winds to give typical open-ocean values for sensible heat, moisture and momentum fluxes, when the measuring instruments were mounted on a 10 m tower, 50 m from the shoreline (Smith and Banke, 1975; Smith *et al.*, 1976). Fluxes measured in this way relate to an area 200 to 1000 m offshore, where in moderate weather the waves do not break. Data collection and processing were carried out using techniques established at the Bedford Institute (Smith, 1974; Dobson *et al.*, 1974).

RESULTS

In this first application of the eddy correlation technique to measuring gas fluxes, there was not the variety in field conditions (Table 1) necessary to try to characterize gas exchange processes. The observed drag coefficients (Table 1) are slightly lower than we might have expected. At 7 m/s wind speed $C_{10} \simeq 1.1 \times 10^{-3}$ (Smith and Banke, 1975) and for stable stratification (z/L = 0.3) a reduction to 0.9×10^{-3} might be expected (Hsu, 1974). The values obtained here, $C_{10} \simeq 0.6 \times 10^{-3}$, may indicate that the beach site chosen (Fig. 3) is not entirely typical of ocean conditions; but they do not lie outside the scatter of Smith and Banke's values. The upward CO_2 fluxes observed are in agreement with our expectations for the seasonal warming of the ocean. In addition, alkalinity analysis of water samples, which for several reasons probably did not give quantitative results, nevertheless indicated that the water was substantially supersaturated with CO_2 relative to the air. Our CO_2 fluxes correspond to the upper range of values recorded for the ocean (Schindler, 1975).

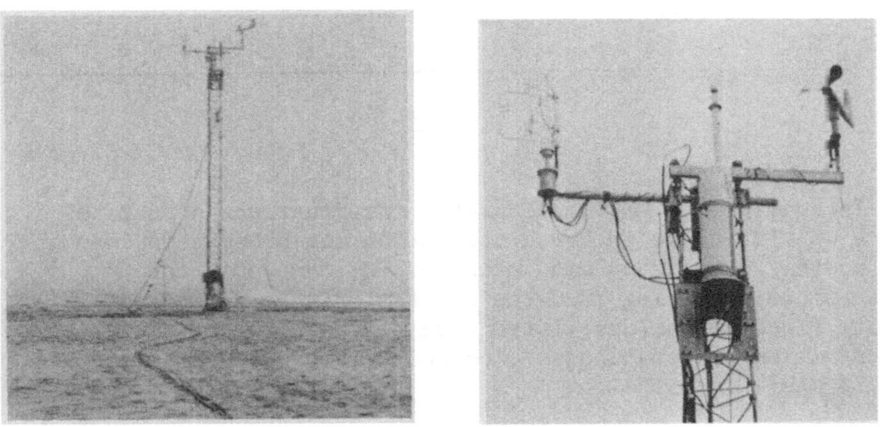

Figure 3. Instrumented 10 m tower on the beach at Sable Island.

Table 1 Summary of Results

Date (1976)	July 8	July 9	July 9
Starting time	2248Z	0130Z	0225Z
Duration (min)	29	42	42
CO_2 flux $\overline{qu_3}$ (μmol/m^{2}s)	0.5	0.8	0.8
Heat flux $\rho c\overline{tu_3}$ (J/m^{2}s)	-13.9	-13.0	-12.9
Momentum flux $\rho\overline{u_1u_3}$ (N/m^2)	- 0.034	- 0.043	- 0.039
Wind speed U (m/s)	6.9	7.5	7.7
Sea temperature T_s (C)	12.65	12.65	-
Air temperature T (C)	14.8	14.7	-
Stability z/L	0.35	0.26	0.28
Drag coefficient $C_{10} = -\overline{u_1u_3}/U_{10}^2$	0.00058	0.00062	0.00053

z = measurement height

$L = -(T+273)(-\overline{u_1u_3})^{3/2}/(gK\overline{tu_3})$, Monin-Obukhov stability length

g = gravitational acceleration, 9.8 m/s^2

u_i = wind velocity fluctuation component: (1) downwind, (2) lateral, (3) vertical

K = 0.4, von Karman constant

ρ = density of air

c = specific heat of air at constant pressure, 10^3 J/kgC

t = temperature fluctuation about the mean

In addition to CO_2 concentration fluctuations associated with a CO_2 flux, the infrared absorption measuring technique used in this experiment also measures CO_2 concentration fluctuations resulting from the air temperature and hence density fluctuations associated with a sensible heat flux. A CO_2 concentration fluctuation, q', caused by a temperature fluctuation, t, is

$$q' = -\frac{t}{T}\,Q \qquad (2)$$

where Q is the average CO_2 concentration and T is the absolute temperature of the air. The correction which must be applied to the apparent measured CO_2 flux, F, (Eq. 1) is

$$- \overline{q'u_3} = \frac{Q}{T} (\overline{tu_3}) \tag{3}$$

The corrected CO_2 flux values (Table 1) are about 30% less than the uncorrected values. Several more data runs taken in similar conditions have been analysed and give results similar to those tabulated.

In the analog chart record for the run starting at 0130 UT, July 9, 1976 (Fig. 4), much of the apparent CO_2 fluctuation may actually be sensor noise. Time series of the products of the digitized CO_2, temperature, and downwind velocities (respectively) with vertical velocity (Fig. 5) give a qualitative indication of

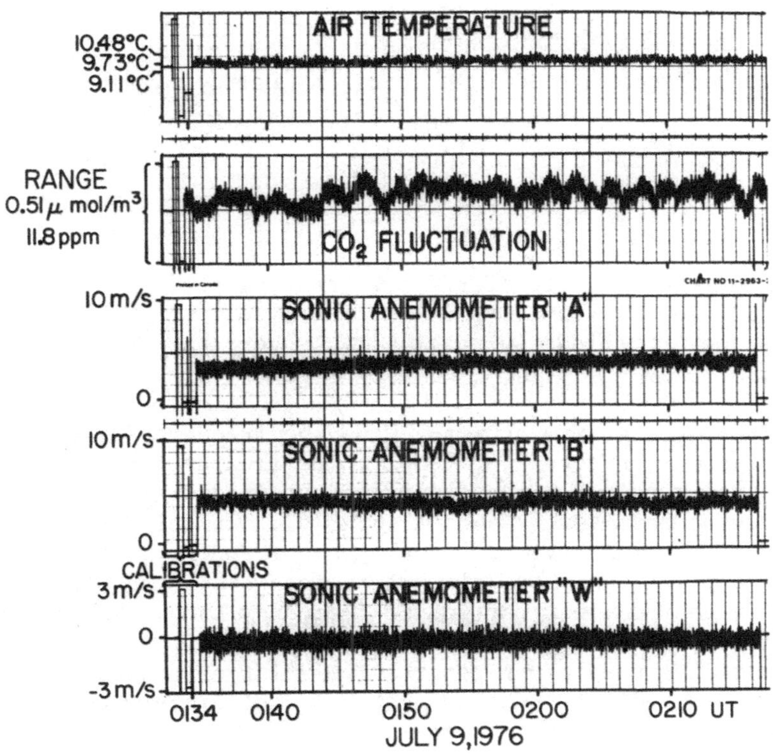

Figure 4. Chart recording of sensor signals: air temperature, CO_2 concentration, and three wind components.

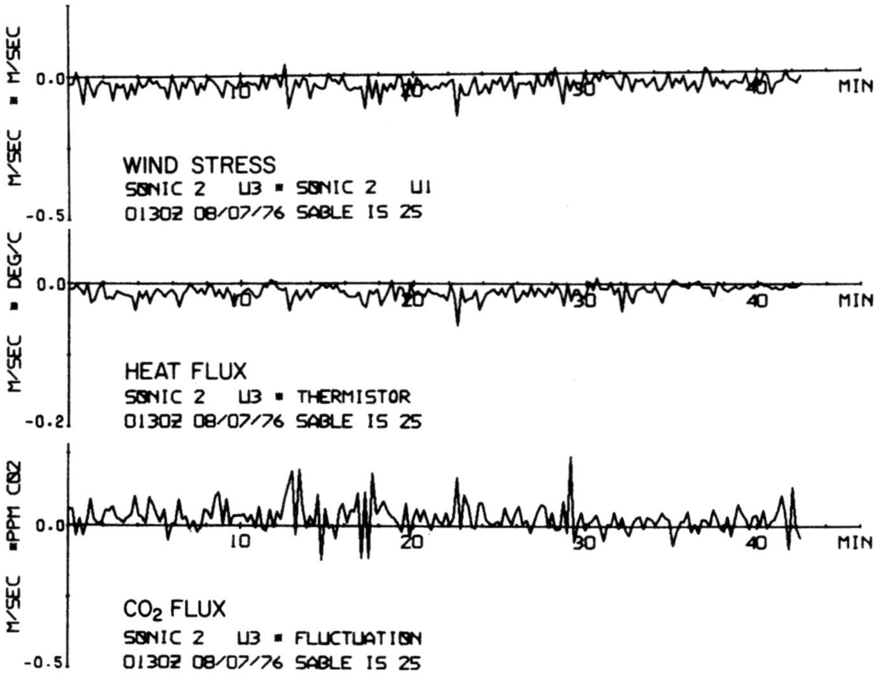

Figure 5. Time series of CO_2 flux, heat flux, and wind stress averaged over intervals of 12.8 sec.

the averaging time required for representative estimates of the fluxes. Individual events typically last 1 to 4 min, and so we must average over periods substantially longer than 4 min, in this case 43 min.

For frequencies from 0.01 to 10 Hz the frequency distribution of CO_2 variance (corrected for temperature fluctuations) for this run is plotted together with noise data obtained earlier with the sampling region enclosed by a hood (Fig. 6). In this plot of (frequency x spectrum) against logarithmic frequency, the area covered in any frequency band is proportional to the contribution to the variance from that particular frequency band. The essentially 'white' sensor noise increases to the right as the logarithmic bandwidths used cover progressively larger frequency intervals. The observed CO_2 spectrum, which is the difference between the spectrum of the observed signal and the noise, shows a broad maximum around 0.2 Hz, tails off to nearly zero at 1 Hz, and becomes lost in the noise above 5 to 10 Hz. The peak at 3 Hz is possibly associated with vibration of the tower.

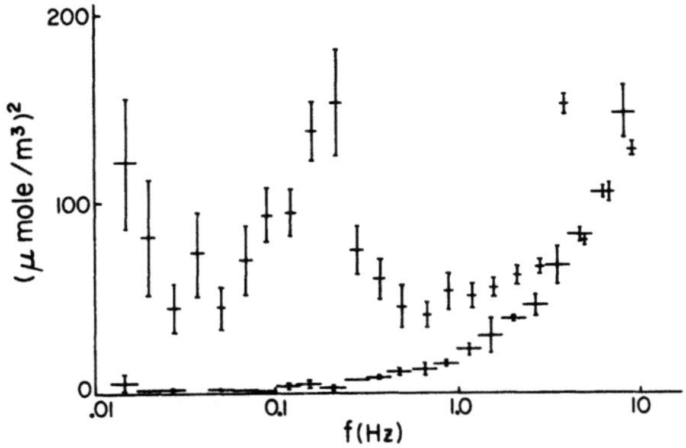

Figure 6. Upper plot: CO_2 concentration spectrum x frequency versus logarithmic frequency (corrected for temperature fluctuations).
Lower plot: Noise of the instrument: Height of symbols gives standard error calculated from variability during data run.

For the same run, the CO_2 flux cospectrum (Fig. 7a) is largely confined to a narrower range of 0.05 Hz to 0.3 Hz than the heat and momentum flux cospectra (Fig. 7b and 7c) which show significant contributions from about 0.05 Hz to 2 Hz. Heat and momentum flux cospectra are negative, indicating downward fluxes. In Figure 7 the area enclosed by the cospectrum is proportional to the contribution to the flux over any particular frequency band.

In this first measurement, we have demonstrated that a direct flux measurement can be made over a short enough time and a small enough area so that environmental quantities expected to be related to gas exchange remain essentially unchanged.

In this particular set of environmental conditions, the CO_2 fluxes were relatively large and the heat fluxes relatively small. Under other conditions, air temperature fluctuation corrections could play a very important role, though the sharply peaked CO_2 fluctuation spectra, if typical, would help in extracting the true value of a CO_2 flux. Future measurements should now be made under more widely varying conditions of wind speed, gas partial pressure differences, air and water temperatures, humidity, and perhaps other environmental quantities to obtain relations which will enable determinations of fluxes of other gases not amenable to direct measurement.

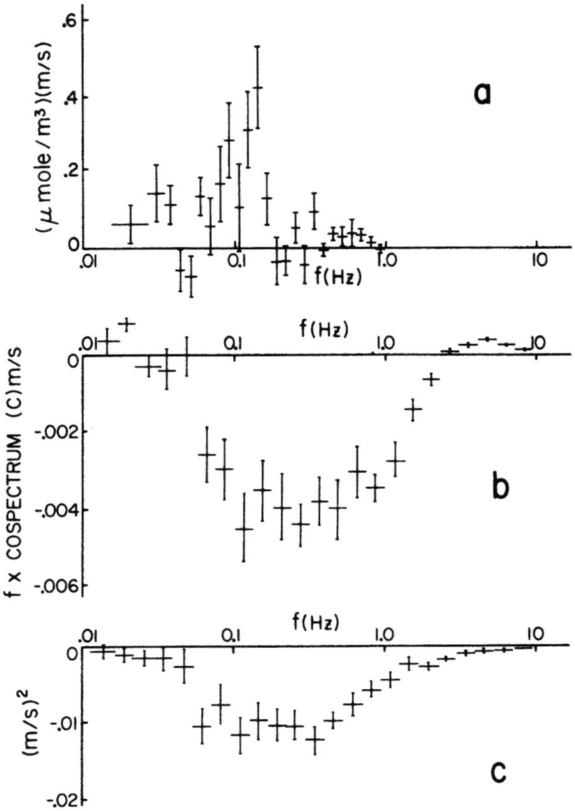

Figure 7a. Cospectrum of CO_2 and vertical wind shows frequency distribution of CO_2 exchange (corrected for temperature fluctuations).

Figure 7b. Cospectrum of heat and vertical wind fluctuations shows frequency distribution of heat flux.

Figure 7c. Cospectrum of downwind and vertical wind fluctuations shows frequency distribution of wind stress.

ACKNOWLEDGEMENTS

The authors wish to thank R.J. Anderson, E.G. Banke, and F. Zemlyak for their able assistance in the field study, and the staff of the Sable Island Upper Air Station for their hospitality. The authors also wish to thank Profs. L. Hasse and S. Bakan for pointing out the need for a temperature correction.

REFERENCES

Broecker, W.S., and T.H. Peng, Gas exchange rates between air and sea, *Tellus*, *26*, 2135, 1974.

Businger, J.A., Interaction of sea and atmosphere, *Reviews of Geophysics and Space Physics, 13,* 720-726 and 817-822, 1975.

Dobson, F.W., R.F. Brown, and D.R. Chang, A set of programs for analysis of time series data including fast Fourier transform spectral analysis, *Bedford Institute of Oceanography Report BI-C-74-2,* 1974.

Hsu, S.A., On the log-linear wind profile and the relationship between shear stress and stability characteristics over the sea, *Boundary-Layer Meteorol., 6,* 509-514, 1974.

Jones, E.P., T.V. Ward, and H.H. Zwick, A fast response atmospheric CO_2 sensor, to be published in *Atmospheric Environment.*

Peng, T.-H., T. Takahaski, and W.S. Broecker, Surface radon measurements in the North Pacific Ocean Station Papa, *J. Geophys. Res. 79 (12)*, 1772-1780, 1974.

Peng, T.-H., and W.S. Broecker, The rate of gas exchange across the sea-air interface in the Atlantic Ocean, *Bull. Am. Meteorolog. Soc., 57,* 144, 1976.

Pond, S., Air-sea interaction, *Trans. Am. Geophys. Union, 53 (6),* IUGG389-IUGG394, 1971.

Pond, S., The exchanges of momentum, heat and moisture at the ocean-atmosphere interface, in *Numerical models of ocean circulation,* Proc. U.S. Acad. of Sciences Symposium, Washington, D.C., 1972.

Schindler, D.W., Factors affecting gas exchange in natural waters, *Limnol. Oceanogr., 20,* 1053-1055, 1975.

Smith, S.D., Program A to D for analog-to-digital conversion and processing of time series data, *Bedford Institute of Oceanography Computer Note BI-C-74-1,* 1974.

Smith, S.D., and E.G. Banke, Variation of the sea surface drag coefficient with wind speed, *Quart. J. Roy. Met. Soc., 101,* 665-673, 1975.

Smith, S.D., R.J. Anderson, E.G. Banke, E.P. Jones, S. Pond, and W.G. Large, A comparison of the air-sea interaction flux measurement systems of the Bedford Institute of Oceanography and the Institute of Oceanography, University of British Columbia, *Bedford Institute of Oceanography Report BI-R-76-17,* 1976.

DISCUSSION

C. A. Friehe

Is the CO_2 device output linear or exponential?

S. D. Smith:

The line strength would decay exponentially along the sensing path. For the small fluctuations seen, nonlinearity is no problem.

H. Lacombe:

Apparently the flux is upward - so, locally and at the moment, the ocean is polluting the atmosphere.

S. D. Smith:

The observations were carried out during a period of warming of the ocean; the CO_2 solubility was therefore decreasing; and CO_2 was released from the sea. This was confirmed by alkalinity measurements of seawater, which confirmed that the seawater was CO_2 saturated.

J. A. Businger:

The cospectrum of CO_2 falls off sooner than the cospectra for momentum and heatflux. Is it possible that the lateral separation between the CO_2 probe and the sonic anemometer is the cause of this fall-off?

S. D. Smith:

The CO_2 and the sonic anemometer were separated by 1 m in a more or less crosswind direction. At 2Hz and 8 m/s we have wavelengths of order 4 m so that if the turbulence is more or less isotropic at 1/4 wavelength we should be all right at all frequencies where we have heat and momentum flux. I shall review existing data on coherence of wind from laterally separated anemometers to check this important point.

M. Coantic:

The only previous attempt to measure CO_2 fluxes by the eddy correlation method I know of is that of DESJARDINS, at Cornell University (see DESJARDINS, 1974, Int. Tour. Biometeorology, 19, pp. 76-83). Could you indicate the relative accuracy of your measuring instrument with respect to the one used by Desjardins?

S. D. Smith:

No, I had not been aware of this work but wish to thank you for the reference. The noise level is measured with the sensing path closed, and the sensitivity is checked by evacuating the sensing path.

GAS EXCHANGE AND EVAPORATION STUDIES IN A CIRCULAR WIND TUNNEL, CONTINUOUS RADON-222 MEASUREMENTS AT SEA, AND TRITIUM/HELIUM-3 MEASUREMENTS IN A LAKE

K. O. Münnich, W. B. Clarke+, K. H. Fischer, D. Flothmann, B. Kromer, W. Roether, U. Siegenthaler++, Z. Top+, and W. Weiss

Institut für Umweltphysik, Universität Heidelberg

INTRODUCTION

This paper reports on gas exchange research activities pursued at this laboratory, with the aim of increasing present knowledge on gas exchange and on the parameters which control it. Laboratory (wind tunnel) experiments as well as field work (with Rn-222 and He-3) are under way. - Two of the present authors (K.O.M. and D.F.) have previously presented [1] a theoretical treatment of the problem. Another, more detailed recent publication by Deacon [2] draws similar conclusions. The most important factor controlling gas exchange is wind speed; at low wind, however, surface cooling by evaporation is also to be considered.

LABORATORY EXPERIMENTS IN A WIND TUNNEL: EVAPORATION AND ISOTOPE FRACTIONATION

The experimental set-up is shown schematically in Fig.1. The circular wind/water tunnel (its air flow has some resemblance to the flow in a hydrodynamic clutch) is clearly a compromise. The advantages are to have a clean and gas tight, medium size system (10 liters of water), allowing basically independent control of wind speed (by rotor frequency), and air moisture (by the rate of nitrogen flux flushing the system). The disadvantage, of course, is that one deals with a rotating system, where Coriolis forces influence the air motion. This primarily results in a stronger coupling of the rotor blade motion to the water surface in the channel, than it would be the case in a linear system.

+McMaster University, Dept. of Physics, Hamilton, (Ontario), Canada
++Physikalisches Institut der Universität Bern (Switzerland)

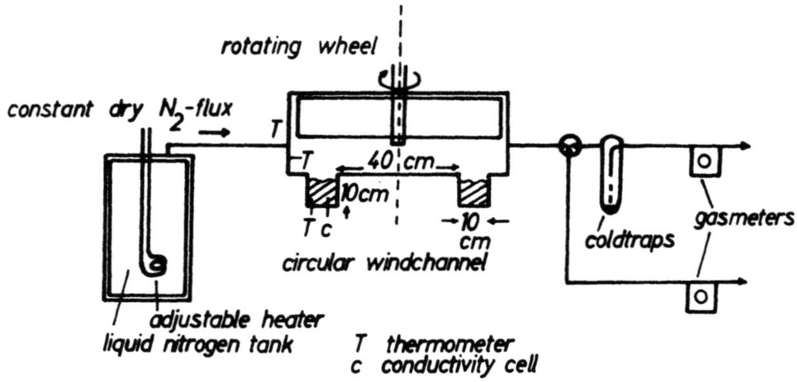

Fig.1 Wind Tunnel Experiments (schematic). - "Wind" is made by a rotating wheel mounted in the chamber top, and momentum is transferred to the water in the annular channel. The chamber is being flushed with dry nitrogen at an adjustable rate. Water vapour can be collected fractionation-free in the cold traps.

The important point with this circular wind tunnel, however, is that, like in nature, momentum is transported into the water from the air above, and that the problem of increasing fetch which makes the state of the water quite varying along the channel length in a linear tunnel does not apply: the water reaches a homogeneous steady state everywhere along the annular channel.

Data on evaporation versus wind speed obtained with the system is shown in Fig.2 (taken from Vogt [3]). The data points fall on the straight line

$$w = 0.5 + 3.5 \cdot 10^{-3} U, \qquad (1)$$

with non-zero evaporation at zero wind. This is presumably due to convection induced by the fact that saturated air close to the water surface may be up to 1% lighter than the dry air above. For comparison with data in nature one can, assuming a log wind profile, tentatively adjust eq. (1) so that it gives w(0,10m) as a function of U(10m). This procedure needs multiplication of the proportionality factor in eq. (1) with $(\approx 0.6)^2 = 0.36$ (compare Fig.6) and brings it very close to $1.2 \cdot 10^{-3}$, which agrees exactly with Laevastu's [4] evaporation formula . This, however, is too good an agreement, since this procedure neglects the Coriolis forces and thus underestimates transfer velocity.

Fig. 3 shows the difference in isotopic composition (the permill depletion ε (O-18) of the isotope ratio ([O-18]/ [O-16]) of the vapour collected in the same experiment. The O-18 depletion in the vapour strongly depends on the air moisture deficiency, but not on

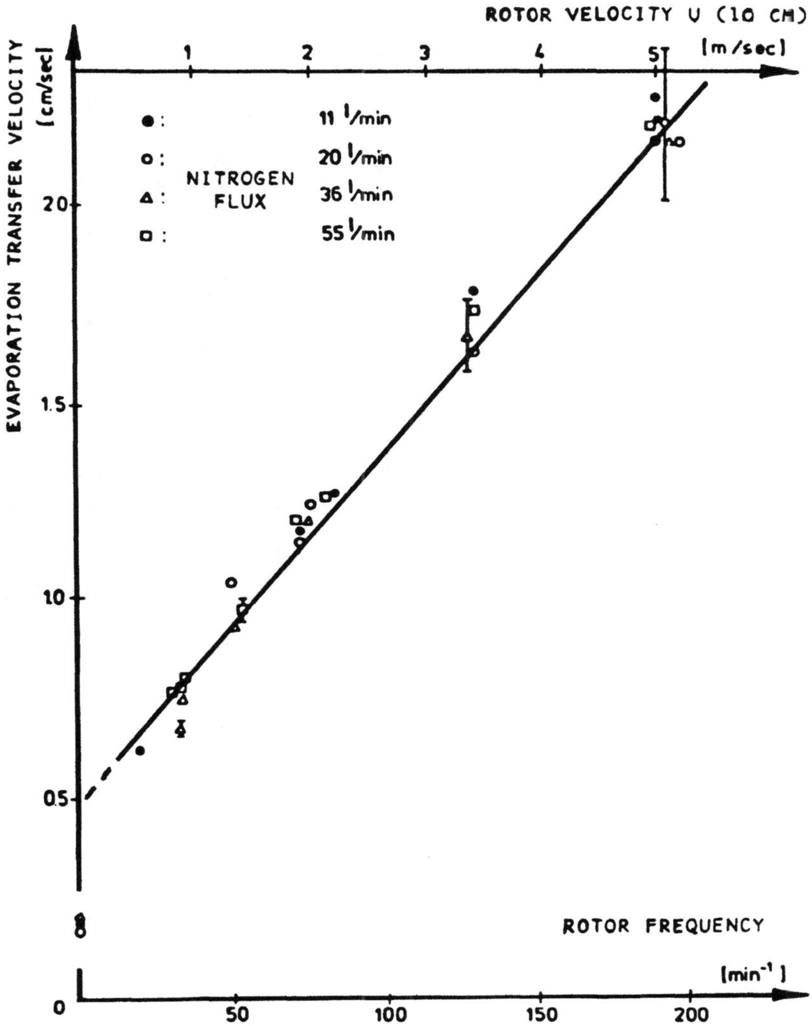

Fig.2: Evaporation transfer velocity w(0,7.5 cm) in the circular wind tunnel (see caption fig.5 for definition of w). - Air moisture is sampled about 7.5 cm above the water surface (chamber outlet). Upper abscissa: rotor velocity, calculated for the center radius of the water channel. Momentum transfer should roughly correspond to a wind speed U (10 meters) of nearly twice the values indicated on the upper abscissa. The error bars shown give the gas meter uncertainty only. Note that points at different nitrogen fluxes (i.e. air moisture) fall upon each other even at zero rotor frequency. Air moisture ranges from 66 to 90% rel. humidity at 75 rpm, and from 57 to 85% at 35 rpm, respectively.

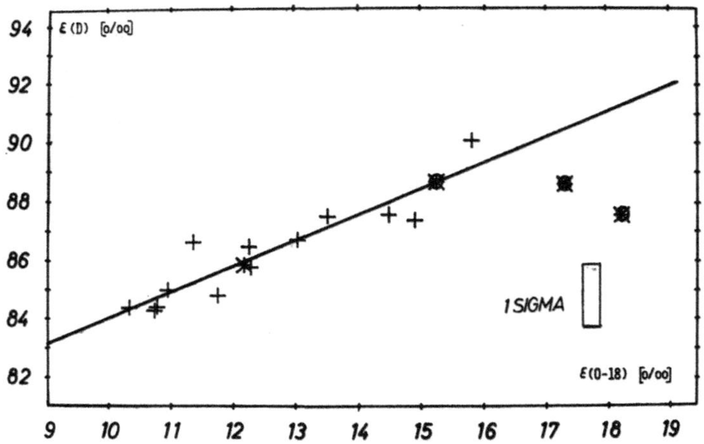

Fig.3 Oxygen-18 fractionation versus moisture deficit. - The circled points represent data obtained at zero wind. The slope is about half the one to be expected for diffusion through a stagnant film. - Figure from Vogt [3] . Due to the usual definition of the fractionation factor α in evaporation $\alpha = 1+\varepsilon = R_{liq}/R_{vap}$ the ordinate rather shows the heavy isotope enrichment in the liquid which, however, is practically identical with the depletion in the vapour (R = [O-18] /[O-16]).

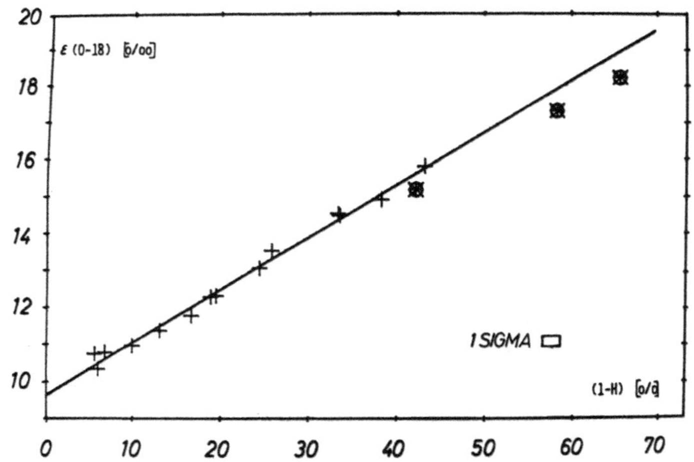

Fig.4 Deuterium versus Oxygen-18-fractionation. - The slope represents the ratio of the deviations ε_o of the diffusion constants HDO and H_2O-18, respectively, from the diffusion constant D of the "normal" water molecule H_2O. The circled points which partly fall off the line were measured at zero wind speed (see text). Figure from Vogt [3].

the evaporation transfer velocity. Extrapolation to air saturated with moisture gives the equilibrium fractionation factor of water in equilibrium with its vapour, while the additional fractionation is kinetic and due to the different molecular diffusion resistances for H_2O and H_2O-18 molecules in the viscous gas sublayer adjacent to the liquid interface. The ratio $D(H_2O)/D(H_2O\text{-}18)$ of the diffusion constants is known [5,6] to be 1.030 which means that the maximum possible depletion of O-18 during evaporation of water into dry air is ε(O-18) = -10-30 =-40‰ and would be reached if the vapour had to diffuse through a stagnant air layer into a well mixed dry space. The kinetic O-18 depletion in the vapour which is observed here, however, is (if extrapolated to zero air moisture) only about -14%, i.e. roughly 1/2 the maximum value of -30%.

This is in fact predicted by models [1,2,5] more adequate to the problem than the stagnant film model. The Danckwerts (residence time) model [1] predicts exactly half the stagnant film diffusion enrichment; other models [2,5] give a somewhat larger value. Usually one defines the ratio of two molecular transport (diffusion) constants which is a dimensionless number (D_x/D_y or specificly ν/D =Sc, the Schmidt number or λ/D = Pr, the Prandtl number etc.) where ν is the molecular diffusion constant of momentum (kinetic viscosity) and λ the one of heat. In general D_x and D_y are the diffusion constants of two different (molecular) species x and y. Then the different models give for the ratio of the transport velocities $w_x/w_y = (D_x/D_y)^n$ with an exponent $0.5 \leq n \leq 1$. In the case of two isotopic species where D_x is only slightly larger than D_y we obtain by linear approximation $w_x/w_y = 1 + n\cdot\varepsilon_0$ where $\varepsilon_0 = (D_x-D_y)/D_y$ The Danckwerts model [1]predicts n = 1/2 if the vapour is sampled at about an elevation $z \approx 70\cdot z_*$ where $z_* = D/U_*$, U_*= friction velocity. At higher elevation the non-fractionating turbulent transfer resistance becomes noticeable, eventually suppresses the apparent fractionation and thus reduces the apparent n. Therefore, our experimental result is only a lower limit for n. As mentioned before, in this system, however, the strong momentum (and matter transport) coupling above the viscous sublayer suggests that the true n is in fact not much larger than 0.5.

If one plots the enrichment ε for the two heavier species HDO and H_2O-18 versus each other (Fig.4) the moisture deficit as well as the exponent n cancels in the ratio and the slope in Fig.4 therefore directly gives the ratio $\varepsilon_0(HDO)/\varepsilon_0(H_2O\text{-}18) = 0.83 \pm 0.1$, in disagreement with the assumption of a simple $1/\sqrt{\mu}$ dependence of the diffusion constant (μ= reduced mass of impact partners) [6] , but in agreement with the findings of Merlivat 1975 (see [5]).

The points measured at zero rotor frequency (Fig.4) definitely fall below the straight line. This is due the transfer resistance in the liquid as has been noticed first by Gat [7] and by Siegenthaler [8] . For the most cases the liquid resistance can be neglected in transfer of tagged water molecules from the liquid to the gas phase. This is due to the fact (fig.5) that the transfer through any diffusion layer is also proportional to the "solubility" of the substance in this layer. The extremely high "solubility" of water vapour in liquid water therefore makes the viscous liquid sublayer extremely permeable for water molecules. Nevertheless, if the viscous liquid sublayer becomes very thick (at zero wind, see below) its resistance becomes noticeable. That deuterium is more strongly influenced is due to the equilibrium fractionation factor being for HDO eight times larger than for 0-18. Therefore a much larger HDO excess is being piled up directly at the interface during evaporation than this is the case for the O-18 tagged water molecules. As soon as the liquid resistance becomes essential the liquid starts to appear more enriched in deuterium than in O-18 and the points in fig.4 fall off the line. A quantitative evaluation of the situation [3] is in agreement with the gas exchange data discussed below.

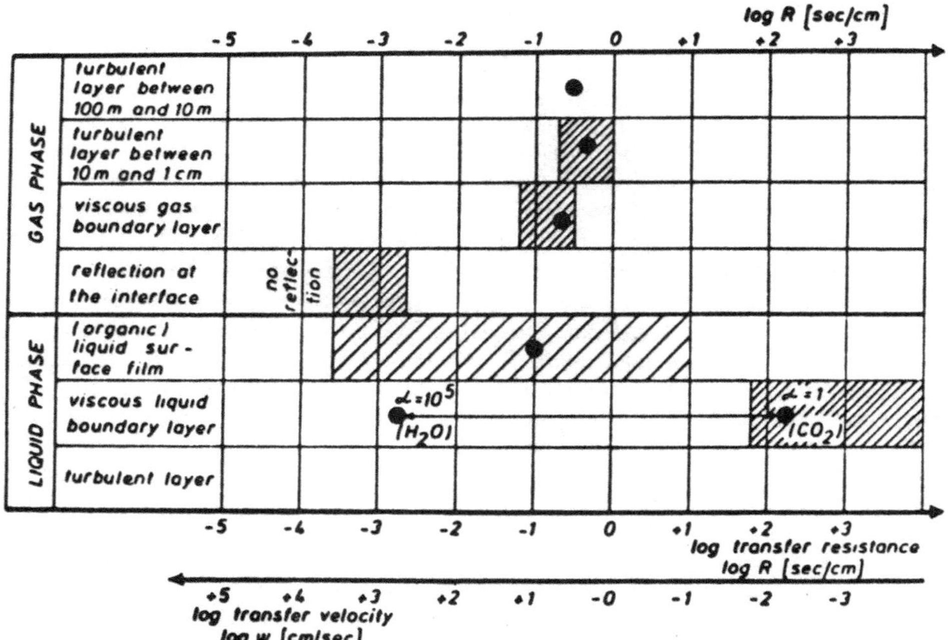

Fig. 5 Schematic representation of the transfer resistances $R_i = 1/w_i = \Delta c_i/j$ (w_i=transfer velocity, Δc_i=concentration difference across layer i; j=substance flux density) of individual layers i on both sides of the air/sea interface. Connecting all resistances in series one sees that due to different values of the solubility α the bottleneck changes from the liquid (for gases, low solubility) to the gas phase (for water vapour, high solubility). Figure from [1]

WIND TUNNEL: GAS EXCHANGE

The circular wind tunnel has primarily been designed for gas exchange studies. The evaporation measurements of the previous section rather are a by-product, and have been undertaken to provide information about the general performance of the system. Up to now most gas exchange data have been obtained on deionized water loaded with CO_2. As can be seen from Fig. 5 and from Liss 1973 [9] or Broecker and Peng 1974 [10] (see also [1]), the transfer velocity of dissolved gases is much slower than the evaporation velocity discussed before. Therefore at the usual nitrogen flushing rates CO_2 concentrations in the gas phase are essentially zero, and the CO_2 one way evades from the water at a rate controlled by the rotor frequency. CO_2 has the prime advantage to be easy to measure in the water samples (titration) taken at intervals from the channel. More recently we installed a conductivity probe in the channel. Through hydration and dissociation the CO_2 dissolved gives a sufficient large conductivity signal (proportional to $\sqrt{[CO_2]}$) which makes the CO_2 concentration direct reading and its exponential decay easy to follow. The observed evasion time τ which is in the order of an hour gives via the relation

$$w = h/\tau \tag{2}$$

(h = water depth in the channel) the evasion transfer velocity w. As has been explained elsewhere 1,9,10 the gas solubility α does not influence relation (2) as long as one can neglect gas phase resistance. Fig. 5 shows that this is the case with CO_2. Despite its chemical reaction with water which allows the conductimetric measurement of its concentration CO_2 essentially behaves like an inert gas under the conditions prevailing here (around 99% of the CO_2 present is inertly dissolved).

The gas evasion velocities obtained with pure water (Lohse 1977 [11]) are presented in Fig. 6 which is taken from Emerson 1974 [12] and contains various other gas exchange data. Our evasion velocity data nicely follow a linear increase with wind velocity and suddenly jump up at a wind tunnel velocity of about 6 meters/sec corresponding to about 10 meters/sec in nature. The jump is characterized by a drastic change in the wave pattern:the waves look more shaky. Unfortunately we don't have a more objective documentation (photographs etc.) of the different wave states as yet. This behaviour - linear increase and sudden change in slope - of the gas exchange rate is interesting in view of the fact that theoretical models [1,2] , predict a linear increase of gas exchange with wind in the moderate velocity range. The prime question then is whether the onsetting stronger increase is primarily due to increasing friction velocity U_* , i.e. to a momentum input jump caused by changing roughness, which after Deacon [2] on the basis of present knowledge seems to be not the case or whether there are other changes going on. In this respect it seems promising that we shall be able to measure momentum input to the water directly via the steady state bulk velocity of the water. The channel walls are smooth and the water flow relaxation

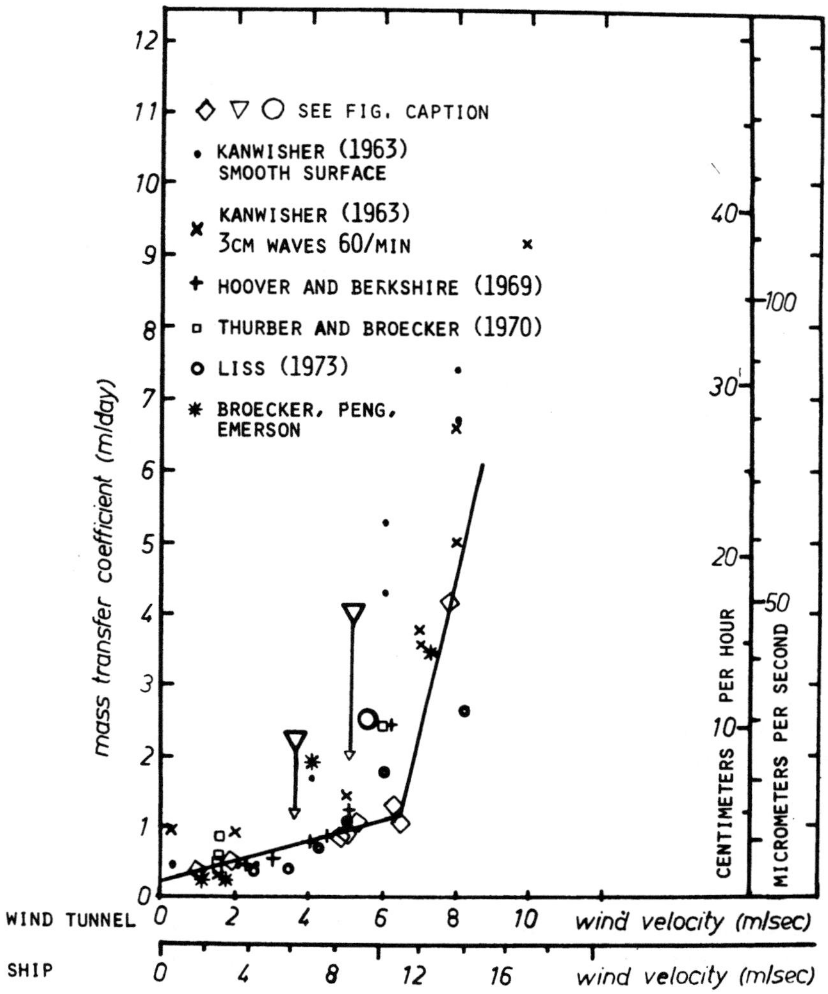

Fig. 6 Summary of gas exchange data available (figure taken from Emerson 1974 [12] and modified). The evasion velocity w on the ordinate is given in several different convenient units. A second abscissa scale has been added to refer to ship's wind data (taken at 10 m). Since wind tunnel velocities are usually at 10 cm above the water surface the two scales have been adjusted to each other with the assumption of a log profile and with a roughness length $z_o \approx .01$ cm. - The tilted squares are our wind tunnel data [11]. The larger circle represents an average of radon-222 data obtained on Challenger 11 (Roether and Kromer 1977 [13]), and the triangles are helium-3 data from Lake Constance 1976/77 (Weiss et al.1977 [14]). The helium-3 evasion velocities have been corrected for the larger diffusion constant with the Danckwerts model. The wind velocities on the lake have been extrapolated from shore data with the help of Mühleisen's 1977 [15] data.

time, being relatively long, should be easy to measure, so that a momentum balance for the channel water can be made.

Another interesting fact which was intrigueing us for a while is that Chr.Bohr 1900 [16] reported CO_2 evasion rates from ethyl alcohol several times greater than the ones he had found with water. We made a measurement with ethanol in the wind tunnel and found this to be basically correct. The point is that the only molecular property of interest here and differing between ethanol and water ist the surface tension which consequently must have an influence on gas exchange rate (capillary waves ?). For safety reasons we made the further measurements not with pure alcohol, but with water to which 10% ethanol were added, this small admixture already reduces most of the difference in surface tension between water and alcohol. The additional advantage is that we can still use the conductometric method to "on line" follow the CO_2 evasion from the liquid. The results are not fully conclusive yet due to the fact that we ran into problems with surface films we had not encountered before. It is, however, quite certain that the ethanol/water mixture shows considerably faster CO_2 evasion than pure water does. Meanwhile we have removed everything but stainless steel from the channel which contains the liquid, in order to minimize the possibility of surface films being formed. It might be possible that these films have a stronger influence on the gas exchange in diluted alcohol than they have in water.

It is obvious from fig.5 that surface films do not have an immediate influence on gas exchange in the sense that they represent a diffusion resistance themselves. They may, however, as the dicussion of the preceding paragraph shows,very efficiently influence gas exchange in an indirect way.

At very low wind speed gas exchange is due to surface renewal caused by evaporation surface cooling, and therefore evaporation produces gas exchange. This is evident from our data in fig.6 which indicate non-zero gas exchange at zero wind. This is even more evidently seen from data not reproduced here with our ethanol/water mixture which were obtained at zero rotor frequency. Meyer 1977 [17] used a homemade resistance thermometer to determine the surface temperature deficit ΔT in our wind tunnel. At zero rotor frequency (N_2-flux 35 liters/sec, 20°C, 35% rel. humidity,evaporation velocity .2cm/sec (see fig.2)) he found ΔT=0.9°C. This value possibly is somewhat too small due to insufficient local resolution of his instrument (see also Katsaros 1977 [18] for comparison). From the data just quoted, the latent heat flux density in our case should have been about j=1.3 mcal/cm^{2}sec. This means a transfer velocity for heat into the water of $w=j/\Delta T=1.5\cdot 10^{-3}$cm/sec. Using Danckwerts' model [1] in which the transfer velocity of CO_2 should be by a factor of 9.5 (the square root of the ratio of the molecular diffusion constants of heat and of CO_2,resp.) smaller one obtains a gas transfer velocity of $1.5\cdot 10^{-4}$ cm/sec. This is in reasonable agreement with the value to be extrapolated from our data in fig.6. The evaporation induced gas exchange presumably does not play a very important role in nature. We nevertheless shall try to study it in this wind tunnel. For this purpose we just have put the whole system into a heat insulated box, where we can run it either at elevated temperature, where the evaporation cooling and the surface renewal induced by it plays an increasing role, or at low temperature down to 4°C where it must vanish due to the decrease of latent heat flux and to vanishing density gradients in the water. Our work with the wind tunnel is closely linked to the field work which will be described in the following two sections.

CONTINUOUS RADON-222 MEASUREMENTS AT SEA

Radon-222 measurements, introduced by W.S. Broecker several years ago, have proven to be an excellent tool for measuring gas exchange under natural conditions. The method [10] is based on the fact that radon-222, the daughter of radium-226 dissolved in sea water, does not come into radioactive equilibrium with its parent if gas exchange through the sea surface is sufficiently intense to open a channel of disappearance from the system, additional to the radioactive decay of radon-222 (radioactive mean life 5.5 days). Under steady state conditions, the depth integral of the radon deficit in the water directly gives the flux density of radon gas through the water surface, and dividing by the radon concentration in the water close to the interface (see fig.8), gives the evasion transfer velocity w.

A problem of the method lies in the fact that the system may not be in a steady state or not be closed. During the period of (exponential) averaging characterized by the radioactive mean life (5.5 days), the gas exchange intensity may have changed; furthermore, the measured radon deficit may not be the correct one, as e.g. internal waves periodically change the deficit integral. A continuously working system to automatically measure radon dissolved in sea water has therefore been designed, which is capable to give a radon concentration profile every two hours [13] . The system degasses about 8 liters of seawater per minute, and the stripped gas replaces one detector chamber volume in 15 minutes. Based on an average radon content of 5 dpm per 100 liters of sea water (see fig.8), the statistical counting error, for the system's 8 detectors combined during a 2 hours scanning period is $\pm$ 1.5%. An overall precision of $\pm$ 2% can therefore be expected. This precision is necessary if one wants to observe the response of the gas exchange rate to changing wind conditions.

The equipment has been tested at sea, the last time this summer on cruise 11 of the RSS "Challenger", approx. 150 miles west of Scotland. Fig.8 shows a typical result, still obtained partly operator-controlled since the computer-controlled winch was not yet available. One notes the wandering of the thermocline during the scan period, which is one reason why the sophisticated equipment is necessary for high precision determinations of the gas exchange rate by the radon

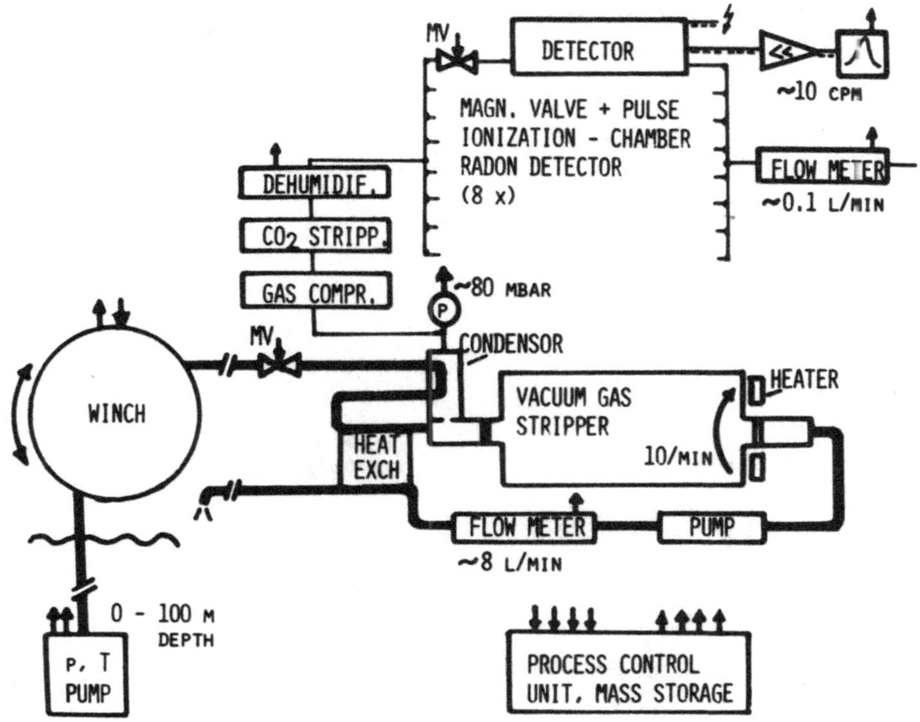

Fig.7 Schematic diagram of automatic radon measuring unit.
══════ water ──── stripped gas ═-═-═-═ detector wiring.
short arrows indicate interfacing with the central process control unit. Water is continuously pumped from depths controlled by a central process control unit, by means of a winch-operated submersible pumping system. The water enters a vacuum stripper in which the dissolved gases are quantitatively removed. The stripped gas is brought to atmospheric pressure and, after removal of CO_2 and dehumidification, flows through one of the 8 radon detectors (slow pulse ionization chambers), again selected by the control unit.

method. The average gas exchange rate obtained on this cruise is included in fig.6. The new radon scanning system will be used during the JASIN (Joint-Air-Sea Interaction) and the FGGE (First Global GARP Experiment) projects 1978/79, and is hoped to give better insight into the parameters controlling gas exchange at sea, particularly since all corresponding air/sea interaction data will then be available as well.

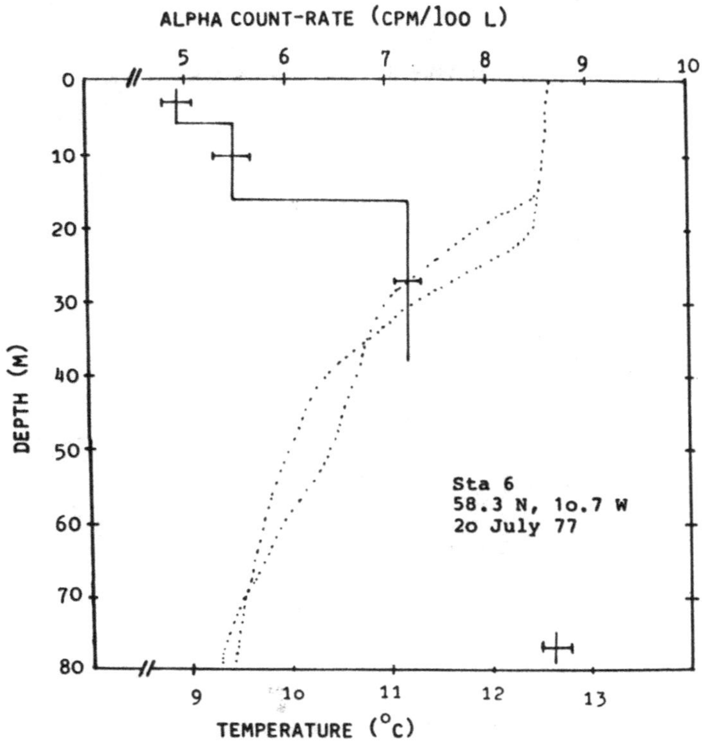

Fig.8 Radon-222 deficit profile, Station 6, Challenger cruise 11 July 1977, west of Scotland. The radon value at 80 meters depth is the radioactive equilibrium value. The close-to-surface value is about 60% of the equilibrium. Note the variation of the temperature profile (dotted) during the scan.

TRITIUM/HELIUM-3 MEASUREMENTS IN LAKE CONSTANCE

Contrary to the radon method, which integrates gas exchange over no more than several days, helium-3, another environmental tracer, is capable of giving average values for a somewhat longer period [20]. The tritium/helium-3 method watches the fate of helium-3, the (stable) product of radioactive decay of environmental tritium in the water, and is well suited to study gas exchange in seasonally overturning inland lakes. The point is that during the period of summer stratification substantial amounts of helium-3 ("excess helium-3") are built up in the hypolimnion, so that the helium-3/helium-4 ratio is very significantly changed from that corresponding to a solubility equilibrium with the atmosphere. During overturn the excess helium-3 mixes to the surface and is then available for evasion into the atmosphere . The present concentration of (nuclear-weapon-produced)

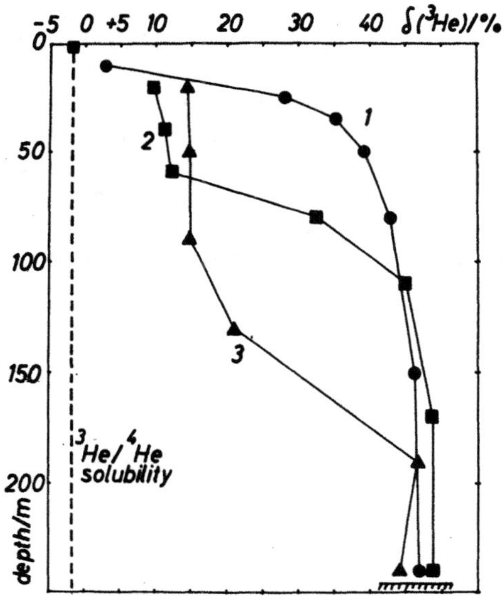

Fig. 9. Helium-3 excess (percent excess of helium-3/helium-4 ratio over atmospheric value) in Lake Constance at different times after the onset of lake overturn (1 = 9 Nov 1976; 2 = 16 Dec 1976; 3 = 13 Jan 1977. - Figure taken from [14]). Evasion of helium-3 into the atmosphere between the dates of profiles 1,3 and 3 is given by the respective differences of the vertically integrated helium-3 excess, making due allowance for new helium-3 production. Dividing these differences by the near-surface helium-3 excess concentration gives the transfer velocities shown in fig. 6 in same way as with the radon-222 method. Continuing helium-3 production shows up as an increase between profiles 1 and 2 below 100 m. Apparently, vertical mixing before the date of profile 2 did not reach depths greater than ≈ 100 m profile 3: ≈ 200 m). The dashed line is solubility equilibrium with the atmosphere [20].

tritium in Lake Constance (Southwest Germany) is, quite uniformly, 180 Tritium Units or 680 picoCurie per liter. Such tritium concentration within two days produces enough helium-3 to change the helium-3/helium-4 ratio in the water to a measurable degree (the helium-3 measurements require a highly specialized mass spectrometer [19]). Fig. 9 shows excess helium-3 observations. Gas transfer rates calculated from these data (cf. caption) are shown in fig. 6. The wind velocity on the lake was extrapolated from shore based data (weather station Friedrichshafen, average between profiles 1 and 2: 3.5 m/sec; between profiles 2 and 3: 2.5 m/sec) with help of observations of Mühleisen [15]. Allowance has to be made for the fact that helium has a much larger molecular diffusivity in water than most other gases (O_2, CO_2, radon) whose coefficients are fairly similar to one another, so that, consequently, helium-3 is lost to the atmosphere considerably faster. To be comparable to "normal" gases the helium results in fig. 6 have been corrected downwards accordingly. The correction is model-dependent, and the one chosen is that from the Danckwerts model [1], which is by the square root of the ratio of the molecular diffusivities. One sees that reasonable agreement with the other data is obtained. The excess helium-3 profiles in fig. 9 additionally give information on vertical mixing.

Acknowledgement: Part of the work presented here is funded by the Deutsche Forschungsgemeinschaft.

REFERENCES

[1] K.O. Münnich, D. Flothmann, Gas exchange in relation to other air/sea interaction phenomena, SCOR Workshop on "Air/Sea Interaction Phenomena" Miami 8-12 Dec 1975 (Background Papers, prepared for the Ocean Sciences Board, Natl.Res.Council, Wash. D.C. 1977)

[2] E.L. Deacon, Gas transfer to and across an air-water interface, Tellus 29 (1977) 363-374

[3] H.J. Vogt, Isotopentrennung bei der Verdunstung von Wasser, Staatsexamensarbeit, Institut für Umweltphysik, Heidelberg 1976

[4] T. Laevastu, I. Clarke, P.M. Wolff, Oceanic part of the water cycle, WMO Report No. 11, Geneva 1969

[5] L. Merlivat, M. Coantic, Study of mass transfer at the air-water interface by an isotopic method, J.Geophys.Res. 80 (1975) 3455-3464

[6] D. Ehhalt, K. Knott, Kinetische Isotopentrennung bei der Verdampfung von Wasser, Tellus 17 (1965) 389-397

[7] J.R. Gat, Environmental isotope balance of Lake Tiberias, in: "Isotope Hydrology 1970", IAEA Vienna 1970, pp. 109-127

[8] U. Siegenthaler, Kinetische Isotopentrennung und Gasaustausch bei der Verdunstung von Wasser, Schlußbericht Schweizer Nationalfonds-Stipendium 1974 (manuscript)

[9] P.S. Liss, Processes of gas exchange across an air-water interface, Deep Sea Res. 20 (1973) 221-238

[10] W.S. Broecker, T.H. Peng, Gas exchange rates between air and sea, Tellus 26 (1974) 21-35

[11] E. Lohse, Bestimmung der Evasionsgeschwindigkeit von CO_2 aus Wasser, Staatsexamensarbeit, Institut für Umweltphysik, Heidelberg, 1977

[12] S. Emerson, Radium-226 and radon-222 as limnological tracers: The CO_2 gas exchange rate; thesis, Columbia Univ., Lamont, N.Y.

[13] W. Roether, B. Kromer, Field determination of air-sea gas exchange by continuous measurement of radon-222, Pure Appl. Geophys. 1977 (in press)

[14] W. Weiss, H. Lehn, K.H. Fischer, W.B. Clarke, Z. Top, B. Kromer, W. Roether, Gas exchange and internal mixing of Lake Constance, Verh.Ges.Ökologie, Kiel, September 1977

[15] R. Mühleisen, Starkwinde an und auf dem Bodensee, Meteorologische Rundschau 30 (1977) 15-22

[16] Chr. Bohr, Die Löslichkeit der Kohlensäure in Alkohol zwischen -67° und $+45^oC$. In- und Evasionscoefficient bei 0°; Annalen der Physik 1 (1900) 244-256

[17] G. Meyer, Konstruktion einer Temperatursonde zur Messung von Temperaturprofilen in der Grenzschicht Wasser/Luft, Staatsexamensarbeit, Institut für Umweltphysik, Heidelberg 1977

[18] K.B. Katsaros, The sea surface temperature deviation at very low wind speeds; is there a limit?, Tellus 29 (1977) 229-239

[19] W.B. Clarke, W.J. Jenkins, Z. Top, Determination of tritium by mass spectrometric measurement of He-3, Int.J.Appl.Radiat. Isotopes 27 (1976) 515

[20] T. Torgersen, Z. Top, W.B. Clarke, W.J. Jenkins, W.S. Broecker, A new method for physical limnology; tritium/helium-3 ages; results for Lakes Erie, Huron, and Ontario, Limnol. Oceanogr. 22 (1977) 181-193

DISCUSSION

H. C. Broecker:

(1) How do you explain the discrepancy between the linear dependence of k with wind speed found by you and the quadratic dependence postulated by Liss, Kanwisher and others?

(2) The sharp increase in k starts at 6 ms^{-1} in your experiments. Others (we too) find that increase at about 2 - 3 ms^{-1}. Can you explain this?

K. O. Münnich:

Simple theories (1, 2) predict linear increase of gas exchange with wind speed or rather with friction velocity. The sharp increase may either mean corresponding increase in friction velocity or even the onset of other mechanisms enhancing gas exchange. Thus the occurrence of the kink needs not to be a sole function of wind speed, it may well depend on other parameters. It seems to me that the <u>square</u> dependence of gas exchange rate on wind speed is just the best we can do at present to fit <u>all</u> data available.

EXPERIMENTAL STUDY OF WIND EFFECTS ON REAERATION

G. E. Mattingly

Visiting Professor, Institut de Mecanique Statistique de la Turbulence, Marseille, France (1973)

Present Address: National Bureau of Standards, Washington, D.C., U.S.A.

INTRODUCTION

The development of water quality prediction models will remain a crucial activity in man's attempt to control pollution. The early modeling efforts of Streeter and Phelps, as well as succeeding efforts, have attempted to elucidate the complexities of reaeration phenomenons. Summaries of these may be found in comprehensive reviews produced by Bennett and Rathbun (2) and Lau (7). It is apparent from both these reviews that considerable effort has been devoted to analysis of the effects of turbulence in dispersion phenomenons. The results have invariably put forth a series of increasingly complex models with which to predict water quality characteristics. These models can be grouped under two headings: (1) phenomenological models*; and (2) computational models.

Phenomenological models, including the present study, attempt to characterize the reaeration processes by means of relationships involving the pertinent large scale quantities. These models are generally empirical, relying on either laboratory or field data to produce a plethora of relationships to predict the reaeration

---

*The phenomenological models, as referred to in this study, are intended to apply to large scale characterizations of gaseous transport into flowing water bodies the case considered here is how the reaeration of a flowing water body depends upon wind speed, water speed and depth. The detailed mechanism through which air passes through the various interfacial layers adjacent to the water surfaces, as exemplified by Merlivat and Coantic (8) is not the direct focus of the present effort.

coefficient. In the past, the flow of wind over water, although a pertinent quantity, has been conspicuously neglected, probably because it remained difficult to properly measure and control. The present effort, conducted in the wind tunnel/water channel facilities at the Institute of Statistical Mechanics of Turbulence (IMST) at the University of Aix — Marseille II, is intended to fill this void.

The successes of the computational modeling efforts, on the other hand, have to date depended upon: (1) The validity of a series of assumptions made to facilitate the analysis; (2) the computational capacity available to the investigator; and (3) the choice of the particular phenomenological model.

EXPERIMENT

The unique IMST facilities in Marseille are designed for the study of wind/water interactions. These wind tunnel and water channel facilities are capable of a wide range of wind and water velocities that provide ideally controlled and reproducible conditions. In addition, both the wind and water flows can be maintained at a temperature of 20°C by individual cooling units in each fluid, a remarkable feature because of the temperature-sensitivity of reaeration phenomenons. Detailed descriptions of these facilities are found in Ref. 4. In the present study, the test section is 8 m long, the water surface is 55 cm wide, and both air and water flows are 27 cm deep.

Air velocity was varied between 0 m/s and a maximum of 15 m/s, in increments of 3.75 m/s. Average velocities of the flowing water were 4.50 cm/s, 9 cm/s, and 18 cm/s. Because of physical constraints, the depth of the water was not varied. Velocities in both the air and water flows were measured with hot wire and hot film anemometry, respectively. The anemometers used were DISA 55D01s without simultaneous linearization. Calibrations preceded and followed several hours of measurement with the smoothed mean calibration curve used to compute both the mean and fluctuating velocity values.

Calibration curves for the hot wire anemometer were very consistent both before and after taking data. Typical time-traces of anemometer signals obtained at various heights above the water surface are shown in Fig. 1. From traces such as these, one is able to detect the approximate thickness of the air boundary layer above the water surface. The top trace exhibits the fluctuations present in the boundary layer adjacent to the ceiling of the wind tunnel.

The two calibration curves for the hot film anemometer were quite different and the various profiles were not entirely consistent; thus, these quantitative results are not presented. The average water velocities are presented as determined via pitot tube

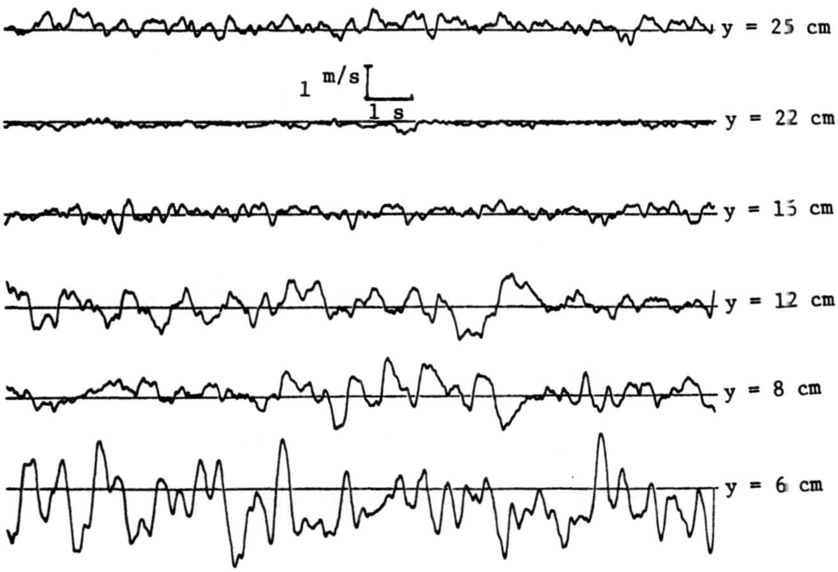

Fig. 1. Wind Fluctuation Velocities above Water Surface (y = Sensor Height above Mean Water Surface

measurements made in the pipe flow which recirculated the water in the channel. Typical, uncalibrated time traces of hot film anemometer signals obtained at various depths below the water surface are shown in Fig. 2. From traces such as these, one is able to estimate the approximate thickness of the water boundary layer below the water surface. The traces labeled y = 1.0 cm exhibit frequent negative spikes, indicating that the water sensor lies in the air space between wave crests. The traces at and below 20 cm exhibit the turbulent fluctuations present in the boundary layer at the bottom of the channel. Water velocity profiles, obtained later by I. Choi in the same facility at IMST are available in Refr (3).

Measurements of the dissolved oxygen concentration in the water were made using the conventional Winkler technique that was facilitated using an electronic titrator. Water samples of 250 ml were taken at recorded time intervals and oxygen concentrations determined within several minutes of the sampling time.

For each test condition, the procedure consisted of chemically deoxygenating the water, using the sodium sulfide method in the presence of the cobalt chloride catalyst (2.6). Although the oxygen decay phenomenons present in this study are chemical rather than biochemical, the interactions of the phenomenons with the water and air are assumed to be identical to biochemical reactions occurring naturally in streams and rivers.

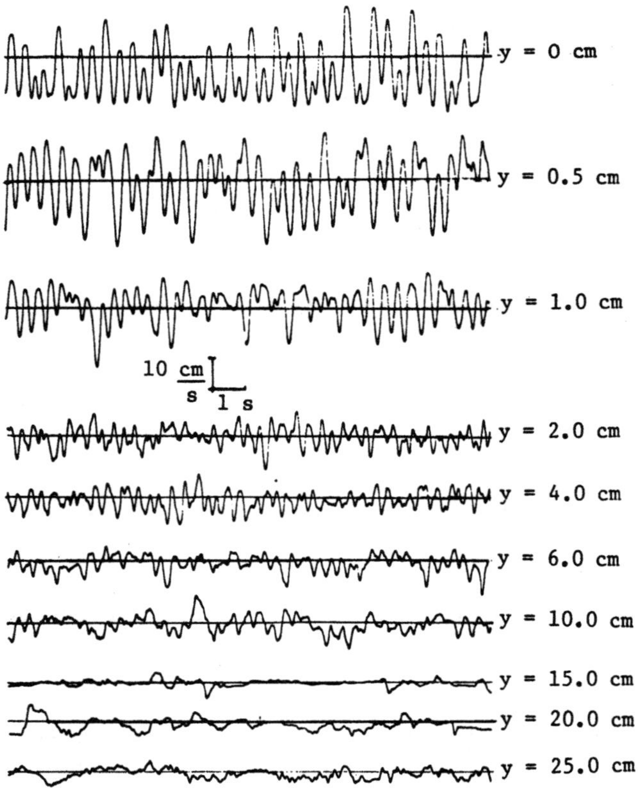

Fig. 2. Water Fluctuating Velocities below Water Surface (y = Sensor Depth below Mean Water Surface)

The amount of sodium sulfide used was 20% in excess of the amount required for complete deoxygenation. Considerable reaction times were found necessary for the deoxygenation to occur. In fact, the procedure leading to the most successful results consisted of titrating a 20°C water sample, then injecting the water with 120% of the dosage plus catalyst required to extract all oxygen, and then slowly circulating this mixture overnight with the pumping system. The following morning, a titration was made to determine the initial oxygen demand of the sodium sulfide residue as well as the approximate initial concentration of the subsequent test run. When this titration indicated that the oxygen concentration was 50% of saturation, then the initial oxygen demand was taken to be 70% of saturation. Through this procedure, the assumption was made that negligible reaeration occurred during the night. As the overnight circulation was done very slowly, the assumption seems reasonable in that a conservative assessment of the overnight reaeration would account for less than 1 ppm dissolved oxygen accumulation. The resulting change in the initial oxygen demand produced a change in the reaeration coefficient of less than 5%.

The next morning, after the cooling systems had brought the air and water to the desired 20°C temperature, both were accelerated to the desired velocity settings, and the test run was begun. Water samples were taken — generally at 5-min intervals initially, with more time allowed between samples as the test continued and successive changes in oxygen concentration diminished. Where possible, i.e., within reasonable time constraints, the asymptotic or saturation value of the oxygen concentration was obtained for the final titrations of the run. During each run, some 15-25 concentration values were recorded to evaluate the pertinent reaeration and chemical reaction rates.

EVALUATION METHOD

To evaluate the pertinent parameters, the reaeration is assumed governed by the classic Streeter-Phelps model (2), using first-order kinetics for both the deoxygenation and reaeration processes. The assumption made in the chemical process through which the water body is deoxygenated is that, in an interval of time, the percentage decrease in oxygen-demanding material is proportional to the time interval. This decrease in oxygen-demanding material is reflected in a corresponding reduction in the amount of dissolved oxygen in the water body.

Equivalently, this reduction can be described by an increase in the dissolved oxygen deficit from the saturation concentration. The dissolved oxygen deficit is decreased by the effects of reaeration. When the present assumption is applied, this portion of the model is identical to that of the oxygen-demanding material. Lastly, sources of dissolved oxygen, as exemplified by photosynthesis, can be described so that the rate at which the dissolved oxygen decreases is a constant with time.

The conventional assumption made in surface water modeling schemes is that the water body is well mixed. Given that assumption, then the concentrations of oxygen-demanding material and dissolved oxygen are spatially uniform, and the only changes occur in time. In the recirculating water channel used in the present study, this mixing is achieved very well by the pump. Thus, water enters the channel with the dissolved oxygen constant and, as it passes down the channel, its surface and upper layers are imparted oxygen by diffusion through the surface and subsequent dispersion into the upper layers of water via the turbulence there. As the water circulates through the pump, it reenters the channel with a uniform but increased oxygen concentration. Therefore, assuming the water is well mixed spatially, the dynamics are written as

$$\frac{dL}{dt} = -k_1L; \quad \frac{dD}{dt} = k_1L - k_2D - P \qquad (1)$$

in which L = biochemical oxygen demand, in parts per million; t = time, in hours; k_1 = deoxygenation coefficient for the dissolved sodium sulfide, in hours^{-1}; P = photosynthetic rate of production of oxygen, in parts per million per hour; D = dissolved oxygen deficit, in parts per million in which $D = C_s - C$ in which C_s = dissolved oxygen concentration; and k_2 = reaeration coefficient, in hours^{-1}. The initial conditions are $D = D_0$ and $L = L_0$, at the test starting time, t = 0. Deposition and the associated chemical oxygen demands (COD) are neglected in the present laboratory scale study. The photosynthetic production of oxygen is neglected.

The initial dissolved oxygen (DO) deficit is checked by extrapolating a line smoothed through the DO data to the time the test was begun. The initial COD, L_0, was found by subtracting the dissolved oxygen concentration of a sample taken just before the test began from the COD of the deoxygenation dosage made the previous evening.

The subsequent test consisted of recording the values of the oxygen concentration as it increased toward the saturation value. With these data, the method of parametric differentiation was used to determine the system parameters, k_1 and k_2, from the data plus the initial values of D_0 and L_0. The method of parameter evaluation, described in detail in Ref. 1, basically fits the data to the system of Eqs. 1 and determines k_1 and k_2 with very rapid convergence according to a least-squares criterion. The resulting "best fit" lines will be presented together with the data in the following section.

EXPERIMENTAL RESULTS

The results of the nonlinear least-square parameter evaluation are presented in dimensional form in Table 1. At this point, it is apparent that wind exerts an extremely significant effect on the reaeration process. The reaeration coefficients are seen to range from 0.477 hr^{-1} - 5.817 hr^{-1} for the tests where water velocity is 18 cm/s. When the water velocity is 9 cm/s, the values for k_2 vary from 0.150 hr^{-1} - 4.762 hr^{-1}. Due to the extremely lengthy time intervals required to conduct the experiment for water velocity of 4.5 cm/s without wind flow, this test was not run. Consequently, the range of values of k_2 for 4.5 m/s wind velocities is only 0.455 hr^{-1}- 4.151 hr^{-1}.

The tabulated values obtained for the deoxygenation coefficient, k_1, are presented in Table 2. Note that ranges and values for k_1 correlate quite closely to those for k_2 and can be estimated to be between one-third and one-half the corresponding k_2 value. In view of these k_1 values, it will be seen that serious errors in determining reaeration coefficients can occur if any significant deoxygenation is not considered concurrently with the reaeration.

TABLE 1. - REAERATION COEFFICIENTS, k_2, in $hours^{-1}$

Air velocity, $\overline{U}_A$, in meters per second (1)	Water Velocity, $\overline{U}_w$, in centimeters per second		
	18.00 (2)	9.00 (3)	4.50 (4)
15.00	5.817	4.762	4.151
11.25	3.770	2.955	2.184
7.50	1.846	1.599	1.150
3.75	1.065	---	0.455
0	0.477	0.150	---

TABLE 3. - WIND SHEAR VELOCITY, U_{*A}, IN METERS PER SECOND

Air velocity, $\overline{U}_A$, in meters per second (1)	Water Velocity, $\overline{U}_w$, in centimeters per second		
	18.00 (2)	9.00 (3)	4.50 (4)
15.00	2.038	1.835	1.044
11.25	1.832	1.031	1.013
7.50	0.943	0.896	0.607
3.75	0.468	---	0.132
0	0.321	0.122	---

TABLE 2. - DEOXYGENATION COEFFICIENTS, k_1, in $hours^{-1}$

Air velocity, $\overline{U}_A$, in meters per second (1)	Water Velocity, $\overline{U}_w$, in centimeters per second			
	18.00 (2)	9.00 (3)	4.50 (4)	0.00 (5)
15.00	1.920	2.000	2.028	2.160
11.25	1.051	1.163	---	1.371
7.50	0.800	0.816	0.944	0.987
3.75	0.160	0.187	0.206	---

TABLE 4. - ROUGHNESS HEIGHT, Y_o, IN MILLIMETERS

Air velocity, $\overline{U}_A$, in meters per second (1)	Water Velocity, $\overline{U}_w$, in centimeters per second			
	18.00 (2)	9.00 (3)	4.50 (4)	0.00 (5)
15.00	4.40	5.30	5.70	6.70
11.25	1.74	2.46	--	4.20
7.50	1.08	1.80	2.16	3.04
3.75	0.017	0.061	0.113	---

The mean distributions of the air velocity above the water surface were surveyed for the corresponding air and water velocities. The data were then fitted to

$$\frac{U_A}{U_{*A}} = \frac{1}{K} \ln \frac{Y}{Y_0} \tag{2}$$

in which U_A = air velocity, in meters per second at the height Y, in centimeters, above the mean water surface. The K is the von Karman constant, 0.4. The fitting process determines the parameter, U_{*A}, the wind shear velocity, in meters per second at the water surface and Y_0, the wind roughness height in centimeters. Values for these two parameters are presented in Tables 3 and 4, with the fitted data shown in Fig. 3. As evident, the air boundary layer is not surveyed immediately adjacent to the mean water surface. For the higher air velocities especially, the profile surveys were abandoned when the frangible hot wire sensor was judged in imminent danger of being struck and broken by small water droplets that were being entrained by the air stream.

A survey was also made of the surface waves produced by the wind. With a vertical grid alined horizontally with the mean water surface, photographs were made of the water surface profile. From these records, wave lengths and wave heights were measured at two fetches, 250 cm and 650 cm for three values of wind velocity. These data, together with wave frequencies obtained near the water surface from traces such as those shown in Fig. 2 enable one to calculate surface wave velocities. Figs. 4(a) and 4(b) show these results where averages are presented as functions of wind velocity.

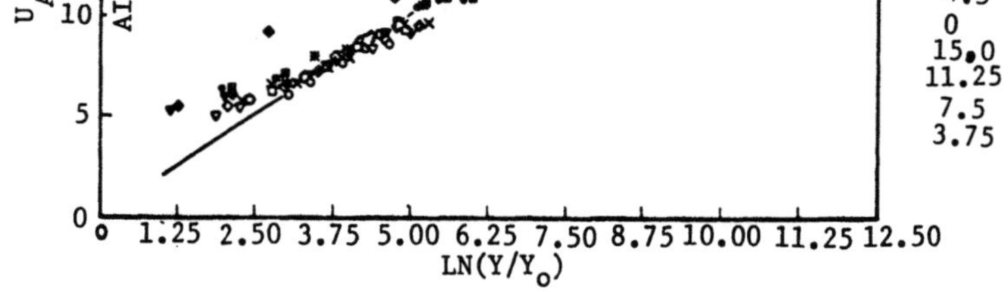

Fig. 3. Wind Velocity Profiles above Water Surface (U_{*A} = Air Shear Velocity; Y_0 = Roughness Height)

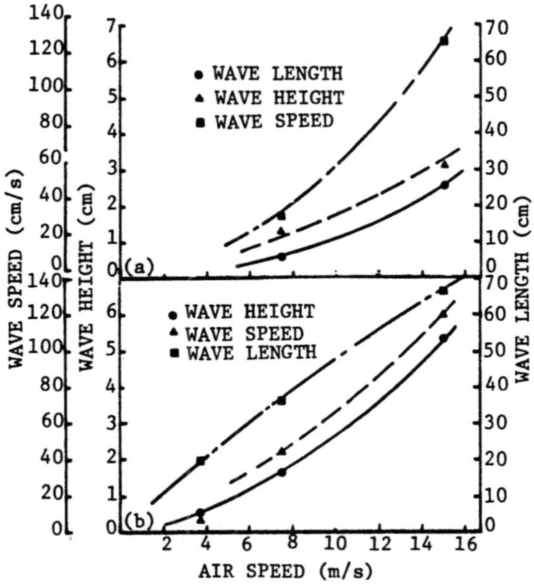

Fig. 4. Wave Characteristics: (a) 250-cm Fetch Length: (b) 650-cm Fetch Length

With the parameters presented in Tables 1 and 2, the best fit lines are computed and plotted with the corresponding data points for constant average water velocity in Figs. 5(a)-5(c). Note that, upon careful examination, all the lines are not straight lines and absolute values of the slopes are not equal to the reaeration coefficients. This is in accordance with the Streeter-Phelps model for dissolved oxygen, in which oxygen decay is present. Accordingly, if this oxygen decay were neglected, the reaeration coefficients obtained via slopes from semilog plots would be less (by k_1L/D) than those presented in Table 1. Since several previous investigators have used the slope method to determine reaeration coefficients, it appears that the accuracy of those results depends critically upon whether the oxygen decay dynamics were completed (5,6). In situations where deoxygenation is achieved solely by means of vacuum degassing, the semilog slope method should give satisfactory results. However, where chemical degassing is used, the time required to complete the reaction is critical to the slope method for determining k_2. From the present study, this reaction time is found to be quite long—of the order of hours depending upon the mixing characteristics prevailing in the water body as well as any reaeration occurring during the degassing period.

Fig. 5(c) includes the results of two tests conducted at the air velocity of 15 m/s. The coincidence of these two results is an indication of the repeatability of the reaeration phenomenons under the closely controlled conditions in the IMST facilities.

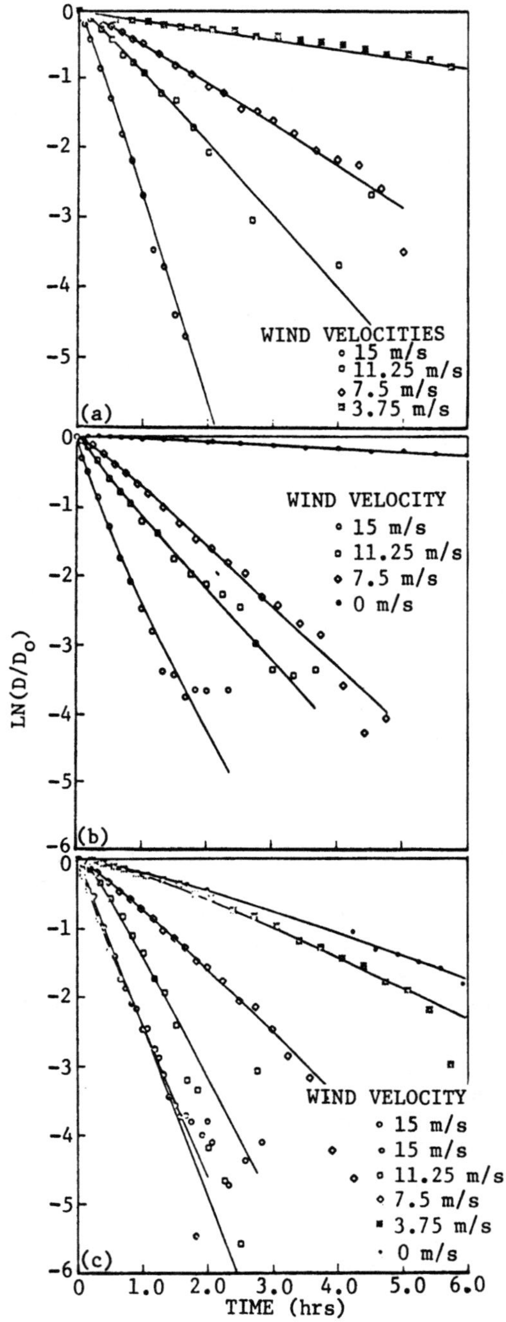

Fig. 5. Dissolved Oxygen Deficit Traces for Constant Average Water Velocity: (a) 4.5 cm/s; (b) 9.0 cm/s; (c) 18.0 cm/s

EMPIRICAL FORMULATION

To formulate the results of the present experimental study into readily useable form, a series of empirical formulas are produced. In these formulas, laboratory values obtained for reaeration coefficients prevailing under windy conditions are presented relative to corresponding values obtained when wind has not been considered. From the results of Bennett and Rathbun, their equation for k_2 as a function of average water velocity and depth is

$$(k_2)_0 = 0.101 \frac{\bar{U}_W^{0.607}}{H^{1.689}} \qquad (3)$$

in which $(k_2)_0$ = the reaeration coefficient, in hours^{-1}; $\bar{U}_W$ = average water velocity, in meters per second; and H = average water depth, in meters. In the empirical formulations that follow, this Bennett and Rathbun result is taken as the reference reaeration cofficient.

Two reasons are provided for this choice of reference. First, the present results contain but one value each for $\bar{U}_W$ of 18 cm/s and 9 cm/s in the absence of wind; the third value for $\bar{U}_W$ = 4.5 cm/s was not obtained owing to the extremely long test time required. Second, and more importantly, the Bennett and Rathbun result is one which has been obtained from a very large data base, which explains why it is widely referred to and used. It has not escaped notice that the windless reaeration coefficients presented in Table 1 differ from the Bennett and Rathbun relation; the disagreement, however, is considered typical in view of the variability found in comparable experimental data. The standard error of estimate of the present results by the Bennett and Rathbun (2) equation is 0.16, which compares favorably with their quoted value of 0.20. Further, the present variation is found to be typical when compared to other laboratory data where the effects of wind are excluded (2).

When the wind-produced excess reaeration is fitted as a function of wind velocity, the result obtained is

$$\frac{k_2}{(k_2)_0} - 1 = 0.2395\ \bar{U}_A^{1.643} \qquad (4)$$

in which k_2 = reaeration coefficient prevailing under windy conditions, in which $\bar{U}_A$ is the wind velocity, in meters per second measured in the free air stream above the boundary layer effects prevailing near the water surface. In the present laboratory experiment, $\bar{U}_A$ is measured at Y = 20 cm above the water surface. For the 11 readings presented in Table 1, the correlation coefficient of r = 0.977 is obtained with the foregoing equation.

When the average water velocity is incorporated, the equation obtained is

$$\frac{k_2}{(k_2)_0} - 1 = 0.2588\ (\bar{U}_A - \bar{U}_W)^{1.618} \qquad (5)$$

in which the only new quantity is the average water velocity, $\bar{U}_W$, in meters per second. With this equation, the correlation coefficient is r = 0.978.

For the values of air shear velocity presented in Table 3, the following empirical equation applies:

$$U_{*A} = 0.0252\ (\bar{U}_A - \bar{U}_W)^{1.645} \qquad (6)$$

in which all velocities are expressed in meters per second. For the 14 values, the correlation coefficient, r = 0.978, is obtained.

From the values given in Table 4, the roughness height in the air boundary layer adjacent to the water surface is empirically expressed as

$$Y_0 = 1.260 \times 10^{-3}\ (\bar{U}_A - \bar{U}_W)^{3.228} \qquad (7)$$

in which Y_0 is in millimeters, and the velocities $\bar{U}_A$ and $\bar{U}_W$ are in meters per second. The correlation coefficient, r = 0.923, is obtained for the 14 values.

ANALYSIS

Reaeration coefficients are determined in this experimental study which indicate that wind-produced excess aeration can reach more than 30 times that predicted by conventionally used empirical expressions. New empirical expressions are developed that describe this significant wind role and correlate well with the experimental values. When wind velocity alone, as measured above the air boundary adjacent to the water surface, is used to account for the excess aeration, the correlation coefficient is 0.977. When both the air and water velocities are used to fit the data, the resulting expression produced a correlation coefficient of 0.978. The difference in the correlation coefficient is insignificant. The ratio of wind-produced aeration to that predicted solely upon water velocity and depth is graphed as a function of wind velocity, in meters per second, in Fig. 6. With the use of Eqs. 1 or 2 in future modeling efforts applied to flowing streams and rivers, perhaps such uncertainties as those described by Yu (9) can be reduced.

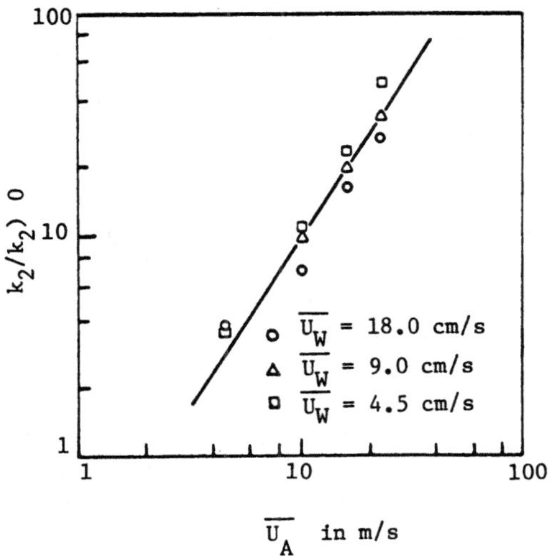

Fig. 6. Ratio of Wind-Produced to Conventionally Predicted Reaeration Coefficients as Function of Wind Velocity, in meters per second

Judging from the significant values determined for the deoxygenation coefficients presented in Table 2, it is essential in laboratory investigations of reaeration phenomenons to perform the data reduction in a manner consistent with the critical degassing step. In the present experiment, the chemical degassing step was found to require very long time intervals for reasonable oxygen deficits to be achieved. Consequently, the deoxygenation process ought not be neglected in the data reduction unless evidence is available which insures that the chemical degassing has been completed.

When degassing is performed by producing a vacuum over the water body analyzed, the deoxygenation process is justifiably neglected once the vacuum is removed and reaeration phenonenons begin. In the corresponding data analysis, the semilog slope determination would give the precise reaeration coefficient provided the remaining assumptions are valid.

CONCLUSIONS

The results of the present laboratory experiment indicate that the influence of wind plays an extremely significant role in the reaeration process. From these results, it is demonstrated that the effect of the wind can produce more than order of magnitude increases in the reaeration coefficient. Detailed profiles for the wind velocity profile over the water surface are given with the corresponding first-order reaction coefficients, k_1 and k_2.

To formulate the results of the present experimental study into a readily useable form, empirical equations are developed which relate wind-produced excess reaeration to wind and water velocities. As well, equations are developed for wind shear velocity and roughness height in the wind stream over the water surface. Of course, it remains to be seen how these equations predict reaeration phenomenons under full-scale conditions. It is recognized that conditions in the field are quite difficult to control and to measure. However, in view of the wide range of reaeration results that are undoubtedly wind-produced in this present study, it is concluded that future efforts in the field should include wind measurement in addition to the usual water speed and depth measurements. In this way, improved modeling efforts can be made along with more realistic economic evaluations of the necessary supplementary means, such as sewage treatment to maintain acceptable water quality in surface streams and rivers.

ACKNOWLEDGMENTS

The present experimental investigation was conducted at the Institute for Statistical Mechanics and Turbulence at the University of Aix-Marseille II, while the writer was on a 6-month sabbatical leave from the Department of Civil Engineering of Princeton University. The writer is especially indepted to A. Favre of the Institute, and his staff — notably Burnage, Bonmarin, Chambaud, Coantic, Dumas, Fulachier, Gaviglio, Quaccia, Ramamon, and Resch. Specific mention should be made of two individuals: J. Minas of the University of Aix-Marseille II, for his generous provision of the necessary chemicals and the use of the electronic titrator; and R. Ezraty, whose generous assistance is gratefully appreciated. The writer's initial interest in reaeration phenomenons was stimulated through conversations with D. J. O'Connor of Manhattan College and through consultant activities with the Water Resources Research Institute of Rutgers University, at which time he worked closely with W. W. Whipple and S. L. Yu. The writer wishes to express his gratitude to the French National Center for Scientific Research (CNRS) for their financial assistance in the form of a subvention forfaitaire.

REFERENCES

1. Beliveau, J. G. and Mattingly, G. E., "Non-Linear Least Square Techniques in Water Quality Problems," *Journal of Environmental Science*, Vol. 4, No. 1, Spring, 1974, pp. 23-39.
2. Bennett, J., and Rathbun, R., "Reaeration in Open Channel Flow," *Geological Survey Professional Paper 737*, U.S. Government Printing Office, Washington, D.C., 1972.

3. Choi, I., "Contribution to the Study of the Physical Mechanism of the Generation of Waves at an Air-Water Interface," Thesis presented to the University of Aix-Marseille II, 24 June 1977.
4. Coantic, M., and Bonmarin, P., "The Air-Sea Interaction Simulation Facility at the Institute of Statistical Mechanics of Turbulence," *Atmospheric Technology*, No. 7, Fall, 1975, pp. 72-79.
5. Eboulaidy, A. F., and Plate, E. J., "Wind Shear-Turbulence and Reaeration Coefficient," *Journal of the Hydraulics Division*, ASCE, Vol. 98, No. HY1, Proc. Paper 8651, Jan., 1972, pp. 153-170.
6. Krenkel, P., and Orlob, G., "Turbulent Diffusion and the Reaeration Coefficient," *Transactions*, ASCE, Vol. 128, Part III, Paper No. 3491, 1963, pp. 293-334.
7. Lau, Y. L., "A Review of Conceptual Models and Prediction Equations for Reaeration in Open Channel Flow," *Technical Bulletin No. 61*, Department of the Environment, Ottawa, Canada, 1972.
8. Merlivat, L. and Coantic, M., "Study of Mass Transfer at the Air-Water Interface by an Isotopic Method," *J. Geophys. Res. Vol. 80, No. 24*, August 1975, pp. 3455-3464.
9. Yu, S. L., "Uncertainties in Water Quality Models: The Case of Atmospheric Reaeration," presented at the December 11-14, 1972, International Symposium on Uncertainties in Hydrologic and Water Resource Systems, held at Tucson, Ariz.

NOTATION

The following symbols are used in this paper:

C = oxygen concentration, in parts per million;
C_s = saturation oxygen concentration, in parts per million;
D = dissolved oxygen deficit, in parts per million;
D_0 = initial dissolved oxygen deficit, in parts per million;
H = mean water depth, in centimeters;
K = von Karman constant;
k_1 = deoxygenation decay rate, in hours^{-1};
k_2 = reaeration rate, in hours^{-1};
k_{2_0} = reference reaeration rate, in hours^{-1};
L = biochemical oxygen demand, in parts per million;
L_0 = initial biochemical oxygen demand, in parts per million;
P = photosynthetic rate of oxygen production, in parts per million per hour;
r = correlation coefficient;
t = time, in hours;
U_A = air velocity, in meters per second;
$\bar{U}_A$ = air velocity 20 cm above mean water surface, in meters per second;

U_{*A} = shear velocity in air at mean water surface, in meters per second;
U_w = mean water velocity, in centimeters per second, in meters per second;
Y,y = vertical distance from mean water surface, in centimeters; and
Y_0 = roughness height in air at the mean water surface, in millimeters or centimeters.

DISCUSSION

E. J. Plate:

Your measurement technique of taking your oxygen sample at one point subjects your data to errors due to inlet and outlet conditions. How did you correct for this?

G. Mattingly:

It is true that our method of taking water samples at a single point could possibly incur inlet and outlet errors should these be present. It is also true that the two point sampling used in your published experiment with Eboulaidy does avoid inlet and outlet errors if these occur due to the presence of an "inlet roller" or "outlet spillways". In response, I would say that the inlet flow condition designed into the facilities at IMST, Luming are well done, meaning (1) the air and water inlets are quite smooth and well-matched (2) there is no "spillway".
There is a tubular beach which may induce some errors. I believe its presence is unavoidable in that reflected wave effects would induce a more severe detrimental effect.

I feel the two point sampling system can suffer if the samples extracted are not true cross-sectional averages at each point.

I also believe that the very long (i.e. "overnight") de-gassing phenomena must be taken into account in the calculation of the reaeration coefficient, $-k_2$. Even using the publicized cobalt catalyst, and after slowly mixing Na_2SO_3 and water overnight, we find the deoxygenation coefficient, k, is approximately 1/3 to 1/2 of k_2.

K. Hasselmann:

If your results are independent of the channel boundaries (top, bottom, fetch) should they not depend only on the difference velocity $u_{air} - u_{water}$, and not on both velocities seperately? If, however, the results do depend on u_{air} and u_{water} separately, would it not be better to express the influence of the boundaries (for example in determining the turbulence level) through an appropriate choice of non-dimensional parameters, rather than the wind speed, in plotting the results?

G. Mattingly:

I have attempted to characterize the wind effects on reaeration by normalizing our present results with respect to a widely used stream reaeration formulation which neglects wind. These normalized results are then correlated with the dimensional wind velocity. I do this to emphasize (expecially to water quality prediction modellers) the effects of wind which while not surprising to some of us have been, to date, neglected by others who look in the area of stream and river modelling.

In presenting the experimental design of the present effort using dimensional quantities I did not mean to convey the fact that the gasous transport depends upon these velocities. I do not believe that reaeration phenomena depend upon the dimensional air and water velocities, individually. This is expecially true in the ocean. Here, reaeration phenomena have to be characterized using velocity differences and (yet to be determined) length scales. With this ultimate goal in mind we have also given reaeration correlations with air-water velocity differences. In addition, air shear velocity and roughness scale are also correlated with air-water velocity differences. I fully expect that the appropriate length scales will emerge from future research efforts on the gaseous transport in the interfacial layers. With these in hand, the proper non-dimensionalisation of the phenomena can be predicted.

MECHANICS OF MULTIPHASE FLUXES OVER THE OCEAN

S. C. Ling, Alice Saad, And T. W. Kao

The Catholic University of America
Department of Mechanical Engineering
Washington, D. C. 20064, U.S.A.

INTRODUCTION

The study of multiphase fluxes of water over the ocean has long been recognized by scientists (Roll, 1965) as being important to the understanding of the exchange process between the ocean and the atmosphere. Recently, Ling and Kao (1976) have successfully obtained a preliminary solution showing how water droplets produced by both the whitecaps and sea sprays play a major role in the transport of energy and moisture from the ocean. The evaporating water droplets produce a large moisture source for the humidity field, while the latent heat of evaporation contributes to a large heat sink for the temperature field. The resultant temperature stratification further influences the stability of the wind field which in turn affects the exchange processes. Thus the problem was solved by a set of coupled equations governing both the wind, the droplet concentration, the humidity, and the temperature fields. In addition, one must further consider the coupling effect between the surface layers of the atmosphere and the ocean. Under high sea states most of the forces produced between the wind and the sea are through the form drag on the waves. This produces a surface (drift) layer of water to move faster than the orbital velocity of the wave crest, thus causing the crest to break. Under equilibrium condition, very little energy is required to maintain the motion of the developed waves, hence a major part of the energy that is derived from the wind is directly dissipated through the action of wave breaking. The mechanics of wave breaking is very similar to that of a low head waterfall. A substantial part of the kinetic energy of this falling sheet of water is converted into the energy of entrainment which in turn suppresses the development of intense micro-scale turbulence and prevents quick dissipation of the energy. Subsequent unstable boiling effect of the entrained air mixture converts the energy of entrainment back into intense large-scale turbulence, which in turn promotes more efficient heat transfer in the ocean. Recently, a detailed study of the structure of such a turbulent field was performed by Ling and Saad (1977). The result shows that the large-scale turbulence has a wide range of inertial subrange which is associated with the unstable two-phase flow. This indicates that air entrainment is an important mechanism which is responsible for the large-scale turbulence in the surface (mixing) layer of the ocean. Groups of large entrained air bubbles tend to rise quickly to the sea surface and form clusters of light foamy heads near the wave crests. These foamy wave crests are easily picked up by the wind and torn into sea sprays. Because small entrained air bubbles rise more slowly to the surface, they are left behind by the moving wave crest to form a long trail of whitecap. Most millimeter size air-bubbles break individually as they broach the water surface. As each air bubble breaks, the surface energy of the bubble is used to create a few jet drops as well as a cloud of submicro film drops (Blanchard, 1963). All these micro droplets are easily diffused and evaporated into the air as a major source of moisture. It is, therefore, important that one should also try to parameterize both the air entrainment and droplet production process as a function of the energy input from the wind to the waves. To do this, one must first obtain sets of consistent field and laboratory data. Recently, we have developed a unique droplet-size concentration measuring system (Leon, 1977). More advanced version of the instrument is now being applied to study the size distribution of droplets produced by the foaming process in a water channel. Also active plans are being made to study, under the Joint Air-Sea Interaction Project (JASIN), the production rate and size concentration of water droplets under high sea states. Hence this paper will represent a part of a continuing effort to obtain more accurate experimental data for updating the preliminary theory of Ling and Kao (1976). In particular, the recent improvement to the theory which includes the effect of both the size distribution of water droplets and their falling velocities will be presented.

STATEMENT OF THE PROBLEM AND GOVERNING EQUATIONS

In the cited paper of Ling and Kao, water droplets were treated as if they had only one effective size. Furthermore, no falling effect of the droplets was considered. As a first step to improve the theory we have now taken these factors into consideration, so that the resultant solution will be more realistic and readily verified by the field data. At this moment we are still awaiting for more accurate field data to update the theoretical model. We have planned to obtain the necessary information during the JASIN project in November 1977. In the experiment, both the vertical profiles of

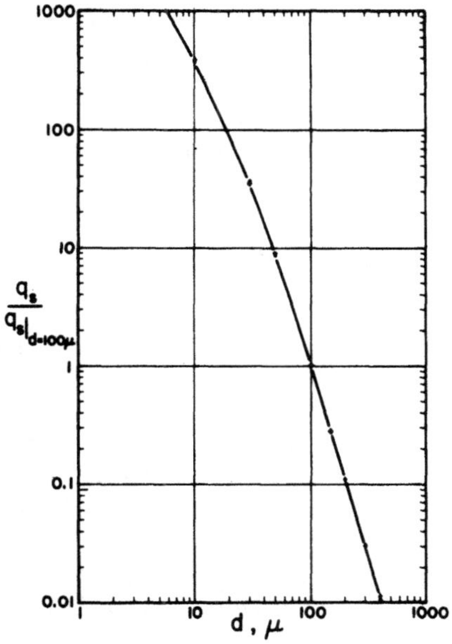

Fig. 1. Normalized production rate of different sizes of droplets by whitecap with different energy levels.

the droplet size concentration, wind velocity, humidity, and temperature will be measured simultaneously. From these data and using the analytical model we should be able to solve for the production rate (surface flux) of the various sizes of droplet under different sea states. Presently, we shall use only the laboratory data as a first trial. The production of droplets by breaking waves was simulated in the laboratory by a low head hydraulic jump created in an open channel 0.6 m by 0.6 m in cross section and 10 m long. An air blower, developing a wind of 4m/sec, was directed at the hydraulic jump and in the direction of the water flow. A droplet size concentration detector was placed a few centimeters above the foamy water surface and 1 m downstream of the hydraulic jump to measure the size concentration of the water droplets. Because the measurement was taken very close to the water surface, the measured size concentration should be closely related to the actual production rate of the droplets by the whitecap; i.e., unaffected by the effect of falling velocities and evaporation. Measurements were taken for hydraulic jump with different energy levels. For each level of energy, the measured production rates for different sizes of droplet q_s were normalized by $q_s|_{d=100\mu}$ and plotted in Fig. 1. One notes that the droplets maintained a universal distribution of sizes, ranging from 1 μ to 400 μ, with more micro size droplets produced than large droplets. This observation has greatly simplified the analysis in that only the flux for one size of droplet needs to be found or defined.

Because of the vast horizontal extent of the ocean compared to the typical height of the atmospheric surface layer, one may justify to treat this layer as being both quasi uniform in the horizontal direction and quasi steady (Sheppard, 1958). As shown previously by Ling and Kao (1976), the governing equation for all transportable quantities θ can be expressed as

$$\frac{\partial}{\partial z} K(z)\frac{\partial \theta}{\partial z} + S_\theta - SK_\theta = 0 , \qquad (1)$$

where z is the vertical coordinate referring to the mean wave trough as the datum, K(z) is the eddy diffusivity, and S_θ and SK_θ are the source and sink functions of θ, respectively. Since we are concerned only with high sea states (whitecaps state) where the value of K is so large compared with any molecular diffusivities, one may safely assume that K is universally applicable to all transportable quantities including the water droplets. However, the falling effect of the water droplets is to be treated separately as a sink effect which is independent of the turbulent diffusivity – a technique which is well proven for the treatment of the transport of sediment (Rouse, 1938). The present study will use the same eddy diffusivity K(z) as in the previous study; i.e., in the lower z level it is dominated by the constant turbulent shear stress produced by the form drag on waves, and in the upper level it is modified by the Monin–Oboukhov stability length parameter (Ling and Kao, 1976).

The equation governing the vertical transport of water droplets is best treated as discrete droplet size groups denoted by subscript n as

$$\frac{\partial}{\partial z} K(z)\frac{\partial Q_n}{\partial z} + S_{Qn} + w_n \frac{\partial Q_n}{\partial z} + S_{tn} = 0 , \qquad (2)$$

where Q_n is the concentration of water droplets having a mean group diameter d_n, S_{Qn} is the droplet source produced by sea sprays and whitecaps, w_n is the terminal fall velocity of the droplets, and S_{tn} is the transfer of droplets from one size group to the other due to evaporation. The term S_{tn} consists of two parts, droplets transferred in and transferred out of the n group. The number rate of droplet transferred out from group n to the smaller size group n-1 can be expressed as the total mass rate of evaporation (Ranz and Marshall, 1959) of the n group droplets divided by the change in mass of a droplet from size d_n to d_{n-1};

$$S_{tn}(\text{out}) = \frac{-2(1+0.25Re_n^{0.5})\pi D_v d_n (H_s - H)\rho_a Q_n}{\frac{\rho_w \pi}{6}(d_n^3 - d_{n-1}^3)} \qquad (3)$$

Similarly, the number rate transferred in from the next larger size group d_{n+1} can be expressed as

$$S_{tn}(in) = \frac{2(1+0.25Re_{n+1}^{0.5})\pi D_v d_n+1(H_s - H)\rho_a Q_{n+1}}{\frac{\rho w\pi}{6}(d_{n+1}^3 - d_n^3)} , \qquad (4)$$

where Re_n is the Reynolds number expressed as $w_n d_n/\nu_a$, ν_a the kinematic viscosity of air, D_v the molecular diffusivity of water, ρ_a and ρ_w the mass density of air and water, and H_s and H the saturated specific humidity and specific humidity, respectively. H_s is a function of the absolute temperature T (Ling and Kao, 1976). Equation (2) may be normalized as

$$\bar{K}\frac{\partial^2\bar{Q}_n}{\partial\bar{z}^2} + \frac{\partial\bar{K}}{\partial\bar{z}}\frac{\partial\bar{Q}_n}{\partial\bar{z}} + \bar{S}_{Qn} + \bar{w}_n\frac{\partial\bar{Q}_n}{\partial\bar{z}} + 12\frac{\rho_a}{\rho_w} D_v \frac{\varepsilon}{U_o} H_s(T)(1-H_o^*)(1-\bar{H}) \times$$

$$\left[\frac{(1+0.25Re_{n+1}^{0.5}d_{n+1}\bar{Q}_{n+1}}{(d_{n+1}^3 - d_n^3)} - \frac{(1+0.25Re_n^{0.5})d_n\bar{Q}_n}{(d_n^3 - d_{n-1}^3)}\right] = 0 , \qquad (5)$$

where ε is the average wave height, U_o the reference wind at $\bar{z}_o = 20$, and q_{s1} the average surface flux of the smallest droplet size group 1. The various normalized variables are defined as $\bar{K} = K/U_o\varepsilon$, $\bar{Q}_n = Q_n U_o/q_{s1}$, $\bar{z} = z/\varepsilon$, $\bar{S}_{Qn} = S_{Qn}\varepsilon/q_{s1}$, $\bar{w}_n = w_n/U_o$, $\bar{H} = (H_o^* - H^*)/(H_o^* - 1)$, and H^* and H_o^* are the relative humidity and the relative humidity at reference height $\bar{z}_o = 20$, respectively.

The governing equation for the normalized humidity $\bar{H}$ can be expressed as

$$\bar{K}\frac{\partial^2\bar{H}}{\partial\bar{z}^2} + \frac{\partial\bar{K}}{\partial\bar{z}}\frac{\partial\bar{H}}{\partial\bar{z}} + \sum_{n=1}^{m} 2(1+0.25Re_n^{0.5})\pi D_v \frac{q_{s1}}{U_o}\frac{\varepsilon}{U_o} d_n(1-\bar{H})\bar{Q}_n = 0. \qquad (6)$$

The governing equation for the normalized temperature $\bar{T}$ is expressed as

$$\bar{K}\frac{\partial^2\bar{T}}{\partial\bar{z}^2} + \frac{\partial\bar{K}}{\partial\bar{z}}\frac{\partial\bar{T}}{\partial\bar{z}} - \sum_{n=1}^{m}\frac{L_h}{C_p|T_w - T_o|} 2(1 + 0.25Re_n^{0.5})\pi D_v \frac{q_{s1}}{U_o}\frac{\varepsilon}{U_o} d_n \times$$

$$H_s(t)(1 - H_o^*)(1 - \bar{H})\bar{Q}_n = 0 , \qquad (7)$$

where $\bar{T} = (T - T_o)/|T_w - T_o|$, T the absolute air temperature, T_w the absolute sea surface temperature, T_o the absolute temperature at the reference height $\bar{z}_o = 20$, L_h the latent heat of evaporation, and C_p the specific heat of air. The last left hand terms of both (6) and (7) represent the total source of humidity and sink of latent heat produced by the evaporation of all water droplets, respectively. The source term $\bar{S}_{Qn}(\bar{z})$ in (5) is related to the surface flux of droplets q_{sn} through a distribution factor within the wave trough. In this paper we shall use the same factors as those used in the cited reference of Ling and Kao. Because there will be no injected surface droplet source above the wave crest, $\bar{S}_{Qn}(\bar{z}) = 0$ for $\bar{z} > 1$. To estimate the size range of droplets which is important to the present problem, we weigh the droplet distribution curve shown in Fig. 1 by a weighing factor which is proportional to both the evaporation rate and the effective evaporated volume. The evaporation rate is approximately proportional to $d^{1.2}$ and the effective evaporated volume is proportional to d^3 times the resident time of the droplet in air divided by the time required for complete evaporation. The weighted function was found to be a minimum at d less than 0.5μ. It increased as d to a maximum at 100μ and then decreased as d^{-2} to a minimum value for $d > 400\mu$. It shows that although there are a large number of submicro size droplets they do not have sufficient net volume to contribute to the moisture and latent heat. On the other end of the droplet size spectrum, the large droplets have a much larger volumetric content and rate of evaporation, but due to their short life time (large terminal velocity) and small population they also have very little contribution to the moisture. From the weighted function one may divide the effective droplet size distribution into $m = 3$ groups; $d_1 = 50\mu$, $d_2 = 150\mu$, and $d_3 = 250\mu$. They represent the weighted average diameters for the following ranges of droplet sizes: 0.5 - 100μ, 100 - 190μ, and 190 - 400μ, respectively. Because the droplet size distribution was found to be universal for a wide range of energy input to the foaming system, one may consider the ratio between the surface droplet fluxes q_{s1}, q_{s2}, and q_{s3} to be constant and expressable as $q_{s2} = 0.032q_{s1}$, and $q_{s3} = 0.006q_{s1}$. Thus for the present solution one needs only to specify the value of q_{s1}. Since both the temperature field and the humidity field are very sensitive to the magnitude of q_{sn}, the parametric value for q_{s1}/U_o can be found by an iterative process until the solution gives the best fit for the known temperature and humidity distribution under various atmospheric and oceanic conditions. We have found the parameter $q_{s1}/U_o = 0.26\text{drops/cm}^3$ to give the best result. Similar to the previous cited studies, the parameter ε/U_o was given a value of 0.4 sec. The coupled set of equations (5), (6), and (7) was integrated numerically under suitable boundary conditions also similar to the previous studies. To start the numerical solution, the distribution of eddy diffusivity K(z) was first estimated through the Monin-Oboukhov stability length parameter based on a Bowen ratio of $B_W = 0.12(T_W - T_o)$, see Ling and Kao, 1976. It was then successively refined with the Bowen ratio based on more accurate vertical fluxes of heat and moisture derived from the calculated solution.

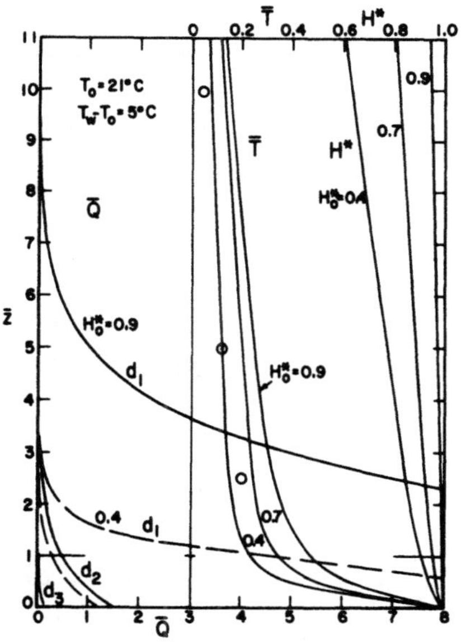

Fig. 2. Distribution of water droplets, temperature, and relative humidity for unstable air, warm sea with $T_w - T_o = 5°C$, and different reference relative humidities H_o^*. Small circles are soundings by Craig (1946, p.46).

RESULT AND CONCLUSION

The calculated solution for the case of warm sea and unstable air is shown in Fig. 2. Both the two larger droplet size groups, $d_3 = 250\mu$ and $d_2 = 150\mu$, give a distribution which is insensitive to the reference humidity H_o^*, while the small size group $d_1 = 50\mu$ is very sensitive to the reference humidity. This indicates that a major part of the large droplets does fall back into the sea. Both the temperature and humidity distributions are very close to those found over a warm sea. The solution for a smaller temperature difference, with $T_w - T_o = 2°C$, is shown in Fig. 3. It represents the most prevalent condition found over the ocean. One notes that the temperature profiles have a strong negative gradient near the sea surface, and they are moved towards the left with temperatures less than T_o. This is clearly caused by the latent heat of evaporation. Thus no amount of artificial modification for K alone can produce such an effect. The solution for the neutral state of air as represented by a sea-air temperature difference of $T_w - T_o = -1°C$ is shown in Fig. 4. The water droplet distributions are not much different from the unstable case. The temperature distribution shows a distinctively strong negative temperature gradient at the bottom level which tends to extract heat from the ocean, and a strong

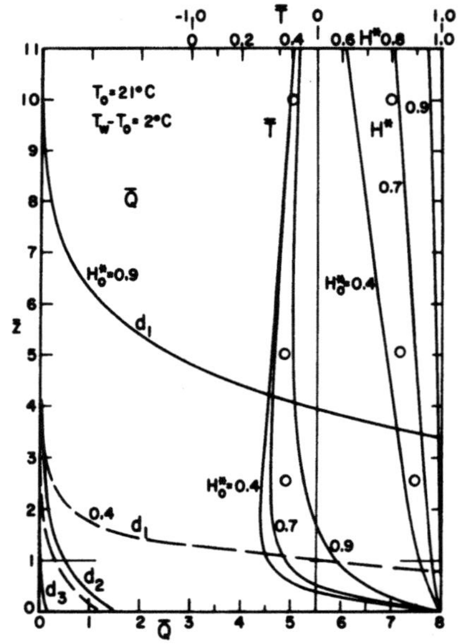

Fig. 3. As in Fig. 2 except $T_w - T_o = 2^oC$. Small circles are soundings by Craig (1946, p. 31).

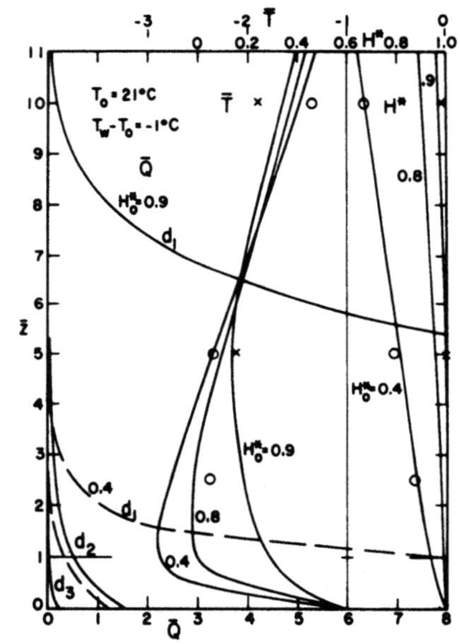

Fig. 4. As in Fig. 2 except for neutral state with $T_w - T_o = -1^oC$. Small circles are soundings by Craig (1946, p. 22), and the crosses by Emmons (1947, p. 68).

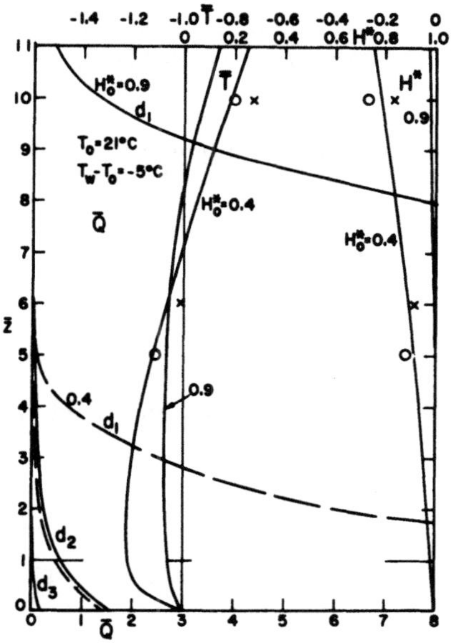

Fig. 5. As in Fig. 2 except for stable air, cold sea with $T_w - T_o = -5^oC$. Small circles are soundings by Craig (1946, p. 20), and the crosses by Emmons (1947, p.61).

inversion layer on the top which also tends to extract heat from the upper air to balance the effect of the latent heat of evaporation. As the reference humidity increases, the effect of latent heat sink tends to take place at a higher level above the sea. In general, the theory predicted the correct temperature and humidity profiles. The case of a cold sea and stable air is shown in Fig. 5. Under low humidity, there is still a small negative temperature gradient next to the sea surface. The air tends to extract heat from the sea even though the sea is cold. This is due to the forever presence of a strong latent heat sink. It is only at very high humidity where the air is totally saturated that there is no effect due to the latent heat near the sea surface. In this case, the population of the small d_1 droplets is greatly increased to form deep sea fog, while the larger d_2 and d_3 droplets remain relatively unaffected. Here again the theory predicted the correct behavior of both the temperature and the humidity profiles.

In conclusion, one may say that the newly refined theory for multiple size droplets gives very realistic results. With more improvements from actual ocean measurements one should be able to greatly refine the analytical model. We are now more confident that the eddy diffusivity K is universally applicable to all transportable quantities. Finally, the latent heat is a powerful mechanism which promotes the transfer of heat from the ocean to the atmosphere - a factor which has to be fully accounted for in any realistic theory of the sea-air interaction.

ACKNOWLEDGEMENT

This work was performed under the Office of Naval Research contract No. N00014-77-C-0017.

REFERENCES

Blanchard, D.C., 1963, Progress in Oceanography, 1, Pergamon Press, 72-202.

Craig, R.A., 1946, Pap.Phys.Oceanogr.Meteor., 10, No. 1

Emmons, G., 1947, Pap.Phys.Oceanogr.Meteor., 10, No. 3.

Leon, R., 1977, 13th Ann.Meeting of Amer.Insti. of Aero. and Astro., paper No. 77-305.

Ling, S.C. and Kao, T.W., 1976, J.of Phys.Oceanogr., 6, 306-315.

Ling, S.C. and Alice Saad, 1977, Physics of Fluids, November.

Ranz, W.E. and Marshall, W.R., 1952, Chem.Eng.Prog., 48, 141-146, and 173-180.

Rouse, H., 1938, Proc. Fifth International Congress of Applied Mechanics, 55, John Wiley, 550-554.

Shappard, P.A., 1958, Roy.Met.Soc. (London) Wuart.J., 84, 205-224.

DISCUSSION

R. L. Street:

In our paper at this conference we report on the fluxes at the interface under spray conditions. Using a detector with a size range of 25 to 1250 mm for droplets, we measured the droplet spectrum and computed the evaporation from droplets by summing first over all droplet diameters at a given height and then by summing over the droplet boundary layer thickness. For the conditions in our facility, which appear to be well related to reported field conditions, we find that the contribution of droplet evaporation to the humidity field is negligible, droplet evapotation being only 2 % of the total measured evaporation. These calculations are similar to and consistent with those of Lai and Shemdin (JGR, 79, 1974) and our estimation method, but not our results, is equivalent to that used by Wu (JGR, 79, 1974). In addition we do not find the temperature decrease below ambient air temperature (reported by the author for near isothermal conditions) in our facility.

E. L. Mollo-Christensen

The apparent discrepancies between the results of Ling and those quoted by Street obviously need to be resolved. That cannot be done here, and since the differences seem both serious and fundamental, I hope that there will be progress in identifying the reasons for the discrepancies. I thank Prof. Street for his comment.

E. L. Mollo-Christensen:

You did not include electric effects on droplet transports. Please comment.

S. C. Ling:

Electric charges on water droplets will definitely increase the effective molecular diffusivity of water. It will be taken into account in future refinement of the theory. At the same time one should also address to the very problem of where is the origin of the atmospheric electricity. Could it be related to the radiation belt of the earth? Indeed, it is most likely so. Stretching one's imagination, we may one day unify space science, atmospheric science, and oceanic science into one unified system.

J. A. Businger:

Using the same eddy transfer coefficient K for all droplet categories may become incorrect when the settling speed becomes large.

S. C. Ling:

This will be the time when the concentration of large droplets is very large such that it is sufficient to affect the density of the air. Improvement to the theory will be made if this effect is proven to be important.

W. Pierson:

In air-sea interaction, quantities such as $q'w'$ and $\theta'w'$ are important in the fluxes. You have discussed also $\Delta T/\Delta z$ at the sea surface determine evaporation. How are these terms and effects treated in your model?

S. C. Ling:

These terms are, of course, included in the basic transport equations but in different forms.

S. D. Smith:

You predicted near the end an upward heat transport in the absence of a sea-air temperature difference. Eddy heat flux measurements, such as those of Carl Friehe or my own, often show such an effect. Can you estimate its magnitude in typical conditions?

S. C. Ling:

The present result can be gainfully used to explain many seemingly strange heat and moisture fluxes that are observed in the ocean under different atmosphere and oceanic conditions. In general, both the vertical heat and moisture fluxes are not constant with height due to the source and sink effects of the physical system. Once the solution for a given boundary condition is found, the fluxes of any transportable quantity can be easily found, see Ling and Kas (1976). It should be noted that the present theory indicates that a sea-air temperature difference of $-1^{\circ}C$, instead of $0^{\circ}C$, is the neutral state.

M. S. Longuet-Higgins:

It is very easy to see that in a cloud of spherical droplets, if a vertical line drawn through the cloud intersects N droplets on average, then the area of the air-water interface is increased by a factor 4N. Hence it is not surprising if, for large N, the fluxes are substantially increased.

K. Katsaros:

I feel inspired to make a comment relating to what Prof. Longuet-Higgins just said, which is also related to Dr. Pierson's comment. I will show this afternoon that the adding of waves with no wind has strong effects on the heat flux, explained in part by the increase in surface area. This might also explain the peaks in flux found by Dr. Chesselet when the sea state changed, while the mean wind remained constant. I believe this corroborates Dr. Ling's development.

S. C. Ling:

The latend heat of evaporation or condensation is such a powerful mechanism in the heat transport process that one should not neglect them in any realistic theory of the sea-air interaction. Micro water droplets do exist over the ocean at all time but not visible to the naked eye (a hazy air means water droplets). Thus any increase of the surface area of evaporation or the surface humidity gradient will greatly increase the effect of latent heat.

V. Cardone:

You have used the mean wave heigth to normalize the vertical length scale in your analysis. Would not the roughness parameter, Z_o, be more appropriate as it more nearly characterizes the intensity of whitecapping and spray production.

S. C. Ling:

This roughness length is a non-physical parameter usefull to represent the roughness of the boundary where the logarithmic wind profile goes to zero. The present theory considers the wind profile within the trough of the wave without terminating it artificially at an artificial roughness height. In other words, the present treatment of the problem is more realistic, in that the waves are considered as moving drag objects within an air space above the mean wave trough. The roughness height is represented by the mean wave height, where the mean wave length to wave height ratio is taken to be essentially constant under all high wind states. In the future, the mechanics of whitecaps and sea spray will be parameterized in terms of the energy input from the wind to the wave.

D. A. Friehe

What are typical ratios of latent heat losses from the ocean by moisture evaporation at the surface to that by spray evaporating in the air near the surface?

S. C. Ling:

From the present solution we can see that this ratio can be very small when the sea is cold and the humidity at the reference height is high, i.e. the humidity at the ocean surface is saturated and thus no surface evaporation. On the other hand, when the sea is warm and the reference wind U_o is small the ratio will be very large, because there will be very little water droplets. Under normal warm sea and high sea states the ratio may run around 0.3 to 0.2 depending on humidity at the reference height.

ON THE DYNAMICS OF STEEP GRAVITY WAVES IN DEEP WATER

M. S. Longuet-Higgins

Department of Applied Mathematics and Theoretical Physics
University of Cambridge, Silver Street, Cambridge and
Institute of Oceanographic Scieces, Wormley, Surrey

1. INTRODUCTION

Ocean waves on all scales play an important role in air-sea interaction, since they strongly affect the transfers of heat, momentum, water vapour, gases (O_2 and CO_2), salt particles and micro-organisms, between ocean and atmosphere. For the transfer of momentum, wave breaking may be crucial. Banner and Melville (1977) have presented evidence that breaking tends to induce separation in the air flow above the surface, with a consequent increase in the local momentum flux from air to water by about two orders of magnitude. Wave breaking must also transfer momentum from waves to surface currents and supply energy to mix the upper layers of the ocean, thus greatly increasing the transfers of heat and gases across the surface.

To calculate the behaviour of the boundary-layers at the surface, we need to know in detail the profile and particle velocity at the interface. Even for free gravity-waves this remains a missing element in our knowledge. Until now, most water-wave theory has treated the free surface as a simple spectrum of sine-waves, each propagated with its own phase speed, or else as a weakly nonlinear system, in which there is but a gradual transfer of energy between different wave components. Notwithstanding, most observations of the sea surface under conditions of wave generation show the surface to be strongly non-sinusoidal. On scales larger than a few centimetres, the wave crests tend to be sharply curved, the troughs more rounded.

Now the form of limiting waves as calculated long ago by Michell (1893) does show a sharp angle at the crest as suggested earlier by Stokes (1880). But the calculation of steep, not limiting, waves, with a still rounded crest, presents technical difficulties which have only recently been overcome by the use of accelerated convergence methods (Schwartz 1974; Longuet-Higgins 1975; Cokelet 1977a) or asymptotic analysis (Longuet-Higgins and Fox 1977a and b). These methods have all yielded unexpected yet consistent results, as will be described below in Section 2.

By itself, the anlysis of steep, symmetric waves is however insufficient to describe the observed pattern of the sea surface, which is highly time-dependent. How is it that breakers and white-caps are actually formed? Again, the mathematical tools to investigate this question have till recently been lacking. In 1976, however, the author and E.D. Cokelet published the results of a new method of caclulating unsteady gravity waves, in which many of the old restrictions and assumptions are absent. As described in Section 3, this technique promises to provide rational estimates of the momentum and energy transferred by plunging breakers.

A complementary line of enquiry is described in Section 4, where we outline a recent investigation into the stability of steep gravity waves, showing that there are at least two important types of instability, subharmonic instabilities of the Benjamin-Feir type, and strong local, or superharmonic instabilies, which lead to plunging breakers.

Lastly we show how their conclusions have been verified, not only by laboratory experiments but also by the quite independent time-stepping method of Section 3.

The present situation therfore offers hope for the accurate calculation and understanding of steep surface waves, and of the concomitant flow in the boundary-layers at the surface.

2. PROPERTIES OF STEEP, SYMMETRIC WAVES

For the first time, profiles of steep progressive gravity waves of arbitrary height, calculated with reliable accuracy, were presented by Schwartz (1974, Figure 9). Not surprisingly, as the highest wave is approached the profile becomes quite non-sinusoidal. Schwartz found (1974, Figure 10) that for waves of given length the Fourier coefficient a_1 of the surface profile was not always an increasing function of the crest-to-trough waveheight, as had previously been assumed. Thus for one and the same value of a_1 it was sometimes possible for two quite distinct waves, of different heights, to exist. This of course implies that a_1, the expansion parameter used by Stokes and many others, is not suitable over the whole range. More suitable is the crest-to-trough waveheight $2a$, or some other parameter which increases monotonically with a.

Even after Schwartz's paper (1974) it was generally thought that the phase-speed c of a steep, progressive gravity wave on deep water always increased with the wave steepness, at given wavelength. However, further calculations (Longuet-Higgins 1975) have shown that the wave speed, considered as a function of the crest-to-trough height 2a, increases to a maximum within the range of possible heights, and then diminishes; the highest wave is not actually the fastest. The reason for this behaviour is associated with the fact that the profile of the highest wave, having a very sharp curvature at the crest, intersects the profile of a wave which is not quite so high. Hence the higher wave lies below it over most of the wavelength (see Cokelet 1977 figure 20). Its potential energy is therefore less, and its kinetic energy also.

On examining the form of the crests of steep gravity waves it is found that the surface profiles are asymptotically self-similar (see Figure 1). The length-scale of the flow, which is proportional to the radius of curvature at the crest, is given by

$$\ell = q^2/g$$

where q is the speed of a fluid particle at the crest, in a frame of reference travelling with the phase velocity, and g denotes the acceleration of gravity.

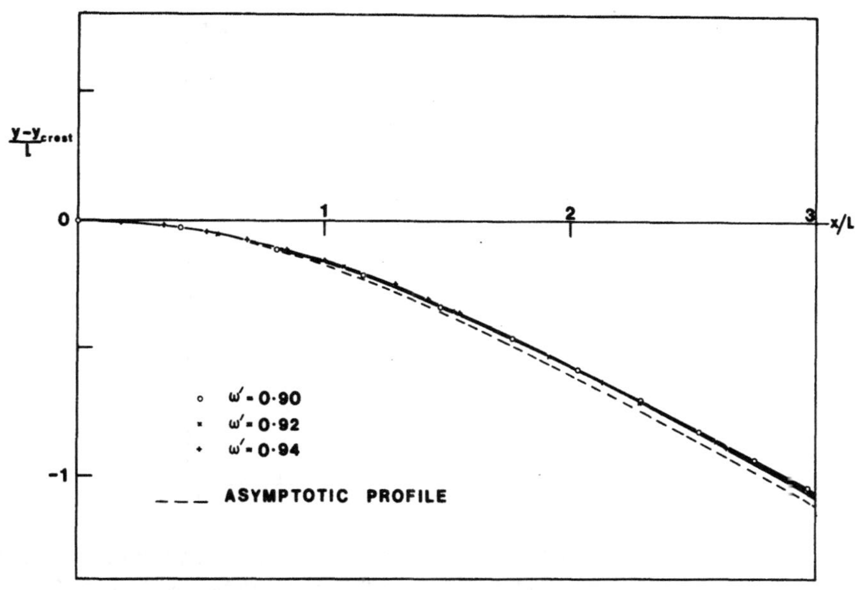

Figure 1 (from Longuet-Higgins and Fox 1977a). Profiles of the crests of steep gravity waves, rescaled by the length $\ell = q^2/g$, showing that the forms are asymptotically similar.

At distances r from the crest such that r/l is large, the free surface is inclined at almost $\pm 30^{\circ}$ to the horizontal. The limiting form of the flow near the crest (shown by the broken line in Figure 1) has been calculated precisely by Longuet-Higgins and Fox (1977a).

The complete form of the wave profile has been calculated by Longuet-Higgins and Fox in a second paper (1977b). The method applies when the scale l of the flow near the crest is sufficiently small compared to the wavelength L, that is when

$$\epsilon^2 \propto l/L$$

is a small quantity. In an inner zone near the wave crests, of radius l the profile is described by the asymptotic flow of Figure 1. In the main part of the wave the flow is as in Michell's highest wave (1893) but now perturbed so as to take account of the finite curvature at the crest. The two expressions are matched over a zone of radius ϵL. To order ϵ^3, the resulting expression for the phase speed c is

$$c^2 = (gL/2\pi)\left[1.1931 - 0.423\,\epsilon^3 \cos\left(2.143 \ln \epsilon + 1.48\right)\right].$$

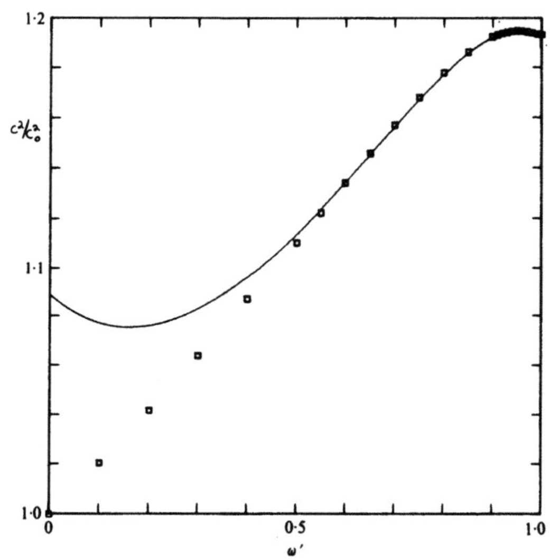

Figure 2 (from Longuet-Higgins and Fox 1977b). The square of the phase-speed c of deep-water gravity waves, as a function of the parameter ω' which increases monotonically with the wave height. Continuous curve represents asymptotic formula given above. Plotted points are from usual expansion in powers of a small-amplitude parameter.

This is compared with the previously calculated values of c^2 in Figure 2. It can be seen that the agreement is close, and that the maximum value of c^2 is very accurately verified.

3. UNSTEADY WAVES

Observations suggest that steep progressive waves have a strong tendency to become steeper on their forward face and to topple forwards, producing either "spilling" or "plunging" breakers. The development of the wave profile, at least up to the point of impact of the jet on the forward face, has been recently followed by a new method described by Longuet-Higgins and Cokelet (1976). Although restricted in the first place to irrotational waves on deep water, the method appears to be capable of extension to waves in the presence of a surface shear current and to water of infinite depth.

The motion is assumed periodic in the horizontal coordinate x with length scale L, though not periodic in the time t. By a simple substitution $\zeta = \exp[ik(x+iy)]$ the free surface is transformed into a simple closed curve C in the ζ-plane, and all points at infinite depth are transformed into a single point in the interior. By a suitable choice of coordinates the complex velocity potential $\chi = \phi + i\psi$ becomes regular and analytic everywhere inside C, at all times. The velocity field in the interior of the fluid depends only on the values of ϕ at the surface, and the computation is carried forward by following the values of x,y and ϕ corresponding to marked particles. At each time-step we compute

$$\frac{dx}{dt} = \phi_x , \qquad \frac{dy}{dt} = \phi_y , \qquad \frac{d\phi}{dt} = \frac{1}{2}(\nabla\phi)^2 - gy - p/\rho$$

where p is the surface pressure. This enables the value of ϕ on $C(t+dt)$ to be found, and hence the tangential derivative $\partial\phi/\partial s$. To find the normal derivative $\partial\phi/\partial n$ is equivalent to solving a Dirichlet problem: how to determine the normal derivative of a function ϕ, given on the boundary and satisfying $\nabla^2\phi = 0$ everywhere in the interior?

Longuet-Higgins and Cokelet (1976) find the solution by means of an integral equation, which is solved numerically, at each time-step. The method involves the neglect of viscosity, compressibility and surface tension, but does not assume the particle acceleration to be small compared to g.

In their first paper (1976) some examples were given of the development of the free surface from a symmetric wave, when the energy level was further raised by the smooth application of a surface pressure distribution, over a short intitial period. After removing the pressure, the free surface curled over and plunged forward, in the way commonly observed (see Figure 3).

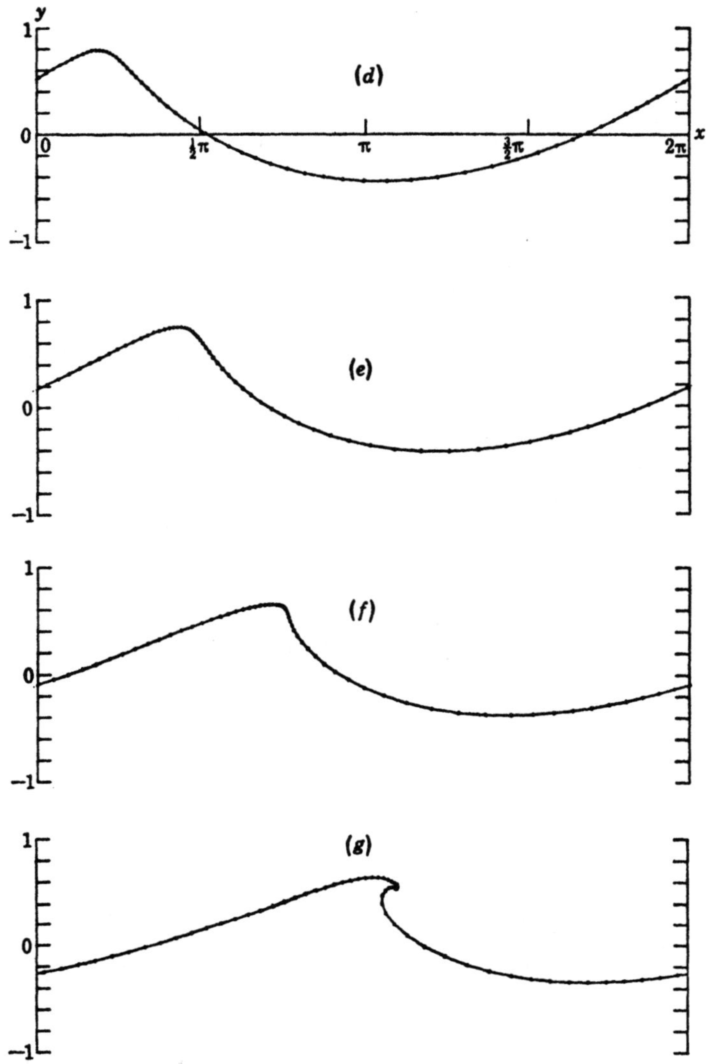

Figure 3 (from Longuet-Higgins and Cokelet 1976) Breaking of a deep-water wave raised to an energy 1.88 times the maximum for steady propagation. The calculation is by numerical time-stepping.

In another set of numerical experiments (1977b) Cokelet has begun with a free wave of exactly sinusoidal form, but moderate or large amplitude and, without applying any pressure at the surface, has followed the subsequent development of the wave profile. Just as in the previous case the wave forms a high-velocity jet and plunges forwards (see Cokelet 1977b, Figure 3). This happens even when the initial energy is less by a factor 2/3 than the maximum energy for a steady, nonsinusoidal wave of the same wavelength (see Figure 4).

From Figure 4 it will also be seen that the momentum in the jet can grow to be a substantial portion of the total momentum of the wave train.

To follow the jet beyond the instant where it impinges on the forward face of the wave requires a different type of analysis, which must be capable of modelling the turbulent shear flow and the entrainment of air and of momentum from the laminar part of the flow. A start in this direction has been made by Longuet-Higgins and Turner (1974).

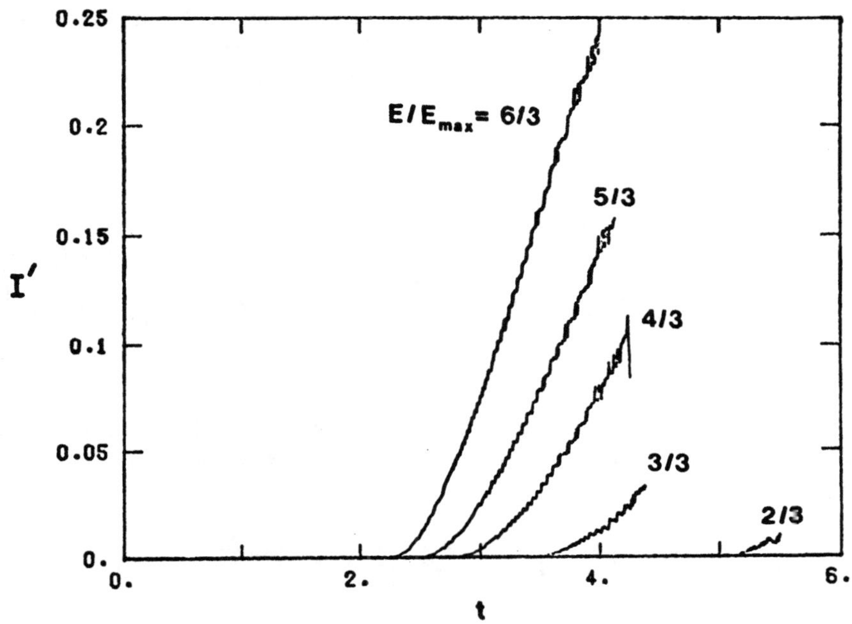

Figure 4 (from Cokelet 1977b). The momentum I' in the jet of a breaking deep-water wave as a function of the time t, for different total energies E. At $t=0$ the surface profile was a pure sine-wave.

4. THE INSTABILITIES OF STEEP GRAVITY WAVES.

To understand the results of numerical computations such as we have just described, a complementary analytic approach is needed. Some light is thrown on the initial stages of wave breaking by a recent investigation of the normal-mode instabilities of a regular train of steep gravity waves (Longuet-Higgins 1977a and b). Among the possible normal perturbations it is found that some are neutral; these are essentially travelling waves superposed on the unperturbed wave train. Some, like the subharmonic instabilities of Benjamin and Feir, grow gradually. There is also a third group which grow much more rapidly and correspond to the initial stages of plunging breakers.

Coordinates are taken relative to an observer travelling with the horizontal phase speed c in the direction of wave propagation. Thus the unperturbed, finite-amplitude wave appears as a steady flow. This flow, in which the surface is the streamline $\psi = 0$, may be accurately calculated by the methods described in Section 2.

On this we now superpose a small perturbation, so that the horizontal and vertical coordinates x and y are, in general, functions of ϕ, ψ and t. Moreover the free surface is given by $\psi = F(\phi, t)$, where F is a function to be determined. The dynamic and kinematic boundary conditions, expressed in terms of ϕ, ψ, t as independent variables, become respectively

$$\left.\begin{aligned} &(y_\phi y_t + y_\psi x_t) - gy(y_\phi^2 + y_\psi^2) = \tfrac{1}{2} \\ &[1-(y_\phi y_t + y_\psi x_t)]F_\phi + (y_\phi^2 + y_\psi^2)F_t + (y_\psi y_t - y_\phi x_t) = 0 \end{aligned}\right\} \quad (4.1)$$

to be satisfied at the surface $\psi = F$. By expanding in a Taylor series in ψ, we replace (4.1) by conditions to be satisfied on $\psi = 0$. Now we write

$$\left.\begin{aligned} x &= X(\phi,\psi) + \xi(\phi,\psi)\,e^{-i\sigma t} \\ y &= Y(\phi,\psi) + \eta(\phi,\psi)\,e^{-i\sigma t} \\ F &= f(\phi)\,e^{-i\sigma t} \end{aligned}\right\} \quad (4.2)$$

where X and Y represent the unperturbed steep wave and ξ, η, f are normal modes with radian frequency σ. Neglecting all squares and products of ξ, η, f we obtain two linear conditions for ξ, η, f on $\psi = 0$, with coefficients depending on the known functions X, Y (for details, see Longuet-Higgins 1977a).

Now X and Y are given by expressions of the form

$$X = -\phi/c + H_1 e^{\psi/c} \sin \phi/c + H_2 e^{2\psi/c} \sin 2\phi/c + \ldots$$

$$Y = -\psi/c + \tfrac{1}{2}H_0 + H_1 e^{\psi/c} \cos \phi/c + H_2 e^{2\psi/c} \cos 2\phi/c + \ldots$$

where $H_0, H_1, H_2, \ldots$ and c are known constants. On expanding the perturbation ξ, η in conjugate Fourier series with unknown coefficients $a_0, a_1, \ldots; b_0, b_1, \ldots$ and similarly f in a Fourier series with coefficients $c_1, c_2, \ldots; d_1, d_2, \ldots$ and equating coefficients of cos $r\phi/c$, sin $r\phi/c$, we obtain an infinite set of equations for the a_s, b_s, c_s, d_s and σ. The system may be solved by truncation of the matrix at successively higher values of r and s, giving the eigenfrequencies σ (in this frame of reference) and the corresponding normal modes. The maximum value N of r and s is raised until numerical convergence is obtained.

Consider first the superharmonics — that is modes which have the same horizontal scale as the unperturbed wavelength L, or less. The calculated frequencies σ are shown in Figure 5, as a function of the steepeness ak of the unperturbed wave (throughout the calculation we choose units so that $k = 2\pi/L = 1$, $g = 1$). At small wave steepnesses the perturbations are free travelling waves with wavenumber n, having n maxima or minima in a horizontal distance L. Positive values of n correspond to perturbations travelling in the same sense as the unperturbed wave. Thus $n = 1$ corresponds to a perturbation which simply shifts the phase of the basic wave, without changing the amplitude. Since the phase-speed is unchanged the frequency of this perturbation is always zero. $n = 2$ corresponds to a perturbation having two maxima and minima per basic wavelength, and so on. Its frequency as $ak \to 0$ is $2 - \sqrt{2} = 0.586$. Generally the limiting frequency is $n - n^{1/2}$ when n is positive. Negative values of n correspond to perturbations travelling in the opposite sense. Their limiting frequency is $|n| + |n|^{1/2}$. Thus $n = -1$ corresponds to a perturbation of the same wavelength as the basic wave- its limiting frequency is 2, since to an observer travelling with the unperturbed wave the frequency of encounter of wave crests is doubled.

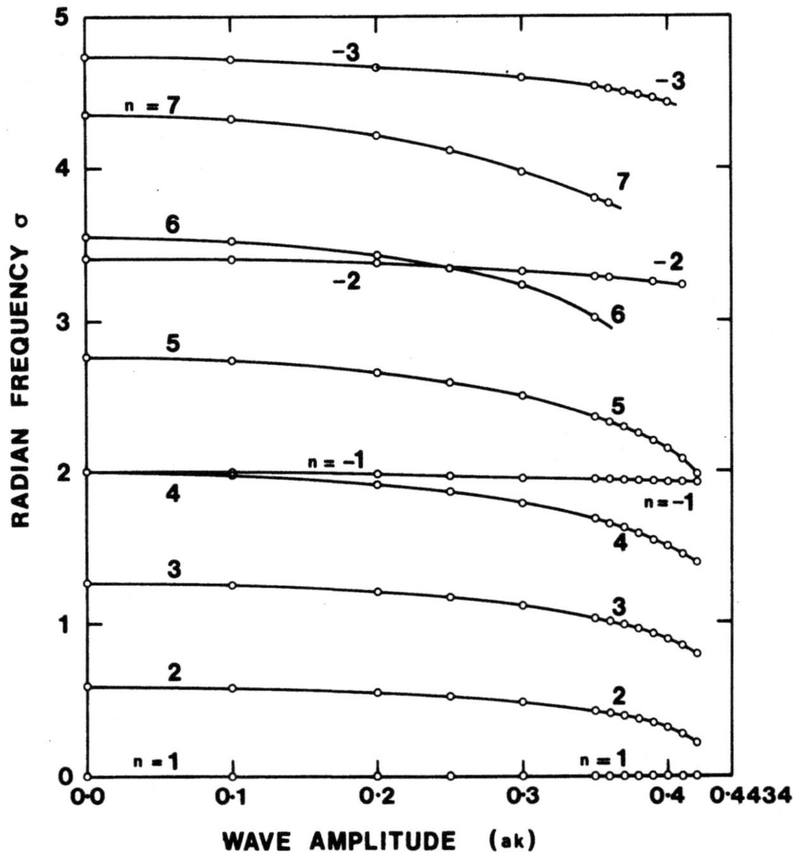

Figure 5 (from Longuet-Higgins 1977a). Frequencies of the normal-mode perturbation of a deep-water wave, as a function of the steepness (dimensionless amplitude) of the unperturbed wave.

As ak increases, the frequency of the perturbations, in this reference frame, always diminishes. The change in frequency, for small ak, is of order $(ak)^2$, and corresponds closely to the change in phase-speed of a short wave due to the presence of a longer longer wave of steepness ak (Longuet-Higgins and Phillips, 1962).

For larger values of ak the frequencies of the modes $n = 1$ and $n=2$ approach one another. On account of practical limitations the calculations have not yet been taken beyond about $ak = 0.42$ (in that case the maximum value of N was 45; the order of the matrices involved is $4N+2$). The frequency σ of the second mode appears to be tending toward zero at the point corresponding not to the highest wave but to the point $ak=0.436$ where the phase speed c is a maximum (see Figure 2). This can be understood by considering a perturbation of the basic wave which increases the height of the wave without changing its phase. At general values of ak the perturbed wave will travel with a slightly different phase-speed from the original, so the perturbation will not be a normal mode; it begins symmetric and becomes asymmetric. But at the value of ak where c is stationary, the perturbed wave travels with virtually the same speed as the unperturbed wave. So we have the possibility of a new normal mode with zero frequency. At such a bifurcation point we may expect a transition to instability.

Confirmation comes from considering the subharmonics, that is to say perturbations having a horizontal scale mL where $m>1$. When m is integral, these may be calculated by a precisely similar method (Longuet-Higgins 1977b). The case $m=2$ (perturbations with repetition distance 2L) is shown in Figure 6. The even modes $n=2/2$ and 4/2 have already appeared in Figure 5 as $n=1$ and 2 respectively. The odd modes display a new phenomenon. At a certain wave steepness $ak=0.228$ the frequencies of the modes $n=1/2$ and $3/2$ 3/2 coalesce and become complex. Those for which $\mathrm{Im}(\sigma)>0$ are unstable. It can be shown (see below) that this is a subharmonic instability of the Benjamin-Feir type (1967) in which the fundamental wave is modulated by side-bands which grow exponentially.

For larger values of ak the rates of growth of the instability (shown in Figure 7) at first increase, but reach a maximum at around $ak=0.32$ before diminishing to zero. At $ak=0.36$ the unstable mode splits again into two neutral modes; the side-band instability disappears, as predicted by Lighthill (1967).

However at ak = 0.41 there is coalescence with the mode n = 2/2 and a new type of instability appears, having zero frequency and a much higher rate of growth (Figure 7). At the critical point itself the perturbation is such as to increase the height of every

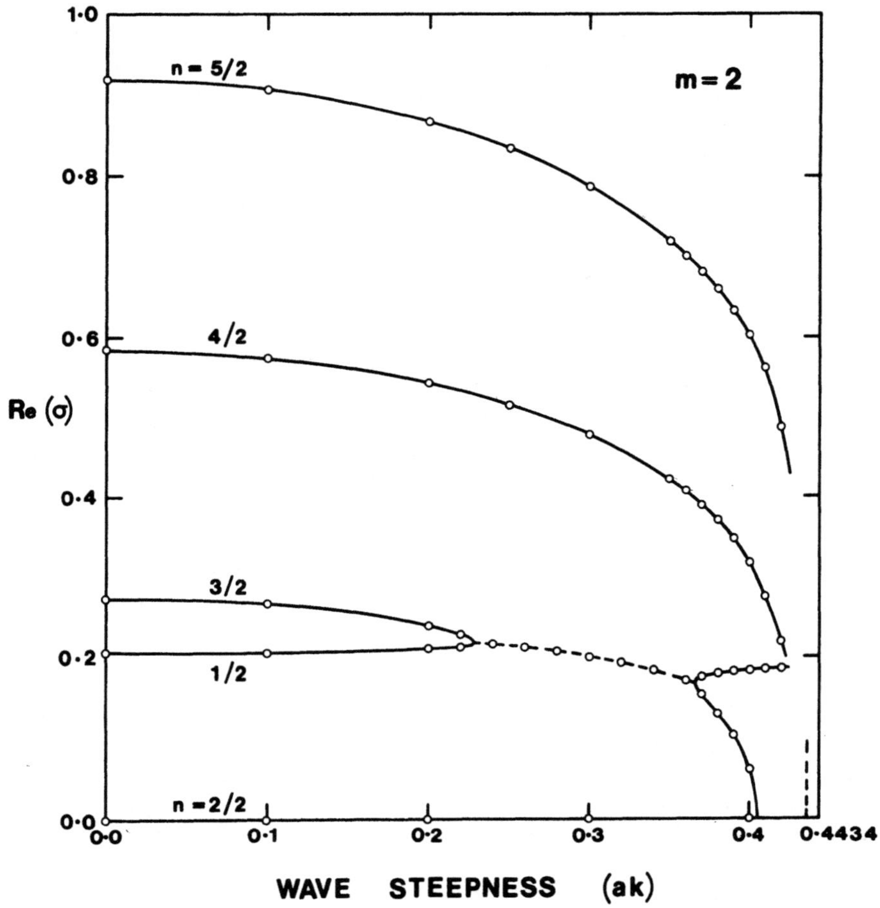

Figure 6 (from Longuet-Higgins 1977b). Frequencies of the subharmonic normal perturbations of deep-water waves in the case $m=2$ (the repetition distance is two wavelengths). The broken curve indicates an unstable mode.

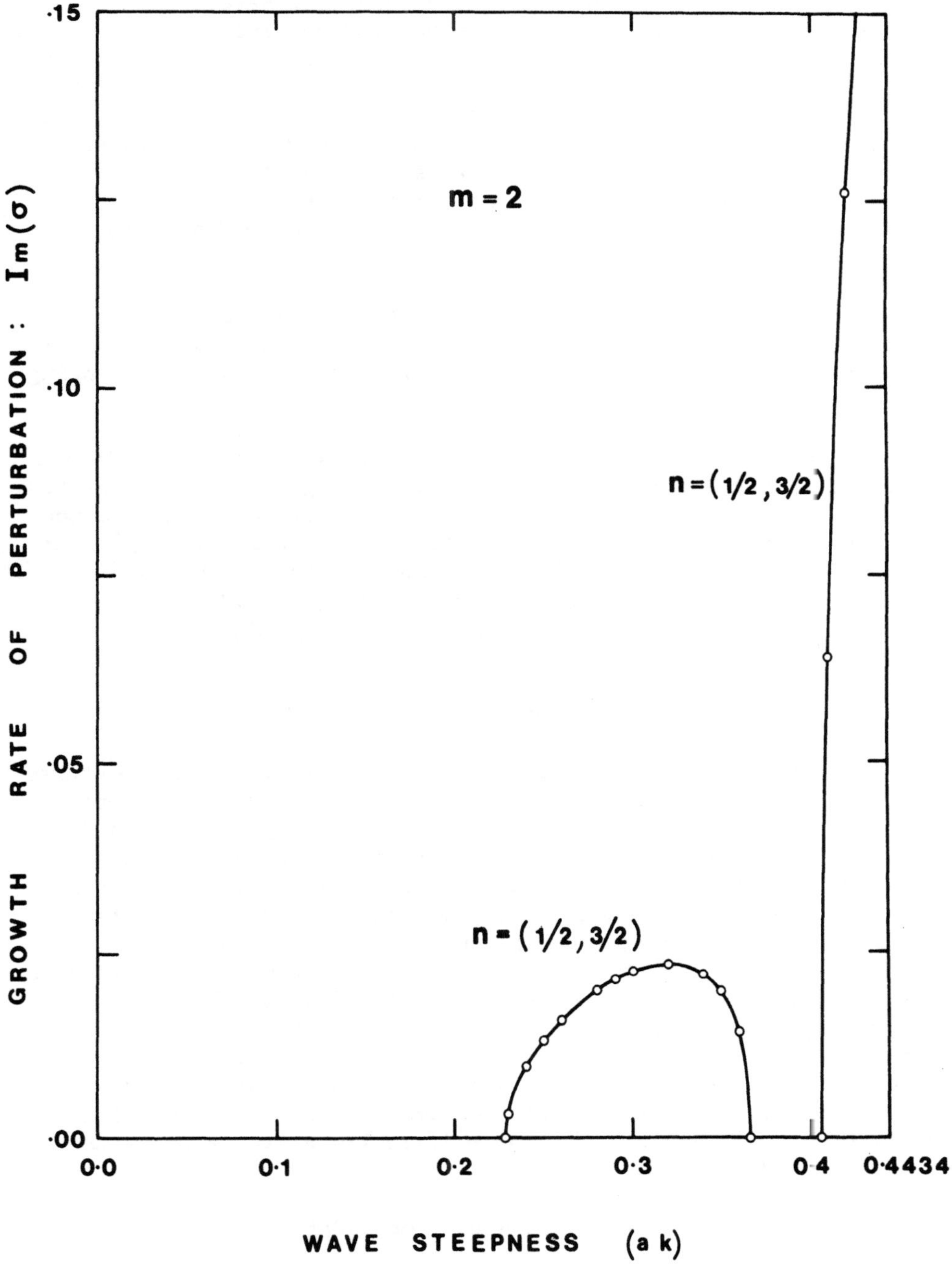

Figure 7 (from Longuet-Higgins 1977b). Rates of growth of the subharmonic modes corresponding to Figure 6. Note the much larger growth rate of the instability which arises at $ak = 0.41$.

alternate wave, at the same time diminishing its wavelength slightly so as to keep the phase speed unaltered; for the intermediate waves, the wave height is diminished and the wavelength slightly increased, with a similar effect. Since the perturbation is an odd one, it contains no component of wavenumber 2/2 = 1 . So it must be possible to increase the height of each individual wave without increasing its first harmonic component. Hence we must be close to the wave steepness ak for which the first harmonic is a maximum. From Figure 10 of Schwartz (1974) this occurs at about $ak = 0.41$ in close agreement with our calculations.

When ak slightly exceeds the critical value 0.41 the form of the instability becomes asymmetrical. From its local character it is evident that this represents the initial stages of a plunging breaker.

By computing subharmonics at other values of m up to $m = 8$ we arrive at the stability diagram shown in Figure 8, where the real part of the perturbation frequency is plotted against the steepness of the unperturbed wave. The boundaries of the unstable region are delineated up to the limit of computation, which is about $ak = 0.42$. The tangent at the origin corresponds to the critical condition $\sigma = \sqrt{2}\, ak$ derived by Benjamin and Feir (1967) on the assumption that ak was small.

5. COMPARISON WITH OBSERVATION

The rates of growth of subharmonic instabilities were measured by Benjamin (1967) in a long wave channel at a constant side-band frequency $\sigma = 0.1$ and for wave steepness ak lying between 0.07 and 0.17 (see Figure 9). These are essentially spatial growth-rates, determined by observing the surface elevation at different distances from the wavemaker. Benjamin and Feir's approximate theory is indicated by the broken curve. In Figure 9 we show also the growth rates estimated from Section 4; the time-rates have been converted to spatial rates of growth by dividing by the linear group-velocity (the errors in this procedure being of order $(ak)^2$. From Figure 9 we see that the present calculations imply a shifting of the critical wave steepness somewhat to the right, so that the theoretical curve luckily passes through the two lowest plotted observations. In general the observations lie a little below the theoretical curve.

6. VERIFICATION BY TIME-STEPPING

To verify the calculations at higher values of ak, Longuet-Higgins and Cokelet (1977) have used the time-stepping technique of Section 3 to test the rates of growth of the normal modes. Thus, in the case $m = 2$ it can be assumed that the motion is space-periodic with repetition distance $2L$. Figure 10 shows a typical example

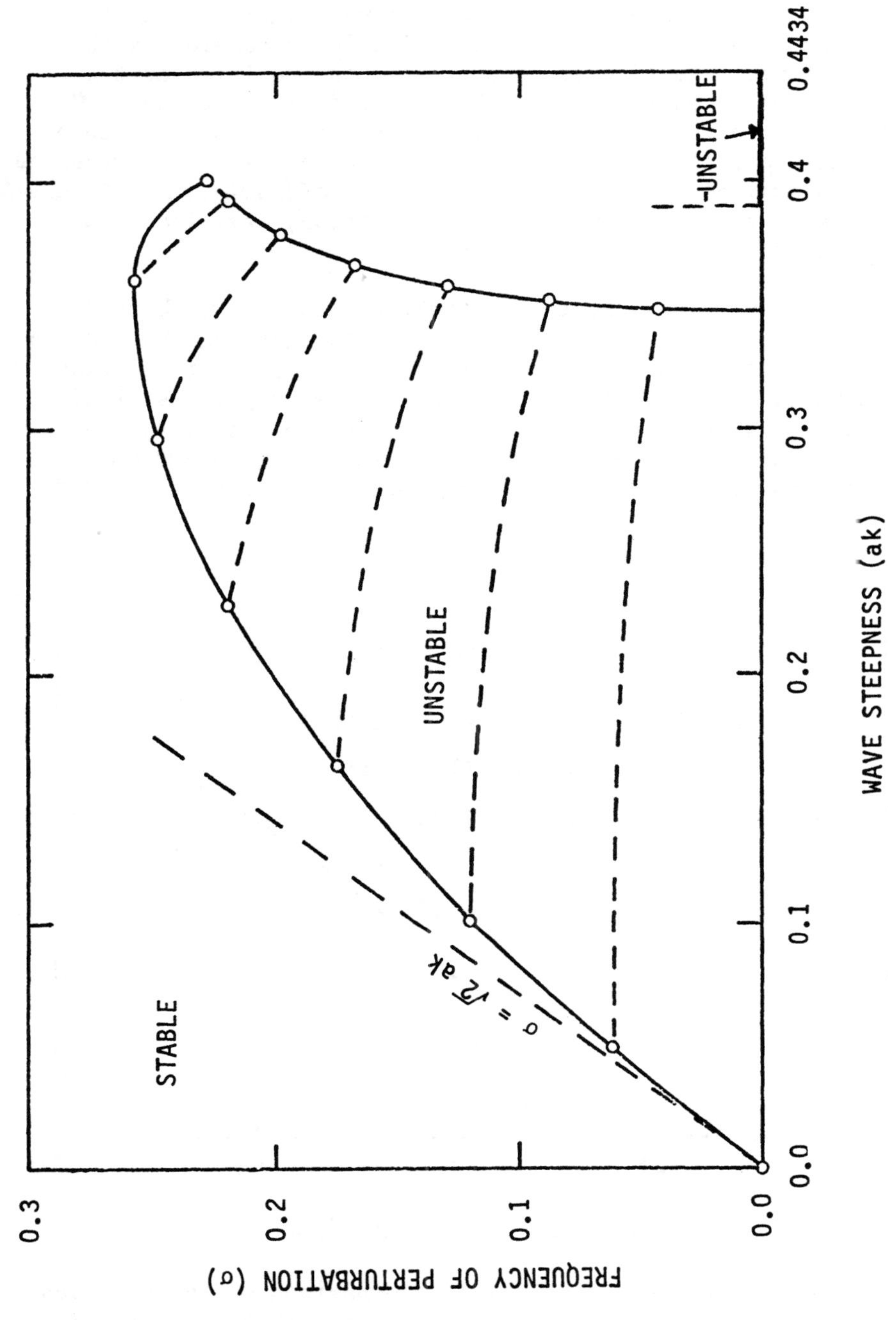

Figure 8 (from Longuet-Higgins 1977b). Stability diagram for gravity waves of finite amplitude on deep water.

in which the initial perturbation at t = 0 corresponds to the unstable mode $n = (1/2, 3/2)$ when $ak = 0.32$. The perturbed wave is shown on the left and the computed perturbation, enlarged ten times in the vertical, is shown on the right. Time advances downwards. At subsequent times $t > 0$ the perturbation is calculated as the difference between the time-stepped profile on the left and the unperturbed profile advanced with the theoretical phase speed c. Since the perturbation has both an in-phase and a quadrature component, after one half-cycle $\frac{1}{2}T$ the perturbation is simply reversed in sign, and mulitplied by constant factor $\exp(\beta T/2)$, where $\beta = \mathrm{Im}(\sigma)$. This in effect transfers a positive perturbation, say, from one wave crest to the adjacent wave crest.

In Figure 10 we see about three half-cycles, in which the perturbation is transfered from one wave to the next and back again about three times, after which it has grown to the point of breaking. The calculated magnification after one half-cycle was 1.57, compared to the value 1.46 given by the normal mode theory. Thus these two quite independent methods are essentially in agreement. It was verified also that the decaying modes did indeed decay and

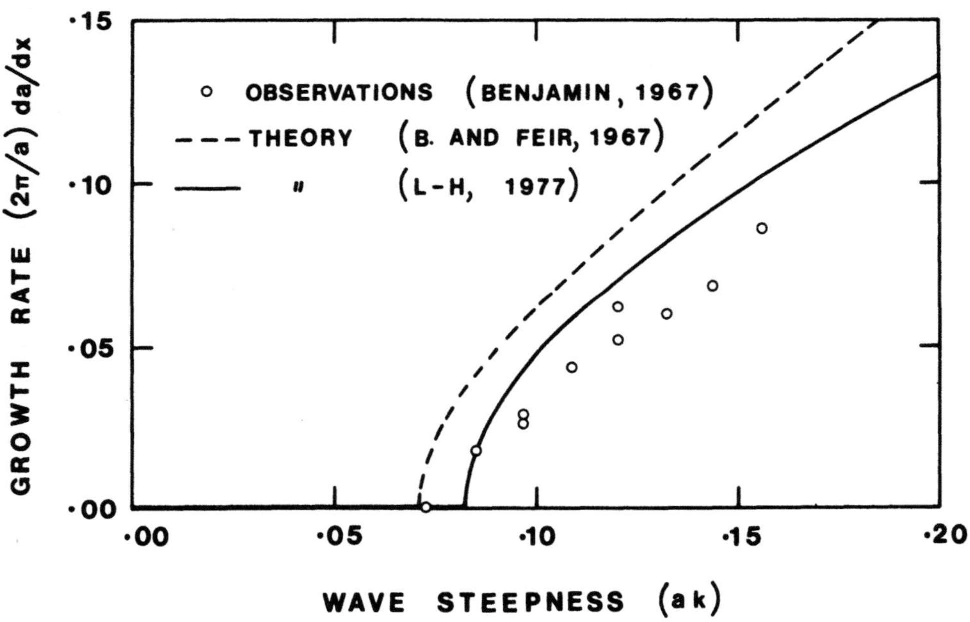

Figure 9 (from Longuet-Higgins 1977b). Comparison of the observed growth-rates of subharmonic instabilities with the theory of Benjamin and Feir (1967) and the present calculations.

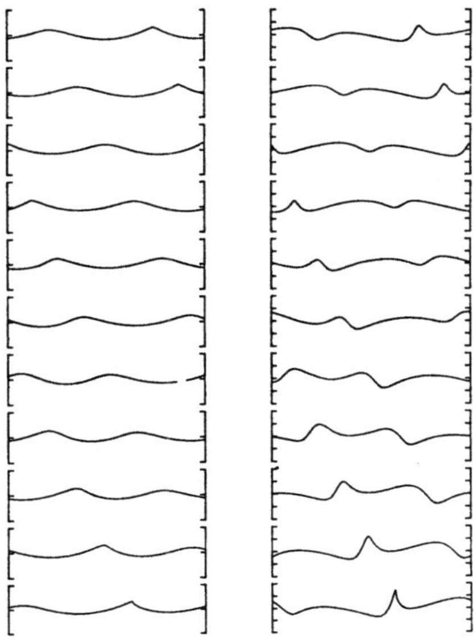

Figure 10 (from Longuet-Higgins and Cokelet 1977). Growth of the subharmonic instability $n = (\frac{1}{2}, \frac{3}{2})$ when $ak = 0.32$. The left-hand column shows the time-stepped surface profiles. The right-hand column shows the resulting "perturbation" enlarged vertically by a factor 10. Time increases downwards.

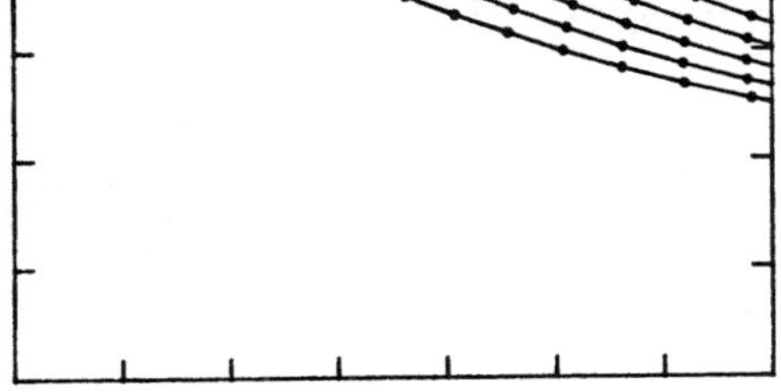

Figure 11 (from Longuet-Higgins and Cokelet 1977). Close-up of the surface profile in Figure 10 near the instant of overturning, as seen by an observer travelling with the speed of infinesimal waves. The interval between profiles is 1/50 of a wave period.

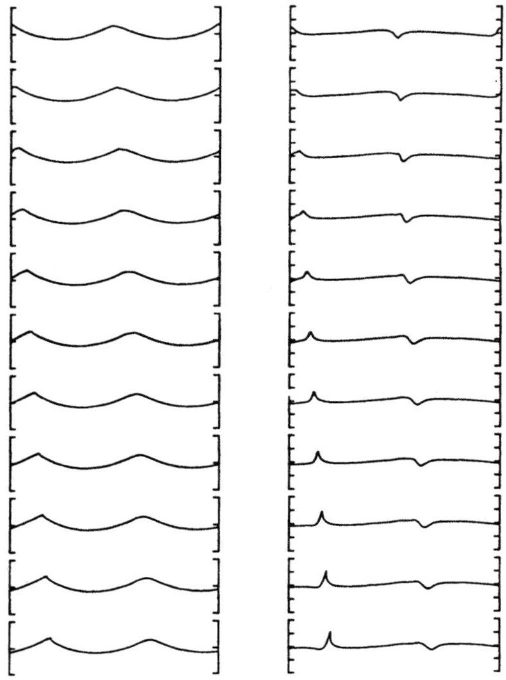

Figure 12 (from Longuet-Higgins and Cokelet 1977). Growth of the local instability $n = (2/2, 3/2)$ when $ak = 0.41$. The left-hand column shows the time-stepped profiles, and the right-hand column the "perturbation" enlarged by a factor 10. Note that the initial rate of growth is more rapid than in Figure 10. The final stages are however very similar.

that the neutral modes at ak = 0.1 and ak = 0.38 indeed had rates of growths not significantly different from zero.

Figure 11 shows an enlargement of the surface profiles near the instant of overturning, in a frame of reference travelling forwards with the linear wave speed $c_0 = (g/k)^{1/2}$. Successive profiles are separated by $1/50$ of a (linear) wave period.

Very similar results were obtained in the case ak = 0.41, but after a much shorter lapse of time (Figure 12). This indicates that the subharmonic instabilities at say ak = 0.32 merely serve to transfer energy at a low rate from each alternate wave to the waves in between. But that as soon as a critical wave steepness is reached (about where $c = c_{max}$) a local instability with much higher growth-rate takes over.

7. CONCLUSIONS

It is reasonable to suppose that the type of surface features revealed both by close observation and by the numerical calculations of Sections 4 to 6 will have a marked effect on the air-flow above the surface, and on the occurrence of flow separation.

In this study we have not so far included the effects of viscosity and capillarity, which will introduce qualitatively new effects; nor have we yet introduced a wind-drift layer at the surface. However it does appear possible to extend the numerical technique so as to include a surface shear at least.

So far we have considered only waves in deep water. In water of finite depth we may expect that the subharmonic instabilities will tend to be suppressed. On the other hand the local, plunging instabilities may very well be enhanced.

REFERENCES

Banner, M.L. and Melville, W.K. 1976 On the separation of air flow over water waves. J. Fluid Mech. 77, 825-842.

Benjamin, T.B. 1967 Instability of periodic wavetrains in nonlinear dispersive systems. Proc. R. Soc. Lond. A 299, 59-67.

Benjamin, T.B. and Feir 1967 The disintegration of wave trains on deep water. Part 1. Theory. J. Fluid Mech. 27, 417-430.

Cokelet, E.D. 1977a Steep gravity waves in water of arbitrary uniform depth. Phil. Trans. Roy. Soc. A 286, 183-230.

Cokelet, E.D. 1977b Numerical experiments on the deformation of steep water waves. Proc. I.M.A. Conf. on Mathematics in Oceanography, University of Bristol, June 1977.

Lighthill, M.J. 1967 Some special cases treated by the Whitham theory. Proc. R. Soc. Lond. A. 299, 28-53.

Longuet-Higgins, M.S. 1975 Integral properties of periodic gravity waves of finite amplitude. Proc. R. Soc. Lond. A. 342, 157-174.

Longuet-Higgins, M.S. 1977a The instabilities of gravity waves on deep water. I Superharmonics.
Submitted to Proc. R. Soc. Lond. A, Sept. 1977.

Longuet-Higgins, M.S. 1977b The instabilities of gravity waves on deep water II Superharmonics.
Submitted to Proc. R. Soc. Lond. A. Sept. 1977.

Longuet-Higgins, M.S. and Cokelet, E.D. 1976 The deformation of steep surface waves on water. I. A numerical method of computation.
Proc. R. Soc. Lond. A. **350**, 1-26.

Longuet-Higgins, M.S. and Cokelet, E.D. 1977 The deformation of steep surface waves on deep water II. Development of normal instabilities.
In preparation, Sept. 1977.

Longuet-Higgins, M.S. and Fox, M.J.H. 1977a Theory of the almost-highest wave. The inner solution.
J. Fluid Mech. **80**, 721-74.

Longuet-Higgins M.S. and Fox, M.J.H. 1977b Theory of the almost-highest wave. II. Matching and analytic extension.
(Submitted to J. Fluid Mech., Sept. 1977).

Longuet-Higgins, M.S. and Phillips, O.M. 1962 Phase velocity effects in tertiary wave interactions.
J. Fluid Mech. **12**, 333-336.

Longuet-Higgins, M.S. and Stewart, R.W. 1960 Changes in the form of short gravity waves on long waves and tidal currents.
J. Fluid Mech. **8**, 565-583.

Longuet-Higgins, M.S. and Turner, S.J. 1974 An entraining plume model of a spilling breaker.
J. Fluid Mech. **63**, 1-20.

Michell, J.H. 1893 The highest waves on water.
Phil. Mag. (5) **36**, 430-437.

Schwartz, L.W. 1974 Computer extension and analytic continuation of Stokes's expansion for gravity waves.
J. Fluid Mech. **62**, 553-578.

Stokes, Sir G.G. 1880 Supplement to a paper on the theory of oscillatory waves.
Mathematical and Physical Papers **1**, 314-326.
Cambridge Univ. Press.

DISCUSSION

P. A. Taylor:

In the computations you have shown of the instability of Stokes waves following an input of energy by surface pressure forces, how much energy was added? Is this likely to be a significant process leading to wave breaking in the sea?

M. S. Longuet-Higgins:

In our 1976 paper (P.R.S.A. 350, 1) the added energies were respectively 0·43, 0·61, 0·80 and 0·93 E_{max}. In Cokelet (1977) no energy was added, but the initial energies were 6/3, 5/3, 4/3, 3/3 and 2/3 E_{max}., the wave being initially sinusoidal. In our latest computations (Longuet-Higgins & Cokelet 1977) the waves began as steady, finite-amplitude waves and virtually no energy was added. The perturbations nevertheless grew to breaking-point. In natural conditions there is probably a slow, through patchy, input of energy from the wind to the steeper waves, accompanied by intermittent loss through breaking. We may soon be able to model this.

W. J. Pierson:

(1) Can these techniques be extended to short crested aperiodic waves?

(2) How big a domain influences the solution to the partial differential equations for each point in the solution?

M. S. Longuet-Higgins:

(1) The investigation of unstable modes by perturbation analysis (Section 4) could quite possibly be extended to three-dimensional motions. However, to do the time-stepping (Section 5) in three dimensions is probably not practicable at the present time.

(2) The solution is governed by an integral equation in which the kernel is equal to $\ln PQ$, where P and Q are points on the boundary of the contour C (P the given point, Q the integration point). So the influence of each point Q falls off rather slowly. Essentially this is because the fluid is incompressible, and what happens at any point on the boundary is felt immediately at other points of the fluid.

M. A. Donelan:

In attempting to produce a single whitecap in a wave tank by sending a short group of steep waves down the tank, I noticed a very strong instability qualitatively similar to your normal mode instability. However $\kappa h < 1{\cdot}363$ i.e. outside the range of the Benjamin and Feir instability. Is your instability for $ak > 0{\cdot}4$ limited to the same "kh" range as theirs?

M. S. Longuet-Higgins:

All my calculations so far have been for deep water ($kh \gg 1$). It should be quite practicable to extend the calculations to water of finite depth, and this is being started already. What one expects is that, in finite depth, the subharmonic instabilities (which involve large horizontal scales) will tend to be suppressed, starting first with those modes which in deep water would have the smallest rates of growth. The instabilities for $ak > 0{\cdot}4$ were, however, local, being concentrated near the wave crest. These will surely have their counterparts in shallow water, but we do not yet know how their growth rates will be affected.

THE DISPERSION RELATION FOR WIND WAVES IN DECAY AREA

H. Mitsuyasu, Yi-Yo Kuo and A. Masuda

Research Institute for Applied Mechanics
Kyushu University, Fukuoka, Japan

Random waves are generated by wind in the first half of a wind-wave flume. The latter half of the flume is kept free from wind to measure the wave characteristics unaffected by the wind and wind-generated current. The waves in the latter (decay) area are measured with a linear array of wave gauges, and their phase velocities and coherences are determined by a usual technique of the cross-spectral analysis. The measured results are compared with the third-order approximation to the theory of two-dimensional random waves, which we have developed recently (Masuda *et al.* 1977). Agreement between the theory and the experiment is satisfactory, and observed characteristics of the phase velocity and coherence of the spectral components can be attributed to the effects of the nonlinearity and angular dispersion of the random waves.

1. INTRODUCTION

To a first approximation, ocean waves can be described as a linear superposition of statistically independent free waves which have random phases and satisfy the linear dispersion relation. Obviously, however, the nonlinearity of the wave motion introduces a number of interesting phenomena, such as generation of trapped waves (Tick 1959), the energy transfer among wave components (Phillips 1960, Hasselmann 1962) and phase velocity effects (Longuet-Higgins and Phillips 1962). In recent years much interest has been paid to the nonlinear dispersion relation in a random wave field (Huang and Tung 1976, 1977, Weber and Barrik 1977, Barrik and Weber 1977). This is perhaps due to the increasing utilization of remote sensing techniques in oceans wave studies. The dispersion relation is urgently needed to relate the spatial characteristics measured with the remote sensors to the temporal characteristics measured by the conventional methods. In the actual wind wave field, however, the phenomenon is not so simple as we might expect even when we confine our attentions to the phase velocity of the random waves: Wind and wind-generated current affect the phase velocity of the wind waves (Shemdin 1972, Kato 1974).

In order to study the purely nonlinear effects of waves on the phase velocities of spectral components we have analyzed, in this paper, the data of laboratory wind waves in decay area where the wind and the wind-generated current are absent. Measured results have been compared with the nonlinear theory of two-dimensional random waves which we have developed recently (Masuda *et al.* 1977).

2. EQUIPMENTS AND PROCEDURE

The measurements were made in a wind-wave flume 13.4 m long, 0.6 m wide and 0.8 m high (figure 1). Water depth was 0.365 m.

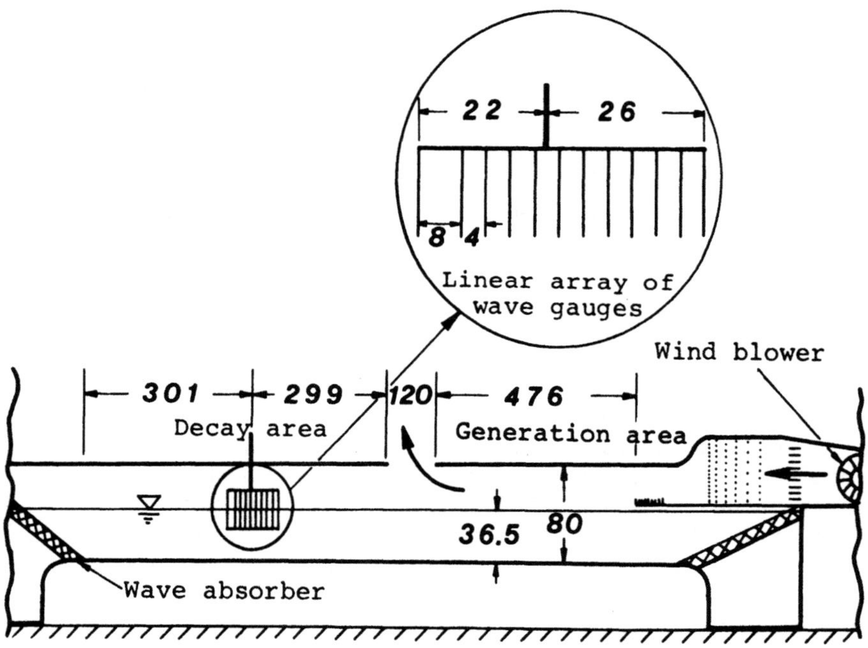

Fig. 1 Schematic diagram of wind-wave flume (unit in cm)

Wind waves were generated by wind in the first half of the flume, and the latter half of the flume was kept free from wind to measure the random wave field unaffected by the wind and drift current. The latter half of the flume was made airtight to eject the air flow smoothly from the ceiling gap at the middle part of the flume. It was confirmed that the drift current attenuates very rapidly in the decay area and is negligible at the location of the wave measurements. Waves were measured by using a linear array of twelve wave gauges aligned equidistantly with 4 cm intervals except for the last gauge (figure 1). Each probe of the wave gauge is of resistance-type, which consists of two parallel platinum wires with a diameter 0.1 mm and 2 mm space.

Winds with three different speeds U: 10, 12.5 and 15 m/sec were used for generating wind waves with different spectral peak frequencies. Here U is a cross-sectional mean speed of the wind in the flume. Wind-generated waves propagate from the generation area to the decay area in the latter half of the flume and finally absorbed by the wave absorber placed at the end of the flume. Waves were measured continuously for 13 minutes at the location shown in figure 1, and wave signals from twelve gauges were recorded simultaneously on a data recorder.

3. ANALYSIS OF THE WAVE DATA

The twelve wave records of each run were digitized simultaneously at sampling frequency 40 Hz. For the convenience of the spectral analysis, each of the wave data was divided into ten subsamples of 51.2 sec, which contained 2048 data points. In order to measure the phase velocity of the spectral components, cross-spectra

$$Co_{1n}(f) - iQ_{1n}(f) \ , \quad n = 2, 3, \cdots 12, \tag{1}$$

were computed through the ordinal fast Fourier transform procedure, by using the first wave data η_1 nearest to the generation area and the succeeding wave data η_n (n = 2, 3, ··· 12). In this way, we can compute the cross-spectra of waves measured at two points of different spacing (r = 4 cm, 8 cm, 12 cm, ···) along the dominant direction of wave propagation. Final data of the cross-spectra were obtained by taking sample mean of ten sub-samples of raw wave spectra and taking moving average of successive seven line spectra. Therefore, equivalent degrees of freedom of the spectra are approximately 140. By using the cross-spectra of waves, the phase lag $\theta(f)$, the coherence Coh(f) and the phase velocity C(f) of the spectral components were determined as

$$\theta(f) = \tan^{-1}[\ Q(f)/Co(f)\ ] \ , \tag{2}$$

$$Coh(f) = \{[\ Co^2(f) + Q^2(f)\ ] \,/\, \phi_1(f)\phi_n(f)\}^{1/2} \ , \tag{3}$$

$$n = 2, 3, \cdots 12$$

and $$C(f) = 2\pi f r/\theta(f) \tag{4}$$

where r is the distance between the wave gauges, i.e., r = 4, 8, ... 48 cm.

4. EXPERIMENTAL RESULTS

Typical examples of the measured results are shown in figure 2. From the top to the bottom, power spectra at two stations, normalized phase velocity C/C_0 and coherence are shown respectively as a function of the frequency f. Here, C_0 is the phase velocity of long-crested linear wave in finite depth.

Power spectra. Since the attenuation of the spectral energy is very small for relatively short distance of propagation ($r < 12$ cm), power spectra at two stations almost coinside with each other. Spectral peak frequency f_m is 2.7 Hz for the waves generated by the wind with U = 10 m/sec, and 1.8 Hz for those generated by the wind with U = 15 m/sec.

Coherence. The coherence of the waves measured at two stations are nearly unity for the frequency components near the spectral peak frequency f_m when the distance between two wave gauges is relatively short (r = 4 , 12 cm). Although the data for very long distance are not shown, the coherence near f_m decreases gradually with the increase of the distance between two wave gauges. The coherence also decreases gradually with the increase of the frequency of the spectral component. In addition to such gradual decreases, very rapid decrease of the coherence can be seen locally near 4.8 and 8.5 Hz for U = 10 m/sec, and near 3.4 Hz for U = 15 m/sec. These frequencies of local drops of the coherence correspond roughly to twice or three times the spectral peak frequency f_m.

Phase velocity. The normalized phase velocity C/C_0 is nearly equal to but slightly larger than unity in a relatively wide frequency range near the spectral peak frequency. That is, the phase velocities of energy-containing components of random waves are fairly close to those for the long-crested linear waves. It will be shown later that slight differences between C and C_0 can be attributed to the effect of angular spreading of random waves. However, the rapid increase of C/C_0 appears at the frequencies where the local decrease of the coherence is observed. Particularly, the increase of C/C_0 is very rapid for relatively large spacing of the wave gauges. These peculiar features of the coherence and phase velocity are attributed to the nonlinear effects of random

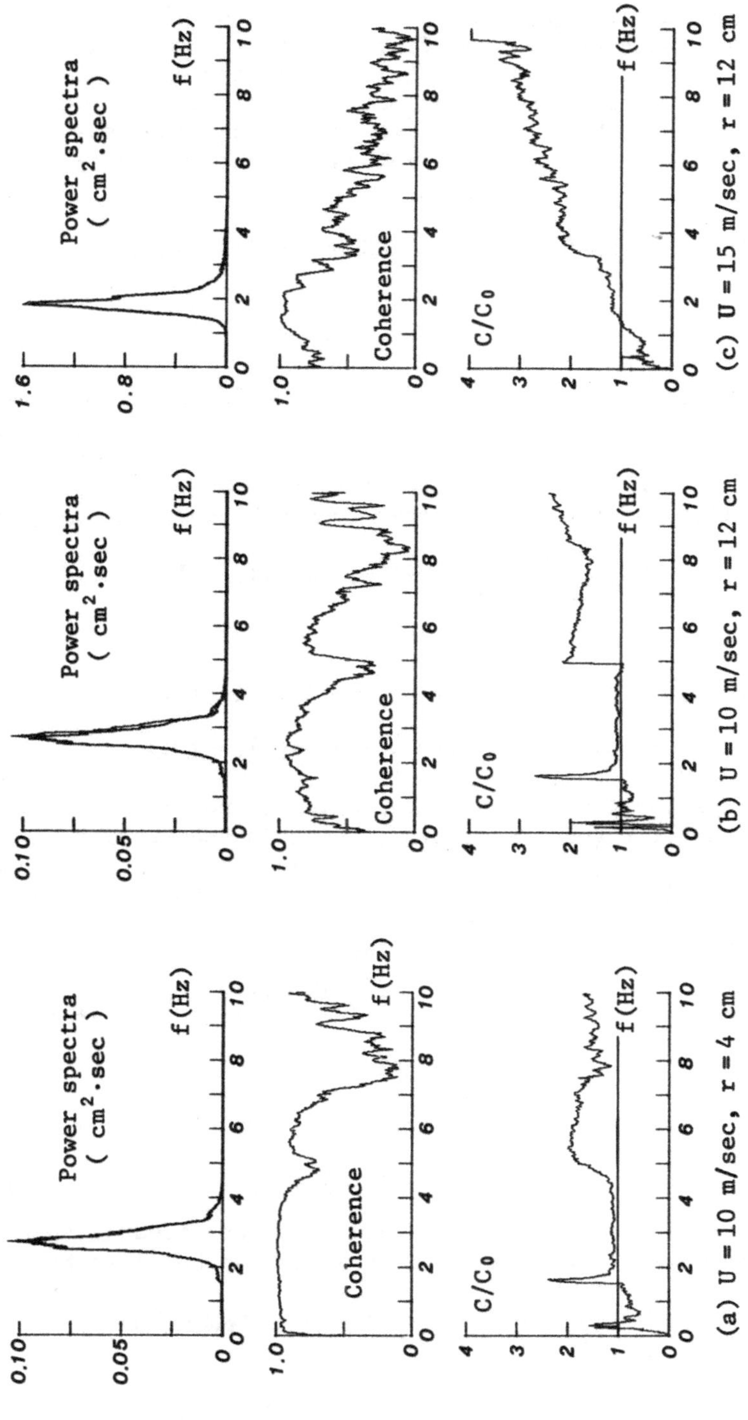

Fig. 2 Power spectra, coherence and phase

wave field, which will be discussed later. Anomalous value of C/C_0 at 1.6 Hz can be attributed to the transverse oscillation of water in the flume. Although the amplitude of the transverse oscillation is very small, it can not be negligible when the spectral density of wind-generated waves is very small at that frequency. Such phenomena are not seen for U = 15 m/sec, where the spectral density of wind-generated wave has moderate values at that frequency.

5. NON-LINEAR THEORY OF RANDOM GRAVITY WAVES

Outline of the Theory

In order to examine the dispersion relation of random gravity waves, we have developed a weakly non-linear theory of random waves to the third order along the same line as that of Phillips (1960) [Masuda *et al.* 1977]. It is shown that the wavenumber-frequency spectrum $\psi_2(\mathbf{k},\omega)$ of the second-order (forced) wave can be given by

$$\psi_2(\mathbf{k},\omega) = \int_{\mathbf{k}_1}\int_{\omega_1} 2\left\{\frac{f_2(\mathbf{k},\omega,\mathbf{k}_1,\omega_1)}{g-\omega^2/|\mathbf{k}|}\right\}^2 \psi_1(\mathbf{k}_1,\omega_1)\psi_1(\mathbf{k}-\mathbf{k}_1,\omega-\omega_1)d\mathbf{k}_1 d\omega_1 , \quad (5)$$

where

$$f_2(\mathbf{k},\omega,\mathbf{k}_1,\omega_1) = \frac{1}{2}\{\omega_1^2 - \omega_1\omega + \omega^2 - \omega_1(\omega-\omega_1)<\mathbf{k}_1,\mathbf{k}-\mathbf{k}_1>$$
$$- \omega\omega_1<\mathbf{k},\mathbf{k}_1> - \omega(\omega-\omega_1)<\mathbf{k},\mathbf{k}-\mathbf{k}_1>\} . \quad (6)$$

$\psi_1(\mathbf{k},\omega)$ is the spectrum of the first-order (free) waves with vector wavenumber $\mathbf{k}$ and frequency ω, and g is the acceleration of gravity. In (6) $<\mathbf{k},\mathbf{k}'>$ denotes the cosine of the angle formed by two vectors $\mathbf{k}$ and $\mathbf{k}'$. In the same approximation, the cross spectrum $Cr(\omega,\mathbf{r})$ of the surface displacements at two points separated by a distance $\mathbf{r}$ is given by

$$Cr(\omega,\mathbf{r}) = \int_{\mathbf{k}} e^{-i\mathbf{k}\cdot\mathbf{r}}\psi(\mathbf{k},\omega)d\mathbf{k} = \int_{\mathbf{k}} e^{-i\mathbf{k}\cdot\mathbf{r}}\{\psi_1(\mathbf{k},\omega)+\psi_2(\mathbf{k},\omega)\}d\mathbf{k}$$
$$= Cr_1(\omega,\mathbf{r}) + Cr_2(\omega,\mathbf{r}) . \quad (7)$$

The frequency spectrum $\phi(\omega)$ is given by

$$\phi(\omega) = Cr(\omega,\mathbf{0}) = \int_{\mathbf{k}}\psi(\mathbf{k},\omega)d\mathbf{k} = \phi_1(\omega) + \phi_2(\omega) . \quad (8)$$

To the third-order of approximation, a non-linear dispersion relation for weakly non-linear random waves can be given by

$$g - \frac{\omega^2}{|\mathbf{k}|} = \int_{\mathbf{k}_1}\int_{\omega_1}\frac{4\{f_2(\mathbf{k},\omega,k_1,\omega_1)\}^2}{\{g-(\omega-\omega_1)^2/|\mathbf{k}-\mathbf{k}_1|\}}\,\psi_1(\mathbf{k}_1,\omega_1)d\mathbf{k}_1 d\omega_1$$
$$+ \int_{\mathbf{k}_1}\int_{\omega_1}\{f_3(\mathbf{k},\omega,\mathbf{k}_1,\omega_1,-\mathbf{k}_1,-\omega_1) + f_3(\mathbf{k},\omega,\mathbf{k}_1,\omega_1,\mathbf{k},\omega)$$

$$+ f_3(\mathbf{k},\omega,\mathbf{k},\omega,\mathbf{k}_1,\omega_1)\}\psi_1(\mathbf{k}_1,\omega_1)d\mathbf{k}_1 d\omega_1 \quad . \tag{9}$$

Here $f_3(\mathbf{k},\omega,\mathbf{k}_1,\omega_1,\mathbf{k}_2,\omega_2)$ denotes

$$\begin{aligned} f_3(\mathbf{k},\omega,\mathbf{k}_1,\omega_1,\mathbf{k}_2,\omega_2) &= \frac{1}{4}(\omega_1^2|\mathbf{k}_1| + \omega_2^2|\mathbf{k}_2|) - \frac{1}{4}(\omega\omega_1|\mathbf{k}_1| + \omega\omega_2|\mathbf{k}_2|) \\ &- \frac{1}{2}(\omega_1+\omega_2)|\mathbf{k}_1+\mathbf{k}_2|\{\omega_1<\mathbf{k}_1,\mathbf{k}_1+\mathbf{k}_2> + \omega_2<\mathbf{k}_2,\mathbf{k}_1+\mathbf{k}_2>\} \\ &- \frac{1}{2}\omega(\omega_1+\omega_2)\frac{|\mathbf{k}_1||\mathbf{k}_2|}{|\mathbf{k}|}<\mathbf{k}_1,\mathbf{k}_2> \\ &+ \frac{1}{2}\omega|\mathbf{k}_1+\mathbf{k}_2|<\mathbf{k},\mathbf{k}_1+\mathbf{k}_2>(\omega_1<\mathbf{k}_1,\mathbf{k}_1+\mathbf{k}_2> + \omega_2<\mathbf{k}_2,\mathbf{k}_1+\mathbf{k}_2>). \end{aligned} \tag{10}$$

The relation (9) reduces to the form derived previously by Longuet-Higgins and Phillips (1962) for the case of a uni-directional wave field (Masuda *et al.* 1977).

Practical Computations

In order to apply the preceding theory to the analysis of the measured waves, it is necessary first to know $\psi_1(\mathbf{k},\omega)$ and $\psi_2(\mathbf{k},\omega)$. Since the measured spectrum is the results of both free waves and forced waves, we must separate them. The separation is possible by the iterative method. Within the present approximation we consider that the measured frequency spectrum $\phi_{obs}(\omega)$ is given by

$$\phi_{obs}(\omega) = \phi_1(\omega) + \phi_2(\omega) \quad , \tag{11}$$

where $\phi_1(\omega)$ and $\phi_2(\omega)$ are the spectra of free and forced waves respectively. We put the zero-th order of iteration as follows:

$$\phi_2{}^0(\omega) = 0, \ \psi_2{}^0(\mathbf{k},\omega) = 0, \ \phi_1{}^0(\omega) = \phi_{obs}(\omega) \quad , \tag{12}$$

and

$$\psi_1{}^0(\mathbf{k},\omega)d\mathbf{k}d\omega = \phi_{obs}(\omega)\cdot G(\omega,\theta)\delta(g|\mathbf{k}| - \omega^2)d|\mathbf{k}|d\theta\, d\omega \quad , \tag{13}$$

where $G(\omega,\theta)$ denotes an angular distribution function of the free waves and δ is the delta function. The following form of $G(\omega,\theta)$ is assumed in the present study;

$$G(\omega,\theta) = \begin{cases} G'(m)\cos^m\theta & |\theta| \leqslant \pi/2 \\ 0 & |\theta| > \pi/2 \end{cases} \tag{14}$$

where $G'(m)$ is a normalizing function. Substituting (13) into (5) we can compute the first-order of iteration $\psi_2{}^1(\mathbf{k},\omega)$ and $\phi_2{}^1(\omega)$. The first-order of iteration for $\phi_1{}^1(\omega)$ can be obtained by

$$\phi_1{}^1(\omega) = \phi_{obs}(\omega) - \phi_2{}^1(\omega) \ . \tag{15}$$

Repeating the same procedures we can compute higher-order of iterations successively. Fortunately, this iterative method yields very rapid convergence and the second order of iteration is sufficient for our present purpose.

Practical computations are summarized as follows; we assume $m = 2, 4, 6$ for the angular distribution function (14), compute $\phi_1(\omega)$ and $\phi_2(\omega)$ by the iterative method mentioned above, and determine the cross spectrum $Cr(\omega,r)$ [$= Cr_1(\omega,r) + Cr_2(\omega,r)$] by substituting $\phi_1(\omega)$, $\phi_2(\omega)$ and $G(\omega,\theta)$ into (7). Finally, in order to compare the theory with the measurements, the coherence and the phase velocity are computed from (3) and (4) by using the cross spectrum. In the third order approximation we consider the non-linear dispersion relation and use

$$\delta(g|\mathbf{k}| - \omega^2 + \epsilon(\omega,\theta)) \ , \tag{16}$$

instead of $\delta(g|\mathbf{k}| - \omega^2)$, where $\epsilon(\omega,\theta)$ is an increment of $g|\mathbf{k}|$ deduced by the non-linear dispersion relation (9).

6. COMPARISON BETWEEN THEORY AND MEASUREMENTS

Effects of angular spreading. First, the effects of angular spreading of waves on the phase velocity and coherence are studied by comparing the measured results with the predictions of linear theory. Since nonlinear effects are expected to be negligible near the spectral peak frequency, such analysis can shed light on the characteristics of the energy-containing components of wind waves. Figure 3 shows an example of the results, where the angular distribution function is assumed as $\cos^m\theta$, $m = 2, 4, 6$. The figure shows that the angular distribution function of the form $\cos^4\theta$ gives the best fit to the data both for C/C_0 and for the coherence near the spectral peak frequency ($f/f_m = 1$). Such a comparison was also done for the data of other cases: different spacings of wave gauges and different wind speeds. It was found that the linear theory can explain the measured data in the following frequency range; $f/f_m = 0.8 - 1.6$ for $U = 10$ m/sec, and $f/f_m = 0.9 - 1.3$ for $U = 15$ m/sec. It was also found that the angular distribution function of the form $\cos^4\theta$ fits the data of the former case and that of the form $\cos^2\theta$ fits the data of the latter case. Figure 4 shows the relation between the coherence of the frequency components at the spectral peak and the separation distance of wave gauges. The distance r is normalized in the form kr, where $k = 2\pi/L$ is the wave number. The relations predicted by the linear theory are also shown in the same figure, where the angular distribution function was assumed again as $\cos^m\theta$, $m = 2, 4, 6$. Figure 4 shows that the angular distribution function $\cos^4\theta$ fits to the data for $U = 10$

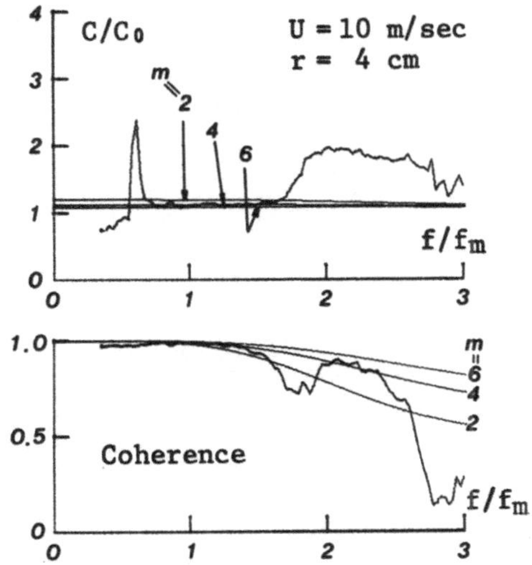

Fig. 3 Phase velocity and coherence compared with the theoretical predictions

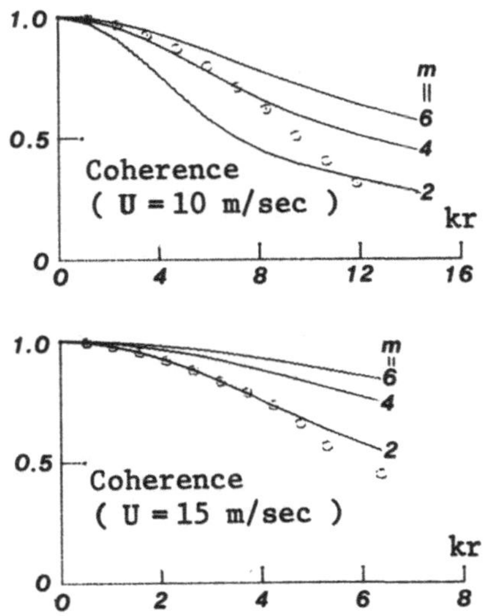

Fig. 4 The relation between the coherence at f_m and the separation distance

m/sec and $\cos^2\theta$ to the data for U = 15 m/sec. At large separation distance of wave gauges (kr > 5), however, the power m changes gradually with the increase of the separation distance. Closer investigation of the original data of the coherence showed that such phenomena are related to the attenuation of spectral energy. With increasing the distance between two wave gauges the attenuation of the wave energy near the spectral peak increases gradually. Slight but anomalous decrease of the coherence was found in a frequency range where the attenuation of the wave energy was considerable.

Nonlinear effects. Although the linear theory of two-dimensional random waves can give fairly good explanation for our measured results, the validity is confined to the dominant part of the wave spectrum near the spectral peak. That is, either rapid increase of the phase velocity or corresponding local decrease of the coherence can not be explained by the linear theory. Therefore, the results are compared, in the next step, with the non-linear theory outlined in 5.

Figure 5 shows the results of such comparison. Each figure shows, from the top to the bottom, the normalized phase velocity C/C_0, the coherence, and the normalized power spectrum $\phi f_m/E$ at the upward station, respectively as the function of f/f_m. Here, E is the total energy contained in the wave spectrum. In each figure, the curve (2) corresponds to the prediction of the second-order theory of two-dimensional random waves. The angular distribution

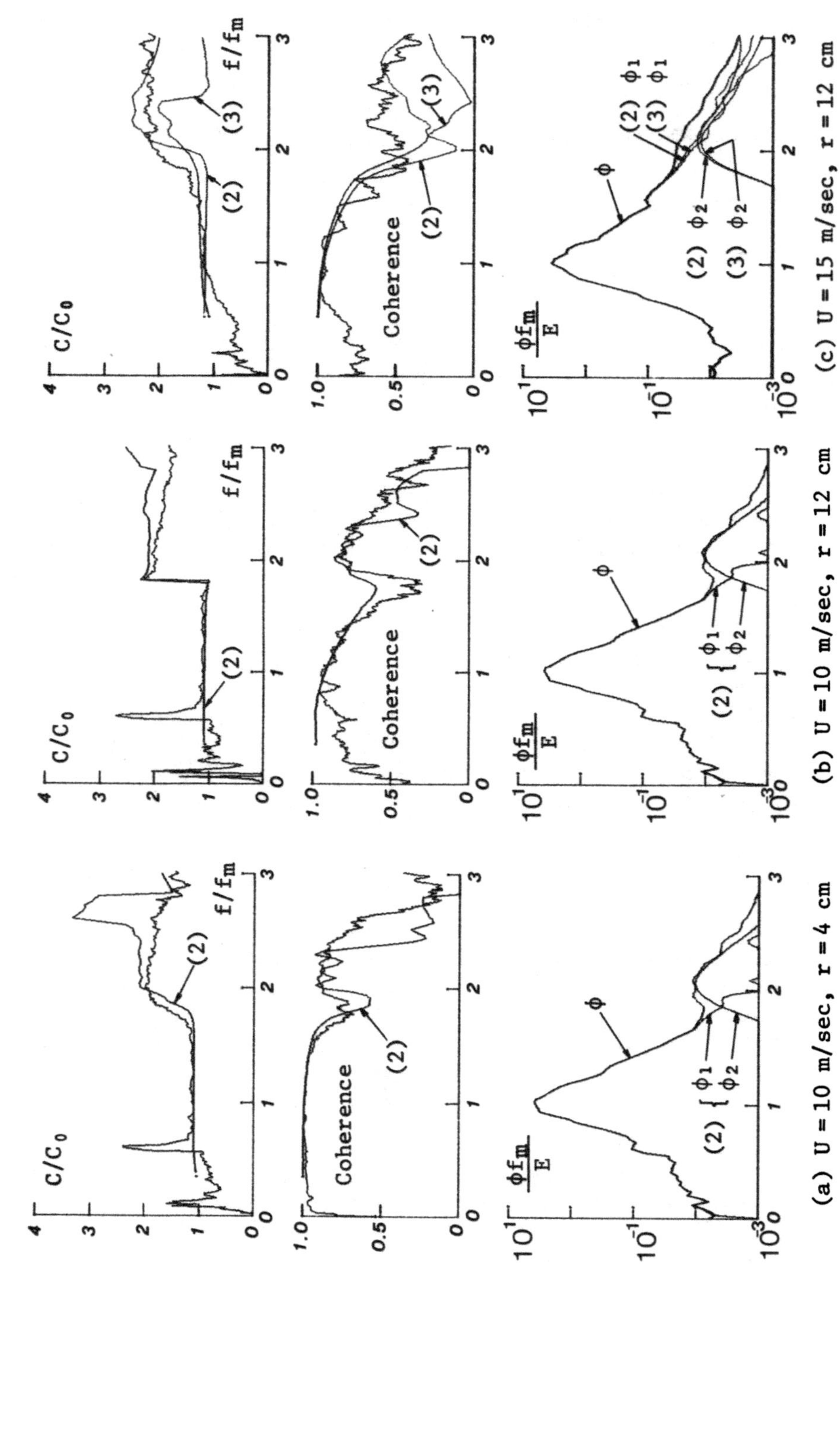

Fig. 5 Phase velocity, coherence and power spectra

function was assumed to be proportional to $\cos^4\theta$ for U = 10 m/sec, and $\cos^2\theta$ for U = 15 m/sec. Figure 5 shows that the second-order theory gives good explanations of our measured results.

For the case of U = 15 m/sec, two theoretical curves, the second-order approximation denoted by (2) and the third order approximation denoted by (3), are compared with the measurements. The third-order approximation shows better agreement to the measurement in the dominant part of the spectrum, but the second-order approximation is better in the high frequency region ($f \gtrsim 2f_m$), although the reason is not clear. From the results shown in figure 5 we can conclude that our measured results can be explained by the non-linear theory of two-dimensional random waves. Furthermore, it can be seen that the spectral energy near the frequency $f/f_m = 2$ can be largely attributed to the nonlinear effects. In other words, they are mainly due to forced waves not to free waves. Very rapid increase of C/C_0 and rapid local decrease of the coherence occur at the frequency where the spectral density of the free waves $\phi_1(f)$ coinside to that of the forced waves $\phi_2(f)$. It is also quite natural that the measured C/C_0 is approximately 2 for the frequency range $1.8 < f/f_m < 3$, because the spectral components in this range are almost due to the forced waves accompanied by the dominant waves in a frequency range $0.9 < f/f_m < 1.5$.

CONCLUSIONS

The conclusions of this study are summarized as follows: For the wind waves in decay area, the phase velocity and the coherence of the energy-containing components near the spectral peak are very close to those given by the linear theory. However, spectral energy in a high frequency region ($f > 1.8\ f_m$) is largely due to the second-order (forced) waves and contributions of the free waves are relatively small. Both the rapid increase of the phase velocity and the rapid local decrease of the coherence occur at the frequency near $1.8\ f_m$, both of which are due to the effects of the second-order (forced) waves coexisting with the free waves.

The authors are indebted to Mr. T. Honda, Mr. K. Eto, Mr. M. Tanaka and Miss N. Uraguchi for their assistance in various stage of this study. They also wish to express their appreciation to Dr. H. Honji for his invaluable comments on the first draft of the paper.

REFERENCES

Barrick, D. E. and Weber, B. L. 1977, J. Phys. Oceanogr., 7, 11-21.
Hasselmann, K. 1962, J. Fluid Mech., 12, 481-500.

Huang, N. E. and Tung, C. 1976, J. Fluid Mech., 75, 337-345.
Huang, N. E. and Tung, C. 1977, J. Phys. Oceanogr., 7, 403-414.
Kato, H. 1974, Rep. Port & Harboure Res. Inst., 13, 3-32.
Longuet-Higgins, M. S. and Phillips, O. M. 1962, J. Fluid Mech., 12, 333-336.
Masuda, A., Kuo, Yi-Yu and Mitsuyasu, H. 1977, (Unpublished).
Phillips, O. M. 1960, J. Fluid Mech., 9, 193-217.
Shemdin, O. H. 1972, J. Phys. Oceanogr., 2, 411-419.
Tick, L. J. 1959, J. Math. Mech., 8, 643-651.
Weber, B. L. and Barrick, D. E. 1977, J. Phys. Oceanogr., 7, 3-10.

DISCUSSION

E. J. Plate:

Is the exponent in the coherency actually an effect of the directional form of the spectrum, or could it be an effect of the "noise" which is gradually decaying?

H. Mitsuyasu:

(1) Our measured data of c/c_o and coherence can be explained consistently by this form of the angular distribution function. Therefore, this is not due to the effect of the noise.

(2) Our independent observation of the angular distribution function by rotatable linear array gives almost the same results.

W. J. Pierson:

(1) Did your pictures show capillary waves in the wind area that died out in the no-wind area?

(2) What caused the spike in c/c_o at 0.7 hertz?

H. Mitsuyasu:

(1) Yes, capillary waves are clearly observed in wind area of our experimental flume.

(2) Anomalous spike in c/c_o at 1.6 Hz (normalized frequency $f/f_m = 0.7$) is attributed to the transverse oscillation of water in the flume. Although the amplitude itself is very small it cannot be negligible when the spectral density of the wind waves is very small at that frequency.

O. Shemdin:

One observes non-linear spectral peaks in the wave heigth spectrum even in the generation area. Have you applied your higher order spectral calculations to wave spectra in the active wind generation region?

H. Mitsuyasu:

Yes we did. As you mentioned, a non-linear spectral peak in the wave spectrum has been clearly observed.

If we take the effect of drift current on the phase speed of free waves into account, our theory gives a fairly good explanation of this phenomenon.

F. Ramamonjiarisoa:

Is the theory different from the theory Huang and Tung?

H. Mitsuyasu:

The initial formulation is the same, but the results are different from their results.

K. Hasselmann:

I wonder if you would like to comment on the relative energy level of free waves and second order nonlinear forced waves in the ocean. I seem to recall that the linear dispersion relation for ocean wind waves - which one can measure, for example, with pitch-roll buoys or HF backscatter - is very well satisfied for frequencies well beyond the double-peak frequency.

H. Mitsuyasu:

According to our wind wave data obtained at the ocean research tower, the contribution of the free waves, in many cases, seems to be larger than that of the second order nonlinear waves (although the latter is not negligibly small). That must be a reason why your data show the linear dispersion relation up to such a high frequency as $f \geq 3$ fm.

MODULATION OF CENTIMETRIC WAVES BY LONG GRAVITY WAVES: PROGRESS REPORT ON FIELD AND LABORATORY RESULTS

Omar H. Shemdin

Jet Propulsion Laboratory-California Institute of Technology

The modulation of short wind waves (centimeter to decimeter long) by long gravity waves is investigated under field and laboratory conditions. The field study employed a wave follower capable of tracking ocean waves with frequencies less than 1.0 Hz and heights less than 2.0 m. A high response laser-optical system was used to detect upwind-downwind and cross-wind slopes of short waves. Cross correlations of short wave slopes with long waves reveal the presence of definite coupling between the two. A well defined peak appears in the cross-correlation function at 45° downwind of the long wave crest, suggesting higher density waves there.

The laboratory investigation was conducted with wind over periodic long waves. The periodicity of the latter allowed determination of statistically meaningful energy spectra of short waves at different phase locations along the long wave profile. The laboratory results appear to be consistent with the relaxation theory introduced by Keller and Wright (1975). The relaxation constant is computed from the data and found to have dependencies on wave frequency and wind speed. Preliminary comparisons between field and laboratory results are discussed.

1. INTRODUCTION

One of the potentially most important developments in the history of ocean wave measurements is the advent of remote sensing using active microwave and HF techniques. These methods can provide information on wave height, length and direction over large spans of the ocean under active weather conditions, regardless of time of day or night. Figure 1 is a wave image obtained with an L-band imaging radar during hurricane Gloria in September 1976. The figure also shows its Fourier transform delineating the dominant wave direction within 180°. The latter can be uniquely resolved by a process known as "focusing" performed on the radar optical correlator. The L-band imaging radar transmits microwaves 23.0 cm long. The return is primarily from short waves of the same length through Bragg scattering. Other wavelengths contribute through geometrical optics. The intensity and source of return is dependent on the radar inclination angle. Similarly, an X-band imaging radar at a high incidence angle derives its return primarily from 3.0 cm waves due to its higher transmitted frequency. Clearly, the short wave structure of the ocean surface and its modulation by long waves is primarily responsible for imaging the long waves. A better understanding of short wave modulation by long waves is of central importance in this investigation.

The importance of short waves in relation to atmospheric transfer from air to water has recently been suggested by Valenzuela and Wright (1976), and Garrett and Smith (1976). In a recent collaborative study in the Bight of Abaco, Snyder et al. (1977) found the magnitude of atmospheric transfer due to normal pressure to be of order 0.2 τ_0 where τ_0 is the surface wind stress. This recent finding implies that 0.8 τ_0 is supported by short waves. Short waves can contribute to the growth of long waves through a maser type mechanism originally suggested by Longuet-Higgins (1969). The strength of the energy transfer from short waves to long waves remains a subject of active research with contributions reported by Hasselmann (1971), Valenzuela and Wright (1976) and Garrett (1976). The most realistic estimates for such a transfer are of order 0.1 τ_0 or less. Assuming this to be accurate and taking note of losses due to wave breaking and turbulence, it would appear that at most 0.7 τ_0 is transferred from wind to surface drift through short waves. Our present knowledge of the details associated with atmospheric transfer to waves and through waves to surface drift is lacking. There is a need for accurate measurements of the capillary and short gravity wave structure in the ocean. Remote sensing potentially offers such a capability, and yet verification of remote sensing techniques requires careful surface truth measurements of the capillary and short gravity wave structure. The recent interest in both remote sensing and

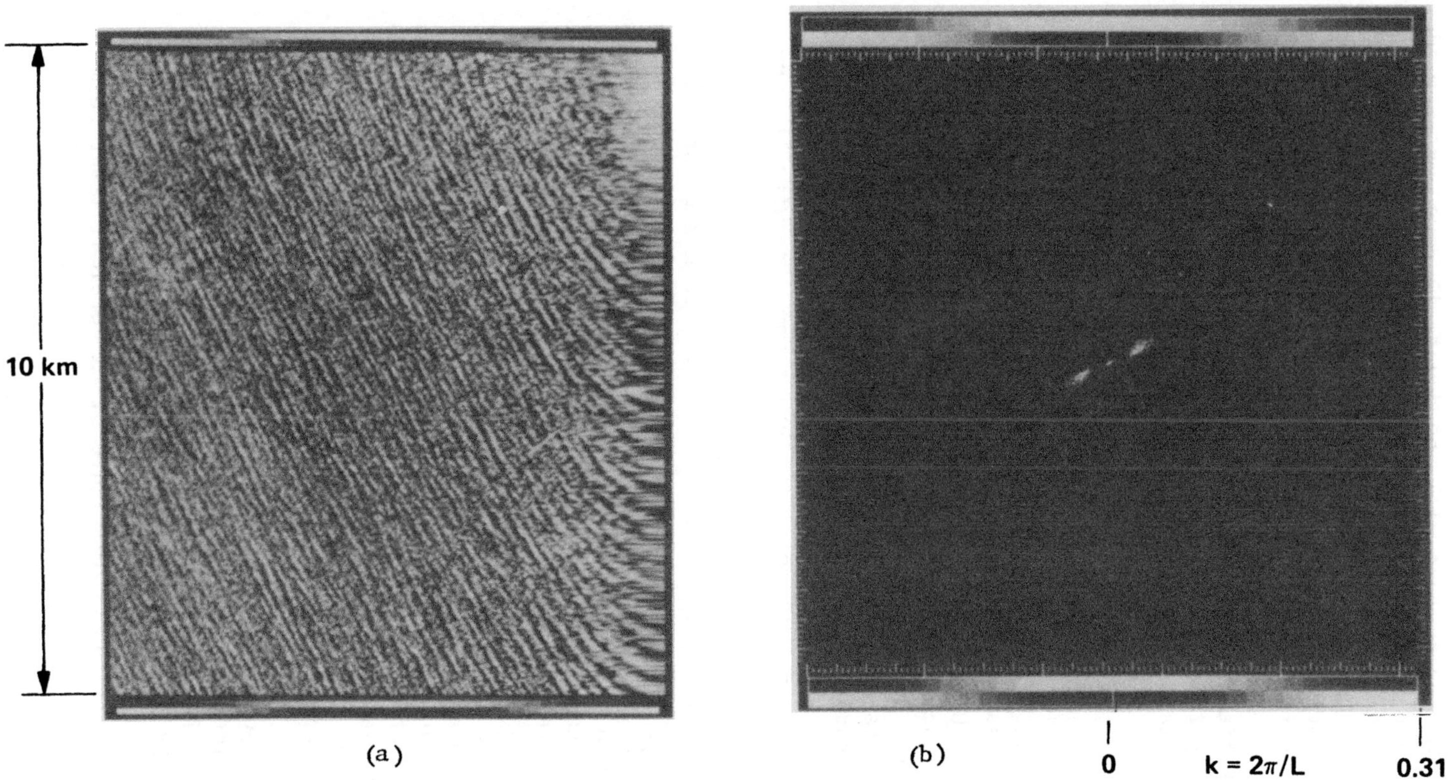

Figure 1. L-band synthetic aperture radar image of hurricane Gloria on 30 September 1976. Aircraft located 94 km ahead of eye in direction of hurricane travel at 9.15 km altitude.
(a) Image showing dominant waves 225 m long.
(b) Two-dimensional Fourier transform of image.

ocean surface microstructure will, no doubt, improve our knowledge of the dynamics of the ocean surface layer.

The present paper constitutes a progress report on the laboratory work conducted at the University of Florida wind and wave tank and on the field work conducted at Marineland, Florida in 1975. The laboratory findings are a key to our ability to analyze and interpret field data. They are discussed here and compared with laboratory radar measurements and also short wave measurements obtained in the field.

2. INSTRUMENTATION

Surface waves in the capillary and short gravity range are generally considered to have wave lengths ≤ 1.0 m and free wave frequencies ≥ 1.0 Hz. Table 1 gives wavelengths and frequencies of standard radars used in remote sensing of the ocean surface. Bragg scattering dominates the radar return signal at high depression angles so that only short surface waves have a direct impact on the radar return signal under such conditions.

Experimental investigations of short waves require the use of high frequency response wave sensors. Capacitance wave gages are used successfully in gravity wave studies, but are limited in frequency response for short wave studies. Sturm and Sorrell (1973) demonstrated that the frequency response of a capacitance gage was flat for waves longer than 8.0 cm (frequencies less than 4.5 Hz) as shown in Figure 2. The instrument response becomes even more important in studies of modulation of short waves by long waves. Here a short wave with wave number, k, measured by a stationary sensor will correspond to a variable encounter frequency along the long wave profile. A typical ocean wave with 1.0 m amplitude and 10 sec period will have an orbital velocity 62.8 cm/sec in an infinite depth situation. An 8.0 cm long short wave will have an oscillating encounter frequency with respect to the stationary sensor in the range 4.5 ± 49.3 Hz. A capacitance gage of the type tested by Sturm and Sorrell will provide a distorted wave height signal over most of the long wave profile.

A high response laser-optical wave slope detector was developed at the University of Florida to facilitate studies of short surface waves. The early version of this sensor was described by Tober et al. (1973) and an improved model was described by Palm (1975). The most recent instrument model used in both wave tank and field studies is shown in Figure 3. A vertical laser beam originating underwater is used. The beam is bent by refraction at the surface when waves are present. The deflection angle is proportional to the surface wave slope and is detected by an

Table 1. Radar frequencies and corresponding Bragg wave frequencies and lengths

Band	Radar Frequency (GHz)	Surface Waves λ (cm)	Surface Waves f (Hz)
Ka	35.0	0.85	30.7
Ku	13.30	2.3	10.0
X	9.30	3.2	8.0
C	5.25	5.7	5.5
L	1.20	23.0	2.6

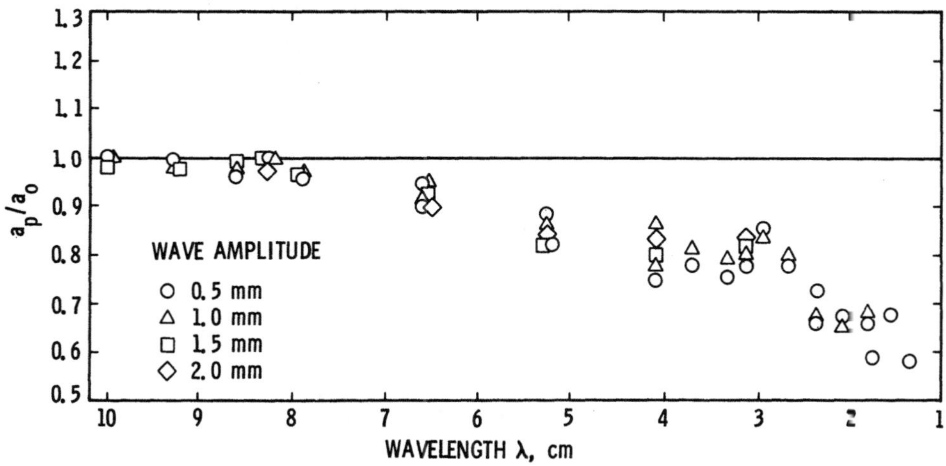

Figure 2. Response of a capacitance wave gage to short surface waves [after Sturm and Sorrell (1973)].

optical receiver with a photo diode which determines the X-Y coordinate values of the deflected beam. The frequency response of the sensor is limited by the size of the laser beam spot on the water surface. The sensor is estimated to have a flat response up to 80.0 Hz. A notable design feature of the laser-optical instrument is its insensitivity to mean water surface displacements to within ± 30 cm. The instrument is especially suitable for the study of modulation of short waves by long waves provided that the long wave amplitude does not exceed 30 cm. Useful laboratory experiments were conducted in this range. The results are discussed in (3).

The amplitude constraint is stringent for field studies where 1.0 m wave amplitudes occur frequently. To offset such a limitation the laser-optical device was mounted on a wave

(a)

(b)

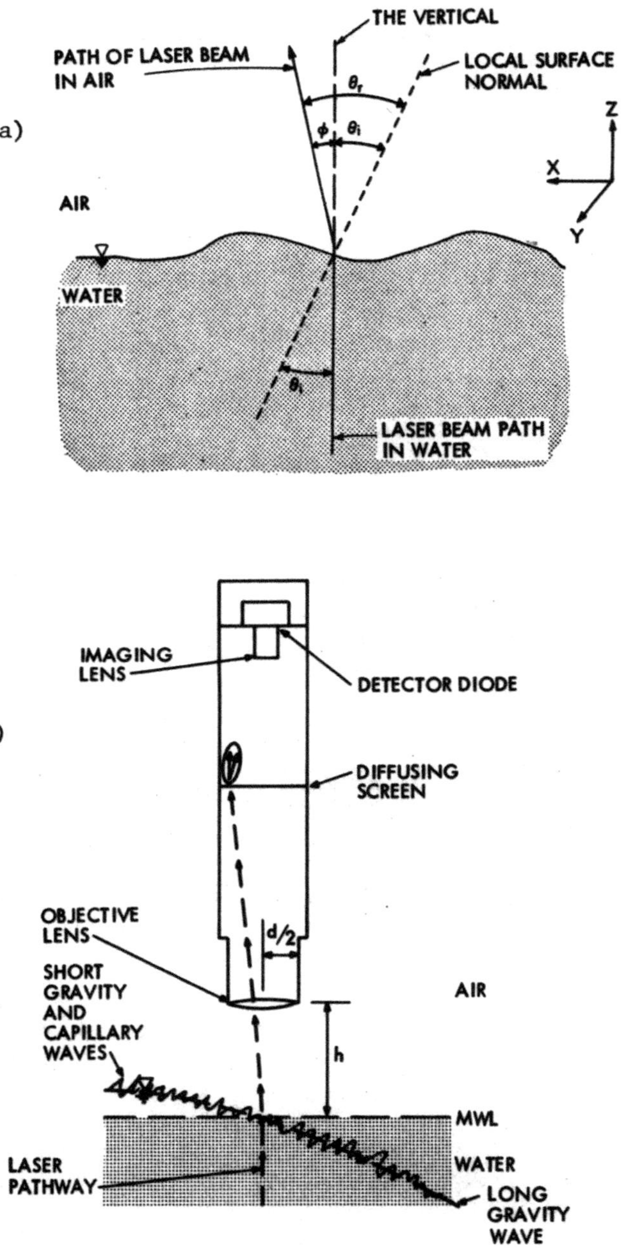

Figure 3. Laser-optical device for detection of short wave slopes.
(a) Geometry of operation.
(b) Schematic design of optical receiver.

following platform capable of tracking long ocean waves with frequencies longer than 1.0 sec to within a tracking accuracy of ± 10 cm. The wave follower design specifications and capabilities were reported by Tober and Shemdin (1977). The laser-optical instrument, when mounted on the wave follower, allows for the measurement of short waves with amplitudes ± 20 cm and their modulation by long ocean waves. Short wave modulation measurements with the wave follower and laser-optical system were obtained offshore from Marineland, Florida. The apparatus, shown in Figure 4, was deployed in 10.0 m water depth. The results are discussed in (4).

3. LABORATORY RESULTS

A detailed account of the laboratory experiment was given by Reece (1976, 1977). Briefly, the laser-optical device was placed in a stationary mode in the wind and wave facility. Periodic long waves were generated mechanically with a 2.0 second period and 10 cm wave height. Wind was then introduced in small increments with a controllable blower. At each wind speed the upwind-downwind and cross-wind components of surface slopes were measured by the laser-optical system and recorded. Simultaneously, long waves were measured with a capacitance wave gage and recorded on the same data acquisition system. The data on short wave intensities and their modulations were then analyzed by digital techniques using the IBM-360 computer.

The periodic nature of the mechanically generated long waves allowed for the segmenting of the continuous data record with respect to phase locations along the long wave profile. It was found by Reece (1977) that eight segments provided sufficient definition of the short wave spectra for frequencies ≥ 5 Hz. Statistical confidence in slope spectral densities were achieved by averaging spectra at corresponding phase locations along consecutive long waves. In Figure 5 the data segments are shown in 5(a). A short wave spectrum measured over the trough is compared to one measured over the crest in 5(b). The comparison reveals a frequency shift in the crest spectrum to higher frequencies compared to the trough spectrum. The latter is attributed to the increase in crest encounter frequency imposed by long wave orbital velocity. To substantiate this the crest spectrum was corrected for the orbital velocity difference between crest and trough of the long wave. The modified crest spectrum, also shown in Figure 5(b), has the same general shape, especially at high and low frequencies relative to the peak, compared to the trough spectrum. The following simple frequency-wave number relationship was used in the transformation.

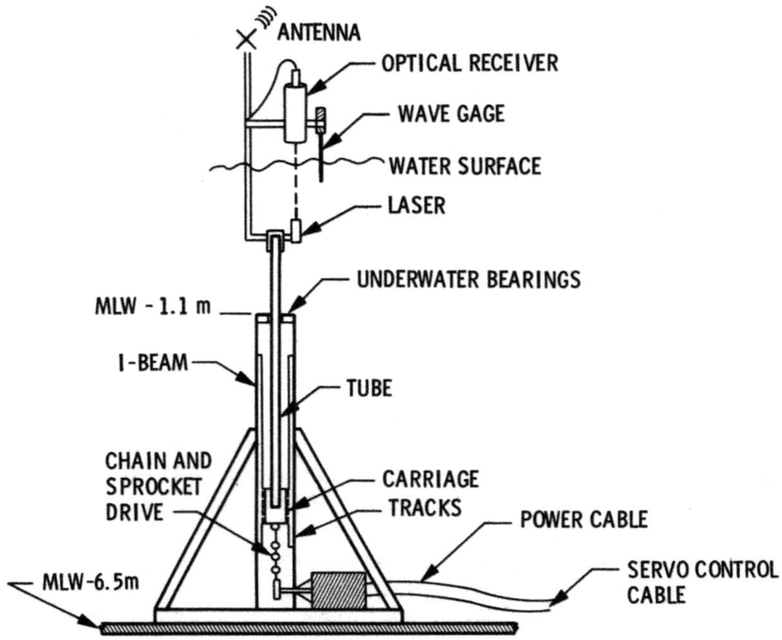

Figure 4. Apparatus for detecting short waves in the ocean.

$$\sigma_s = \sigma_o + k_s U_s + k_s U_o \, , \tag{1}$$

where σ_o is the free short wave frequency defined by the dispersion relationship in terms of wave number, k_s, gravity, g, surface tension, T, and water density, ρ.

$$\sigma_o^{\,2} = g k_s + \frac{T k_s^{\,3}}{\rho} \quad . \tag{2}$$

In Eq. (1) U_s and U_o are the wind-induced drift and orbital velocity of long waves, respectively. Eq. (1) provides the type of transformation needed to compare short wave slope time histories and spectral densities to wavelength information and cross-section return obtained with radars and photographic techniques. The important consequence of Eq. (1) is that the orbital component of encounter frequency, $k_s U_o$, can be significantly greater than the mean frequency, σ_o, especially under field conditions. Demodulation of short waves becomes necessary if space and time measurements are to be correlated.

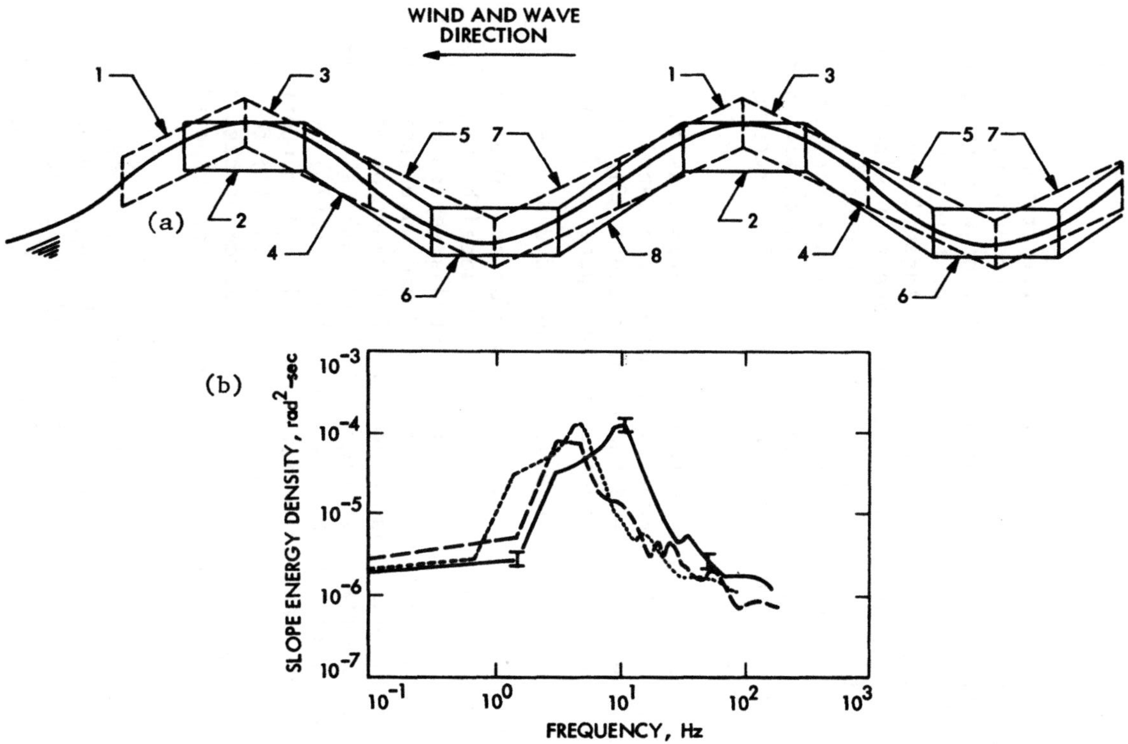

Figure 5. (a) Segments of the long wave profiles used for calculating short wave spectra.
(b) Crest spectrum (———) corrected for orbital velocity to correspond to an equivalent trough spectrum (--------) and compared to measured trough spectrum (— — —) [after Reece (1977)].

It is convenient to distinguish between short waves in terms of their free frequencies, σ_o, or wave numbers, k_s, as given by Eq. (1). The orbital velocity components of a modulated short wave may be averaged out to yield mean intensities over the long wave profile. The corresponding mean frequency is

$$\sigma_s = \sigma_o + k_s U_s , \tag{3}$$

so that a wave number, k_s, will have an encounter frequency dependent on wind speed due to surface drift. Accounting for this effect it is possible to obtain the variation of mean slope intensities at different wind speeds in terms of free wave frequencies, σ_o. The mean upwind-downwind slope intensities, ϕ_o, are graphed in Figure 6 and show linear dependence on wind speed. The apparent saturation at $f_o = \sigma_o/2\pi = 6.25$ Hz cannot be explained and is not consistent with similar measurements reported previously (Shemdin, 1974). The linear increase in ϕ_o with wind speed is consistent with positive correlation of SCAT return signal with wind speed.

A normalized modulation amplitude, m, is defined as the ratio of modulation amplitude density to mean slope density, ϕ_o. The dependence of m on wind speed and σ_o is shown in Figure 7. A monotonically decreasing dependence on wind speed is observed at frequencies, $f_o = \frac{\sigma_o}{2\pi}$ equal to 9.4, 15.6 and 20.0 Hz. At $f_o = 5.0$ Hz the m values show a peaking behavior at 6.0 m/sec suggesting a resonance type phenomenon such as noted by Plant, et al. (1977). Keller and Wright (1975) conducted a series of experiments in the University of Florida wind and wave facility. They operated their X-band radar over a similar range of wind speeds and mechanical waves. In Figure 8 their results are compared to wave slope intensities obtained by Reece (1976) at $f_o = 9.4$ Hz. The agreement is satisfactory and lends support to the proposed demodulation technique suggested by Eq. (1) for transforming frequency to wave numbers. The agreement also lends support to the Bragg scattering theory used to infer properties of short surface waves from radar measurements.

The phase of the slope intensity modulations is shown in Figure 9 for different f_o and wind speeds. Although trends can be discerned at each f_o, it can only be stated confidently that the phases are in the range 40° - 140° downwind of the long wave crest.

The results shown in Figure 7 highlight the important revelation that the normalized modulation, m, can be on the order of 1.0 at low wind speeds. It is one order of magnitude larger than predicted by the two-scale hydrodynamic theory of Longuet-Higgins (1969)

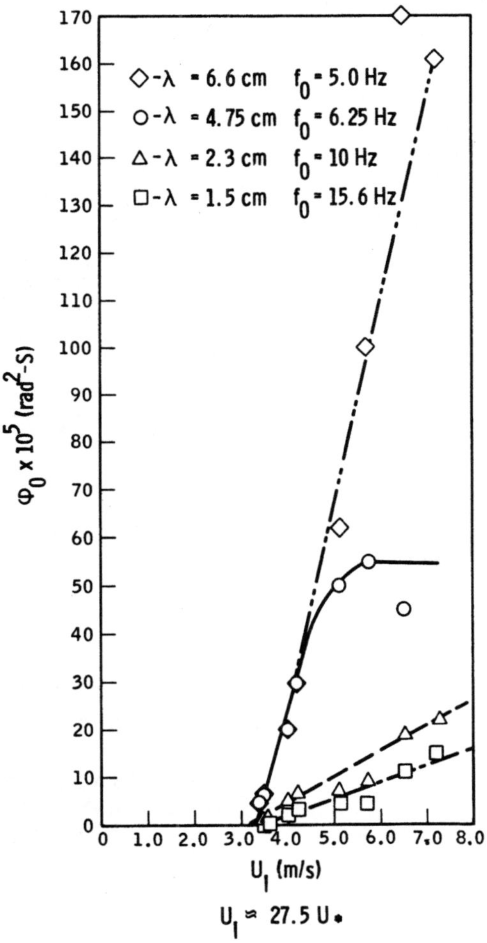

Figure 6. Mean upwind-downwind slope intensity as a function of wind speed for different short wave free frequencies, f_o.

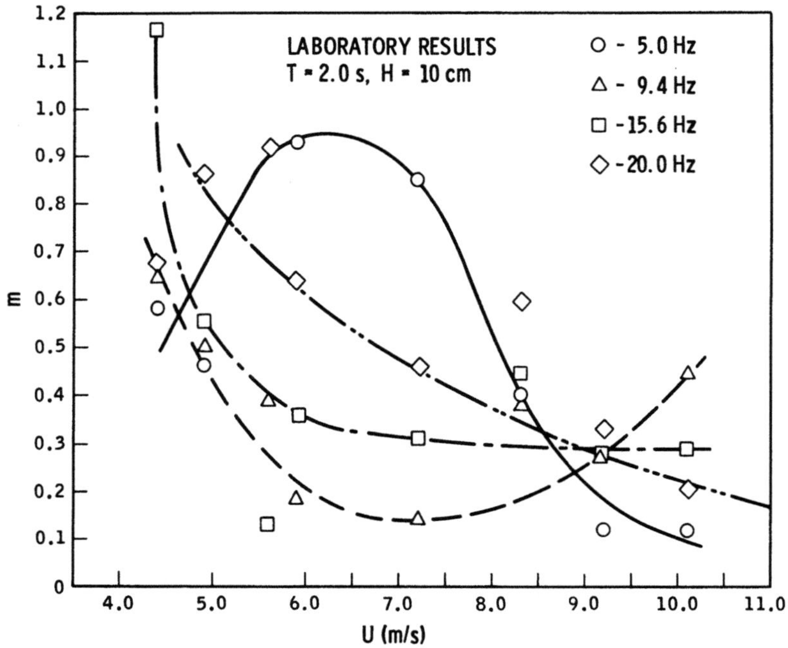

Figure 7. Normalized modulation amplitude, m, as a function of wind speed for different free wave frequencies, f_o.

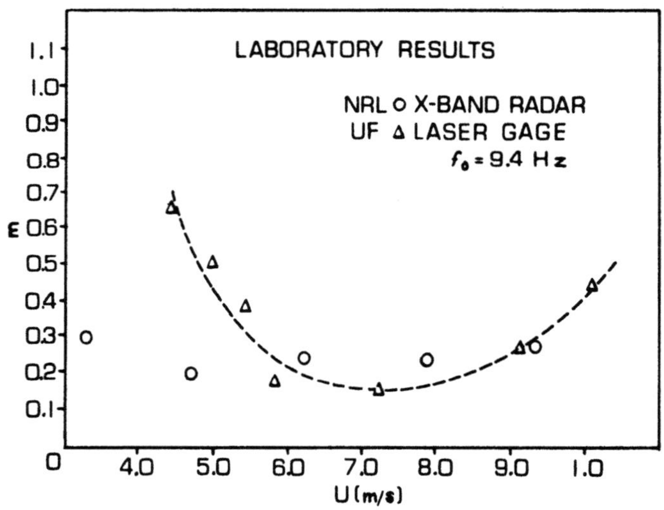

Figure 8. Comparison between radar and laser measurements of short wave modulation, m, in the laboratory.

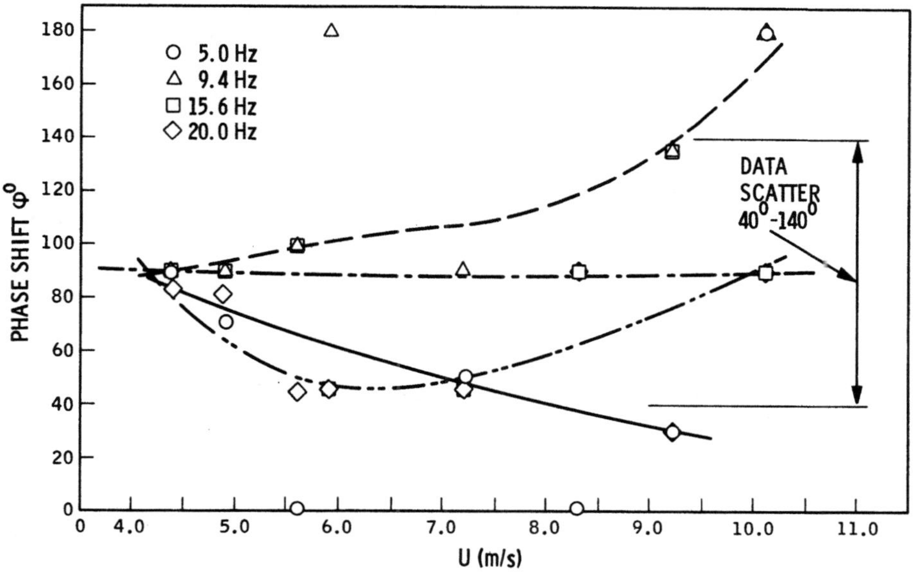

Figure 9. Phase shift of short waves down crest of long waves for different free wave frequencies, f_o.

which also predicted peak short intensities at the long wave crest. Reece (1976) compared the laboratory results to the relaxation theory proposed by Keller and Wright (1975) and found an acceptable agreement in modulation amplitude and phase provided that a proper choice of the relaxation constant was made. The latter was found to depend on wind speed, f_o, and long wave properties. The relaxation theory provides valuable insight towards understanding wave modulation of short waves. Its shortcoming lies in the fact that the relaxation constant is not a "constant" but depends on wind speed and long wave properties.

4. FIELD MEASUREMENTS AND DATA ANALYSIS

In the field tests conducted at Marineland, Florida in 1975 simultaneous measurements were obtained and recorded for upwind-downwind and cross-wind wave slopes, long wave surface displacement and wind speed. The wave slopes were measured by the laser-optical device operated in the wave following mode. The long wave surface displacement was obtained with a pressure transducer mounted on the wave follower I-beam at 4.90 m below the mean water level. The wave slope and surface displacement were recorded at 400 samples per second. Long wave time series were obtained by averaging 30 consecutive points in approximately 10 minute records. Power spectra of wave height were obtained using the Fast Fourier Transform technique.

The slope data records contain significant information at high frequencies up to 100 Hz. Two different analysis procedures were pursued.

a. RMS Cross-Correlation

The simple and most obvious cross-correlation between wave height and short wave slope gives the coupling between wave height and rms intensity of short waves. This calculation was carried out by direct integration

$$R(\tau) = \int s^2(t)\ \eta(t+\tau)\ dt, \qquad (4)$$

where $R(\tau)$ is the correlation function, s and η refer to slope and surface displacement, respectively. Alternately, $R(\tau)$ can be computed by the Fourier transform technique

$$C(\sigma) = S(\sigma)\ H^*(\sigma)\ , \qquad (5)$$

where C, S and H refer to the Fourier transforms of R, s, and η, respectively, and H* is the complex conjugate of H.

To avoid aliasing the wave slope data the slope squared signal, s^2, was averaged over 150 consecutive data points. This provided a good measure of the s^2 envelope over the long waves. The wave height spectrum for the Marineland data has a peak at 0.012 Hz. Only insignificant energy remains above 0.5 Hz. The cross correlation between s^2 and η contains most of the significant contributions across the spectrum as can be inferred from Eq. 5. The cross-correlation function, $R(\tau)$, was calculated by using both Eqs. (4) and (5) independently. The agreement between the two was quite reassuring.

A possible source of error in the rms cross-correlation is the contribution of the mean slope of long waves. This error can be eliminated by subtracting long wave slope from the total signal before computing the cross-correlation function. The dominant slope intensities of waves are primarily in the higher frequency range so that the error cannot be expected to be very large. Also, it is eliminated in the second and more detailed calculation procedure discussed in 4(b). The cross-correlation function, $R(\tau)$, is shown in Figure 10. A distinct peak appears at 1.0 sec

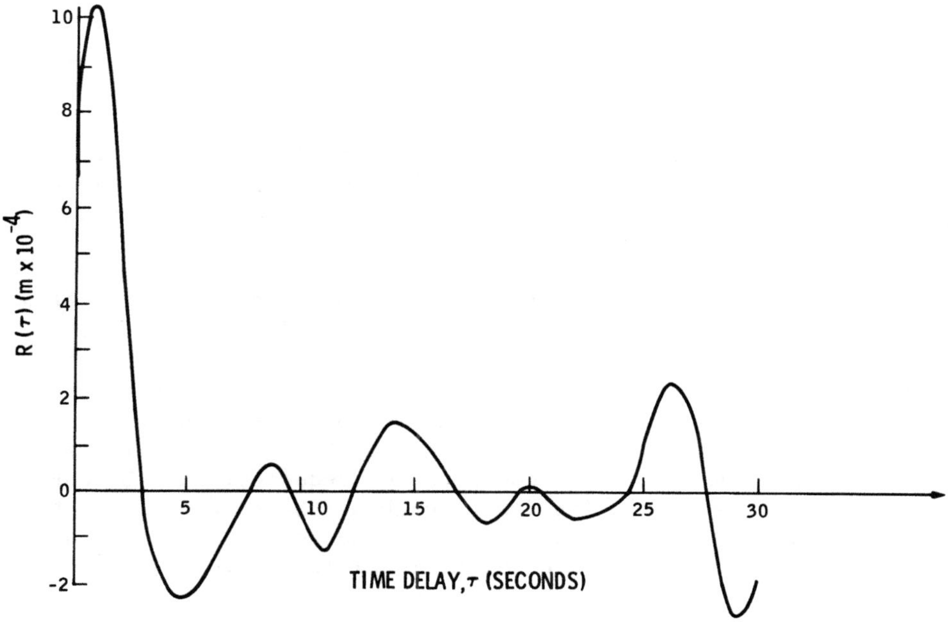

Figure 10. Cross-correlation function of slope squared signal with long wave surface displacement. Peak is located at 1.0 sec delay time.

time lag. The dominant wave period was 8.0 sec, suggesting that the small scale wave structure is more intense on the forward slope of long waves.

b. Demodulation, Bandpassing and Cross-correlation

As a consequence of the laboratory investigation it is essential that the orbital velocity of long waves be considered in the dispersions equation relating k_s and σ_s of short waves. Eq. (1) may be written in terms of phase speeds

$$c = c_o + U_s + U_o \ , \qquad (6)$$

where

$$\sigma_s = k_s c$$

and

$$c_o^2 = \frac{g}{k_s} + \frac{Tk_s}{\rho}. \tag{7}$$

The surface drift, U_s, depends on the wind speed U_1 and is normally 2 - 4% of U_1. Most field and laboratory experiments yield $U_s = 0.03\ U_1$. The latter was used in the demodulation calculations.

The periodic nature of long waves in the laboratory permitted demodulation in the frequency domain. In the field it is only possible to demodulate in the time domain. The time history of short waves at a fixed station exhibits a relatively high encounter frequency in the vicinity of a wave crest and vice versa near a wave trough. A short wave with wave number, k_s, has a higher frequency on the long wave crest compared to frequency, σ_o. Demodulation in the time domain implies stretching the time scale of the measured signal near the long wave crest and compressing it near the long wave trough. For a unidirectional long wave, η,

$$\eta_\ell = \sum a_i \cos (k_i X - \sigma_i t), \tag{8}$$

where the subscript i refers to different long wave components. The orbital velocity, U_o, at the surface is

$$U_o(t) = \sum \sigma_i a_i \coth\ k_i h \cos (k_i X - \sigma_i t) , \tag{9}$$

where h is the local water depth.

In the Marineland Experiment the most useful measurements were obtained on December 15, 1975 during a dominant swell situation. The water depth at Station I was 10 m and a directionally narrow wave height spectrum was observed. It was inferred that Eq. (9) provided a reasonable approximation of the orbital velocity in the Marineland situation. More calculations to quantify errors due to directional spreading are presently in progress. The validity of Eq. (9) allows demodulation in the time domain through a time-dependent scale factor, F(t).

$$F(t) = \left[1 + \frac{U_o(t)}{c_o + U_s}\right], \tag{10}$$

where U_o and c_o are as specified in Eqs. (7) and (9), and $U_s = 0.03\ U_1$. A more general random sea surface is directional. Demodulation in such a situation must incorporate wave directionality as is being done in analyzing the recently acquired data during the West Coast Experiment, described by Shemdin et al. (1977).

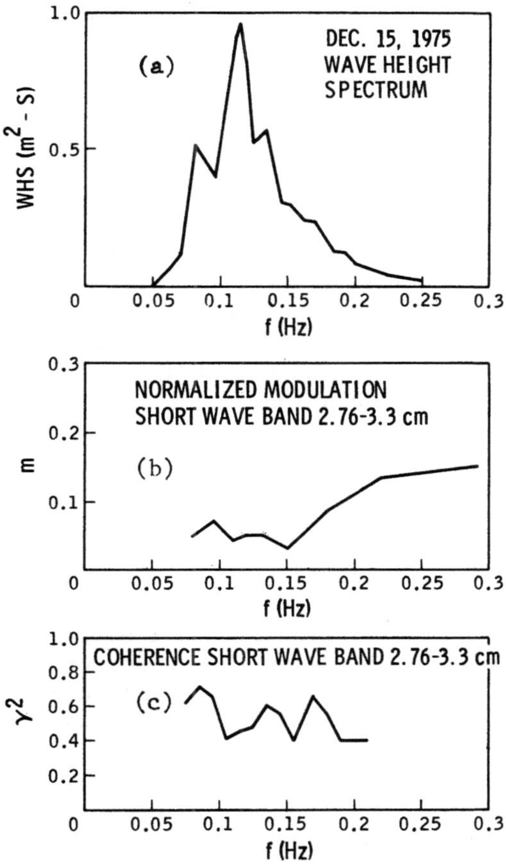

Figure 11. Results of ocean measurements at Marineland, Florida. (a) Long wave height spectrum. (b) Normalized modulation, m. (c) Coherency function.

The demodulation procedure produces a modified time history which maintains a dispersion relationship described in Eq. (3) so that for a given k_s there corresponds a fixed σ_s. The demodulated time history can then be band-passed for a given σ_s or k_s and transformed back to the time domain for cross-correlation with the long-wave height-signal. The cross-spectrum of the demodulated and band-passed signal at k_s can also be obtained to determine the strength of amplitude modulation of k_s at different spectral frequencies of the long waves. The latter is equivalent to what a real aperture radar gives when cross-correlating the signal return with a long-wave height-signal.

For the random and near unidirectional condition encountered in the Marineland Experiment a normalized modulation factor, m, is defined, for a short wave band, as the ratio of the cross-spectrum at a particular long wave frequency to the cross-correlation function, $C(\tau=0)$. The wave height spectrum is shown in Figure 11. The corresponding m value and coherence function of short waves in the band 2.76 - 3.30 cm are also shown in Figure 11. The m values constitute the first estimates of this type and are preliminary in some sense. There is a noticeable lack of a peak in the m-spectral values near the peak frequency of wave height spectrum. The wind speed was 5.0 m/sec from 60° north. The swell direction was from 90° north. The m values corresponding to the peak of the wave height spectrum are of order 0.1 suggesting only 10% modulation.

5. SUMMARY OF RESULTS AND CONCLUSIONS

In this paper we present results from laboratory and field experiments on the modulation of short waves by long waves. It is found that long waves significantly modulate the short wave dispersion by their orbital velocity and that demodulation is necessary if the data collected by remote sensors are to be compared to surface penetrating devices. The laboratory results indicate modulation to be up to 100% of mean value and with the proper choice of the relaxation constant, the results are consistent with the relaxation theory predictions. It is also found that short wave intensities are maximum on the forward slope of long waves in the range 40° - 140° downwind of the crest. The field data on short waves can only be demodulated in the time domain. The results to date are preliminary and only available for a narrow range of wind speeds. The modulation level is weak for wave lengths in the range 2.76 - 3.30 cm.

ACKNOWLEDGMENT

This work was supported by the NOAA-Space Oceanography Program under Grant NG-29-72 and by the National Science Foundation under Grant OCE-7602713. The collaborative aspect of the Marineland Experiment was supported by the NASA Office of Applications through the Seasat-A Surface Truth Program. The cross-correlation calculations for field data were carried out by D. Evans.

REFERENCES

Garrett, C. and Smith, J. (1976), "On the Interaction between Long and Short Surface Waves," J. Phys. Oceanog., 6, 925-930.

Hasselmann, K. (1971), "On the Mass and Momentum Transfer between Short Gravity Waves and Large Scale Motions," J. Fluid Mech., 50, 189-205.

Keller, W. C. and Wright, J. W. (1975), "Microwave Scattering and the Straining of Wind-Generated Waves," Radio Science, 10, 139-147.

Keller, W. C. and Wright, J. W. (1976), "Modulation of Microwave Backscatter by Gravity Waves in a Wave Tank," NRL Report 7968, Naval Research Laboratory, Washington, D.C., 1-20.

Longuet-Higgins, M. S. (1969), "A Non-linear Mechanism for the Generation of Sea Waves," Proc. Roy. Soc., A, 311, 371-389.

Palm, C. S. (1975), "An Instrument to Measure the Two Dimensional Wave Slope Spectrum of Ocean Capillary Waves," Doctoral Dissertation, Univ. of Florida, Gainesville,

Plant, W. J., Keller, W. C., and Wright, J. W. (1977), "Modulation of Coherent Microwave Backscatter by Shoaling Waves," private communication.

Reece, A. M. (1976), "Modulation of Wind Generated Waves by Long Gravity Waves," Ph.D. Dissertation, University of Florida, Gainesville.

Reece, A. M. (1977), "Modulation of Short Waves by Long Waves," J. Bound. Layer Meteo., in press.

Shemdin, O. H. (1972), "Wind-Generated Current and Phase Speed of Wind Waves," J. Phys. Oceanog., 2, 411-419.

Shemdin, O. H. (1974), "On the Remote Sensing of Directional Wave Spectra and Surface Winds," Seasat-A Scientific Contributions, NASA, Washington, D.C.

Shemdin, O. H., Inman, D. L., and Blue, J. E. (1977), "West Coast Experiment Test Plan," JPL Report 900-765, Pasadena, 69 pages.

Snyder, R. L., Long, R. B., Dobson, F. W., and Elliot, J. A. (1977), "The Bight of Abaco Pressure Experiment," Proc. NATO Symposium on Turbulent Fluxes through the Sea Surface, Wave Dynamics and Prediction, Bendor, France, Plenum Publishing Corp.

Sturm, G. V. and Sorrell, F. Y. (1973), "Optical Wave Measurement Technique and Experimental Comparison with Conventional Wave Height Probes," Applied Optics, 12, 1928-1933.

Tober, G., Anderson, R. C., and Shemdin, O. H. (1973), "Laser Instrument for Detecting Water Ripple Slopes," Applied Optics, 12, 4, pp. 788-794.

Tober, G. and Shemdin, O. H. (1977), "Design and Performance of an Oceanic Wave Follower," JPL Report, in preparation.

Valenzuela, G. R. and Wright, J. W. (1976), "The Growth of Waves by Modulated Wind Stress," J. Geophys. Res., 81, 5795-5796.

DISCUSSION

O. Phillips:

I have a question concerning your de-modulation procedure. When short waves ride over long ones, there are two kinematical effects. The short waves are convected by the orbital velocity of the long one, as you said, giving a modulation of the apparent frequency. But they are also strained by the long waves - the wave-number of an energy packet is modulated also. Consequently, an instrument like a back scattering radar, which responds to a given wave-number, will be detecting different energy packets on crests and troughs. Since the spectral densities vary rapidly with wave-number, could not a considerable bias be introduced?

O. Shemdin:

At the outset of the experiment we did anticipate difficulties with wave-number modulation in a packet. When we analyzed the results we found that spectral extremities over the trough could be made to match those over the crest if a transformation in the frequency domain was made taking account of the orbital motion. The difference in the demodulated spectra we attributed to straining imposed on the short waves by the long wave. In this sense we have assumed the short waves to be free waves advected and strained by the long wave. We have not explored wave-number straining in a packet sense but plan to pursue the idea.

K. Hasselmann:

1. In demodulating your short-wave signal to remove the Doppler shift induced by the long-wave orbital velocity do you not have a problem in the ocean? It is not possible to reconstruct the horizontal velocity in a two-dimensional spectrum from a single-point wave height signal. Did you find that the coherency was significantly increased after demodulating as compared with a straight, uncorrected cross spectral analysis of long-wave amplitude and short-wave energy?

2. You mentioned that Wright's relaxation time theory contained a free parameter (the relaxation time) as a function of frequency which enabled one to fit almost any modulation data. Doesn't the theory in fact imply a relation between amplitude and phase of the modulation transfer function which could be tested?

O. Shemdin:

1. In the Marineland Experiment the wave follower and sensors were stationed in 10 m depth. The waves were narrow band (swell)

and reasonably well organised in direction with respect to shore. The velocity field derived with a single point wave height measurement can give reasonable results, at least compared with a broad and highly directional wave field. We also have two slope measurements which we plan to use to improve accuracy over single point wave height measurement.

2. Yes, Wright's theory does imply a relation between amplitude and phase as suggested. We find Wright's theory able to predict modulation height and phase for a given relaxation constant. But we find that the relaxation constant depends on wind speed and short wave frequency. The theory does not predict the resonance type phenomenon we observe at 5Hz free frequency.

P.A. Taylor:

It would be interesting to see slope spectra for your field experiment covering both gravity and capillary wave ranges. As far as I am aware there are few (or no?) such spectra in the published literature?

O. Shemdin:

There are few high frequency spectra obtained with high response wave gages. We have such spectra and will plan to include them in future publications.

R. Long:

The bottom scattering process examined by myself has not been given equal time here because of the difficulty of parameterizing the effect. It remains a possibly important process in some situations. However, some data on the bottom in the JONSWAP area has become available and, while not definitive, supports the notion that the process is not important there. In any event, detailed directional spectra are being measured in the JONSWAP area at this time, and we will soon have a chance to look for the scattered background field which should arise in the presence on an increasing swell beam if this scattering process is active there.

MEASUREMENT OF THE PRESSURE AND VELOCITY FIELD BELOW SURFACE WAVES

L. Cavaleri*, J. A. Ewing+ And N. D. Smith+

*Consiglio Nazionale delle Ricerche, Venice
+Institute of Oceanographic Sciences, Wormley, England

Observations of the wave elevation, pressure and three components of velocity have been made at different depths from a tower in the Adriatic Sea near Venice in a water depth of 16 m. The spectral analysis of the pressure measurements show significant differences from linear wave theory which depend on the wavelength and depth of the transducer. The velocity field was measured with two electromagnetic current meters in the same vertical line. The amplitudes of the vertical and horizontal velocities were about 10% less than expected on the basis of linear wave theory. The downward transport of horizontal momentum, estimated from the co-spectrum of the vertical and horizontal velocities, was found to be much greater than the total atmospheric stress. The large values of the momentum flux are associated with deviations in the phase of the horizontal velocity component with respect to the wave elevation from the value expected using linear wave theory.

INTRODUCTION

Linear theory has been used in many engineering and scientific studies to predict the pressure fluctuations and orbital velocities below the surface due to waves. It is therefore important to know if measurements show any significant departure from the theory.

The early experiments of Draper (1957) and Tsyplukhin (1963) suggested that the bottom pressure in shallow water was more severely attenuated for the longer period waves than was predicted by linear theory from a knowledge of the surface wave height. In contradiction subsequent work using digital power spectral analysis techniques by Bowden and White (1966) showed that there was evidence for greater pressures within 1 m of the bottom especially for the longer period waves while Simpson (1969) did not find any significant departure from linear theory in the range where the coherence between surface elevation and pressure was high. Homma et al. (1966) observed that the attenuation of pressure with water depth varied with wave frequency but they did not detect any variation of pressure attenuation with the depth of the transducer. However, in a study using two pressure transducers Esteva and Harris (1970) detected a greater deviation from the linear law for the deeper instrument.

Comparisons of measured horizontal velocities compared to the pressure at the same depth below waves have also been made by Bowden and White (1966) and Simpson (1969). Both studies indicate that the velocities were generally greater than expected on the basis of linear theory compared with pressure measurements but no direct comparisons were made with the surface elevation. Thornton and Krapohl (1974) found that the horizontal velocity spectra were about 12% higher than expected theoretically from the surface wave spectrum while the phase angle between one velocity component and wave elevation was in good agreement with linear wave theory.

Measurements of the transfer function of vertical velocity compared with wave elevation have been made by Yefimov and Khristoforov (1969), Taira (1971) and Thornton and Krapohl (1974). Both Yefimov and Khristoforov and Thornton and Krapohl report good agreement between the amplitude response compared with linear wave theory but, in the case of the phase difference between vertical velocity and wave elevation, Yefimov and Khristoforov detect significant departure from linear theory while Thornton and Krapohl's measurements of phase are close to theory. In contradiction the work of Taira (1971) indicates that the amplitude of the vertical velocity is about 10% smaller than expected on the basis of linear theory.

The most unexpected and scientifically important deviation from linear theory is however concerned with the measurement of large Reynolds stresses first reported by Shonting (1964) and confirmed subsequently by Yefimov and Khristoforov (1969). The measurements of these authors show that the vertical and horizontal velocities are not in quadrature thus indicating the existence of a large vertical momentum flux greater than the total atmospheric stress.

The principal aim of this work was to try to resolve some of the differences reported by previous investigators using an off-shore tower located near Venice from which it is possible to make a series of reliable measurements over a period of time.

Theory

Linear theory (Cartwright, 1962) assumes the sea surface can be represented by the sum of an infinite number of sinusoids each with a particular amplitude, frequency, wavenumber and direction. The phases of the sinusoids are randomly distributed. The wave-number k and frequency σ (= $2\pi f$) are given by the dispersion relation $\sigma^2 = gk \tanh kH$ where H is the mean water depth and g is the acceleration due to gravity. The advantage of this representation is that an energy spectrum can be defined by adding linearly the variances of the component waves over defined frequency intervals. Furthermore any linear response arising from the wave field can also be represented by a spectrum which is related to the wave spectrum through a transfer function.

In this study we are mainly concerned with the transfer functions between the amplitudes of the pressure p, vertical velocity w and horizontal velocities u and v to the wave amplitude h. On the basis of linear theory these transfer functions are given by (see for example Nagata (1964))

$$K_p = \frac{\rho g \cosh k(H-z)}{\cosh kH}, \quad K_w = \frac{g \sinh k(H-z)}{c \cosh kH}, \quad K_u = K_v = \frac{g \cosh k(H-z)}{c \cosh kH}$$

where the phase velocity $c = \sigma/k$ and z is the vertical co-ordinate measured positively downwards from the mean level of the free surface.

In random waves with a frequency spectrum E_h the spectra of the horizontal velocities E_u and E_v depend on the form of the directional distribution of wave energy. In this case we must combine the velocity spectra and compare measurements of $(E_u + E_v)/E_h$ with the function $K_U \equiv K_u$.

Linear theory predicts that the pressure and two components of horizontal velocity are in phase with the wave and that the vertical velocity lags the wave elevation by 90°.

Measurements

The measurements were made from the oceanographic tower of C.N.R. situated 13 km off the coast of Venice in the Adriatic Sea. The average water depth at the tower is 16 m; this increases offshore with a bottom slope of 1:1000. In the vicinity of the tower the sea bed is a smooth sand-mud mixture with bottom contours running almost parallel to the coast. The tower has a square section 5 m wide with four supporting legs. Three floors are available for storage, power, generators, living accommodation for four people and instrumentation.

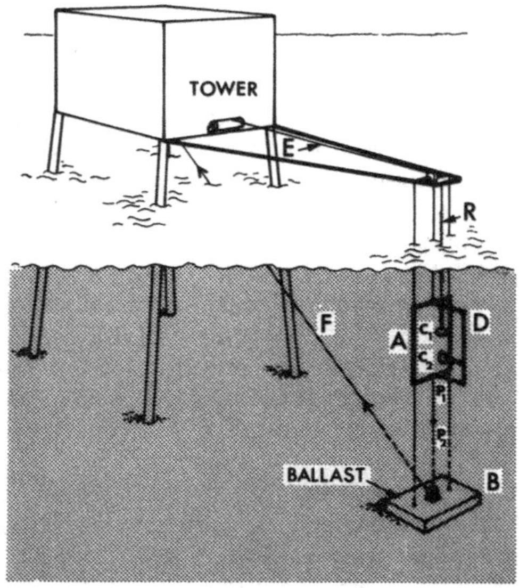

Fig. 1 Schematic diagram of tower and instrumental arrangement

Meteorological and tidal data are recorded routinely on the tower. The tidal range during the measurements was less than 0.7 m with a maximum change of 6 cm during any 30 min. record.

A platform (Figure 1) which extends 7 m horizontally away from the south-east side of the tower was used for the study. The measurements were made from a cart A which slides on two parallel vertical wires fixed under tension to a heavy ballast weight B on the bottom. The wire guides were situated 7 m from the nearest leg of the tower of diameter 0.6 m; measurements made from the cart were thus considered free of interference effects caused by the tower for waves approaching Venice from directions between 045° and 225°. (For all measurements described here the wind was from the north-east where the maximum fetch to the tower is about 100 km). The depth of the cart could be controlled accurately using another wire E, F whose length together with a tide gauge reading allowed the depth of the cart to be determined within 2 cm. Strong tension exerted in the wire E, F ensured that the motion of the cart during measurements was negligible.

Two electromagnetic current meters C_1, C_2 and two pressure transducers P_1, P_2 were fixed on the cart in the same vertical line. A wave staff R was also placed immediately above the other instruments.

Instrumentation

The electromagnetic current meters were developed by I.O.S. (Tucker et al., 1970). Their accuracy is better than 1% up to 1 m/s with a resolution of 0.001 m/s. Calibration of the current meters was carried out in a towing tank. The current meters were placed 0.57 m apart in the same vertical line and at right angles to each other on a rotating arm, D, of the cart. The rotating arm was adjusted so that the vertical plane of the lower current meter C_2 was aligned with the visually observed predominant wave direction. Thus current meter C_2 measures the two components w and u. (Hereinafter we refer to u as the velocity in the predominant wave direction). The upper current meter C_1 in the horizontal plane measures the components u and v.

The pressure transducers were manufactured by Bell and Howell and their accuracy is better than 0.2% of the maximum pressure oscillation. The average pressure was removed using a back-off voltage. The dynamical effect of water motion was cancelled by covering the transducer with a small, soft, water-filled plastic bag (Cavaleri, 1973a). The two pressure transducers were placed at distances of 1.1 m and 4.3 m below the current meter C_1.

Wave elevation was measured with a resistance wave staff made at C.N.R. for which linearity was better than 0.5% of the overall excursion. Experimental tests (Cavaleri, 1973b) showed that errors due to water clinging to the staff were less than 2%.

Analysis

The signals from the transducers were recorded in digital form using a data-logging system developed by C.H. Clayson. (An early form of the digital logging system has been described in Clayson and Smith, 1970). The measurements from the different channels were sampled simultaneously at 0.25 s intervals.

Spectra and cross-spectra were computed using the Fast Fourier Transform algorithm. The total record length of 30 min. duration was divided into 14 non-overlapping sections each of length 128 s. Spectra were evaluated by taking the Fourier transform of each section and averaging over the 14 sections. This procedure yields spectral estimates at $^{1}/_{128}$ Hz interval with 28 degrees of freedom.

The confidence limits for the amplitude and phase of transfer functions estimated using spectral analysis techniques can be determined in terms of the number of degrees of freedom and coherence of the process (Jenkins and Watts, 1968). In this study we are usually concerned with the comparison of measured and theoretical transfer functions when the squared coherence, γ^2, is greater than 0.98. For 28 degrees of freedom this leads to 95%

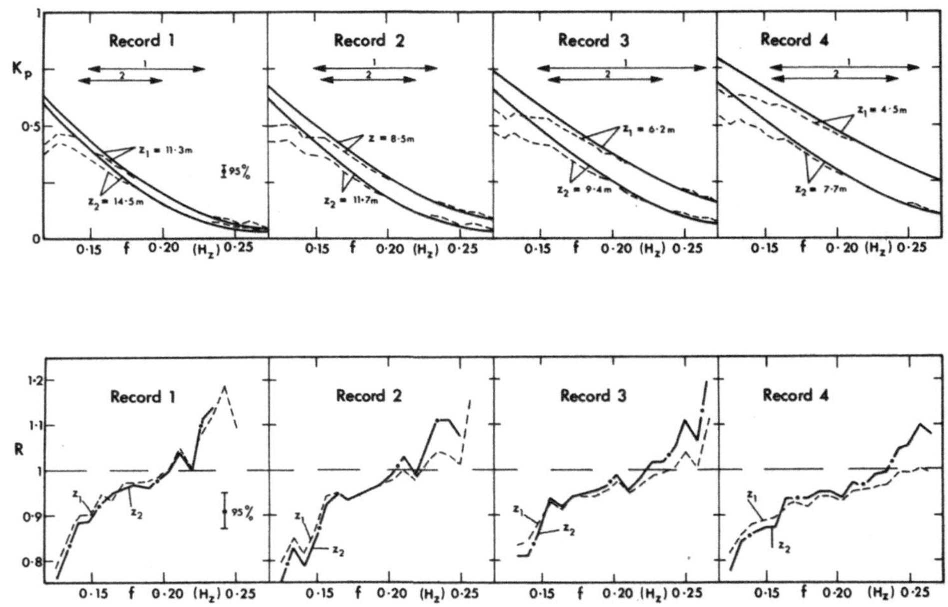

Fig. 2 Upper part: Comparisons of experimental (— — — —) and theoretical (———) amplitude response functions for pressure measurements at depths z_1 and z_2.
Lower part: Comparisons of the ratio R from measurements at depths z_1 and z_2 (— — — —, —·—·—·—) with theory (R=1).

confidence limits of $\pm 7\%$ and $\pm 4^{\circ}$ for the amplitude and phase response functions.

Comparison of pressure measurements with linear theory

The results are presented in two forms. Firstly the experimental amplitude responses of the two pressure transducers P_1, P_2 (at depths z_1 and z_2 below the mean water level) are compared with linear wave theory in the upper half of Figure 2). Secondly, the ratios $R = \sqrt{(E_p/(K_p{}^2 E_h))}$ for the same records are shown in the lower half of this figure. Details of the records are given in Table 1. The range of squared coherence greater than 0.98 is shown by the arrowed lines. 95% confidence limits for the amplitude response and R are shown for one record only; other records having very similar confidence bands.

Comparisons of K_p over the range $\gamma^2 > 0.98$ show that the measured pressure amplitude response is consistently less than expected on the basis of linear theory for both instruments over a range of depths from 4.5 m to 14.5 m. Differences between theory

Table 1

Record	Time (GMT) 11 Mar 1976	Depth of pressure transducer P_1 (m)	Mean wind speed (m/s) and direction (deg.)	Significant wave height (m)	Peak of wave spectrum (Hz)
1	0257	11.3	11; 070	1.42	0.18
2	0343	8.5	11; 070	1.38	0.18
3	0433	6.2	12; 070	1.36	0.19
4	0513	4.5	11; 070	1.40	0.20

Table 2

Record	Time (GMT) 10,11 Mar 1976	Depth of current meter C_2 (m)	Mean wind speed (m/s) and direction (deg.)	Significant wave height (m)	Peak of wave spectrum (Hz)
5	2238	3.7	15; 080	1.74	0.16
6	2330	10.4	14; 080	1.63	0.17
7	0031	7.9	14; 080	1.84	0.18
8	0122	5.6	13; 080	1.56	0.18

Table 3

Record	Time (GMT) 10,11 Mar 1976	Depth of current meter C_2 (m)	Mean wind speed (m/s) and direction (deg.)	Significant wave height (m)	$\overline{u^2}$	$\overline{v^2}$	$\overline{w^2}$	$\overline{uw}$	$\overline{vw}$
					$(cm/s)^2$				
5	2238	3.7	15; 080	1.74	559	90	662	172	18
6	2330	10.4	14; 080	1.63	101	25	52	22	6
7	0031	7.9	14; 080	1.84	204	36	188	71	22
8	0122	5.6	13; 080	1.56	260	42	277	85	11
9	0620	10.6	11; 070	1.28	40	6	23	9	1
10	0848	3.8	8; 070	1.07	210	33	247	55	14
11	0949	8.0	8; 070	0.90	33	5	31	10	3
12	1032	5.6	7; 070	0.79	51	12	62	15	2

and experiment are greatest at lower frequencies where the deviation exceeds the 95% confidence limits. The measured phase difference between pressure and wave signals was always close to and within the 95% confidence limits of $\pm 4^{\circ}$; phases are therefore not shown in the diagrams.

The variation of pressure with depth and frequency is exhibited more clearly in terms of the ratio R shown in the lower half of Figure 2. (If R = 1 then experiment and theory agree. Values of R < 1 imply a greater attenuation of pressure than would be expected from linear theory). The lower diagrams show that at low frequencies there is a 10% greater attenuation of pressure with depth than expected from linear theory; the difference between experiment and theory decreases as the frequency increases and R approaches 1. Furthermore the slope of R with frequency is less for pressure measurements near the surface. This latter observation is confirmed by noting in the individual diagrams that the experimental ratio R for the pressure transducer at depth z_2 ($z_2 > z_1$) always lies below the curve R for pressure measured at z_1 at lower frequencies and conversely at higher frequencies.

The analysis of other records (not shown here) taken during conditions when the wave heights (and spectral amplitudes) were considerably less than Records 1 to 4 show the same quantitative features. Because of this, and also since the variation of R continues into the low amplitude, long wavelength part of the spectrum, we conclude that the deviation of the pressure attenuation from linear theory is not a finite amplitude effect.

Comparison of velocity measurements with linear theory

During the early part of this investigation it was found that measurements of the horizontal velocity ratio K_u exhibited considerable scatter far exceeding the expected confidence limits while measurements of the vertical velocity K_w showed a smooth variation. Other measurements (not shown here) of the common velocity component in the main wave direction, u, available from both current meters also showed a large degree of variability. We believe that most of the scatter in velocity measurements made with an electromagnetic sensor in the horizontal plane is due to the occurrence of stalling at certain angles of incidence to the plane of the head. Soulsby (1971), Tucker (1972) and P.G. Collar (personal communication) have confirmed the existence of stalling from tests in still water but the effect cannot be easily predicted for a random velocity field due to surface waves although it seems reasonable to assume spurious energy is introduced into the velocity spectrum mainly at higher frequencies. In this study we have therefore estimated the horizontal velocity spectrum by using the current meter C_2 (in the vertical plane) for estimates of horizontal velocity u and current meter C_1 for v. Since current meter C_2 is aligned in the predominant wave direction stalling from the transverse component v should be very small for a wind-generated sea with no swell or low frequency energy from directions other than that of the wind). Thus we have compared measurements of $(E_u(z_1) + E_v(z_2))/E_h$ with the theoretical ratio K_U assuming (with small error) v is at the same depth as u. (The variance of the velocity in the transverse direction was less than 20% of the variance of velocity in the main wave direction for the measurements shown in this paper).

Table 2 gives details of the records which are plotted in Figure 3. The four records, taken within a period of about 3 hours, cover depths of measurements from 3 m to 10 m.

The upper diagrams of Figure 3 compare the experimental and theoretical ratios for the vertical velocity K_w. There is always a high coherence between vertical velocity and wave elevation as shown by the arrowed lines for $\gamma^2 > 0.98$. In general the experimental values for amplitude response are about 10% lower than linear wave theory while the phase angle, Φ_w, (only shown for Record 6) is close to theory.

The middle diagrams of Figure 3 show comparisons for the horizontal velocity spectrum are also about 10% less than predicted by linear theory with some indication of greater deviations for measurements near the surface (Record 5). The level of squared coherence for the component u and wave elevation was less than that for the vertical velocity. The most marked deviation from linear theory is associated with the phases of the component velocity, Φ_u, with respect to wave elevation which was significantly different from the theoretical value of 0° (see Record 6; other records have similar characteristics).

Lastly the lower diagrams show that the ratio of vertical to horizontal velocities is closer to theory than the individual velocity ratios.

Estimates of the downwards momentum flux associated with surface waves

The previous section of this paper showed deviations in the phase angle of the horizontal component u (in the predominant wave direction) with respect to the wave elevation from linear theory. In contrast the phase of the vertical velocity agreed with linear wave theory. The implications of this phase difference is seen in direct computations of the phase angle and co-spectrum between u and w. Positive values of this co-spectrum imply a downward flux of horizontal momentum.

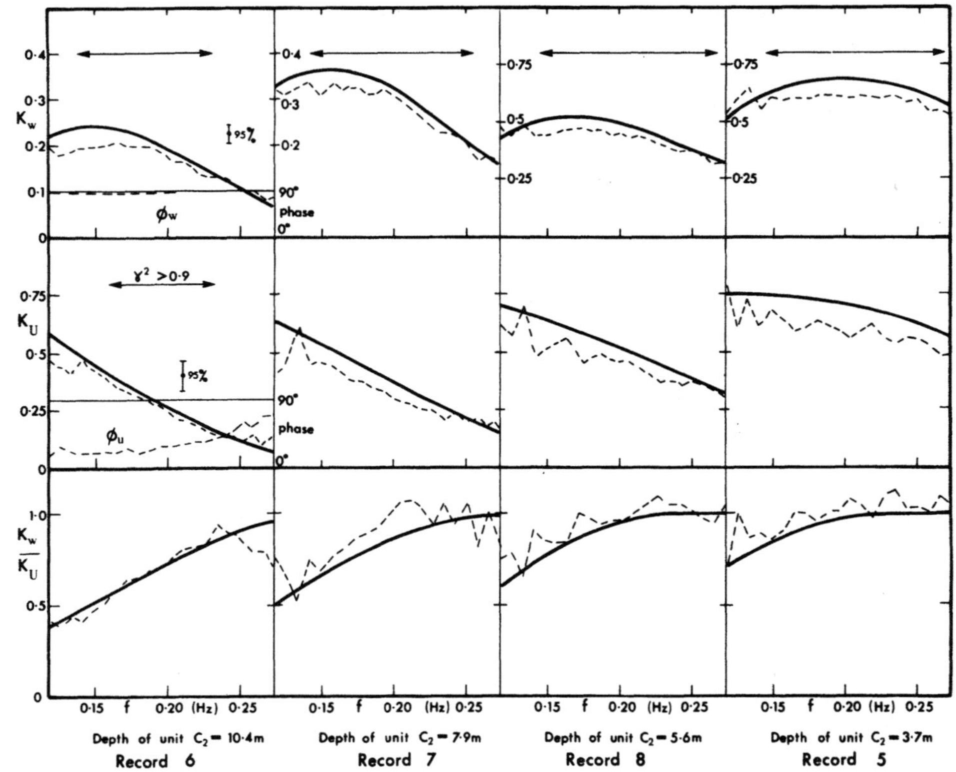

Fig. 3 Comparisons of experimental (– – – –) and theoretical (———) transfer functions:-
Upper diagrams: Vertical velocity
Middle diagrams: Horizontal velocity
Lower diagrams: Vertical velocity/Horizontal velocity

Table 3 shows details of eight records (out of a much larger set) taken during conditions of active wind-wave generation where wave breaking was observed. Wave heights during the first four records were greater than those in the remaining records. Since the density of water is unity in the c.g.s. system of units $\overline{uw}$ is in dynes/cm² when velocities as expressed in cm/s as in Table 3. (uw was evaluated over the frequency range associated with waves up to 0.3 Hz). The atmospheric stress τ at the surface can be estimated to range from about 0.5 to 2 dynes/cm² (using a drag coefficient of 10^{-3}) for the records given in Table 3. It is clear that all measurements of $\overline{uw}$ greatly exceed τ. The transverse component of momentum flux was also estimated from the correlation between v on head C_1 with w on head C_2. As noted earlier measurements from head C_1 in the horizontal plane are influenced by stalling nevertheless the integrated statistic $\overline{vw}$ should not be greatly affected. Values of $\overline{vw}$ are about an order of magnitude smaller than $\overline{uw}$.

We now consider possible instrumental sources of error before discussing these results in further detail. A number of checks have been made on the electromagnetic current meter system. First, we have checked the frequency response of the current meter system with known inputs and we have confirmed that, for frequencies less than about 0.5 Hz, there is unit amplitude response and zero phase lag. Two dynamic tests have also been carried out in a tank in the laboratory. First, by fitting the head to a rotating arm (in the vertical plane) and comparing the output from the current meter with the known features of the rotating arm, and second, by making measurements of the component velocities in mechanically generated waves. Both these tests showed that the phase angle between u and w was within 5° of the theoretical value of 90°. Another possible source of error in measurements of the co-spectrum is in the alignment of the sensors in the vertical plane and internally in the head itself. While these errors are important in turbulence measurements in the boundary layer (see for example Heathershaw, 1976) where angular misalignments of about $\pm 2^{\circ}$ are critical, in our measurements phase shifts as large as 30° clearly cannot be accounted for in this way.

Table 3 shows that $\overline{uw}$ is greater near the surface and decreases rapidly with depth. A comparison of the co-spectrum of u and w with the surface wave spectrum shows that individual peaks of the wave spectrum are seen in the co-spectrum (as noted previously by Yefimov and Khristoforov, 1969, 1971). The momentum flux was found to vary approximately as $(h/z)^2$ where h is the significant wave height and z the depth of the measurement. (Due to the limited range of observed wave conditions it was not possible to detect any variation with wave period).

Deviations of the phase angle between u and w from theory also imply that the particle orbits below waves (in deep water) are not circular but are asymmetric. In a laboratory study of waves and particle velocities Morison and Crooke (1953) showed that, in deep water, the particle paths are close to circles (centred about the mean particle position) even when the waves are very steep. Only in very shallow water do finite depth effects change the shape of the orbit from that expected on the basis of higher order wave theories. The range of wavelengths present in this study in relation to water depth extends from deep water waves to wavelengths in water of intermediate depth; we can not therefore explain our measurements of the phase shift on the basis of shallow water effects or on the basis of finite amplitude effects.

Kitaigorodskii and Miropol'skii (1967) suggested that the

measurements of Shonting (1964) were made in a relatively narrow frequency band which did not include small scale turbulence with the result that $\overline{uw}$ exceeded τ. It is difficult to see how the neglect of high frequency turbulence could reduce the total momentum flux since the momentum flux due to turbulence would have to be upwards towards the free surface. Navrotskii (1967), in a theoretical study of waves and turbulence in the surface layer, discussed the direct effect of the wind in producing a phase shift of the horizontal component of orbital velocity u and estimated that wave stresses at least one order of magnitude greater than τ could be generated by this means. However this effect is confined to a region near the surface and is unable to explain the large values of $\overline{uw}$ at depths several times greater than the wave height.

Yefimov and Grishin (1972) considered the combined effect of nonlinearity and turbulence (characterized by an eddy viscosity) on wave perturbations in the surface layer and were able to compute the Reynolds wave stresses and phase differences between the orbital velocities u and w for different values of the viscosity. They showed that the vertical velocity component conformed more closely to linear wave theory than the horizontal component. The phase difference between u and w was greatest near the surface when it reached values of about 10° and 25° for the first and second harmonics of a periodic wave. The measured phase differences of about 30° in this study in random waves are greater than this theory predicts.

The measurements of Thornton and Krapohl (1974) were made in swell-like conditions during light winds; these authors did not find any difference between the amplitude and phase of velocity components compared with linear wave theory. In the laboratory tests it was also found that mechanically generated waves compared closely with linear theory. This strongly suggests that the velocity field due to wind waves under conditions of active generation are influenced by the turbulence near the surface. We have not yet been able to make any measurements in swell waves to support this view.

Conclusions

Measurements of the pressure fluctuations due to surface waves have been compared with linear wave theory. The analysis of these measurements show that there are differences from linear wave theory which depend on the wavelength and the depth of the measurements. These differences are greatest at low frequencies and increase with the depth of the transducer.

The amplitudes of the vertical and horizontal velocities were about 10% less than expected on the basis of linear wave theory. The phase of the vertical velocity agreed with theory but there were significant deviations in the phase of the horizontal velocity in the predominant wave direction compared with theory. As a result there exists a downwards momentum flux which was found to be much greater than the total atmospheric stress. The differences between measurements and theory do not appear to be caused by finite amplitude effects but rather by the influence of turbulent velocity fluctuations near the surface.

Acknowledgements

We are grateful to Mr M.J. Tucker for his advice during this work. Many of our colleagues made this study possible especially Messrs Braithwaite, Clayson, Curiotto and personnel on the C.N.R. tower.

References

Bowden, K.F. and R.A. White, Measurements of the orbital velocities of sea waves and their use in determining the directional spectrum, Geophys. J. Roy. Astron. Soc., 12, 33, 1966.

Cartwright, D.E., Analysis and statistics. In "The Sea", Volume 1, Interscience, 846 pp., 1962.

Cavaleri, L., Misure di onde con sensori di pressione. C.N.R. Tech. Report No. 79, 1973a.

Cavaleri, L., Ondametro a resistenza-critica, miglioramenti e precisione ottenibile. C.N.R. Tech. Report No. 78, 1973b.

Clayson, C.H. and N.D. Smith, Recent advances in wave buoy techniques at the National Institute of Oceanography. In Electronic & Radio Engineers, 1970.

Draper, L., Attenuation of sea waves with depth. La Houille Blanche, 6, 1957.

Esteva, D. and D. Lee Harris, Comparison of pressure and wave gauge records. Proc. Twelfth Coastal Engineering Conference, 101, Washington, D.C., 1970.

Heathershaw, A.D., Measurements of turbulence in the Irish Sea benthic boundary layer. In "The Benthic Boundary Layer", Plenum, 323 pp., 1976.

Homma, M., K. Horikawa and S. Komori, Response characteristics of underwater wave gauge. Proc. Tenth Coastal Engineering Conference, 99, Tokyo, 1966.

Jenkins, G.M. and D.G. Watts, Spectral analysis and its applications. Holden-Day, 525 pp., 1968.

Kitaigorodskii, S.A. and Y.Z. Miropol'skii, The theory of turbulent exchange in the upper boundary layer of the ocean. Izv. Atmos. and Oceanic Phys. 3, 704, 1967.

Morison, J.R. and R.C. Crooke, The mechanics of deep water, shallow water, and breaking waves. Beach Erosion Board, Tech. Memo. No. 40, 1953.

Nagata, Y., The statistical properties of orbital wave motions and their applications for the measurement of directional spectra. J. Oceanogr. Soc. Japan, 19, 169, 1964.

Navrotskii, V.V., Waves and turbulence in the ocean surface layer. Oceanology, 7, 755, 1967.
Shonting, D.H., A preliminary investigation of momentum flux in ocean waves. Pure and Appl. Geophys., 57, 149, 1964.
Simpson, J.H., Observations of the directional characteristics of sea waves. Geophys. J. Roy. Astron. Soc., 17, 93, 1969.
Soulsby, R.L., The effect of an axial component of current on the output of an e/m log. University of North Wales, Menai Bridge Science Laboratories Oceanography Report No. 70-1, 1970.
Taira, K., Wave particle velocities measured with a Doppler current meter. J. Oceanogr. Soc. Japan, 27, 218, 1971.
Thornton, E.B. and R.F. Krapohl, Water particle velocities measured under ocean waves. J. Geophys. Res., 79, 847, 1974.
Tsyplukhin, V.F., The results of instrumental study of wave attenuation within the deep sea (in Russian). Okeanologiya, 5, 833, 1963.
Tucker, M.J., N.D. Smith, F.E. Pierce and E.P. Collins, A two-component electromagnetic ship's log. J. Inst. Navig., 23, 302, 1970.
Tucker, M.J., Electromagnetic current meters; an assessment of their problems and potentialities. Proc. Soc. Underwater Technology, 2, 53, 1972.
Yefimov, V.V. and G.A. Grishin, Dynamics of surface waves with allowance for the turbulent nature of the motion. Izv. Atmos. and Oceanic Phys., 8, 28, 1972.
Yefimov, V.V. and G.N. Khristoforov, Some features of the velocity field in the layer of wind-driven swell. Izv. Atmos. and Oceanic Phys., 5, 597, 1969.
Yefimov, V.V. and G.N. Khristoforov, Spectra and statistical relations between the velocity fluctuations in the upper layer of the sea and surface waves. Izv. Atmos. and Oceanic Phys., 7, 841, 1971.

DISCUSSION

W. Rosenthal:

If your measurements could be explained by linear theory, one would have to introduce momentum sinks probably near or at the bottom.

Question: is stratification present or is the bottom out of a material with perhaps a special elasticity modulus (mud/clay)?

L. Cavaleri:

Stratification was not present during our measurement. The

bottom was a sand/mud mixture; we have no information on the elasticity.

H. Mitsuyasu:

Please tell us something about the spectral peak frequency of your data. At some frequency, the correction factor, say R, for the case of pressure, crosses the line R = 1. Does that frequency correspond to the spectral peak frequency?

L. Cavaleri:

No. The spectral peak for these measurements was at about 0.18 Hz and therefore below the crossing point R = 1.

W. Pierson:

Could estimates of transfer function effects have influenced the results?

L. Cavaleri:

Our estimates of the transfer function were reported only for squared coherence greater than 0.95 and can therefore be considered reliable.

E. L. Mollo-Christensen:

Since the occurrence of high Reynolds stress may be due to wave interaction with rotational flow did you observe indications of the presence of topographic Rossby waves, edge waves, mean shear flow or other disturbances?

L. Cavaleri:

The tide-gauge records did not show any long-period oscillations. Our measurements were taken in a wide variety of conditions and showed the same systematic difference from theory which we have not been able to explain by other "disturbances".

K. Hasselmann:

Apart from the puzzles your results present regarding the local wave dynamics, is there any way one can balance the large vertical divergences of momentum you obtain as required by a negative second flux divergence? The vertical flux input by the wind and the horizontal momentum divergence of the attenuating waves both appear to be too small by order of magnitude.

L. Cavaleri:

We are unable to explain our present results by any known physical mechanism. We shall be making further measurements.

W. Pierson:

Could the instrumentation itself have been shedding vortices in the field measurements and not in the lab measurements?

L. Cavaleri:

In the field experiments the momentum flux sensor in the vertical plane was aligned with the predominant wave direction; in this case vortex shedding would certainly be small as confirmed by the high measured coherence.

R. L. Snyder:

I wonder if the authors have made any correction for mean currents, which might be a factor at the higher frequencies?

L. Cavaleri:

The local currents were small at about 10 - 15 cm/s. The currents were variable in direction and unlikely to be the cause of any consistent discrepancy between theory and experiment.

WHITECAPS AND MOMENTUM TRANSFER

M. A. Donelan

Canada Centre for Inland Water
867 Lake Shore Road
Burlington, Ontario

INTRODUCTION

When wind blows over a still water surface the momentum so transferred appears in the form of surface waves and wind drift currents. The flux of momentum τ is generally partitioned into that due to skin friction τ_s (viscous drag) and that due to the form drag of the waves τ_f. A part τ_w of the form drag adds momentum to the surface waves; the rest τ_c acts through the waves to alter the underlying currents. Finally, some of τ_w is continually being lost by the waves through whitecapping τ_b; the rest τ_η produces a net increase of wave momentum.

$$\tau = \tau_s + \underbrace{\underbrace{\tau_\eta + \tau_b}_{\tau_w} + \tau_c}_{\tau_f}$$

The momentum, having been transferred across the air-water interface, is diffused downwards. This paper attempts to explore the effect of whitecapping on the flux of momentum beneath the interface and on the mixing of the upper layers.

BACKGROUND

Several recent measurements of mean and fluctuating velocities beneath a water surface subjected to wind stress have indicated that the wave orbital velocities play little or no part in the flux of momentum. Shemdin (1972) found close adherence of the velocity profile to the logarithmic law for constant stress turbulent boundary layers. The constant stress condition was achieved in his wind-wave tank by recirculating the drift current to maintain the mean water level constant. There was no evidence in the velocity profiles of the presence of waves, and Shemdin pointed out that the calculated Stokes mass transport was an order of magnitude less than his observed drift currents. Jones and Kenney (1977) argue that the subsurface boundary layer has many of the essential characteristics of the turbulent boundary layer over a solid surface and so could be expected to scale in much the same way; i.e. with the turbulent friction velocity u_* and depth Z rather than with scales associated with the waves. They demonstrate that the spectra of horizontal velocity fluctuations scale with u_* and Z in the frequency band between wind waves and inertial currents and suggest that the wave orbital velocities may be "inactive" in the transport of momentum. Phillips (1966) argues that the momentum flux to the longer (phase speed $c>5u_*$) waves τ_W is only a small fraction of the total; whereas the energy flux to the waves, of order $\tau_W c$, may be comparable to the direct energy flux to the currents from the wind, of order $(\tau-\tau_W)u_*$. Whitecapping could therefore be expected to be responsible for a significant proportion of the near surface turbulence, but very little of the momentum transfer.

The foregoing is not meant to be a comprehensive review of previous work on the velocity profile and turbulence beneath white-capping waves, but rather to set the stage for the exploratory measurements which follow.

MEASUREMENTS AND RESULTS

The measurements being reported here are conveniently divided into three groups: (a) a slowed motion film of dye-marked flow beneath a single whitecap; (b) laboratory and field in situ measurements of the turbulence beneath wind-generated wave fields; (c) laboratory and field in situ measurements of mean profiles beneath wind-generated wave fields. These measurements were made with a variety of measuring apparati which included a large wind-wave tank, an open-lake field site, two miniature current meters, two capacitance wave staffs and various atmospheric turbulence sensors.

Measuring Apparatus

The wind-wave tank is 4.5 metres wide and 83 metres long with a maximum water depth of 1.5 metres and an overall height of 3 metres. It is equipped with an hydraulically driven piston-type wave generator, and closed circuit wind system capable of free-stream speeds of about 16 metres per second. The field site, described fully by

Birch et al (1976), is a bottom-mounted tower at the western end of Lake Ontario in 12.5 metres of water. Both sites are equipped with mini-computer controlled 12 bit analog-to-digital converters with inherent accuracy of 0.05%.

One of the miniature current meters is a commercially available (Marsh-McBirney Model 522) electromagnetic current meter. The other was developed expressly for the measurement of three-dimensional turbulence embedded in an highly oscillatory flow (Donelan and Motycka, 1977). It is a small sphere, the drag force on which is transmitted via a thin stalk to an internal strain-gauged support - see Figure 1.

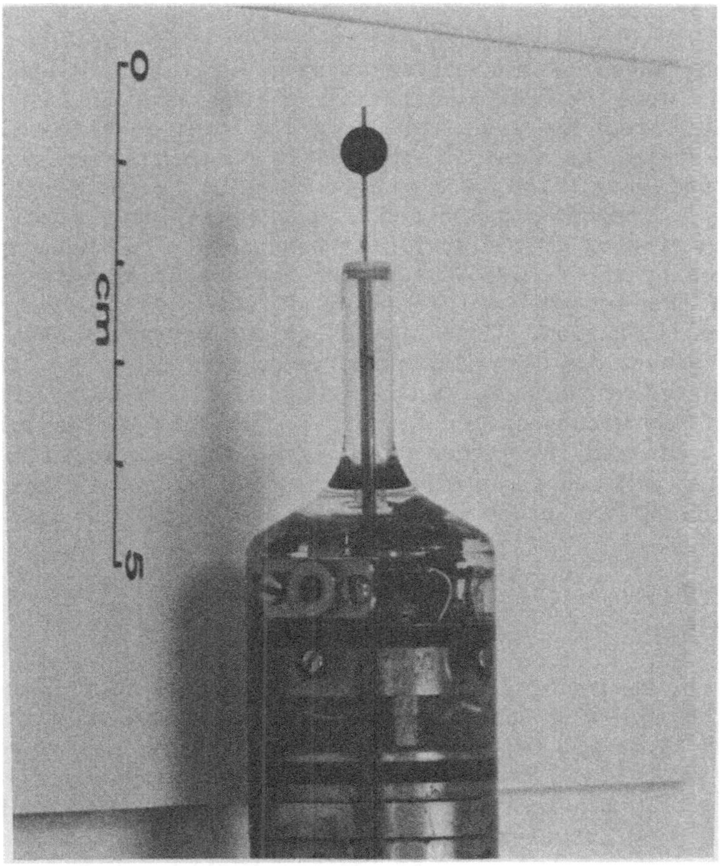

Figure 1.
The miniature drag sphere velocity probe.

Film

An attempt to reveal the way in which a whitecap forms, persists and finally disappears was made by filming the passage of a steep wave train in the wind-wave tank. The fluid was illuminated from above in a 20 cm wide vertical slice 3 m long in the direction of wave propagation and the camera was mounted to the side of the tank and set to run at five times the normal speed. The appearance of a single whitecap in the camera's field of view was achieved by generating a short group of steep short waves followed, an appropriate time later, by a short group of steep long waves. As the long waves pass through the short waves, gentle wave breaking (spilling) ensues. In all cases the amplitude of the short waves is less than 10% of the amplitude of the long waves. So that the short waves can be regarded as a catalyst locally increasing the steepness of the long waves, which then break in an apparently similar manner to natural whitecaps.

Long waves of two different wave lengths, 4 m and 8 m, were filmed. Just before the arrival of the wave trains a line of rodamine B red dye was laid along the centre of the illuminated surface strip. As a result, the gently heaving surface appears as a bright red line which is finally broken up by the frothy bubbling whitecap. Longuet-Higgins and Turner (1974) have given an excellent description of the surficial behaviour of a whitecap which is confirmed by the film. In addition the deeper effects are clearly seen in the intermittent plunging of the dye and the associated turbulent diffusion. Some time after the passage of the whitecap, the turbulence has been dissipated and the extent of vertical mixing is indicated by the wedge-shaped patch of diffuse dye. The orientation of the wedge - point forward - reflects the abrupt start of the whitecap and its rather more gentle demise. Significantly the dye patch, and hence the direct mixing due to the whitecap, extends to depths of the order of the wave height rather than the wave length.

Turbulence

Measurements of fluctuating velocities in the laboratory were made with the drag sphere mounted in the centre of the tank at a depth of 33 cm and fetch of 50 m. Its stalk was normal to the mean orbital plane of the waves. The waves were generated by an equivalent 10 m height wind of 18 m/s in a mean water depth of 1.25 m. After 90 minutes, to ensure steady state conditions, all channels were sampled at 20 Hz for 13.6 minutes. Figure 2 shows the measured spectra of surface elevation η, horizontal velocity u, vertical velocity w and the spectrum of vertical velocity deduced from linear theory applied to S_η (f). There are several striking

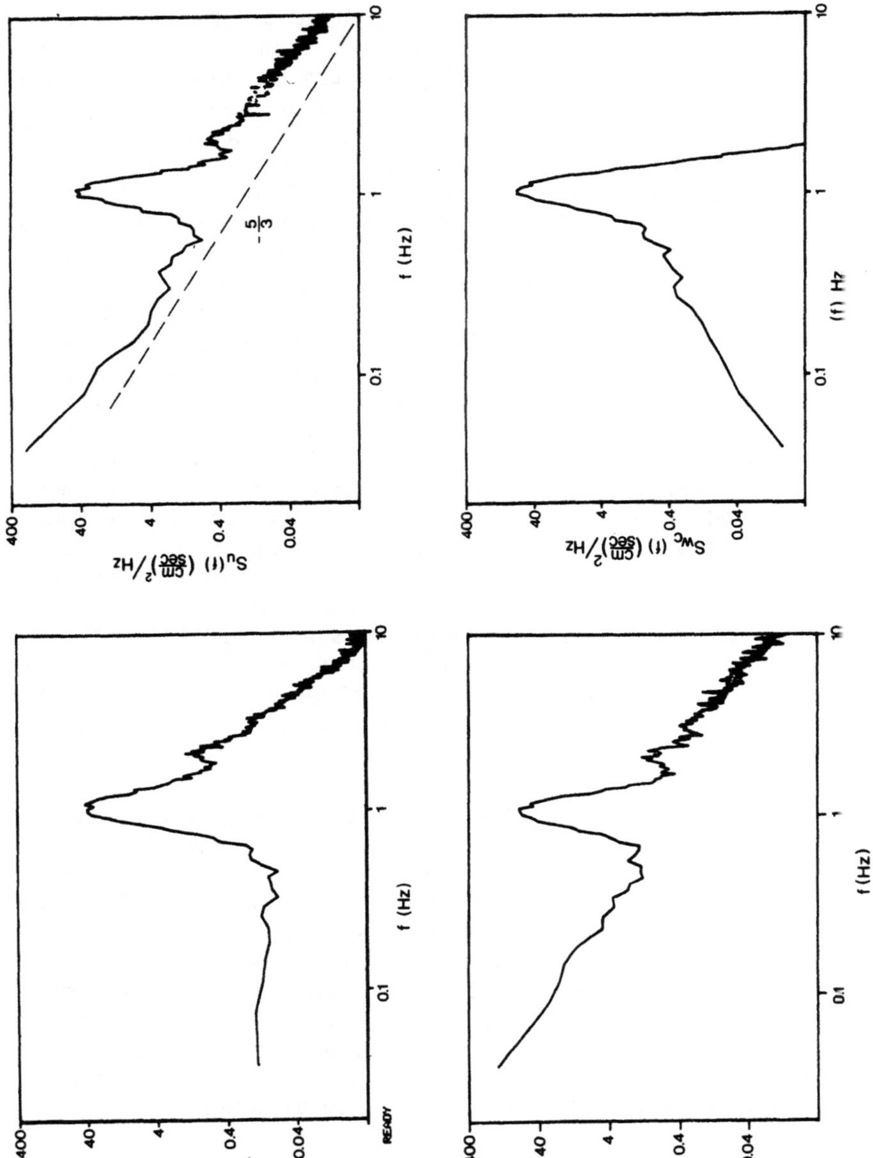

Figure 2.

Laboratory spectra of surface elevation η , longitudinal velocity u, vertical velocity w and vertical velocity computed from S_η (f). The axes are logarithmic.

similarities and dissimilarities among these spectra: (1) They all exhibit the strong wave-related peak, and, while S_w and S_{wc} are in remarkable agreement around the wave peak, S_u and S_w show an increase at frequencies below the peak and a slower fall-off above the peak than S_η or S_{wc}. (2) The ubiquitous -5/3 law is manifest in the low frequencies of both S_u and S_w and suggests that there is a transfer of turbulence energy towards high frequencies from the lowest frequencies shown. The velocity spectra of Jones and Kenney (1977) also conform to the -5/3 law. (3) If we subtract S_{wc} (f) from S_w (f) we are left with two lines of more or less constant but different slope: -5/3 below the wave peak and -5/2 above the wave peak. A similar treatment of S_u (f) with S_{uc} (f) (not shown) yields the same line slopes but with the, perhaps significant, difference that the ordinate separation of the beginning of the -5/2 slope from the extended -5/3 slope is larger in S_w (f) than in S_u (f). The elevation of the start of the high frequency slope above the extended low frequency slope indicates a source of turbulence energy near the wave peak. Further, the higher spectral energy density in S_w (f) at the start of the high frequency slope implies that the turbulence injected near the peak is richer in vertical velocity fluctuations. Lemmin et al (1974) discuss the change in character of the turbulence beneath breaking waves from two-dimensional, associated with a -3 spectral power law, to three-dimensional, associated with a -5/3 spectral power law. Whitecapping would be expected to preferentially produce vertical and longitudinal turbulent fluctuations over lateral fluctuations. However, the directional spread of the waves would ensure production of some lateral fluctuations in a fixed frame of reference aligned with the mean wave direction. Therefore, the -5/2 slope is consistent with the interpretation of quasi two-dimensional turbulence produced by breaking waves.

Figure 3, on linear axes, illustrates the very distinct spectral gap between wave orbital velocities and the lower frequency turbulence. Note that the turbulent part of S_u (f) is larger than that of S_w (f), while the reverse is true of the orbital peaks. The momentum transferred at wave frequencies is essentially zero, while the area under the quadrature spectrum Q_{uw} reflects the out-of-phase character of the wave orbital velocities. The momentum transferred at low frequencies is about 80% of that delivered by the wind τ . Figure 4, on a logarithmic frequency scale, demonstrates that all the momentum is transferred at frequencies at least an order of magnitude less than the surface wave frequencies.

The spectra from a single 13.6 min. field observation, with 10 m wind speed of 10 m/s, are displayed in Figure 5. In this case the electromagnetic current meter was used. It was mounted at a depth of 2.8 m and sampled at 5 Hz. The wave and turbulent regions, separated by a distinct spectral gap, are again evident in S_u (f) and S_w (f). The spectra of Figure 5 show roughly the same

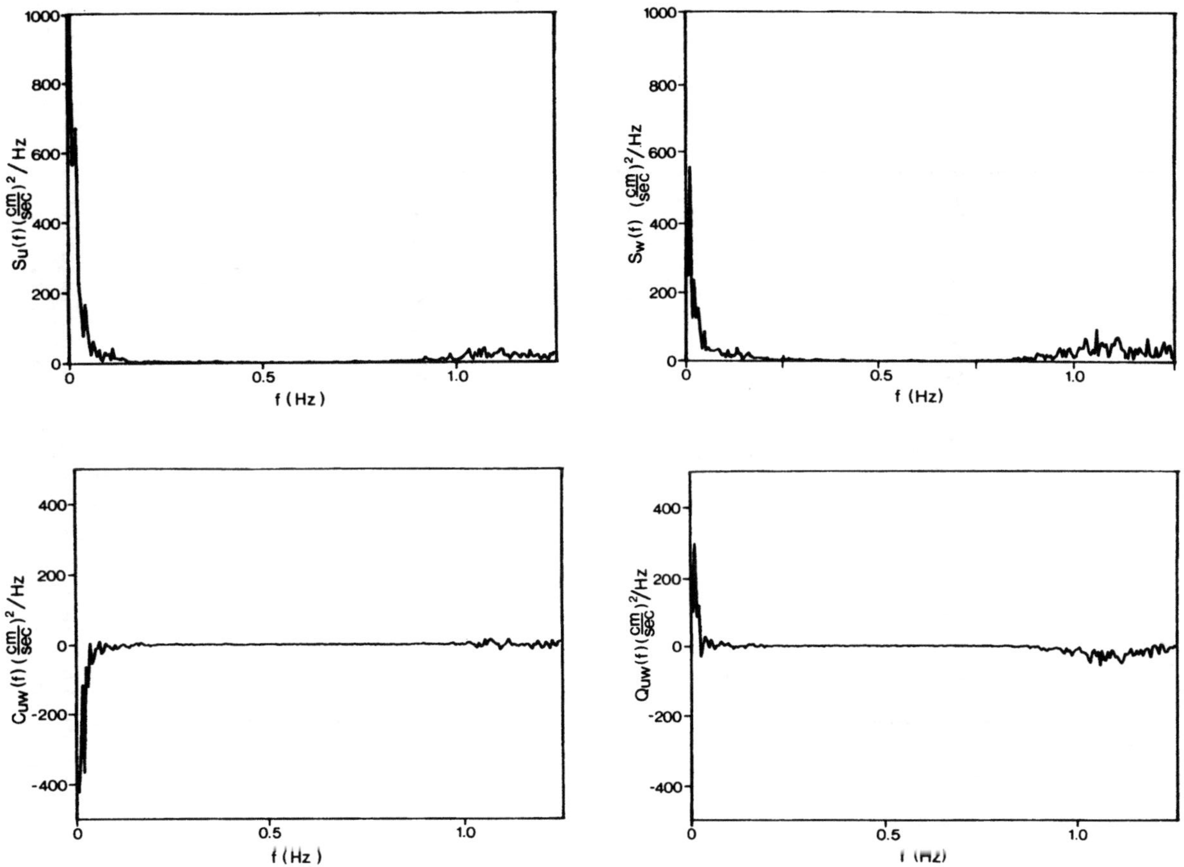

Figure 3.

Spectra and cross-spectra of velocity components for the same data set as in Figure 2. The axes are linear and the spectral gap is evident.

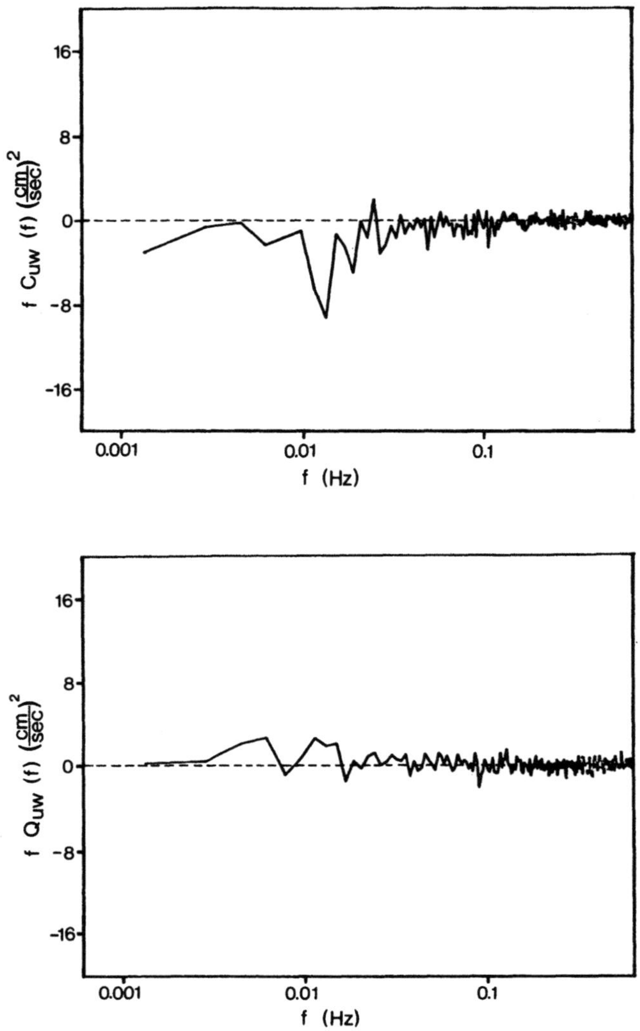

Figure 4.
Cross-spectra of the velocity components on logarithmic frequency axes. The data set is the same as in Figure 2.

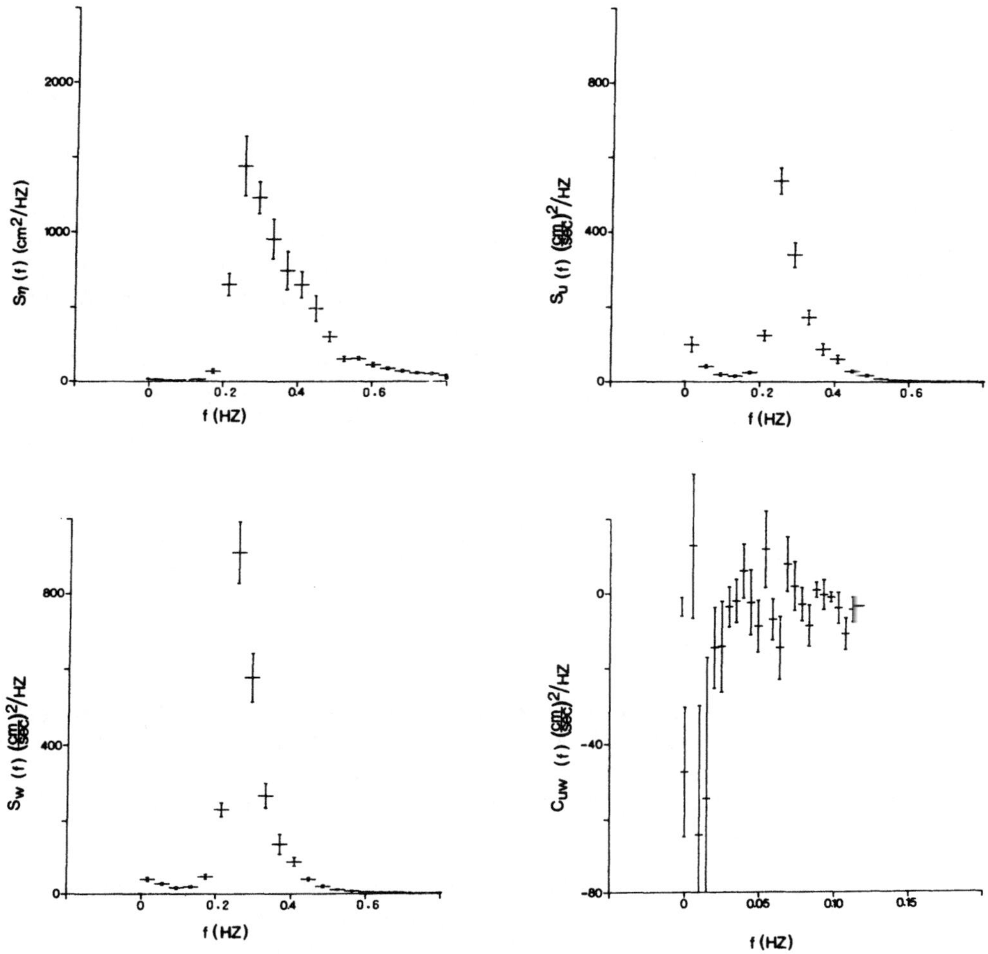

Figure 5.

Spectra and cross-spectra of surface elevation and velocity components for field data. The axes are linear but the cospectral frequency axis is greatly expanded to show the location of the momentum flux peak.

relative magnitudes of the peaks of $S_u(f)$ compared to $S_w(f)$ as in Figure 3. Finally, the momentum flux $C_{uw}(f)$ is entirely at frequencies one tenth or less that of the wave spectral peak and is about one half the measured wind stress τ.

Velocity Profiles

The velocity profiles of Figure 6 were obtained with the electromagnetic current meter. The field observations are 13.6 minute averages, while the laboratory observations are 6.8 minute averages. Both profiles are normalized by the total depth H and the friction velocity $u_* = \sqrt{\tau/\rho_w}$. τ was measured directly using an x-wire anemometer in the laboratory and an anemometer bivane in the field. The laboratory observations collapse onto a single curve showing that the downwind drift is concentrated in the upper 10% of the tank. Here, unlike the experiment of Shemdin (1972), the fluid is entirely contained within the tank and therefore the horizontal pressure gradient, produced by the wind set-up, is reflected in a linear shear stress gradient. Assuming that the shear stress τ_z is zero at the depth of vanishing velocity gradient, we see that the bottom stress amounts to 30% of the surface stress. There is no apparent effect of the waves on the mean velocity profile even though the waves differ markedly in peak wavelength and rms wave height. In both cases the mass transport velocity, integrated over the spectrum

$$U_s = \frac{16\pi^3}{g}\int f^3\, E(f)\; e^{-8\pi^2 f^2 Z/g}\, df$$

is less than 25% of the drift velocity at the surface and drops to 10% or less at the depth of zero velocity of the measured profiles. So that the influence of the wave mass transport velocity is not evident in the laboratory profiles. Here f is wave frequency and E(f) is the one-dimensional spectrum.

The field profiles, on the other hand, tell a quite different story. The open circles, representing very short fetch observations, fall on the normalized laboratory curve. Here again the computed wave mass transport would be buried in the strongly sheared downwind drift. The solid dots, on the other hand, suggest a somewhat less strongly sheared surface current, and do not collapse onto the normalized laboratory profile. In this case, although the wave mass transport velocity is still relatively small at the surface, it decays less rapidly with normalized depth and affects the measured profile in the manner shown.

The measured momentum transfer decreased with depth, but the scatter was too large to allow any firm conclusions. We hope to make more refined flux measurements in the spring of 1978 using the drag sphere.

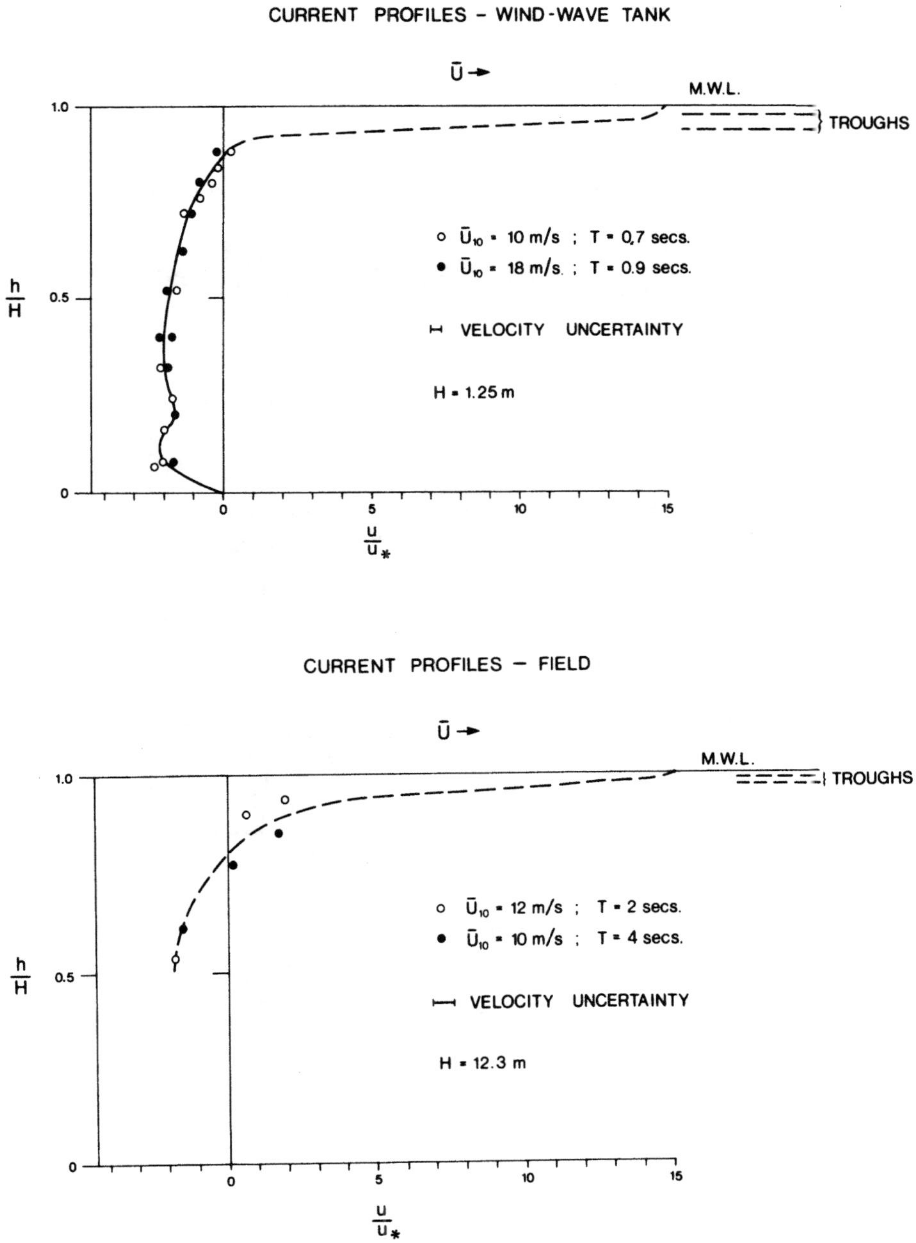

Figure 6.

Normalized current profiles for two laboratory runs and two field runs. The dashed curves are drawn to yield no net mass flux on the assumption that the surface drift velocity is $15u_*$ or approximately 2.5% of $\overline{U}$, as frequently observed. The lines labelled 'troughs' mark the significant trough levels for the two runs in each case. T is the period of the peak of the wave spectrum.

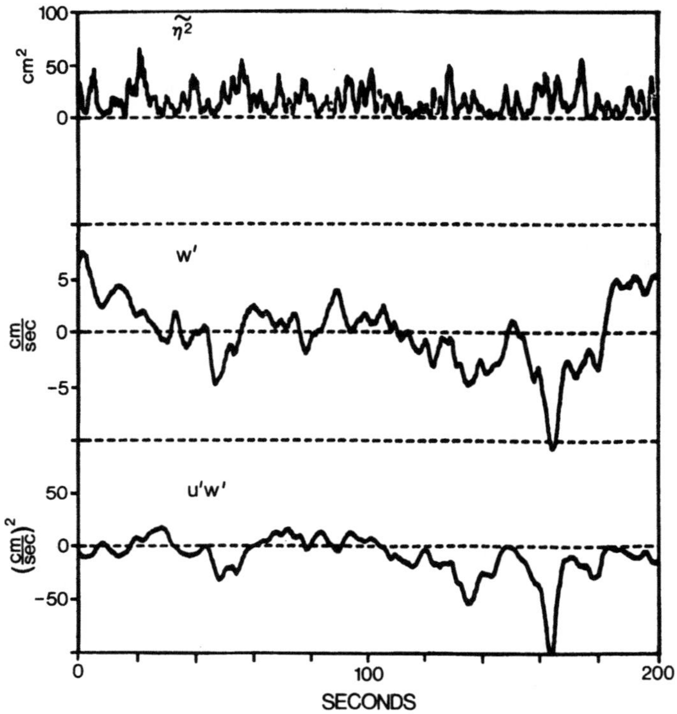

Figure 7.

Coincident time series of the running mean (over one period) variance of surface elevation $\widetilde{\eta^2}$, the low-pass filtered vertical velocity w' and the momentum flux computed from low-pass filtered velocity components u'w'. The data set is the same as that of Figure 2.

CONCLUSIONS

The following specific points are revealed by the observations reported here:

1. The immediate mixing of surface waters due to turbulence produced by a whitecap penetrates to depths of the order of the wave height.

2. The form of the velocity spectra suggests that turbulent fluctuations are produced at low frequencies - possibly through the shear of the mean profile - and at frequencies above the wave peak - probably due to whitecapping. The momentum flux is entirely due to the low frequency fluctuations.

3. The mean velocity profile is very strongly sheared near the surface. The shear may be significantly affected by the wave mass transport if the waves are sufficiently long.

From these points a general picture of the effects of whitecapping emerges. The stress of the wind on the water produces a very strongly sheared downwind current near the surface. When a wave breaks the downward pulse of fluid acts on the velocity shear to produce a downward pulse of momentum. Though each individual pulse of momentum may be quite brief, the intermittent nature of this phenomenon is reflected in the very low frequencies of the momentum flux.

These intermittent pulses of momentum are diffused as they propagate downwards and the appearance of the instantaneous momentum flux is as in Figure 7. The orbital motions have been filtered from the velocity components using an exponential low pass filter with cut-off of 0.1 Hz. The running variance of the wave height is also shown to indicate the rough correspondance in periodicities of the variations of momentum flux and wave groupiness.

ACKNOWLEDGEMENTS

I am grateful to Messrs. T. Nudds, D. Beesley and J. Carew for assistance in the collection of field and laboratory data.

REFERENCES

Birch, K.N., E.J. Harrison and S. Beal, 1976. "A computer-based system for data acquisition and control of scientific experiments on remote platforms." Proc. Ocean '76 Conf., Wash.,D.C. p. 25B-1.

Donelan, M.A., and J. Motycka, 1977. "A miniature drag sphere velocity probe." Submitted to Rev. Sci. Instr.

Jones, I.S.F., and B.C. Kenney, 1977. " The scaling of velocity fluctuations in the surface mixed layer." J. Geophys. Res., 82, 1392-1396.

Lemmin, U., J.T. Scott and U.H. Czapski, 1974. "The development from two-dimensional to three-dimensional turbulence generated by breaking waves." J. Geophys. Res. 79, 3442-3448.

Longuet-Higgins, M.S., and J.S. Turner, 1974. " An 'entraining plume' model of a spilling breaker." J. Fluid. Mech. 63, 1-20.

Phillips, O.M., 1966. "The Dynamics of the Upper Ocean.". Camb. Univ. Press. 261 pages.

Shemdin, O.H., 1972. "Wind-generated current and phase speed of wind-waves." J. Phys. Oceanogr., 2, 411-419.

ABSTRACT

This paper describes field and laboratory experiments aimed at assessing the manner in which wind-generated waves release momentum to the underlying currents. The primary measuring tools are a miniature drag sphere velocimeter designed expressly for this purpose, and a miniature electro-magnetic current meter. The momentum transfer process is considerably more intermittent beneath the waves than it is in the air, being affected by local instabilities of the surface. Far enough (a few wave heights) beneath the surface the sudden downward pulses of momentum are quickly diffused, and the momentum transfer occurs at frequencies an order of magnitude less than that of the peak of the surface wave spectrum.

A short slowed-motion film illustrates the depth of the disturbance caused by a single "whitecap".

DISCUSSION

M. S. Longuet-Higgins:

We have made observations of turbulence due to breaking waves in our wave channel at I.O.S. Wormley, which is 6 ft. wide and about 8 ft. deep. Starting from rest, we found that the depth to which the turbulence penetrated increased with time. After waves of period 1 second had been running for several minutes, almost the whole tank was stirred up.

O. Phillips:

Remember, though, that under natural conditions the breaking is very intermittent. The scale of turbulence is small enough (of order of the wave height) and the turbulence will decay rapidly and may have disappeared before the next breaking event.

M. Donelan:

I think that the film supports Phillips' comment, i.e the turbulence, as indicated by the dye spreading, seems to die out very quickly (perhaps in 5 of 10 seconds).

M. S. Longuet-Higgins:

In our tank experiments it appeared that some small-scale turbulence was carried to greater depths by larger-scale convective motions associated with groups of breaking waves.

LABORATORY EXPERIMENTS ON THE INTERACTION BETWEEN SURFACE WAVES AND FREE CONVECTION IN WATER*

K. B. Katsaros, W. J. Shaw and K. Drummond

Department of Atmospheric Sciences
University of Washington
Seattle, Washington, U.S.A.

INTRODUCTION

Laboratory experiments have been performed to explore the effects of interaction between surface waves and free convection. This work is an extension of an earlier set of experiments on high Rayleigh number (10^9) free convection in water. In particular the temperature drop across the thermal boundary for convection below a free upper surface was related to the net upward heat flux and a relation of the form

$$N = AR^{1/3} \tag{1}$$

was found (Katsaros *et al.*, 1977), where A is a constant. N is the Nusselt number and R the Rayleigh number defined through

$$N = \frac{Q}{k(\Delta T/D)} \tag{2}$$

$$R = \frac{\alpha g \Delta T D^3}{\kappa \nu} \tag{3}$$

where Q is net upward heat flux
k thermal conductivity
ΔT temperature drop across the aqueous thermal boundary layer at the air water interface
D depth of the fluid

---

* Contribution Number 451, Department of Atmospheric Sciences, University of Washington

α thermal expansion coefficient of water
g acceleration due to gravity
κ thermometric difussivity and
ν is kinematic viscosity
Temperature profiles and the statistics of the fluctuating temperature within the boundary layer were also measured (*ibid.*).

With a free upper surface, waves can easily be superimposed on the convective motions. Such a study should have relevance to geophysical phenomena since similar superposition of buoyancy driven motion and waves occur both in the ocean and at internal boundaries of the atmosphere. The effect of waves on heat flux or temperature drop across the aqueous thermal boundary layer has been predicted by O'Brien (1967) and Witting (1971) under the assumption that surface temperature or heat flux remains constant and that the molecules constituting the boundary layer remain constant. Both authors found substantial effects of the order 38% to 900%. The extreme values were obtained for Crapper's (1957) capillary waves. A measure of the "wave effectiveness" W.E. was defined as follows

$$W.E. = \left[\frac{Q_{wavy}}{Q_{flat}}\right]_{\Delta T = const.} = \left[\frac{\Delta T_{flat}}{\Delta T_{wavy}}\right]_{Q = const.} \quad (4)$$

Witting's results for various wave types are given in Table I.

Table I: Witting's (1971) calculated values for "wave effectiveness" as a function of wavenumber, k, and amplitude, a, of the waves.

Wave Type		Wave effectiveness formula	Maximum ka	Maximum W.E.
Capillary		$1 + (ka)^2 + \frac{1}{8}(ka)^4 + 0(ka)^6$	2.29	9.0
Gravity	Stokes*	$1 + (ka)^2 + 7(ka)^4 + 0(ka)^6$	--	--
(irrotational)	limiting	--	0.45	1.38
Gerstner (rotational gravity)		$1 + (ka)$	1.00	2.00

*The Stokes expansion of gravity waves fails to converge at values of ka much less than those appropriate for the wave of limiting amplitude.

One test of the effect of waves on convection included in this study aimed at determining W.E. for high frequency waves. This can be obtained from the NR versus R relationship with and without waves since NR is a non-dimensional heat flux and R a non-dimensional ΔT. A change in ΔT(R) for constant Q(NR), or a change in Q(NR) for constant ΔT(R) can be read as W.E. from such a graph.

Observing tracers in the water with and without waves demonstrated that a change in scales of the convection occurs when waves are present (see Fig. 1).

The scale change of the motion was then quantified by obtaining time-series of temperature at different depths using a resistance film probe in a bridge circuit. Fast Fourier transform spectra were produced from these data.

EXPERIMENTAL SET-UP

The convection tank used for the experiments is seen in Fig. 2. It is of 0.75 m by 0.50 m horizontal cross-section. Water depth used was 0.45 m. The net upward heat flux is simply measured by calorimetry taking heat loss through the tank walls into account ΔT across the aqueous thermal boundary layer was measured with a Barnes PRT5 radiometer. The interface temperature, T_o, is recorded continuously by the radiometer. At regular intervals (30 min) a burst of bubbles from below destroys the boundary layer and the temperature of the bulk of the water, T_{bulk}, is recorded.

Temperature within the water was measured with a resistance film probe (manufactures by Thermosystems Inc.). The sensor is a 1.2 mm long platinum covered quartz fiber of 25 μm diameter, and the platinum is insulated with another quartz layer against the conducting fluid. The probe resistance is measured with a sensitive bridge circuit (for further details see Katsaros *et al.*, 1977). The temperature data was filtered at 20 Hz and recorded on magnetic tape with a 40 Hz digitizing rate.

Waves on the water surface were produced by oscillating a thin wedge up and down across the center of the tank. It was found that waves of 5-6 Hz which traveled without much change in amplitude to the ends of the tank could be produced. Slower waves set-up standing oscillations and faster ones developed crosswaves or did not present a uniform wave field to the radiometer. The wave frequency was measured with a fine resistance wire, whose output was high pass filtered with the low frequency cut-off at 3 Hz. From the frequency, ω, a wave number, k, was calculated using the relation (e.g., Kraus, 1972)

$$\omega^2 = k\, g\, (1 + \frac{\gamma k^2}{\rho_w g}) \qquad (5)$$

where γ is surface tension. The amplitude of the waves could not be accurately determined from the wave-gauge response. Changes in

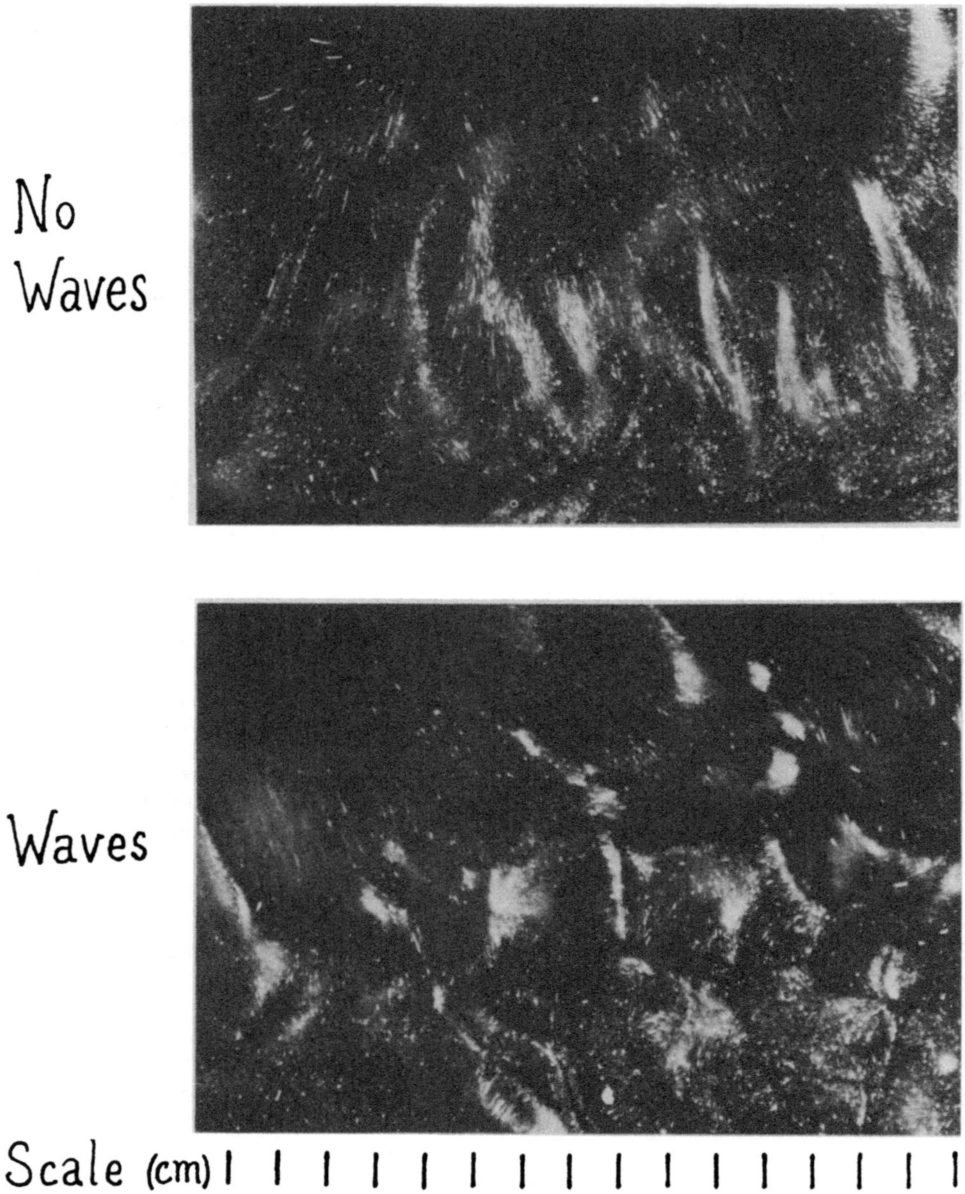

Fig. 1. Horizontal pattern of convection below the free surface of a deep water tank losing heat to the air above with and without waves. The dark lines are regions of cold sinking water. The tracers used were Kalliroscope crystals illuminated by a helium-neon laser light spread in a thin sheet by a cylindrical lens.

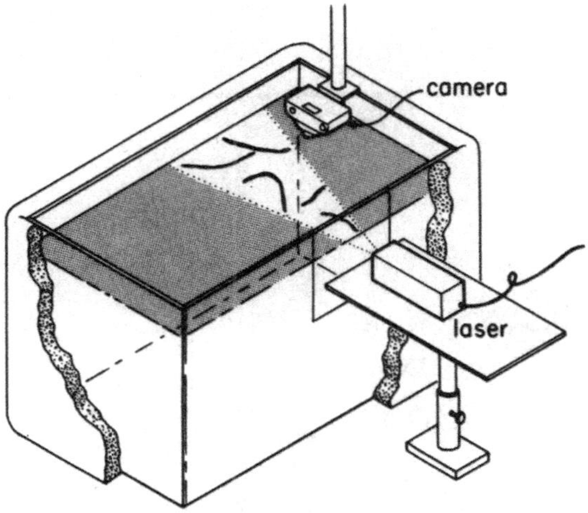

Fig. 2. The laboratory tank.

temperature and conductivity produced drifts. However, photographs taken through the sides of the tank allow an approximate determination of the wave amplitude.

The heat flux out of the tank is simply determined by the temperature and humidity conditions of the room. A thermostat controls the temperature but humidity varies considerably with season. The heat flux is, therefore, not simply related to the surface temperature of the water.

RESULTS

Fig. 3 shows the relationship between NR and R for experiments with and without waves. It can be seen that when waves are added the slope remains about 1/3 in (1), while the coefficient changes by a factor of about 2.0. This is in reasonable agreement with Witting's prediction, although the assumptions of his development are not fulfilled. For instance, the effect is not simply a result of the stretching and distorting of a thin layer of water, the boundary layer, since we know from our earlier study (Katsaros et al., 1977) that the boundary layer is periodically drained off in the convective plumes. The increase in surface are explained by Witting as one of the contributions to W. E. holds, of course, in any case.

It is of interest to consider to which extent the two parameters, Q and ΔT, are responsible for the observed W.E., if we start the experiments with and without waves having the same initial conditions of water and air temperature and room humidity. Fig. 4 and 5 show the relationship of Q versus T_{bulk} and ΔT ver-

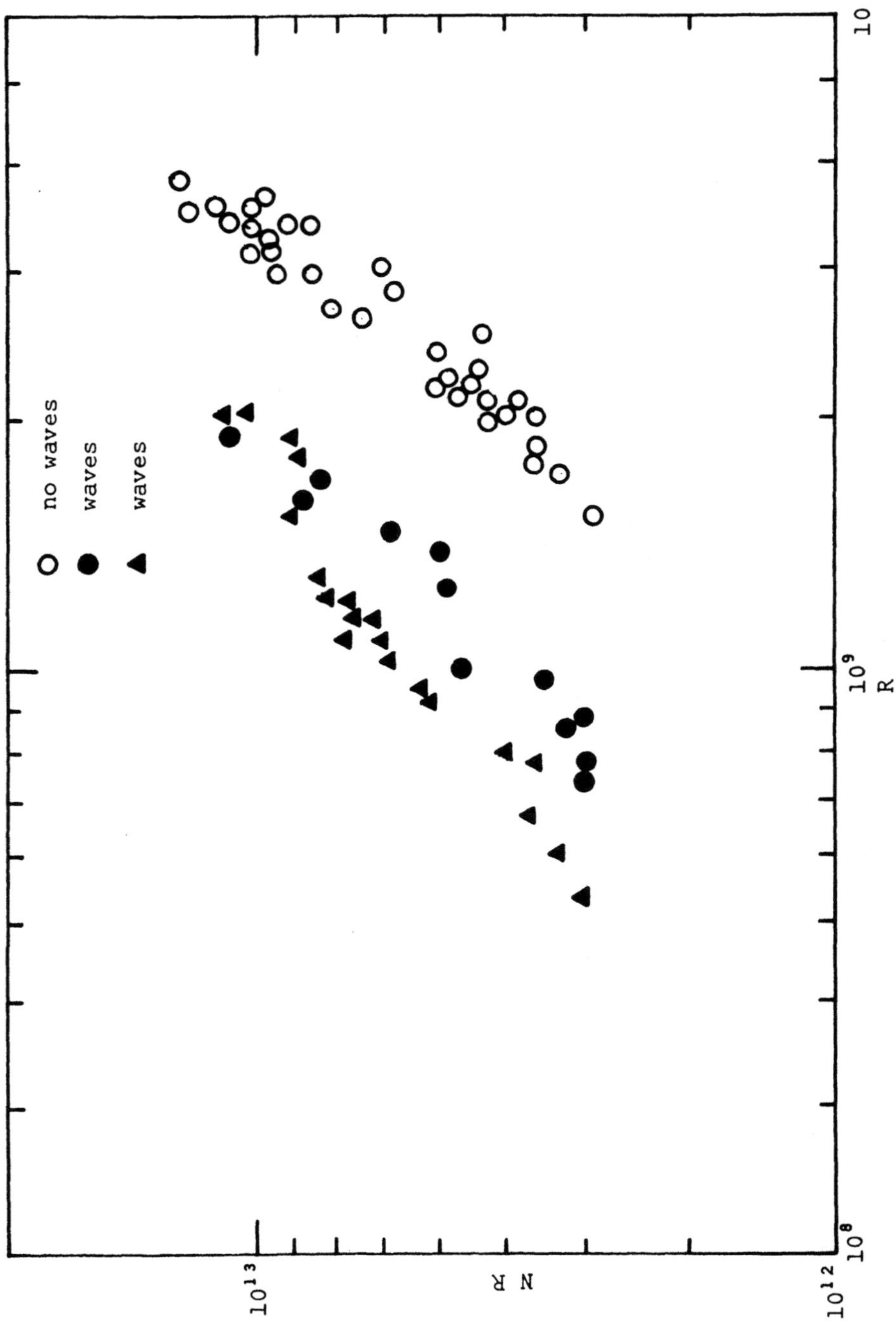

Fig. 3. The NR versus R relation with and without waves. The "wave effectiveness" is given by the relative change along one coordinate axis (see eq. 4).

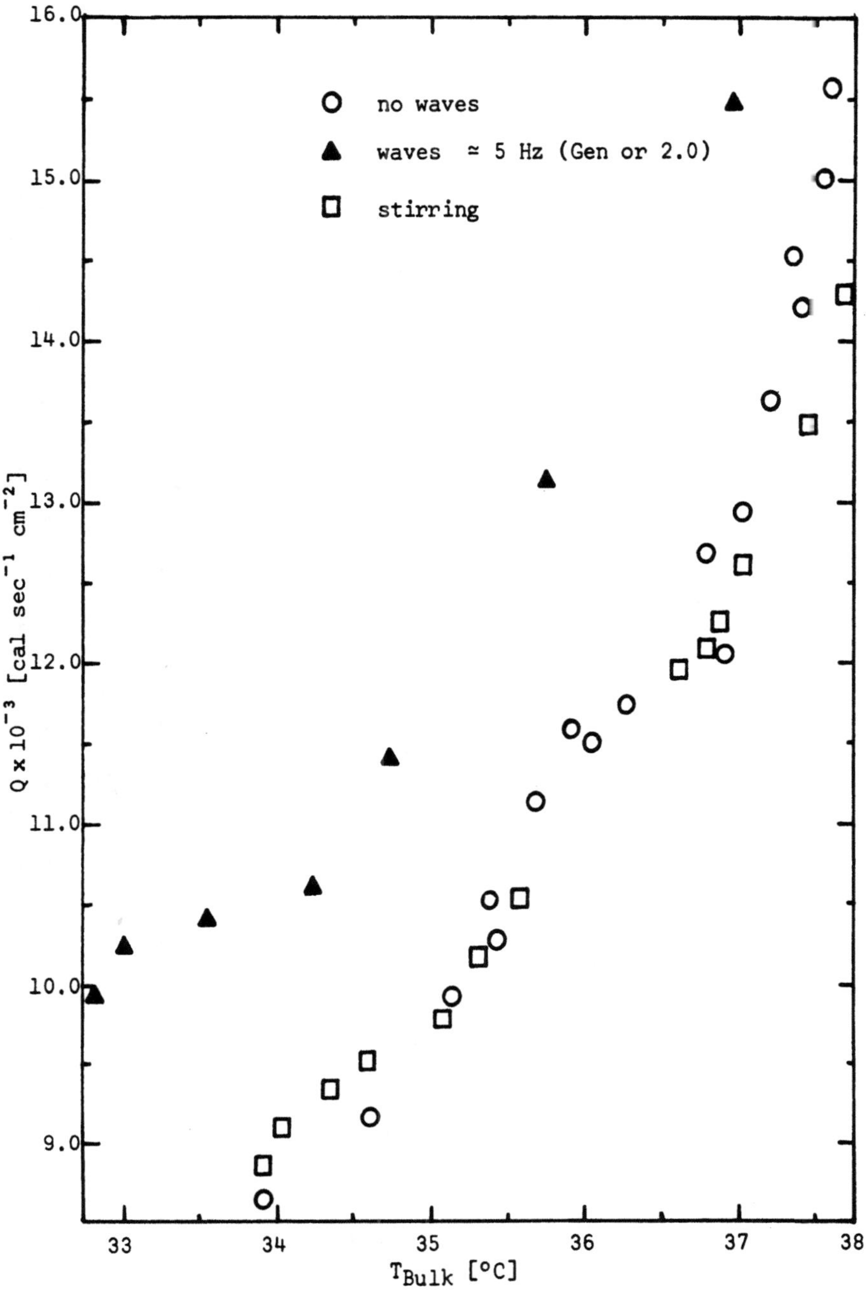

Fig. 4. Observed change in heat flux when waves are generated on the convection tank; all initial conditions remaining fixed. Note that no change is observed in the heat flux, when propellers stir the water below the interface.

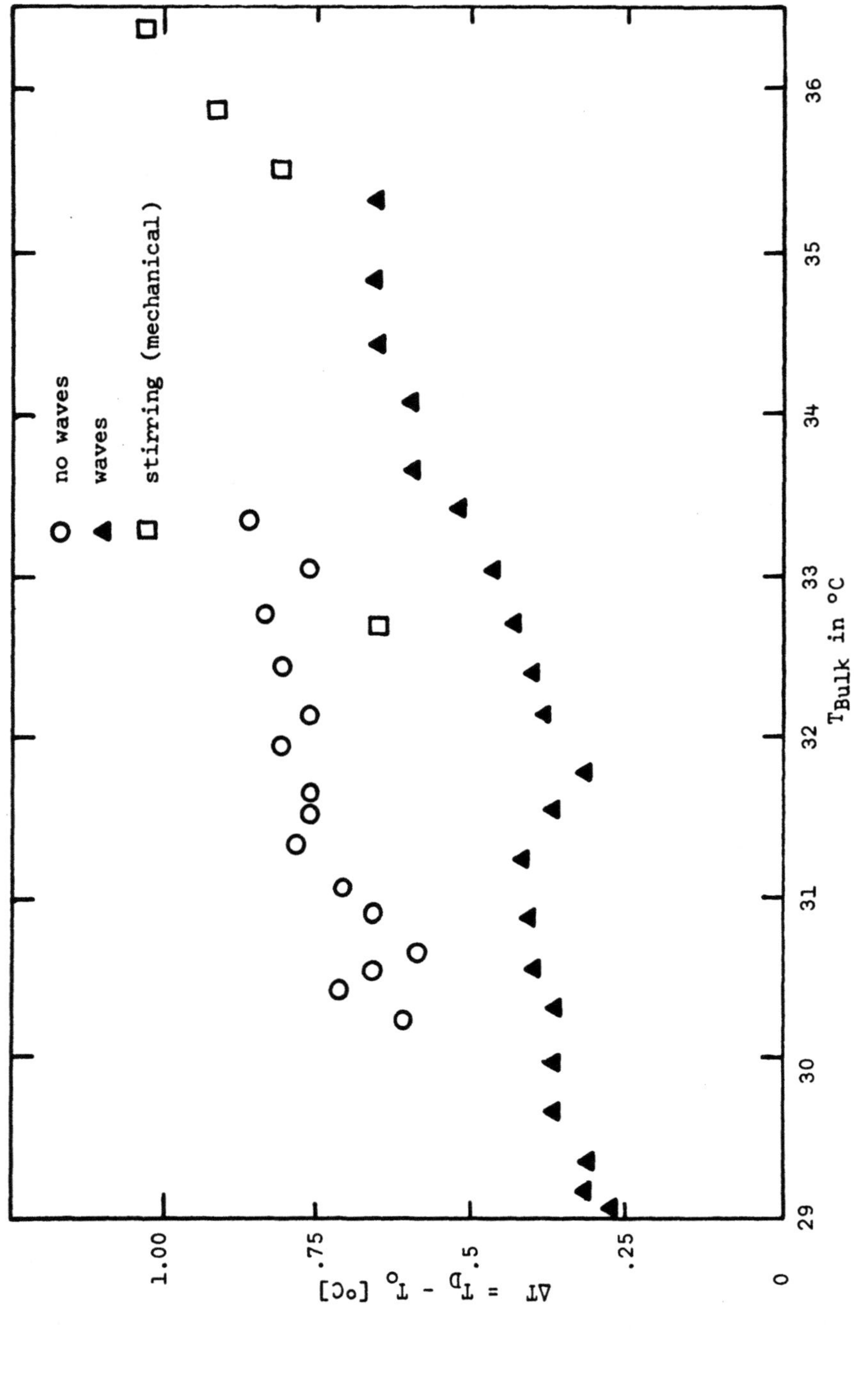

Fig. 5. Observed change in ΔT, when waves are generated on the convection tank; all initial conditions remaining fixed. Vigorous stirring of the bulk water by propellers turning at a depth of 10 cm is seen to cause a substantial decrease in ΔT (No waves were generated during stirring).

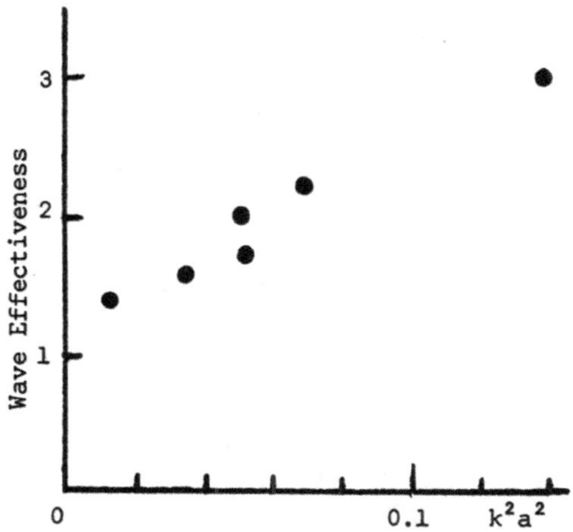

Fig. 6. Relation of W.E. to k^2a^2. The a's are obtained by measuring the amplitude of the waves against the side of the tank with a ruler.

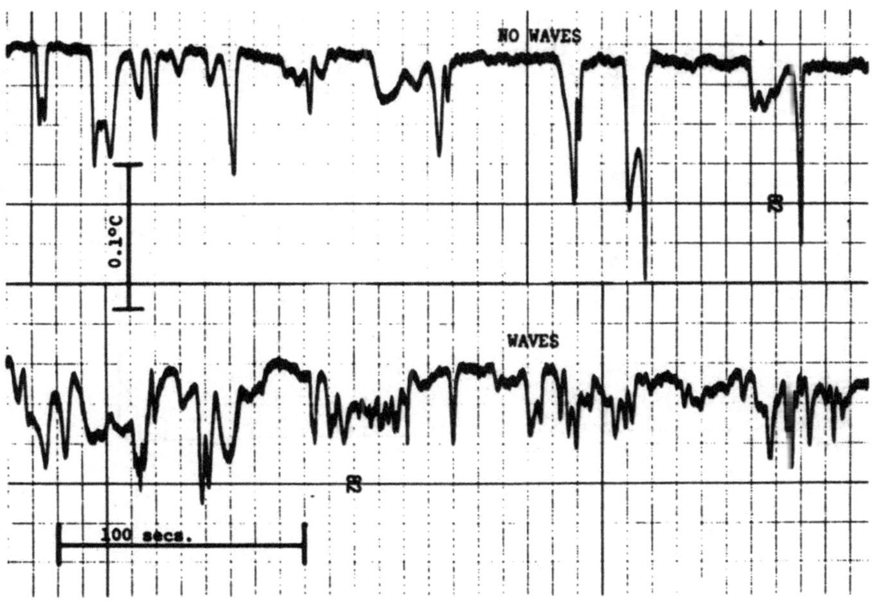

Fig. 7. Time-traces of temperature obtained at 1 cm depth with the resistance film probe.

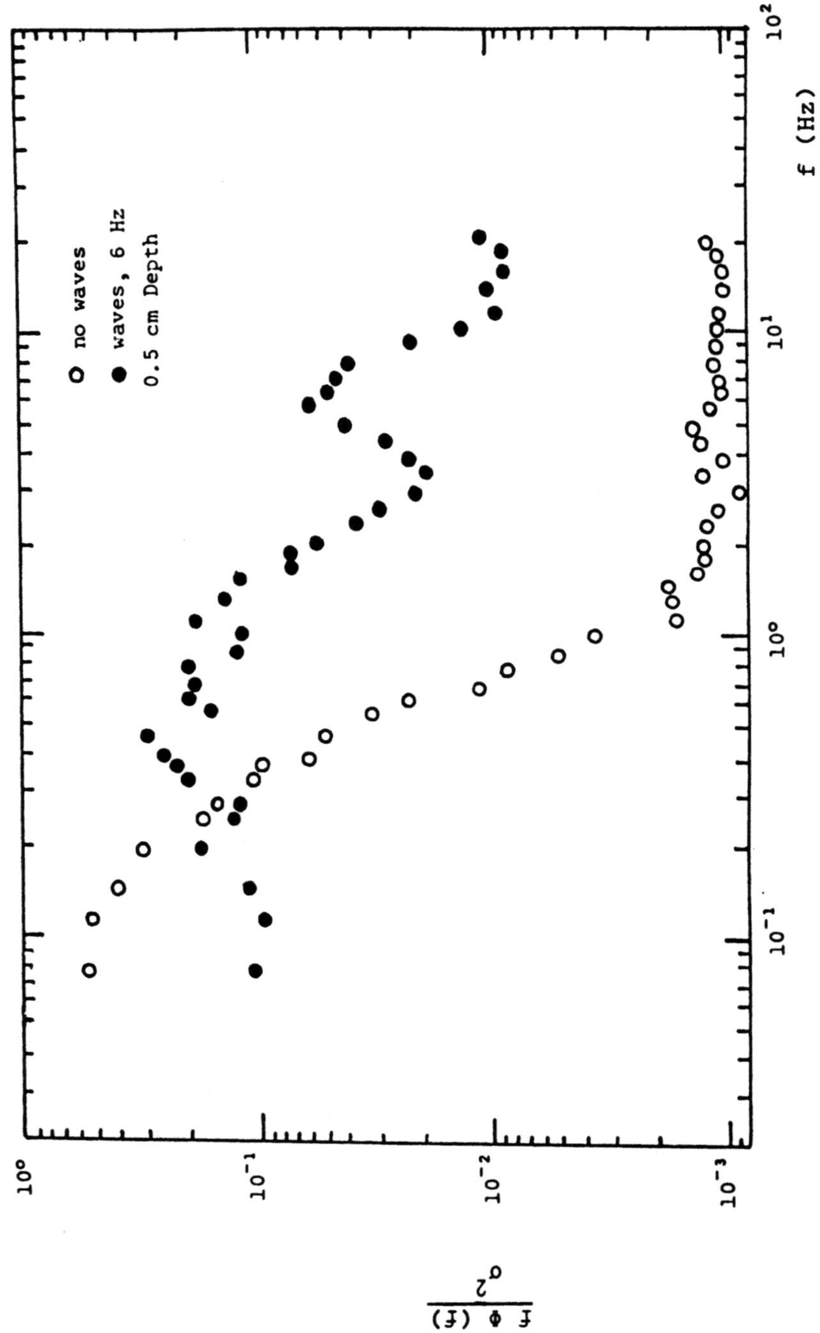

Fig. 8. Frequency spectra of temperature at 0.5 cm depth.

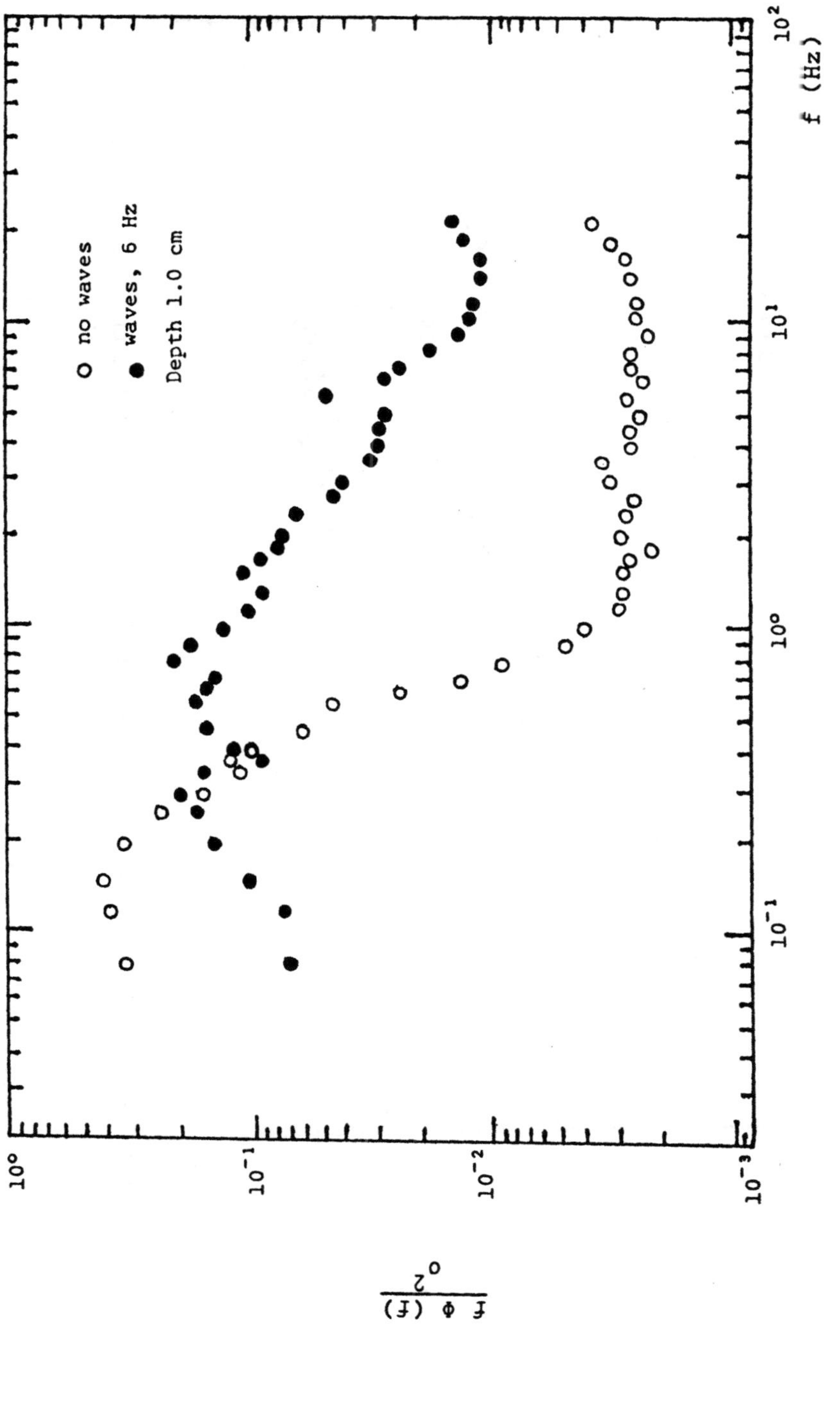

Fig. 9. Frequency spectra of temperature at 1.0 cm depth.

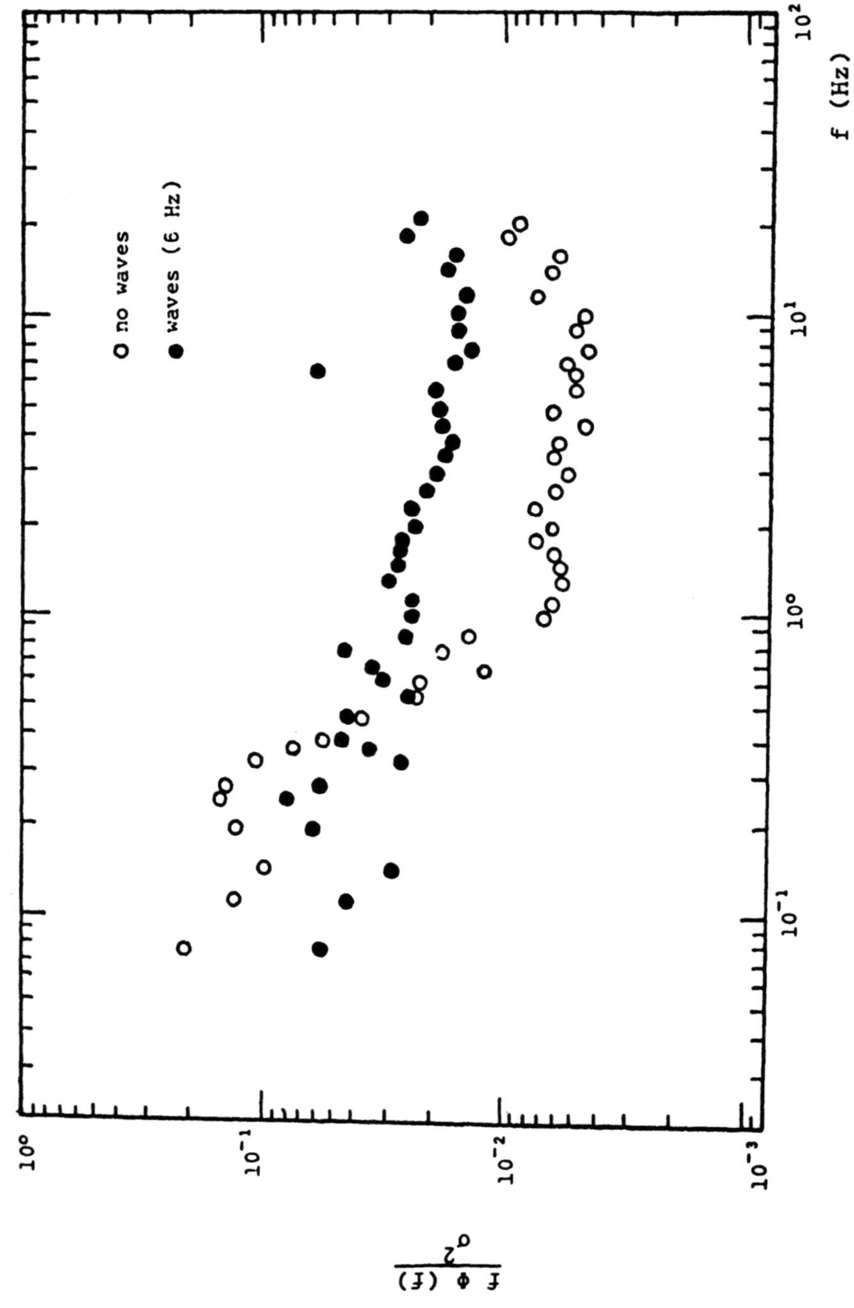

Fig. 10. Frequency spectra of temperature at 3.0 cm depth.

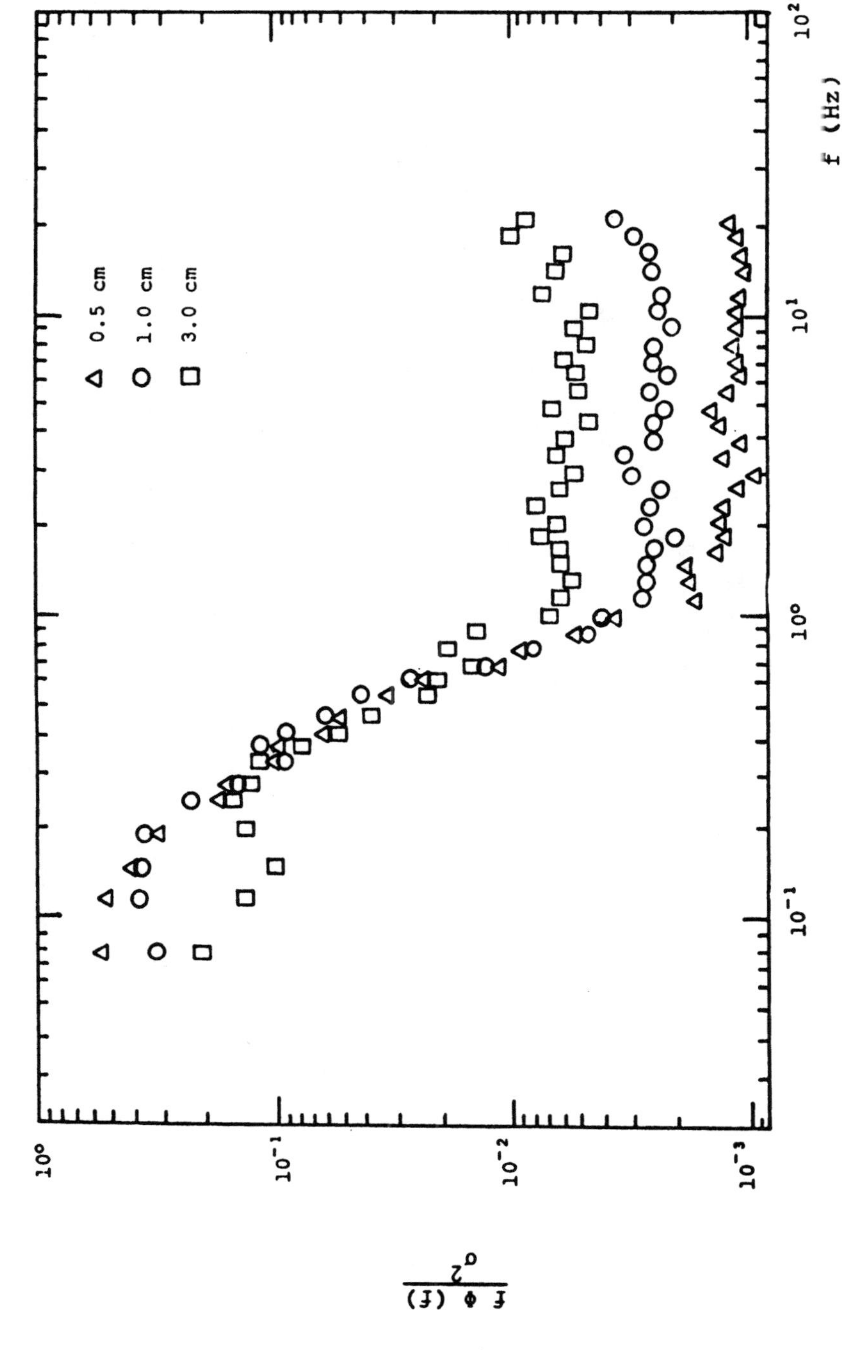

Fig. 11. Temperature spectra as a function of depth without waves.

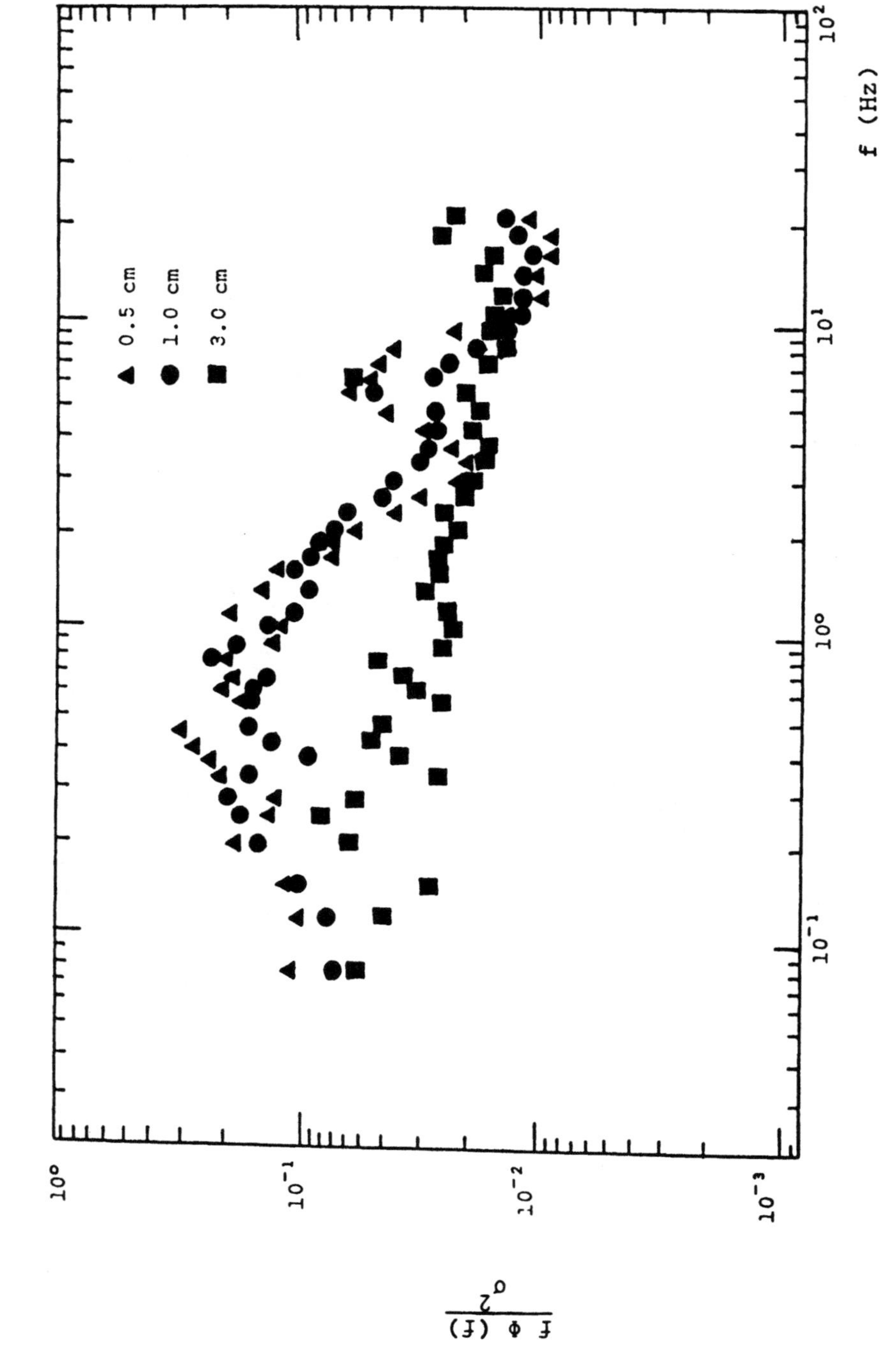

Fig. 12. Temperature spectra as a function of depth with 6 Hz waves.

sus T_{bulk} obtained on the same day (external conditions constant). Both quantities are obviously changed as waves appear.

There was some concern that the wave generator may produce some turbulence not related to the waves themselves, which could effect Q or ΔT. To test this a separate experiment was run with two propellers stirring the water vigorously under the surface. These data runs are marked on the graphs. It is seen that Q was not affected, as expected. ΔT was reduced, of course, but the stirring was of such an intensity and over the whole tank, that we are certain any drag caused by the wave generator would only cause a very small reduction in ΔT.

Work by Hill (1970) and Chang and Wagner (1975) performed in a wind-wave tank also tend to support Witting's theory. From the crude estimates of a and the calculated k's an almost linear relation between k^2a^2 and W.E. was found, Fig. 6. MacIntyre (1971) measured increased gas flux in the presence of high frequency waves.

The effect of the waves on the scales of the convection, seen in Fig. 1 can also be identified in the time-traces of Fig. 7. The cold plumes are much reduced in amplitude but occur more frequently. Fig. 8, 9, and 10 show frequency spectra of the temperature obtained at 3 depths. The effect of the 6 Hz waves is to broaden the spectra near the surface. Comparing the spectral energy distribution as a function of depth in Fig. 11 without waves and in Fig. 12 with waves, one sees the general decrease in spectral energy and broadening of the spectra with depth in both cases.

SUMMARY

Surface waves in the capillary-gravity range of frequencies are seen to interact strongly with deep water free convection. The so called "Wave effectiveness" in increasing heat flux at constant temperature is of the order of 2 for 6 Hz low amplitude waves. The scales of the motion both in time (and space) become shifted to higher frequencies (and wave numbers). Adequate theories to explain the latter effect have not been found.

It is thought that this interaction between surface waves and convective motion may represent an analogy to similar interactions in the upper layers of the ocean, and possibly also in the upper portions of the planetary mixed layer of the atmosphere.

REFERENCES

Chang, J. H. and R. N. Wagner, 1975: Laboratory measurements of surface temperature fluctuations induced by small amplitude surface waves. J. Geophys. Res., 80, 2677-2687.

Crapper, G. D. 1957: An exact solution for progressive capillary waves of arbitrary amplitude. J. Fluid Mech., 2, 532-540.

Hill, R. H., 1970: Laboratory measurements of heat transfer, wind velocity, profiles and temperature structure at an air/water interface. Naval Res. Rept., 7212, pp. 31.

Katsaros, K. B., W. T. Liu, J. A. Businger and J. E. Tillman, 1977: Heat transport and thermal structure in the interfacial boundary layer measure in an open tank of water in turbulent free convection. J. Fluid Mech., (in press).

Kraus, E. G. 1972: Atmosphere-Ocean Interaction. Clarendon Press, Oxford, pp. 275.

MacIntyre, F., 1971: Enhancement of gas transfer by interfacial ripples. Phys. Fluids, 14, 1596-1604.

O'Brien, E. E., 1967: On the flux of heat through laminar wavy liquid layers. J. Fluid Mech., 28, 295-303.

Witting, J., 1971: Effects of plane progressive irrotational waves on thermal boundary layers. J. Fuid Mech., 50, 321-334.

Acknowledgement: We are grateful to Messrs. J. E. Tillman, H. Lahore, M. Brengle, F. Weller, A. Weikert and M. Steere, who contributed to the technical development of the laboratory set-up. This research was sponsored by the Office of Naval Research under Contract N00014-75-C-0502-MOD P00008 Project NR 083-012.

DISCUSSION

M. Coantic:

I believe that the results concerning the respective changes in temperature-drop and heat flux are the consequence of the experimental situation, where a large temperature difference has been set between water and air. In the field, where that difference is generally small, the heat flux should probably be much affected when the temperature-drop changes.

K. Katsaros:

Yes, it is true that since the heat flux (sensible, latent and radiative) is mainly controlled by the air-water temperature difference, which is large (of the order of 10° C in our experiments) it would be very little affected by a change in surface temperature(or in Δ T) caused by the propeller stirring. If

(T_{air} - T_{water}) is smaller, reducing the boundary layer thickness may have a relatively larger effect on heat flux.

H. Charnock:

The flow visualisation - to some extent the spectra - show structures on bigger scales than the waves. Do you think these are more important than those treated theoretically?

M. Coantic:

The theories referred to by Dr. Katsaros establish the heat-flux increase due to waves with respect to a no-flow purely conductive situation. Here, the initial situation corresponds to well-developed convective motions so that the study reveals in fact the effects of the nonlinear interactions between those convective structures and a surface wave motion.

K. Katsaros:

Yes, there exists from the start a separation in the wave-number of frequency scales between the free convection (0.1-1.0 Hz) and the surface waves (6 Hz). The result of the interaction is that the waves smear the scales of the convection over a much broader range of frequencies, where there is a small 6 Hz peak near the surface. This can easily be understood, since the source layer for the convective plumes or sheets is a few mm thick boundary layer at the interface, of the same order as the amplitude of the waves. However, it would be nice to observe the interaction between convection and waves at a wide range of relative scales, particularly at comparable scales. What would happen there I do not know off hand.

LONG WAVE - SHORT WAVE INTERACTION

Dieter Hasselmann

Meteorological Institute, University Hamburg,
Bundesstrasse 55, 2000 Hamburg 13

1. INTRODUCTION

We shall consider the interaction of a long wave (Z_1, k_1,c_1) with a short wave (Z_2, k_2,ω_2), σ^2=kg,for irrotational waves, surface tension will be neglected and all waves shall be in deep water. We shall consider the case of strong coupling $\varepsilon=k_2Z_1>1$. There are essentially three approaches to this problem.

a) Direct treatment of the partial differential equations.
b) A treatment of the Fourier transformed equations, that is then essentially a weak interaction treatment.
c) A WKBJ treatment which for this case can make use of the fact that a Lagragian formulation exists, so that the theory of adiabatic invariants as developed by Whitham (see Whitham(1974) for an extensive list of references) can be employed.

It seems that for our case only the third approach has been worked out (Garrett and Smith, 1976). The WKBJ approach relies on premises which are nearly exactly the opposite to those met in the weak interaction approach. WKBJ theory envisages the short wave as a nearly free wave, free relative to the local environmental parameters, adjusting wavenumber, frequency and mean energy to the local parameters. In interaction theory we envisage a wave which is essentially nonlinear, so strongly nonlinear that the local frequency and wavenumber are subject to large variations as we move from crest to trough of the long wave.Since the interaction is so strong the prevailing view seems to be that an interaction treatment presents at best a formidable task, and at worst that it is doomed to fail. On the other hand, since the direct treatment a) is really quite a simple matter, and since b) is only the Fourier treatment of a) there is some hope to make progress with b). It should be possible to obtain solutions with a) to any desired accuracy and this will shed some light on the reliability of the WKBJ approach, which lies somewhere in between a) and b), retaining the Fourier approach for the short waves, and the local approach for the long wave.

2. THE WKBJ RESULTS OF GARRETT AND SMITH(1976)

In the system moving with the phasespeed c_1 of the long wave the long wave is given as

$$\zeta_1(x) = 2Z_1 \cos k_1 x \tag{2.1}$$

and the velocity is given as

$$u_1(x) = -c_1(1-2(k_1Z_1)\cos k_1 x) \tag{2.2}$$
$$u_3(x) = -c_1(2k_1Z_1)\sin k_1 x \tag{2.3}$$

An observer O_F who is watching the small waves as he is floating in the stream registers variations of gravity, due to the accelerations of his system,

$$g(x) = g(1-2k_1Z_1\cos k_1 x + O((k_1Z_1)^2)) \tag{2.4}$$

and he expects to find small waves satisfying the dispersion relation

$$\omega = [k_2(x)\cdot g(x)]^{1\ 2} = \sigma_2(x) \tag{2.5}$$

and will define the local short wave energy

$$E_2(x) = g(x)\cdot Z_2^2(x) \tag{2.6}$$

and action

$$S_2(x) = \frac{E_2(x)}{\sigma_2(x)} \tag{2.7}$$

He will observe wave groups propagating with

$$\dot{x}_\ell = c_{gr} = \frac{\partial\sigma}{\partial k} \tag{2.8}$$

and expects conservation of action

$$\frac{\partial}{\partial t} S_2 + \frac{\partial}{\partial x}(c_{gr}\cdot S_2) = 0 \tag{2.9}$$

Finally, he knows that the other, stationary, observer O_s staying at a point $x=x_0$, is observing a stationary system, so that the frequency of the wavegroup is conserved

$$\Omega = \sigma_2(k,x) + k_2(x)\cdot u_1(x) = \text{const} \tag{2.10}$$

Thus, the stationary observer, having been informed on the results, puts his equations together

$$\dot{x} = c_{gr}(x,k) + u_1(x) = \frac{\partial\Omega}{\partial k} \tag{2.11}$$

$$\dot{k} = -\frac{\partial\Omega}{\partial x} \tag{2.12}$$

$$\frac{\partial}{\partial t} S_2(x,t) + \frac{\partial}{\partial x}\{(c_{gr}(x,t)+u_1(x,t))S_2(x,t)\} = 0 \tag{2.13}$$

Thus S is conserved along the trajectory of the group. With $k_1Z_1 \ll 1$ the trajectories are easily evaluated and with Garrett and Smith we obtain, with $\alpha = \sigma_1/\sigma_2 = c_2/c_1$,

$$k_2(x) = k_2(1 + 2\varepsilon\alpha^2\cos k_1x) \tag{2.14}$$

and from conservation of action we obtain for the local high frequency amplitude

$$Z_2(x) = Z_2(1 + 2\varepsilon\alpha^2 \cos k_1x) \tag{2.15}$$

These results should be valid for all $k_2 \gg k_1$, provided $k_1Z_1 \ll 1$, regardless of the strength of the coupling k_2Z_1. Of course, WKBJ theory has its limits. The wavepackets must have a length L, so that $k_1L \leq 1$, since otherwise we cannot localize the position of the group. Thus the wavenumber resolution we can obtain is always $\Delta k_2 \geq k_1$, so that the variation in wavenumber predicted by (2.14) $\delta k_2 = 2\varepsilon k_1$ becomes meaningless for $\varepsilon < 1$. At least in this case weak interaction theory can take over, but as we shall see it can also handle all cases for $(\varepsilon\alpha) < 1$.

Furthermore, the condition $k_1L \leq 1$ seems to be somewhat artificial. Either short wave packets separated by some distance interact significantly with each other, in which case the theory is somewhat incomplete, because interaction distances do not enter anywhere; or they do not interact, in which case it should be possible to have a theory with infinitely long groups yielding the same results.

3. THE FOURIER FORMULATION

3.1 The Variational Principle

Defining $a(k,\omega)$ as the Fourier transform of $\zeta(x,t)$

$$a(k,\omega) = \frac{1}{(2\pi)^3} \int e^{-i(kx-\omega t)} \zeta(x,t)\, dx\, dt \tag{3.1}$$

it can be shown (D.Hasselmann, 1977a, hereafter referred to as (I)), that the equations of motion are equivalent to an integral equation for the $a(k,\omega)$.

Writing β for (k,ω) we obtain, in the system with vanishing mean current,

$$(-\omega^2+\sigma^2)a(-\beta_1) = \sigma_1^2 \sum_{n=3}^{\alpha} \int D_n \delta(\beta) a(\beta_2)..a(\beta_n) d\beta_1 ... d\beta_n \tag{3.2}$$

Here $\beta = \sum_{i=1}^{n} \beta_i$

and the coupling terms D_n are functions of $\beta_2 ... \beta_n$, coupling the product $a(\beta_2)...a(\beta_n)$ into $a(-\beta_1)$. Thus,

$$D_n = D_n(\beta_1;\beta_2...\beta_n) \tag{3.3}$$

It has been shown in (I) that the coupling is symmetric:

$$D_n(\beta_1;\beta_2...\beta_n) = D_n(\beta_{\nu_1};\beta_{\nu_2}...\beta_{\nu_n}) \tag{3.4}$$

for $\nu_1 ... \nu_n$ any permutation of $1...n$ provided the following assumption is true: The wave energy and momentum are stationary if the products $Z_k(t)Z_k^*(t)$ are stationary at all k. (Z_k are the spatial transforms of ζ.)
Thus if we define

$$b(\beta) = \frac{a(\beta)}{\sigma(\beta)} \tag{3.5}$$

the equations of motion take a completely symmetric form and can be obtained from the spectral formulation of the Langrangian principle

$$\delta_b L = 0 \tag{3.6}$$

with respect to all variations of $b(\beta)$ and

$$L = L_2 + L_3 + ... \tag{3.7}$$

$$L_n = \frac{1}{n} \int d\beta_1 .. d\beta_n (\sigma_1 .. \sigma_n) D_n(\beta_1 .. \beta_n) \delta(\beta) b(\beta_1) .. b(\beta_n) \tag{3.8}$$

and specifically

$$D_2 = \sigma^2(\beta_1) - \omega^2(\beta_1) = (\sigma_1+\omega_1)(\sigma_2+\omega_2) \tag{3.9}$$

Thus the linear solutions are obtained as

$$b(\beta) = \lambda(\beta)\delta(\omega_\beta-\sigma(\beta)) \tag{3.10}$$

with any arbitrary well behaved function $\lambda(\beta)$, which may still have $\delta(k-k_o)$ contributions. Classical weak interaction theory is well understood and can easily be developed from (3.6). This is exactly equivalent to K.Hasselmann's formulation (K.Hasselmann, 1966; K.Hasselmann, 1967, and references cited there) and will not be discussed here. We mention only that the theory relies on a multiple expansion in many slopes ε_i and thus becomes suspect if the largest slope exceeds unity. The largest slope is given as

$$\varepsilon_{max} = k_{max}\cdot a_{max} \tag{3.11}$$

k_{max} being the highest occupied wavenumber and a_{max} being the largest encountered Fourieramplitude. Thus for our long wave - short wave case, $\varepsilon_{max} = \varepsilon$

$$\varepsilon \quad = k_2Z_1 > 1 \tag{3.12}$$

and we can only hope that classical weak interaction theory converges for $\varepsilon > 1$.

The author believes that the potential of (3.6) has not yet been fully exploited and that probably much can be learned from other fields of science. The next section provides an example, but is hopefully only a first step.

3.2 The Eigenvalue Condition for Short Waves

If we wish to find infinitely long wave groups of constant frequency (to the stationary observer), then all Fourieramplitudes must lie on a line C

$$\beta = (\omega,k) = (\omega_2,k_2) + s\cdot(\sigma_1,k_1) \text{ , } -\infty < s < +\infty \tag{3.13}$$

and must in fact have a δ-function singularity there. The frequency ω_2 is an eigenvalue to be determined, and our stationary observer sees the frequency

$$\Omega_2 = \omega_2 - k_2c_1 \tag{3.14}$$

and the distribution of amplitudes $a(\beta)$ along C appears to him as the wavenumber distribution, since the transformed line C' lies parallel to the wavenumber axis. We shall now also require that the solution is periodic in $k_1\cdot x$, so that our stationary observer sees

$$\zeta_2(x) = e^{i(k_2x-\Omega_2t)}\cdot P(k_1x) \tag{3.15}$$

with

$$P(k_1x) = P(k_1x+2\pi) \tag{3.16}$$

It can in fact be shown that this requirement introduces no loss of generality (D. Hasselmann, 1777b,hereafter referred to as (II)), but this will not be discussed here.

This then means that the distribution in wavenumber space along C' is just a sum of δ-functions located at

$$k_n = k_2 + nk_1, \qquad n = 0,\pm 1\ldots \quad .$$

Now this is in agreement with our spectral equation (3.2).

Our highfrequency wave (3.15) is to be infinitesimal, the long wave is just

$$a_1(\beta) = Z_1[\delta(\beta-\beta_1) + \delta(\beta+\beta_1)];\beta_1 = (k_1,\sigma_1) \tag{3.17}$$

and if we assume that the highfrequency wave has a contribution $\delta(k-k_2)\delta(\omega-\omega_2)$, then of course the interactions D_3, $D_4\ldots$ generate $\delta(\beta)$ at $\beta_n = \beta_o + n\beta_1$. Thus, if we now only consider the contribution D_3 in (3.2) we have

$$(-\omega_n^2+\sigma_n^2)a_n = \sigma_n^2\{D_3(-\beta_n;\beta_{n-1},\beta_1)a_{n-1}+D_3(-\beta_n,\beta_{n+1},-\beta_1)a_{n+1}\}Z_1 \tag{3.18}$$

with $a(\beta_n) = a_n\cdot\delta(\beta-\beta_n)$.

We can continue in this manner with terms D_4, D_5 which become increasingly more complex. Details of the calculation are to be presented in (II), but we mention that the series of higher order coupling terms is probably structured as

$$D_2 \approx \sigma_2\sigma_1$$

$$D_3Z_1 \approx \varepsilon\cdot\sigma_2\sigma_1$$

$$D_4Z_1Z_1 \approx \varepsilon(\varepsilon\alpha)\sigma_2\sigma_1$$

$$\vdots$$

$$D_nZ_1^{(n-2)} \approx \varepsilon\cdot(\varepsilon\alpha)^{n-3}\sigma_1\sigma_2 \tag{3.19}$$

These orders of magnitude have been checked to n=4, but are speculative beyond that. Our limitation to $(\varepsilon\alpha)<1$ stems from (3.19). Once the structure of the problem has been recognized it is just a matter of calculation to carry it to an end. The eigenvalue condition is that the recurrence following from (3.18), including also second order coupling Z_1^2, has a nontrivial solution under the side condition

$$\sum_{n=-\infty}^{+\infty} a_n a_n^* < \infty \tag{3.20}$$

The simplest way to solve the problem is to transform the recurrence into a differential equation for the function

$$\chi(k_1x-\sigma_1t) = \sum_{n=-\infty}^{+\infty} a_n e^{in(k_1x-\sigma_1t)} \tag{3,21}$$

but operator methods are also feasible. We then finally obtain a Mathieu equation for χ and the solution is easily found and identical to the one obtained in the next section.

The most interesting aspect of this approach seems to lie in the demonstration that methods for solving strong coupling are available sometimes and that a thorough and systematic investigation of the problem (3.6) may be worthwhile. The connection between this approach and the WKBJ approach is still not clear, all that we now see is that the results are identical for $(\varepsilon\alpha)<1$.

The real question is: In which circumstances can Whitham's principle $\delta\bar{L}=0$ for the averaged Lagrangian $\bar{L}$ be identified as an asymptotically valid approximation of the exact variational principle $\delta L=0$, either as formulated in (3.6) or by Whitham (1967).

4. THE DIRECT TREATMENT

4.1 The Coordinate Transformation

We shall assume that we had solved the long wave equation, in the absence of short waves to any desired degree of accuracy. This shall be possible since $\varepsilon\alpha^2 << 1$. We are working in the system in which the long wave is stationary, lets its phasespeed be c_1 and assume that c_1 was already corrected for higher order effects.

Then introduce

$$\xi = -\frac{\Phi_1(x,z)}{c_1} \tag{4,1}$$

and

$$\eta = -\frac{\Psi_1(x,z)}{c} \tag{4,2}$$

Φ_1 is the potential function and Ψ the streamfunction of the low frequency solution. Thus

$$\frac{\partial\eta}{\partial z} = \frac{\partial\xi}{\partial x} \; ; \; \frac{\partial\eta}{\partial x} = -\frac{\partial\xi}{\partial z} \tag{4.3}$$

Choose the origin of η so that $\eta = 0$ coincides with $z-\zeta$
We shall not need the solution Φ_1 to any higher than the first order, but the nice thing about our transformation is that we can easily formulate linear boundary conditions for the high frequency field, all coupling terms appearing in Jacobian terms.

4.2 The High Frequency Equations

First of all we note, $\Delta = \partial\xi\xi + \partial\eta\eta$,

$$\Delta\Phi_2(\xi,\eta) = 0, \tag{4.4}$$

since $\Delta\Phi_1 = 0$ holds. The only really questionable approximation is introduced now, we assume that the condition $\Phi_2 \to 0$ for $z \to -\infty$ is met by choosing

$$\Phi_2 = \Phi_2(\xi - i\eta) \tag{4.5}$$

which implies that since we centre our solution on $k_2 >> k_1 > 0$, that sidebands with $k_2 < 0$ will be so weakly occupied that we may neglect them.

We are left then with the boundary conditions on $\eta=0$, and the kinematical boundary condition transforms to, writing $Z(\xi) = \zeta(x)$,

$$\dot{Z} - \Phi_\eta \cdot \frac{J}{\eta_z} - c_1 \cdot Z_\xi \cdot \left(\frac{J}{\eta_z}\right)^2 = 0 \tag{4.6}$$

while the dynamical boundary condition yields

$$\dot{\Phi} + gZ - Jc_1\Phi,_\xi + \frac{1}{2}c_1^2 J_z Z = 0 \tag{4.7}$$

Thus we see that all of our coupling terms, which normally present a problem, have transformed to Jacobian and related terms, and these can be calculated with any desired accuracy. Because of (4.5) we may also replace Φ_η by $-i\Phi,_\xi$, and so for constant frequency we are now left with two coupled ordinary differential equations for the function $F(u)$ and $G(u)$ defined by

$$k_2 Z = e^{ik_2\xi} \cdot F(u)$$

$$\frac{k_2\Phi}{c_2} = e^{ik_2\xi} \cdot G(u) \qquad \text{on } \eta=0$$

$$u = k_1\xi \tag{4.8a}$$

and we further introduce the eigenvalue

$$\Lambda = \frac{\Omega_2 + k_2 c_1}{\sigma_2} \tag{4.8b}$$

to be determined from the requirement that F(u) and G(u) be 2π-periodic. We now introduce

$$\frac{J}{\eta_z} = 1 + P_1$$

$$\left(\frac{J}{\eta_z}\right)^2 = 1 + P_2$$

$$J = 1 + Q_2$$

$$\frac{1}{2} J_{\dot{z}} = \alpha^2 k_2 \tilde{Q}_1 \tag{4.9}$$

noting that all P_i, $\tilde{Q}_i$ are of order $\varepsilon\alpha^2$ and further abbreviate

$$\frac{\partial}{\partial\xi} = k_2(i+\alpha^2\partial u) = k_2 N \tag{4.10}$$

and also introduce the operator

$$M = i\Lambda + \alpha\partial u \tag{4.11}$$

The kinematical boundary condition then yields the equation

$$-MF - \frac{1}{\alpha} P_2 NF + i(1+P_1)NG = 0 \tag{4.12}$$

and the dynamical boundary condition becomes

$$-MG + F(1+\tilde{Q}_1) - \frac{1}{\alpha}\tilde{Q}_2 NG = 0 \tag{4.13}$$

Solving (4.13) for F we obtain

$$F = (1+Q_1)MG + \frac{1}{\alpha} Q_2 NG \tag{4.14}$$

with

$$1 + Q_1 = (1+\tilde{Q}_1)^{-1} \tag{4.15}$$

and

$$Q_2 = \tilde{Q}_2 \cdot (1+\tilde{Q}_1)^{-1} \tag{4.16}$$

Inserting (4.14) into (4.12) we arrive at a second order differential equation for G, with periodic coefficients. The condition of 2π-periodicity then determines the eigenvalues Λ_n. The lowest eigenvalue and eigensolution is what we are interested in.

The details will again be presented in (II) and we only give the final results for the case $(\varepsilon\alpha) < 1$.

The eigenfrequency is

$$\Lambda = \frac{\Omega_2 + k_2 c_1}{\sigma_2} = 1 + 0(\varepsilon^2\alpha^4) \tag{4.17}$$

G is obtained as

$$G = \tilde{G} \cdot e^{-i(1/\alpha - 1/2)u} \cdot e^{4i\varepsilon \sin u} \tag{4.18}$$

with $\tilde{G}$ satisfying the perturbed Mathieu equation

$$\tilde{G}'' + \{(\frac{1}{\alpha} - \frac{1}{2})^2 - 8\varepsilon \cos u - 4i(\varepsilon\alpha)\sin u\}\, \tilde{G} = 0 \tag{4.19}$$

yielding

$$\tilde{G} = e^{i(1/\alpha - 1/2)u} \cdot e^{-2\varepsilon\alpha^2 \cos u} (1 - 8\varepsilon\alpha^2 \cos u)^{-1/4} \tag{4.20}$$

or

$$G(u) = e^{4i\varepsilon \sin u} \tag{4.21}$$

From (4.14) we then obtain an expression asymptotically equivalent to

$$F(u) = iG(1 - 2\varepsilon\alpha^2 \cos u)^{-1} \tag{4.22}$$

which , finally transformed back into (x,z) coordinates yields

$$\zeta_2(x) = e^{ik_2 x} e^{2i\varepsilon \sin(k_1 x)} (1 - 2\varepsilon\alpha^2 \cos k_1 x)^{-1} \tag{4.23}$$

which is in agreement with all of the results of Garrett and Smith. The calculations for $(\varepsilon\alpha) > 1$ are currently being rechecked. (First results were surprising, and are not yet clearly understood).

5. CONCLUSIONS AND SUMMARY

It has been shown that for the case

$\varepsilon\alpha^2 \ll 1$, $(\varepsilon\alpha) < 1$, $\varepsilon = k_2 Z_1 > 1$, $\alpha = \sigma_1/\sigma_2 \ll 1$

all three approaches a), b), c) mentioned in the introduction yield the same results. It is not yet obvious whether this applies also to the case $\varepsilon\alpha^2 \ll 1$, $\varepsilon\alpha > 1$. For the case $\varepsilon\alpha < 1$ we see that the wave-packet approach is not a necessary ingredient for the validity of the WKBJ approach. The modulation properties are properties of the eigenfunctions, and the propagation properties of wavegroups follow from the properties of superpositions of eigensolutions.

An extension of approach a) to cases in which the long wave consists of a few neighboured spectral bands should be possible, and this should give us information how to extend the approach b) to this case.

In any case it has been demonstrated that the spectral formulation does have some potential even in the case of strong coupling.

The basic invariance that we have does not seem to be conservation of action, but rather the conservation of the amplitudes, or if we prefer the conservation of the square of the modulus of our invariant amplitudes of the eigenfunctions.

Thus, the entire high frequency field can be decomposed into

$$\zeta(x,t) = \int B_\lambda \cdot e^{i(k_2 x - \omega_2 t)} \cdot e_\lambda d\lambda \tag{5.1}$$

the e_λ denoting our eigenfunctions. To every eigenfunction there exists the complex conjugate $e^*_\lambda = e_{-\lambda}$, with $B^*_\lambda = B_{-\lambda}$, and we have conservation of the B_λ. If the highfrequency field has a statistical character and the B_λ's become rapidly varying, essentially undeterminable functions on the basis of our limited information, then averages $B_\lambda B^*_\lambda \, d\lambda$ are our conserved quantities. How these averages slowly change if the long wave is not a pure wave with its higher harmonics, but a mixture of such waves is what needs investigation.

REFERENCES

1.) Garrett, C. and Smith, J. (1976):
On the interaction between long and short surface waves.
J.Phys.Oceanogr. 6, 925

2.) Hasselmann, D.E. (1977a):
Wind wave generation by energy and momentum flux to the forced components of a wavefield. (Submitted to J.Fluid Mech.)

3.) Hasselmann, D.E.(1977b):
Long wave-short wave interaction - or - weak interaction theory for surface gravity waves - how weak is weak? (in prep.)

4.) Hasselmann,K.(1966):
Feynman diagrams and interaction rules for wave-wave scattering processes. Rev.Geophysics 4, 1

5.) Hasselmann, K. (1967):
Nonlinear interactions treated by the methods of theoretical physics. Proc.Roy.Soc.A, 299, 77

6.) Whitham, G.B. (1967)
Variational methods and applications to water waves. Proc.Roy. Soc.A 299, 6

7.) Whitham, G.B. (1974):
Dispersive waves and variational principles.
in:Nonlinear Waves. (Ed.Leibovich, S.and Seebass,A.R.) Cornell. (Ithaca and London)

DISCUSSION

A. Favre:

As Prof. Longuet-Higgins said a moment ago, at the IMST, using the space-time correlation method, we measured the phase velocity of the components of the wind waves and found that it is nearly constant in the prediction phase, and equal to that of the main component. How could you use that experimental result in the theory of long wave-short wave interaction?

D. E. Hasselmann:

I do not think that the present theory would predict that effect, but one has to be rather careful in defining what sort of a phase speed one is measuring there, and I actually believe a higher spatial resolution is necessary to check the theory. I shall also stress that the theory in its present form can only handle the case of one single long wave and a spectrum of high frequency waves. It is not at all clear how to extend the theory to a spectrum of low frequency waves plus a spectrum of high frequency waves.

D. J. Webb:

If you repeat the WKB calculation - but using the amplitude-dependent shape changes, described earlier in this meeting by Longuet-Higgins - would your disagreement disappear?

D. E. Hasselmann:

In fact it turns out that the disagreement was due to an error of mine, so that for $k_1 Z_1 \ll 1$ and $k_1 \ll k_2$ weak interaction theory and $\delta L=0$ theory always yield the same result, even if $(k_1Z_1) . c_1/c_2 \gg 1$.

For $(k_1Z_1) \ll 1$ my calculations will certainly agree with those of Longuet-Higgins. It would be interesting to check at which values of (k_1Z_1) my treatment begins to deviate significantly from the exact calculations, with regard to both the eigenfunctions and eigenvalues, but this has not been done yet.

ON THE NONLINEAR TRANSFER OF ENERGY IN THE PEAK OF A GRAVITY WAVE SPECTRUM

M. J. H. Fox

Department of Applied Mathematics and Theoretical Physics, Silver Street, Cambridge

1. INTRODUCTION

Accurate evaluations of the nonlinear transfer of energy within a continuous spectrum of gravity waves are difficult to obtain due to the complicated form of the transfer equation. Computations that have been carried out, for instance by Sell and Hasselmann (1972) and by Webb (1977) have required quite large amounts of computer time to find values of the transfer rate at each point in the wavenumber plane. Thus it has not been practicable to produce high-resolution contour plots of this quantity for a range of different spectral forms.

In this paper we consider a simplified model appropriate to calculating the transfer in a narrow spectrum. Use is made of an interaction equation derived by Longuet-Higgins (1975) for this case, and special model forms for the action spectrum are chosen which enable the multiple integrals appearing in the equation to be reduced to a single integral. This integral may be evaluated very accurately and rapidly - in the first case presented the transfer rate can be evaluated at 200 points in the wavenumber plane in one second of computer time, with an accuracy of better than one percent.

A contour plot of the transfer rate in this case, in which the spectrum is modelled as a single normal curve, is presented in section 2. The results are also integrated to obtain energy transfer rates as a function of frequency.

In section 3 the spectrum is modelled as the sum of two

normal curves, of different amplitudes, variances and mean positions, so that the effect of asymmetry can be studied.

Finally, we see in section 4 how by taking the sum of three normal curves we may obtain close representations of two of the JONSWAP spectra (Hasselmann et al 1973), and we compare the computed transfer rates as functions of frequency with those obtained by Sell and Hasselmann.

2. A SYMMETRIC NORMAL SPECTRUM

The transfer equation given by Longuet-Higgins (1975) for the rate of change of the action density $N(\underline{\kappa})$ at wavenumber $\underline{\kappa}$ near the spectral peak is

$$\frac{\partial N_1}{\partial \tau} = 4\pi \int \cdots \int \left[(N_1 + N_2) N_3 N_4 - (N_3 + N_4) N_1 N_2 \right] \times \delta(\underline{\kappa}_1 + \underline{\kappa}_2 - \underline{\kappa}_3 - \underline{\kappa}_4)\, \delta(\omega_1 + \omega_2 - \omega_3 - \omega_4)\, d\underline{\kappa}_2\, d\underline{\kappa}_3\, d\underline{\kappa}_4 \tag{1}$$

in non-dimensional units, where $N_i \equiv N(\kappa_i)$ and $\delta(\quad)$ denotes the Dirac delta function. ω_i and $\underline{\kappa}_i$ are related by $\omega_i = -\frac{1}{8}(\lambda_i^2 - 2\mu_i^2) + 2M$ where $(\lambda_i, \mu_i) = \underline{\kappa}_i$ and M is a constant.

If we suppose initially that the action density and hence also, in this approximation, the energy density has the normal form

$$N(\underline{\kappa}) = R \exp\left(-\tfrac{1}{2}P\lambda^2 - Q\mu^2\right) \tag{2}$$

then (1) may be reduced to

$$\frac{\partial N}{\partial \tau} = 32\pi^2 R^3 \exp\left(-\tfrac{3}{2}P\lambda^2 - 3Q\mu^2\right) \times \int_{-\frac{\pi}{2}}^{\frac{\pi}{2}} \exp\left(-W + \frac{V^2}{4U}\right) \frac{d\theta}{\sqrt{((P+Q)^2 - (P-Q)^2 \sin^2 2\theta)}} \tag{3}$$

+ three similar expressions where U, V, W are functions involving only positive powers of $\cos\theta$ and $\sin\theta$ (Longuet-Higgins, appendix B; Fox 1976). The integrand in (3) is then a smoothly varying function of θ, and the integral may be satisfactorily evaluated using Simpson's rule. Convergence is rapid as the number of points is increased, 21 giving an estimated accuracy of at least three significant figures.

The value of $\frac{\partial N}{\partial \tau}$ found by this method at the spectral peak agreed precisely with that computed by Longuet-Higgins by a different method, which involved the numerical evaluation of a double integral. As a further check we may compute from our results the values of the integrals on the left hand sides of the two conservation laws

$$\left.\begin{aligned} \int \frac{\partial N}{\partial \tau}\, d\underline{\kappa} &= 0 \\ \int (\lambda^2 - 2\mu^2)\frac{\partial N}{\partial \tau}\, d\underline{\kappa} &= 0 \end{aligned}\right\} \qquad (4)$$

which represent the constancy of wave action and energy respectively. These integrals were in fact found to be very small in comparison with typical values of the integrands taken over the characteristic width of the spectrum. A third conservation law, representing the constancy of wave momentum

$$\int \underline{\kappa}\, \frac{\partial N}{\partial \tau}\, d\underline{\kappa} = 0 \qquad (5)$$

is satisfied automatically by symmetry in this case. It was checked, however, for the asymmetric spectra presented in the next two sections.

Figure 1a) shows a contour plot of the function $\frac{\partial N}{\partial \tau}$ in the case when P = Q = 1. The functions $N(\underline{\kappa})$ and $\partial N(\underline{\kappa})/\partial\tau$ may be integrated along lines of constant frequency, which are parallel to the μ-axis in this approximation, to obtain the frequency spectrum $N(\omega)$, and $\partial N(\omega)/\partial\tau$. These are plotted in figure 1b).

The form of the contour plot of figure 1a) may be explained in terms of the basic resonant interactions contributing to the integral (1). Longuet-Higgins showed that to satisfy the resonance conditions the four wavenumbers $\underline{\kappa}_1$, $\underline{\kappa}_2$, $\underline{\kappa}_3$, $\underline{\kappa}_4$ must lie at the corners of a parallelogram, but must also lie on the same hyperbola, centred at the mean of the four wavenumbers, with asymptotes in directions $\lambda = \pm\sqrt{2}\,\mu$. Longuet-Higgins also showed that for wavenumbers $\underline{\kappa}_1$ outside the peak, but near one of these characteristic directions from the peak the dominant terms in (1) would produce a positive transfer. This effect may be clearly seen in figure 1a).

The negative transfer on the λ and μ axes may be explained in a similar manner (Fox 1976), although this is a weaker effect.

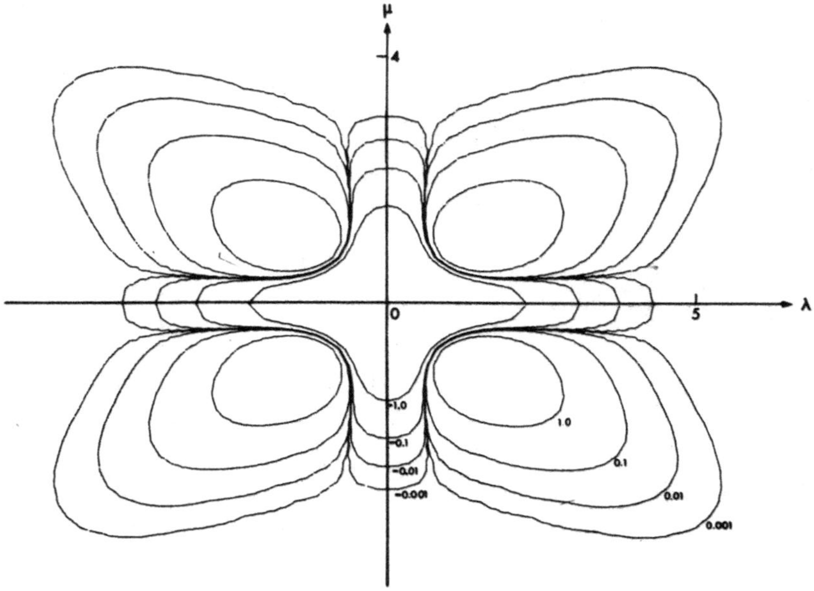

Figure 1(a) Contours of the transfer function $\frac{\partial N}{\partial \tau}$ in the k plane, when the wave spectrum has the asymmetrical normal form (2), in which P = Q = 1 ; R = $\sqrt{(2/PQ)}$

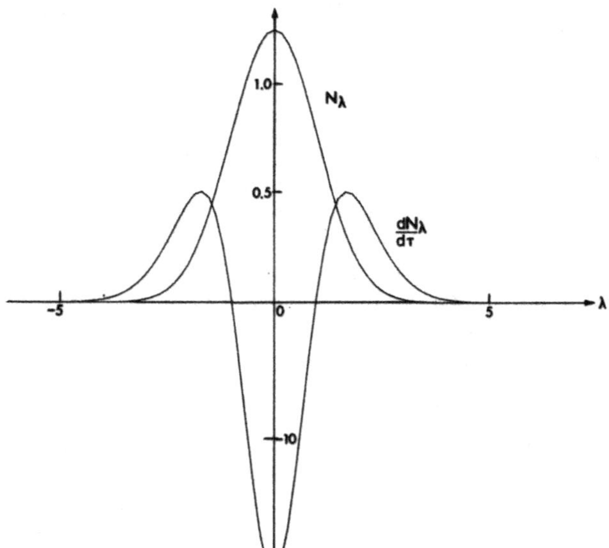

Figure 1(b) The μ-integrated spectrum N_λ and transfer function $dN_\lambda/d\,T$ corresponding to figure 1(a)

3. ASYMMETRIC SPECTRA

We may apply the same method of calculation to narrow wave spectra consisting of the sum of several normal distributions

$$N(\underline{k}) = R \sum_p a_p \exp\left[-\tfrac{1}{2}P_p(\lambda-\lambda_p)^2 - Q_p\mu^2\right] \tag{6}$$

where a_p, P_p, Q_p, λ_p are arbitrary constants (a_p, P_p, $Q_p > 0$). The transfer integral (1) then reduces to a triple summation of terms of the form of (3), the index of each sum running over the number of normal distributions taken.

We are now in a position to investigate the effects on the transfer rate of asymmetry in the spectrum. In all the following examples the spectra have been normalised so as to have a total non-dimensional action of π, by taking

$$R = \left[\sum_p a_p \sqrt{2/P_p Q_p}\,\right]^{-1}$$

Figure 2b) shows the form of spectrum that may be obtained from the sum of two normal distributions having the same width in the μ-direction ($Q_1 = Q_2 = 1.0$), but different widths in the λ-direction ($P_1 = 1.0$, $P_2 = 0.3$). The ratio of the amplitudes is 0.8, and the distance between centres is 1.25. Also shown in figure 2b) is the computed form of the transfer rate as a function of frequency, which may be seen to be appreciably larger on the steeper (low frequency) face of the spectrum than on the high-frequency side. At the peak of the spectrum the transfer is still strongly negative. The contours of transfer rate in the wavenumber plane are shown in figure 2a).

Figures 3a), 3b), 4a) and 4b) show the corresponding results for spectra that have the same structure in the λ-direction as in the previous case, but are successively broader in the μ-direction. In figures 3a), 3b) $Q_1 = Q_2 = 0.25$, and in figures 4a), 4b) $Q_1 = Q_2 = 0.1$, the other parameters being as above. We see that the broadening of the spectrum has little effect on the frequency dependence of the transfer rate, but in the wavenumber plane tends to diminish and even reverse in sign the negative transfer on the λ-axis. This transition to positive transfer takes place first on the steeper side of the spectrum.

This effect may also be explained in terms of the basic resonant interactions. As we have seen the strongest such interactions are those of the configuration suggested by Longuet-Higgins, tending to radiate energy in directions $\lambda = \pm\sqrt{2}\mu$ from regions

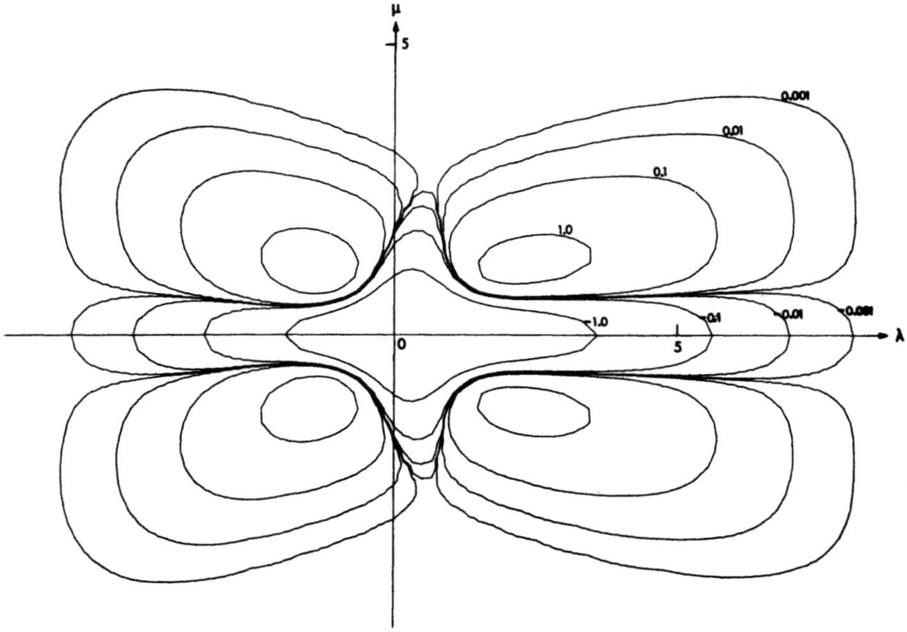

Figure 2(a) Contours of the transfer function ∂N/∂T for the asymmetic spectrum given by (6), in which $P_1 = Q_1 = 1$; $P_2 = 0.3$; $Q_2 = 1$; $a_2/a_1 = 0.8$, $\lambda_1 = 0$, $\lambda_2 = 1.25$

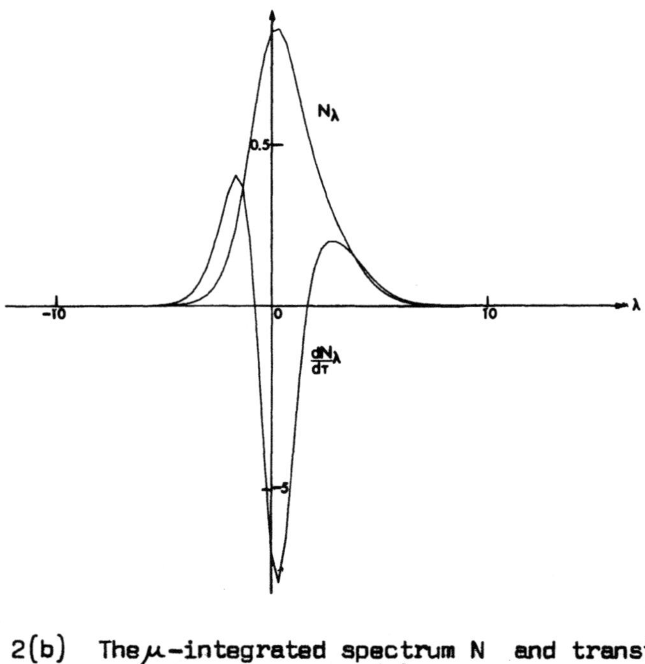

Figure 2(b) The μ-integrated spectrum N and transfer function dN /dT corresponding to figure 2(a)

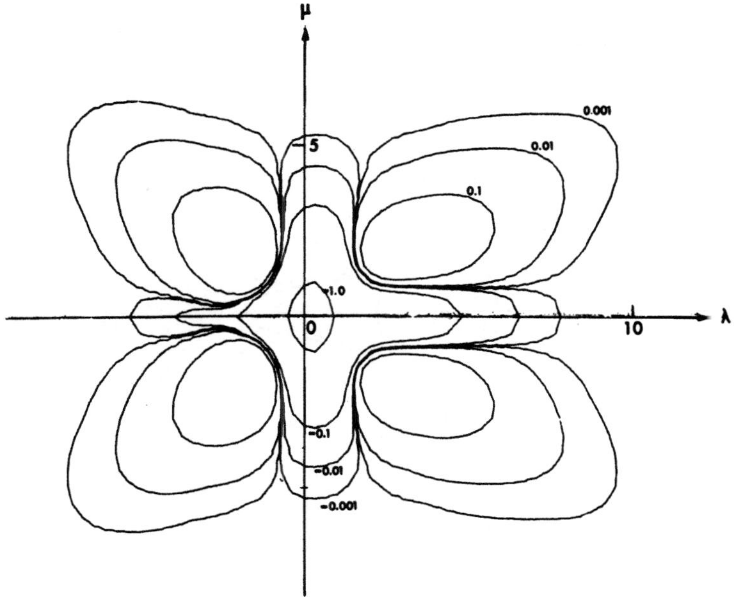

Figure 3(a) As in figure 2(a), but with $Q_1 = Q_2 = 0.25$

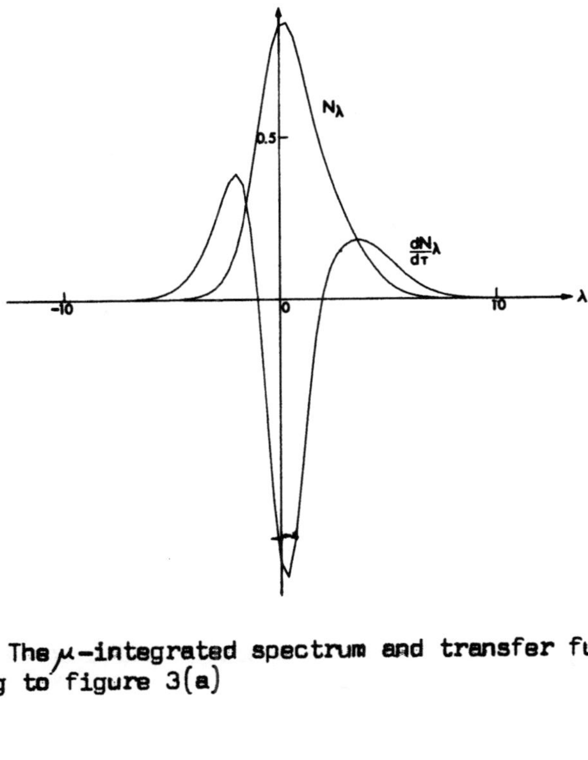

Figure 3(b) The μ-integrated spectrum and transfer function corresponding to figure 3(a)

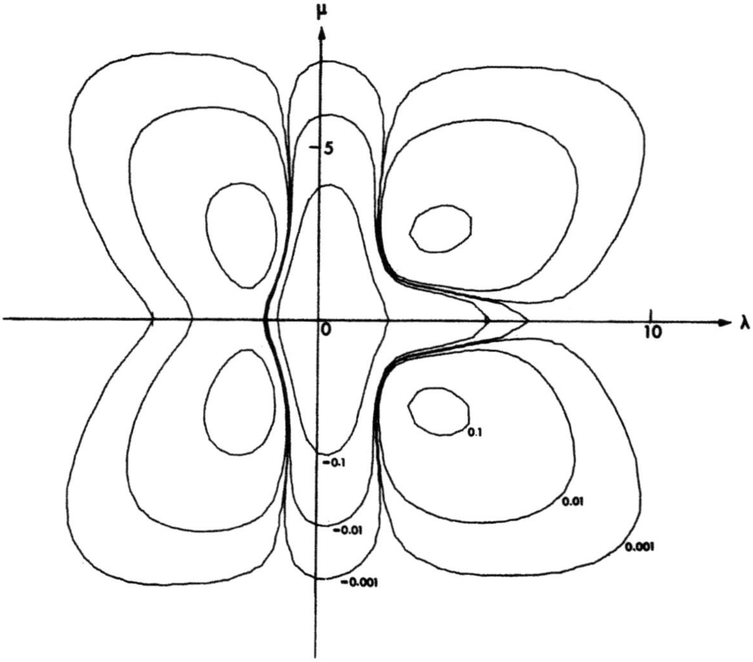

Figure 4(a) As in figure 2(a), but with $Q_1 = Q_2 = 0.1$

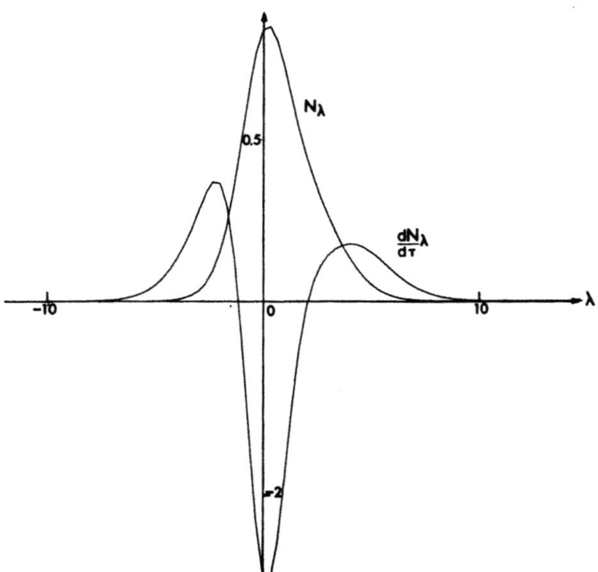

Figure 4(b) The μ-integrated spectrum and transfer function corresponding to figure 4(a)

of high energy (or action) density in the wavenumber plane. For a peak that is broad in the μ-direction a substantial portion of the λ axis will receive a positive transfer through lying in these directions from the extremities of the peak. Also, the transfer is proportional to $(N_1+N_2)\,N_3\,N_4$ and since, in this configuration, $\underline{k}_1$ and $\underline{k}_2$ are on opposite sides of the peak the positive transfer is relatively greater on the steeper side of the spectrum, tending to even out the asymmetry in the λ- direction. Hence it is on this side that the negative transfer first becomes blotted out.

4. A COMPARISON WITH NORTH SEA SPECTRA

We now consider model spectra consisting of the sum of three normal curves. Figure 5a) shows a comparison between such a spectrum, with parameters given in table 1, and the 'mean JONSWAP spectrum' JN5 of Sell and Hasselmann (1972). The Q_p have been chosen to reproduce the effect of the $\cos^2\theta$ spreading factor assumed in the JONSWAP spectrum. In figure 5b) are shown the contours of the transfer function $\partial N/\partial\tau$. These have a form rather similar to that of figure 4a), but as the spectrum is still broader in the μ-direction regions of positive transfer have appeared on the positive, as well as on the negative, λ axis.

In figure 5c) we see a comparison between the μ-integrated transfer rate computed here, converted to physical units, and the results of Sell and Hasselmann, with the frequency scale converted from Hz to rad/s . The order of magnitude is the same in both cases, but the curve found here is much more symmetrical than that of Sell and Hasselmann. This is probably due to asymmetry in the nonlinear coupling coefficient, taken to be constant in equation (1). In particular these results show a tendency for the nonlinear transfers to smooth out the peak in the spectrum. The computations of Sell and Hasselmann indicates that the principal effect of the transfers is to move the peak towards lower frequencies, although greater resolution would be required near the point of transition from positive to negative transfer to determine the rate at which this occurs.

For a narrower spectrum, such as the 'sharp JONSWAP spectrum' R3C of Sell and Hasselmann we expect the narrow band approximation to be better. Figure 6a) shows a comparison between this spectrum and a model spectrum with parameters given in table 2. Figure 6c) shows that the agreement between the computed transfer rate curves is now much closer, both the central minimum and the two side maxima being reproduced fairly well. Again, higher resolution would be required in the computations of Sell and Hasselmann to determine the exact position of maximum negative transfer.

Table 1 Parameters corresponding to the spectrum of Figure 5

p	Pp	Qp	ap	λp
1	1.00000	0.03	1.0000	0.000
2	0.08000	0.03	0.2000	2.000
3	0.01563	0.03	0.0625	7.656

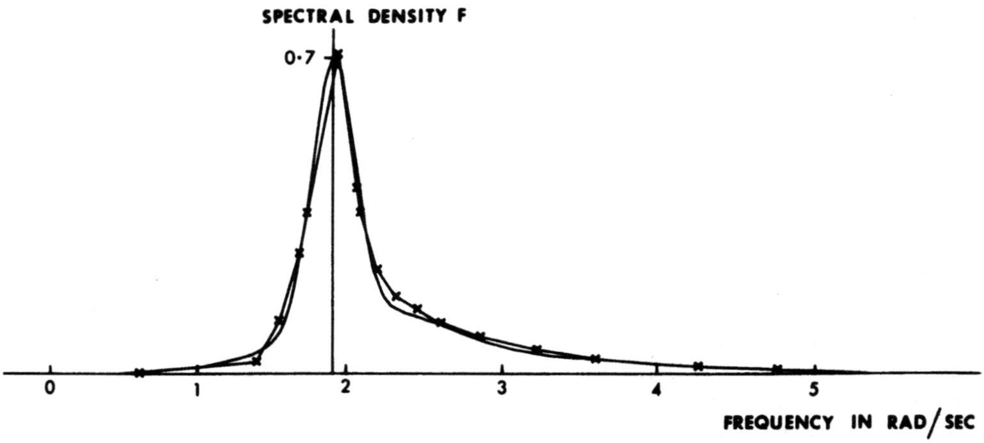

Figure 5(a) The frequency spectrum corresponding to equation (6) (——) and the JONSWAP spectrum JN5 (x – x)

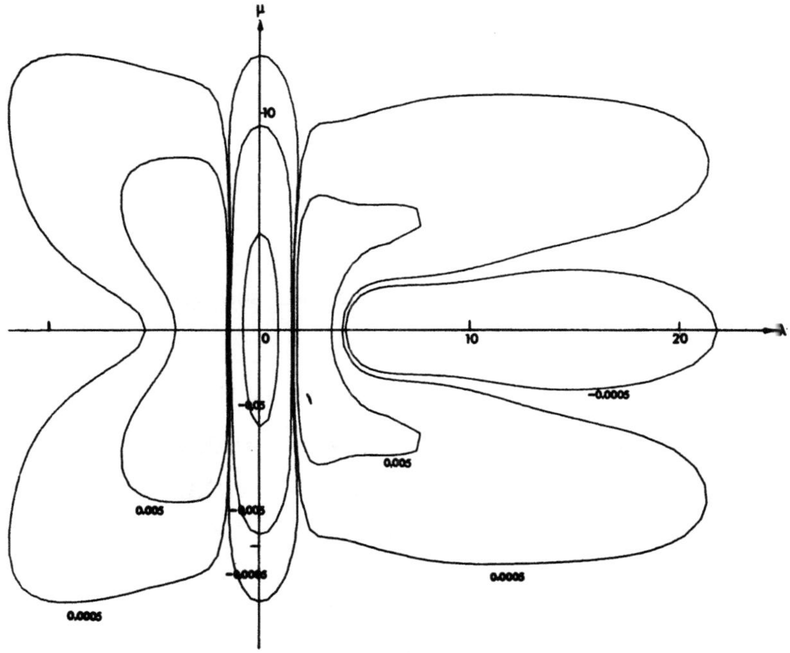

Figure 5(b) The computed transfer function ∂N/∂T corresponding to the spectrum of equation (8)

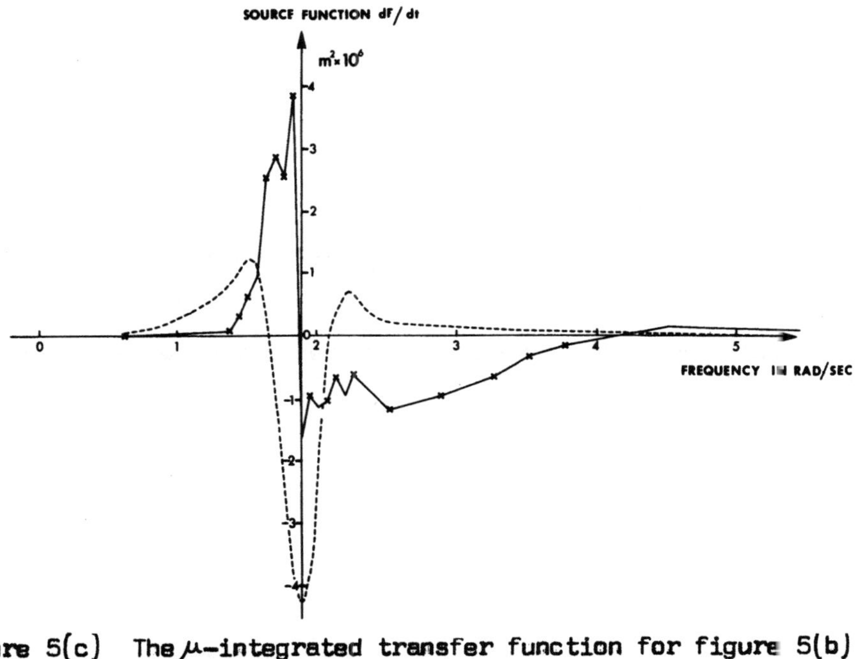

Figure 5(c) The μ-integrated transfer function for figure 5(b) (——) compared with the numerical computations for JN5 by Sell and Hasselmann (1972) (x – x).

Table 2 Parameters corresponding to the spectrum of Figure 6

p	Pp	Qp	ap	λ p
1	1.00000	0.03	1.000	0.000
2	0.08000	0.03	0.110	2.000
3	0.01563	0.03	0.045	7.656

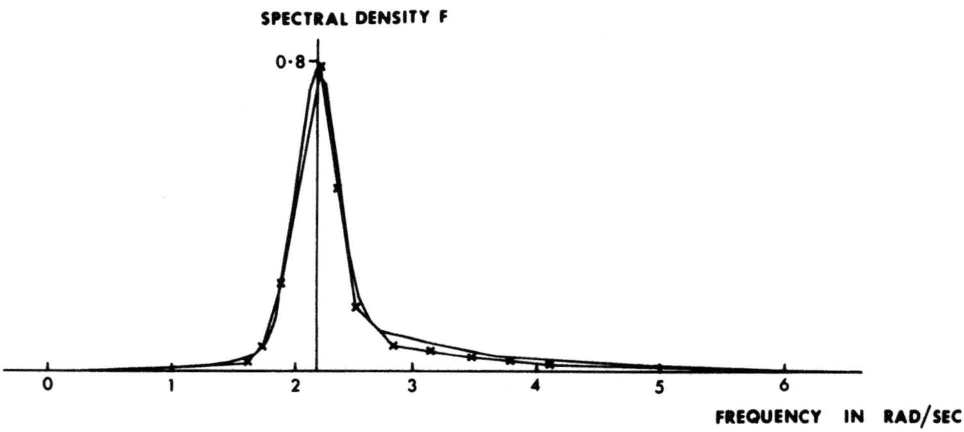

Figure 6(a) The frequency spectrum corresponding to equation (b) (——) and the JONSWAP spectrum R3C (x – x)

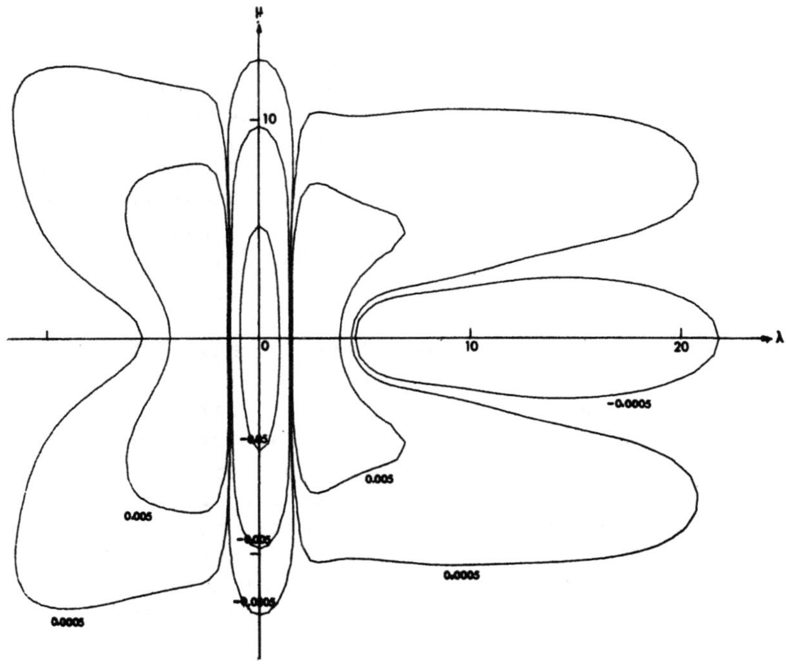

Figure 6(b) The computed transfer function ∂N/∂T corresponding to the spectrum of equation (6)

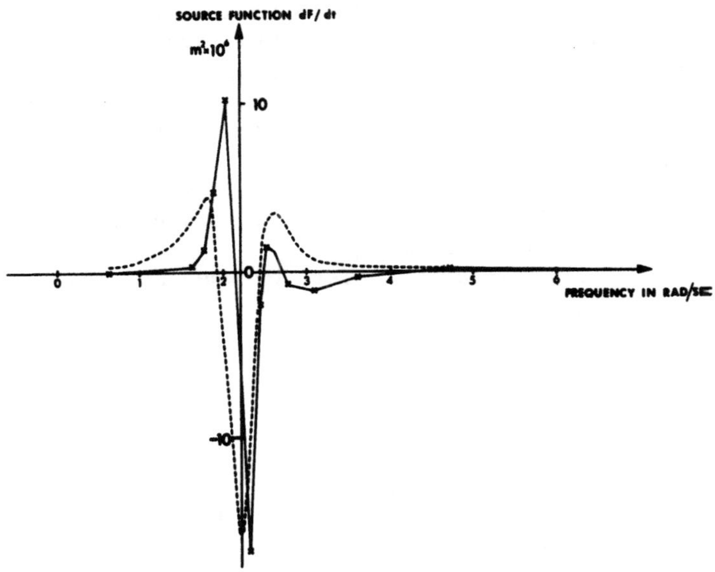

Figure 6(c) The μ-integrated transfer function for figure 6(b), (——) compared with the numerical computations for R3C by Sell and Hasselmann (1972) (x – x)

5. CONCLUSIONS

By adopting a certain analytic description of the energy spectrum we have been able to reduce the energy transfer integral to a manageable form, and hence, for a sufficiently narrow spectrum, to compute the energy transfer precisely and with high resolution in the wavenumber plane. For the 'sharp JONSWAP spectrum' we find that the integrated transfer function agrees well with the computations of Sell and Hasselmann (1972) for which the transfer becomes more symmetric as the spectrum is narrowed.

The spectral forms used here have an exponential decay at large wavenumbers rather than the more realistic algebraic decay usually assumed, and so underestimate the energy present in the high-frequency end of the spectrum. This, together with the rapid increase of the magnitude of the coupling coefficient with wavenumber (Webb 1977), will make these results, in which the coefficient is assumed constant, inapplicable to broader spectra. Thus these computations show a tendency for the spectral peak to become broader, and give no indication that a peak might be built up or moved towards lower frequencies by weak nonlinear transfers.

REFERENCES

Fox, M.J.H. 1976 On the nonlinear transfer of energy in the peak of a gravity-wave spectrum. II. Proc. R. Soc. Lond. A348, 467-483.

Hasselmann, K. etc XV al. 1973 Measurements of wind-wave growth and swell decay during the joint North Sea wave project. Ergänzungsheft zur Deutsch. Hydrogr. Z. A12, 95pp.

Longuet-Higgins, M.S. 1975 On the nonlinear transfer of energy in the peak of a gravity wave spectrum. Proc. R. Soc. Lond. A347, 311-328.

Sell, W. and Hasselman, K. 1972 Computations of nonlinear energy transfer for JONSWAP and empirical wind-wave spectra. Rep. Inst. Geophys., Univ. Hamburg.

Webb, D.J. 1977 Non-linear transfers between sea waves. Deep Sea Res. to appear.

DISCUSSION

K. Hasselmann:

This is really a very nice technique for analysing the structure of the non-linear transfer in the limiting case of a very narrow peak. Herterich and myself have applied it to finite-depth water waves - where the exact integral is particularly difficult to evaluate because the coupling coefficients and dispersion relation are no longer homogeneous in the wave numbers - and found that the finite-depth wave case can be reduced to the infinite-depth case by a straightforward scale transformation. This implies that the stabilisation of the spectral shape and the shift of the peak towards lower frequencies which was found for deep water waves should apply also in the finite-depth case. However: the transfer rates are considerably stronger in shallow water than for infinite depths. A figure is given in the paper by Shemdin et al. in this session.

Perhaps I should comment briefly on the differences between Fox's asymptotic narrow peak calculations and the computations of the complete scattering integral by Sell and myself.

For the JONSWAP spectrum (peak enhancement factor $\gamma = 3,3$) the interactions in the neighbourhood of the peak give much of the story, but interactions of the peak with the higher frequency part of the spectrum are nevertheless important, contributing particularly to the asymmetry of the transfer relative to the peak. Thus I suspect that the differences here are due to limitations of the narrow peak approximation. For a flatter spectrum, such as the Pierson-Moskowitz spectrum ($\gamma = 1$), the narrow peak approximation must necessarily become rather inaccurate, as pointed out by Fox. The difference between Fox's and our results for a sharp spectrum ($\gamma = 7$), however, may be due partially to inaccuracies of our calculations, which really required a higher resolution than we used for this case. It is a useful feature of the Longuet-Higgins-Fox technique that it is most accurate for cases which require the most resolution for the complete integration.

D.E. Hasselmann:

I hope the differences between your calculations and those of Dr. Fox for a narrow peak spectrum, are not mainly attributable to contributions from very high frequencies, because here weak interaction theory certainly fails.

K. Hasselmann:

Although the resonant weak interaction theory is not applicable to interactions involving very short waves, in practice this does not affect the computations of the non-linear transfer by Webb, Sell and myself significantly. The contributions from the very short waves where the theory is no longer applicable are negligible.

H. Mitsuyasu:

According to our observations, in many cases, the maximum negative transfer occurs on the high frequency side of the spectrum (f > spectral peak frequency). Did you compute the transfer for any case that shows the maximum negative transfer at high frequency side?

M.J.H. Fox :

No. In all cases considered the maximum negative transfer occurred at the spectral peak frequency. It seems that a displacement of this negative peak to one side or the other of the spectral peak must be an effect associated with broader spectra.

L. Cavaleri :

How sensitive is the non-linear energy transfer function to small changes in the shape of the spectrum?

M.J.H. Fox :

Experience with the various spectra used in the computations suggests that the transfer function is not very sensitive to small changes in the spectral form, and that quite large changes may be made without qualitatively altering the results.

THE WAVE-WAVE INTERACTION MACHINE

D. J. Webb

Institute of Oceanographic Sciences,
Wormley, Godalming, U.K. GU8 5UB

INTRODUCTION

In this paper* I hope to show that when sea waves interact or scatter off each other they act like a machine. I use the word machine in a broad sense, to mean a system which creates order somewhere at the expense of disorder elsewhere. Thus a machine may create a car at the expense of producing waste heat and pollution elsewhere. In the case of sea waves, an enhanced peak of the spectrum is produced at the expense of creating disorder amongst the very short waves.

In the JONSWAP series of experiments (Hasselman et al., 1973) which studied the development of a sea wave spectrum due to an offshore wind, the sea wave spectra were found to have a very enhanced peak. This enhanced peak had also been noticed by previous experimenters (Barnett and Wilkerson, 1967).

However Hasselmann and his collaborators in the JONSWAP experiments were able to show, at least qualitatively, that the enhanced peak was the result of non-linear transfers between the sea waves. They estimated the non-linear transfer by using Hasselmann's equation (Hasselmann, 1962).

But this appeared to be wrong. Hasselmann's equation was derived from a statistical theory and in statistical theories

---

*A fuller report of this work is being published in Deep-Sea Research (Webb, 1978).

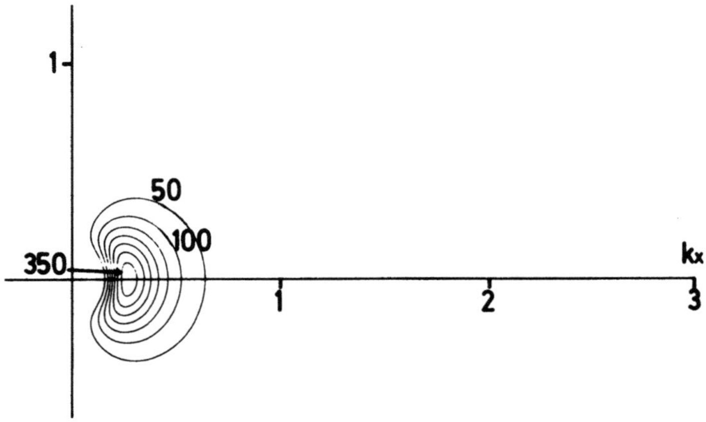

Figure 1 The action density spectrum. This is a Pierson-Moskowitz spectrum with a $\cos^2\theta$ spreading factor, a Phillips constant of 0.01 and a maximum variance at 0.3 Hz.

things always become more disordered with time. But here was a statistical theory increasing the height of a spectral peak by a factor of two or three and thus apparently increasing order.

HASSELMANN'S EQUATION

First action. For oscillating systems, action is an adiabatic invariant. That is, it is conserved if the system changes by only a small amount in each period. Thus for waves moving slowly into shallow water or for waves moving slowly into a current, action is conserved.

If $n(\underline{k})$ is the action density at wavenumber $\underline{k}$, then,

$$n(\underline{k}) = E(\underline{k})/\omega.$$

$E(\underline{k})$ is the energy density, ω the frequency. We expect the total action N, the momentum $\underline{P}$ and the energy E of the system to be conserved during non-linear interactions. This implies

$$N = \int n(\underline{k})\,d\underline{k},$$

$$\underline{P} = \int \underline{k} n(\underline{k}) d\underline{k},$$

$$E = \int \omega n(\underline{k}) d\underline{k},$$

are all conserved.

Hasselmann's equation is based on a statistical theory and it describes the evolution of the mean properties of an ensemble of oceans. If n_1 is the expected action density at wavenumber $\underline{k}_1$, then Hasselmann's equation has the form,

$$dn_1/dt = \int C\ \delta(\underline{k}_1 + \underline{k}_2 - \underline{k}_3 - \underline{k}_4)\ \delta(\omega_1 + \omega_2 - \omega_3 - \omega_4)$$

$$(n_2 n_4 (n_3 - n_1) + n_1 n_3 (n_4 - n_2))\quad d\underline{k}_2 d\underline{k}_3 d\underline{k}_4.$$

The delta functions are there to ensure conservation of energy and momentum. In this, the simplest theory of non-linear interactions, four waves have to be used, because three waves alone cannot satisfy the conservation laws. The term containing products of action densities has to ensure conservation of action density and it should also have the same symmetry as the delta functions. The above term appears to be the simplest function including triple products which satisfies these conditions.

There appears to be no simple physical constraints on the form of the coefficient C. In practice it has a symmetric but otherwise very complex mathematical form (Hasselmann, 1962, Webb, 1978). It increases very rapidly with wavenumber (roughly as k^6).

INTEGRAL

Using a Pierson-Moskowitz spectrum with a $\cos^2\theta$ spreading factor (fig. 1), I used Hasselmann's equation to calculate dn/dt. The result is shown in figure 2. In agreement with Sell and Hasselmann (1972), I found large positive values of dn/dt at the peak of the spectrum. There are also two secondary positive peaks at high wavenumbers at an angle of roughly 45 degrees to the downwind direction and a negative region at high wavenumbers in the downwind direction.

THE MAIN PEAK

The primary problem was to find out how the positive values of dn/dt near the peak of the spectrum arose. I tried a number of lines of attack and noticed that the delta functions in Hasselmann's equation physically represented the scattering of waves k_1 and k_2 to give waves k_3 and k_4, plus the reverse process.

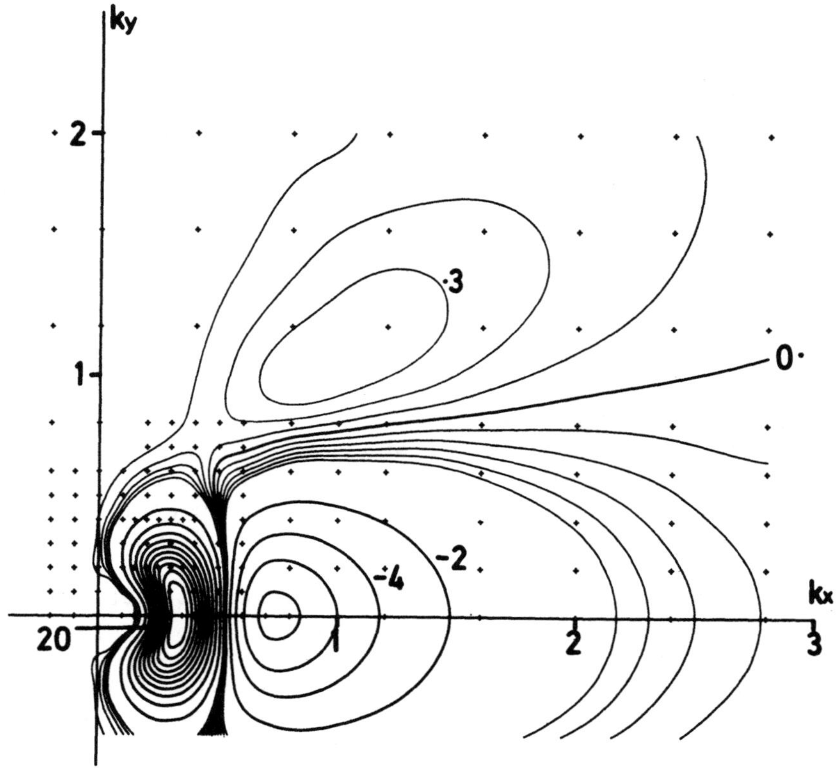

Figure 2 The non-linear transfer dn/dt as a function of wave-number. The contours are marked in units of 10^{-3} M.K.S. units.

In addition, the combination of terms $n_2 n_4 (n_3 - n_1)$ could be taken physically to represent a diffusive flow between waves k_3 and k_1, catalysed by waves k_2 and k_4. As far as n_1 and n_3 are concerned this diffusive flow should smooth out the spectrum. Thus I tried calculating,

$$(dn_1/dt)\,d = 2\int C\delta(k_1+\ldots)\ \delta(\omega_1+\ldots)\ n_2 n_4 (n_3-n_1) dk_2 dk_3 dk_4 .$$

The factor 2 arises because I limited the integral over k_4 so that $k_1 - k_3 < k_1 - k_4$.

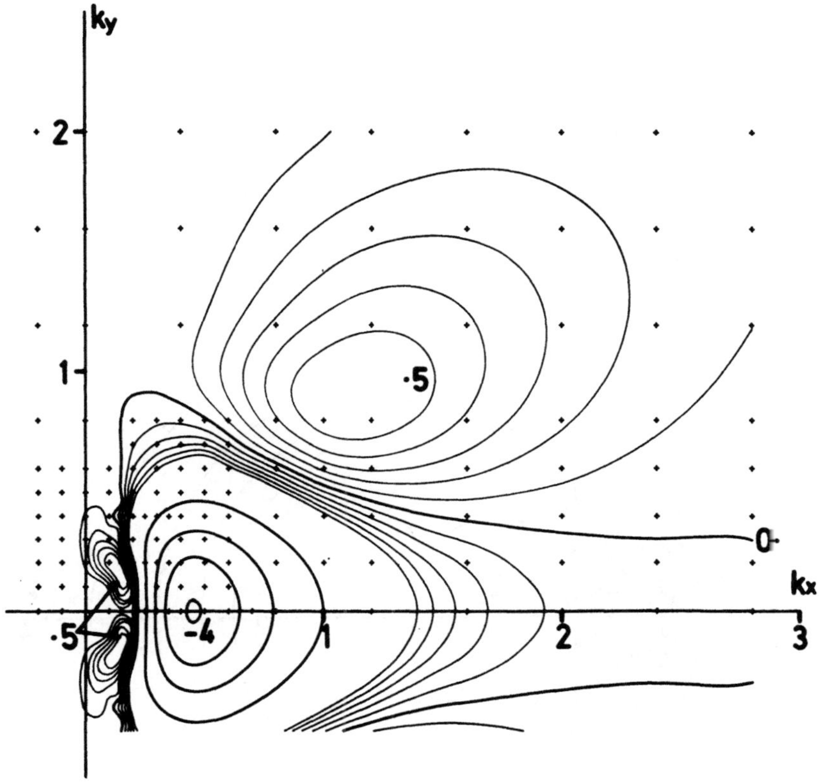

Figure 3 The diffusive contribution to dn/dt. The small crosses represent the points at which the function was evaluated.

The result is shown in figure 3. This does indeed show a flow away from the peak of the spectrum to higher wavenumbers. There is also a flow down the steep low wavenumber face of the spectrum. But this flow is very small, in part because of the low values of C at low wavenumbers.

Because four waves are involved, the diffusive transfer between k_3 and k_1 described by $n_2 n_4 (n_3 - n_1)$ must also produce a transfer between k_2 and k_4. Note that for k_2 and k_4 the direction of transfer does not depend on their relative amplitudes so far as k_2 and k_4 are concerned, the transfer is pumped by the diffusive

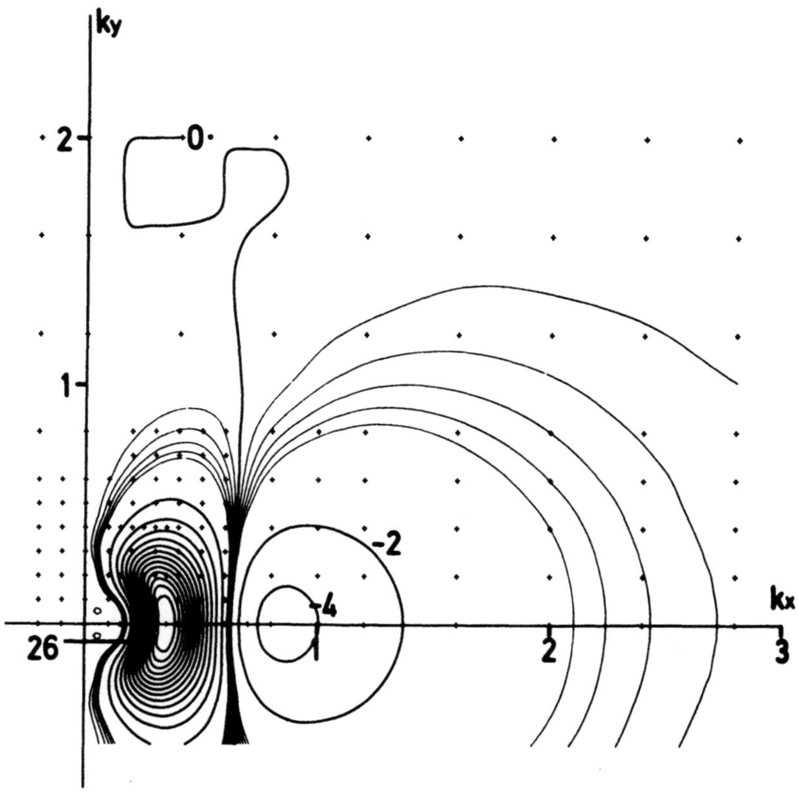

Figure 4 The pumped contribution to dn/dt.

transfer between k_3 and k_1.

The term $n_1n_3(n_4-n_2)$ can now be seen as describing the pumped transfer between k_1 and k_3 due to the diffusive transfer between k_4 and k_2. The contribution to figure 2 from this term is seen in figure 4. We see that it is the pumping term which is enhancing the peak of the spectrum.

ENTROPY

Now this enhancement of the main peak, this apparent creation of order seems unreasonable. Experience has shown that the best measure of order is entropy H. (Note increases in H correspond to more disorder). In statistical mechanics, entropy is defined in terms of the probability f(p), that the system is in a particular part of phase space.

$$H = -\kappa \int f(p) \ln f(p)\, dp.$$

κ is Boltzmann's constant and for systems of waves if p_i defines the amplitude and phase of wave k_i, then $dp = \prod_i dp_i$. If all the information we have is an estimate of the mean action density at each wavenumber then (Webb, 1978),

$$H = \kappa \int \ln n(k)\, dk.$$

The rate of change of H is given by,

$$dH/dt = \kappa \int (1/n)(dn/dt)\, dk.$$

Thus the rate of change of entropy depends only on the relative change in action density. Similarly transfer of action density from regions where it is large to regions where it is small increases disorder.

This partly answers the problem. Although the rate of change of action density at high wavenumbers is small, the relative rate of change can be very great and more than compensate for any apparent increase in order at low wavenumbers.

Hasselmann (1963) showed by substituting his equation into the one for dH/dt that,

$$dH/dt = \kappa \int dk_1 dk_2 dk_3 dk_4\ C\ \delta(k_1+\ldots)\delta(\omega_1+\ldots)$$
$$(\tfrac{3}{4})\ n_1 n_2 n_3 n_4 (1/n_1 + 1/n_2 - 1/n_3 - 1/n_4)^2$$
$$\geqslant 0.$$

Thus for the system as a whole, disorder always increases. For just the diffusive term one finds by the same method,

$$(dH/dt)_d = \kappa \int dk_1 dk_2 dk_3 dk_4\ C\ \delta(k_1+\ldots)\delta(\omega_1+\ldots)$$
$$(\tfrac{3}{4})\ n_1 n_2 n_3 n_4 (\ (1/n_1 - 1/n_3)^2 + (1/n_2 - 1/n_4)^2).$$
$$\geqslant 0.$$

The pumping term is the difference of the above quantities.

$$(dH/dt)_p = \kappa \int dk_1 dk_2 dk_3 dk_4\ C\ \delta(k_1+\ldots)\delta(\omega_1+\ldots)$$
$$(\tfrac{3}{2})\ n_1 n_2 n_3 n_4 (1/n_1 - 1/n_3)(1/n_2 - 1/n_4).$$

The diffusive term always increases disorder and the pumping term may increase or decrease disorder.

In conclusion the increase in height of the main peak is not unphysical because the order created by the pumping term is always less than the disorder created by the diffusive term. In the present case the amount of order created near the peak of the spectrum will be less than the disorder created at high wavenumbers. In the ocean the ordering of the long waves may be noticed because they can travel large idstances and be observed as swell. The growth of the short waves is not so noticeable because they soon saturate the spectrum, break and their energy is lost.

TRANSFER

We have shown that the growth of the peak of the spectrum is not unphysical but we have yet to show how it arises. Two points have to be made. The first is the very non-linear nature of short waves. In the integrals this shows up in the rapid increase in C for large wavenumbers and the much larger region of the k plane associated with short waves. In figure 3 it shows up as the very small diffusive transfer of action density down the forward face of the spectrum, despite the fact that the gradients of action density are large in that region.

The second concerns the conservation laws. In theories of strong turbulence where often the only conserved quantity considered is energy, one is usually happy to let the energy cascade from one region of the spectrum to another. However as soon as there is more than one conserved quantity, and in the present case we have four (counting the two components of momentum), one has to be careful.

Considering just the conservation of action density and energy, then if we transfer a unit of action density to a higher wavenumber in one part of the spectrum, there must be a compensating transfer to a lower wavenumber in another part of the spectrum.

If we look at the integrals in more detail, we see that the largest contributions come from scattering events in which, one of the initial pair of waves and one of the final pair of waves lie in the tail of the spectrum, and the other two waves lie near the peak of the spectrum. The two waves in the tail of the spectrum mean that the coefficient C is very large. Also if k_1 and k_3 are the waves in the peak of the spectrum then in the expression

$$n_1 n_3 (n_4 - n_2) + n_2 n_4 (n_3 - n_1)$$

the term $n_1n_3(n_4-n_2)$ is very large because the product n_1n_3 is large. The other term $n_2n_4(n_3-n_1)$ can be neglected by comparison.

Now $n_1n_3(n_4-n_2)$ represents the diffusive transfer from k_4 to k_2, the transfer being to higher wavenumbers as in figure 3. The transfer from n_3 to n_1, occurring near the peak of the spectrum, will naturally be to lower wavenumbers, to conserve energy. But it cannot be beyond the low wavenumber face of the spectrum, because the product n_1n_3 then drops off rapidly. In fact because of the additional reduction in C as the wavenumber is reduced one finds that the maximum value of dn/dt occurs very near to the peak of the spectrum itself.

To recap, an anisotropy in the diffusive transfer is found due to the very non-linear nature of short waves. The pumped transfer required to conserve both energy and action density, transfers action density from intermediate to low wavenumbers. But it cannot transfer action density much past the peak of the spectrum because the spectrum itself drops off rapidly. The peak of the spectrum itself thus tends to grow.

Of course if there was much swell or other wave energy at very low wavenumbers, some of the pumped action density could be transferred to these very low wavenumbers at the expense of the growth of the peak.

CONCLUSION

The non-linear interaction between sea waves is seen to act as a machine in which the creation of disorder at high wavenumbers, helps to enhance the peak of the spectrum and thus create order at low wavenumbers.

REFERENCES

BARNETT, T.P. and WILKERSON, J.C. (1967) On the generation of ocean wind waves as inferred from airborne radar measurements of fetch-limited spectra. Journal of Marine Research, 25 (3), 292-328.

HASSELMANN, K. (1962) On the non-linear energy transfer in a gravity-wave spectrum. 1: General Theory. Journal of Fluid Mechanics, 12, 481-500.

HASSELMANN, K. (1963) On the non-linear energy transfer in a gravity-wave spectrum. 2: Conservation theorems, wave-particle correspondence, irreversibility. Journal of Fluid Mechanics, 15, 273-281.

DISCUSSION

M. S. Longuet-Higgins:

The two papers that we have just heard (by Fox and Webb) together go far towards clarifying the physical processes underlying Hasselmann's transfer equation. I remember that when some years ago (Review Lecture, R. Society, 1973) I pointed out the need for such clarification, Adrian Gill suggested that we should offer a prize for the first person to succeed. I propose that prize now be awarded jointly to these two authors!

It should however be realised that weak interaction theory is valid only asymptotically, for wave steepnesses within a certain range. It cannot directly describe strong nonlinearities like the overturning of the free surface. Perhaps there is even an intermediate range of wave steepnesses for which the possibility of soliton solutions should be taken into account. Thus, the Davey-Stewartson equations for the envelope of a slowly-modulated wave train in deep water do admit solutions in the form of solitons, which have actually been reproduced experimentally by Yuen and Lake. These are one-dimensional solutions. They correspond to discrete, complex wavenumbers in the generalised nonlinear transform, whereas the ordinary sine-wave components correspond to elements of a continuous spectrum. In our statistical ensemble perhaps we could allow for the existence of a certain number of discrete solitons, on average, per unit horizontal distance, in one dimension.

In two horizontal dimensions such a model is less plausible, because of the probable instability of solitons to two-dimensional disturbances. Nevertheless, one can conceive a statistical model with a certain number of solitons, on average, crossing a given horizontal area. Where solitons intersected they would undergo a certain phase-shift, depending on the amplitudes and the angle of intersection.

K. Hasselmann:

I agree it would be interesting to consider solitons interacting with a continuous background. However, this would be an appropriate model only if the surface wave spectrum did indeed consist of a few essentially δ-function spikes (e.g. swell trains) and a continuum. It is not possible to split up a continuous spectrum into solitons plus background. The interactions within a spectral peak can be treated either as a soliton, if the spectral peak is sufficiently narrow, or by continuous weak interactions, if the peak is not too narrow. The regions of applicability of the two theories are mutually exclusive. I suspect that in the real ocean the continuous interaction theory will be valid even for rather narrow swell peaks, as the necessary condition that the nonlinear transfer time scale is large compared with the swell beat period is normally satisfied.

M. S. Longuet-Higgins:

I would like to explain the reasons for my suggestion. If seen from the viewpoint of linear Fourier analysis, Professor Hasselmann's comments appear to be correct. But for the non-linear Davey-Stewartson equation (which is a special case of the complex Schroedinger equation) the appropriate analytical tool is the inverse scattering method of Zakharov and Shabat (1972) which is simply a nonlinear analogue of the ordinary Fourier transform. In the nonlinear transform representation, the soliton appears as a discrete entity which exists independently of the continuum of sine-wave solutions, and cannot be expressed in terms of them. For a complete description, in general both solitons and sine-waves are required. At low wave amplitudes solitons will be rare, but at higher amplitudes the density of solitons must increase. The solitons do not interact with the continuum or with one another, in the transform space. The best way to construct a statistical ensemble of solitons and sine-waves in the transform space remains to be worked out. Only experience will show whether this gives a more accurate representation of the sea surface than we have at present.

NONLINEAR AND LINEAR BOTTOM INTERACTION EFFECTS IN SHALLOW WATER

O. Shemdin*, K. Hasselmann**, S.V. Hsiao* and
K. Herterich**

*Jet Propulsion Laboratory-Caltech; **Max-Planck-Institut für Meteorologie

ABSTRACT

Spectral wave transformation in shallow water is investigated by examining nonlinear and linear bottom interaction effects. The effect of nonlinear wave-wave interaction in shallow water is investigated by including the depth dependent dispersion relationship in the nonlinear calculations. Dissipative mechanisms examined are bottom friction, percolation within the sand layer, and wave motion in the mud layer induced by hydrodynamic forces acting at the mud line. Comparisons with observations suggest that bottom motion can be one order of magnitude more pronounced than friction or percolation when soft mud occupies the top layer such as found in the Gulf of Mexico. In the North Sea (JONSWAP area) coarse sand with mean grain diameter ≥ 0.3 mm is found in the top sediment layer. Here swell energy dissipation can be explained by the linear percolation mechanism. When bottom sand is fine (mean grain diameter ≤ 0.4 mm), such as found offshore of Panama City and Marineland, Florida, nonlinear bottom friction is found to explain swell dissipation adequately. A nonlinear bottom scattering mechanism was investigated by Long (1973) who found the effect to be possibly important in the JONSWAP area but required detailed directional wave measurements to derive conclusive results. This paper examines five different data sets on wave transformation in shallow water and offers explanations in terms of bottom interaction mechanisms.

1. INTRODUCTION

The growing exploitation of the natural resources of the nearshore and offshore areas of the ocean has generated increased interest in shallow-water wave prediction models. Existing wave models deal primarily with deep water where the energy balance is controlled by atmospheric input, nonlinear wave-wave transfer and wave dissipation by breaking or turbulence. In shallow water, additional processes need to be considered; shoaling and refraction become important, and the wave-wave transfer is enhanced by the stronger nonlinearities of finite-depth waves. Dissipation induced by friction, percolation and bottom motion can also become significant processes, as will be discussed.

In this paper we first review in (2) the various available shallow-water data sets. We then explore in (3) different linear and nonlinear transfer and dissipation mechanisms. In (4) we interpret the various data sets in terms of prevailing mechanisms at the respective sites. The incorporation of different processes in a predictive shallow-water model is outlined in (5).

2. AVAILABLE DATA SETS ON WAVE TRANSFORMATION IN SHALLOW WATER

JONSWAP: North Sea

The transformation of swell in shallow water was reported by Hasselmann et al. (1973) for the JONSWAP area, shown in Figure 1. The damping coefficient was found to be independent of tidal current, suggesting a linear mechanism for energy dissipation rather than a nonlinear one such as bottom friction. Long (1973) investigated the effective damping of waves by scattering of bottom irregularities in the JONSWAP area but was unable to arrive at firm conclusions from the data available to him. He also expressed the need for accurate directional wave-height information at a minimum of two stations before final assessment of this mechanism could be made. In a recent study Rosenthal (1977) estimated the wave dissipation rate due to bottom motion from simultaneous measurements of wave height, bottom pressure and vertical velocity of the bottom at station "PISA". A typical data set is shown in Figure 2. The damping due to bottom motion was found to be one order of magnitude smaller than the total damping calculated from the wave-decay rate.

Mississippi Delta Region: Gulf of Mexico

Large decay rates of waves approaching shore in the delta region of the Gulf of Mexico were reported recently by Tubman and

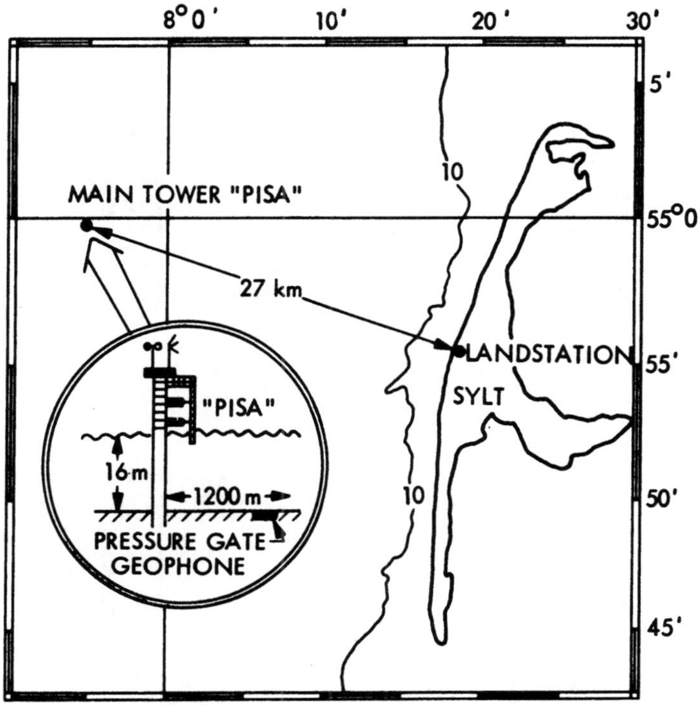

Figure 1. Geographical location for the JONSWAP test site.

Suhayda (1976). Simultaneous measurements of wave height and bottom vertical acceleration were reported in spectral form and are shown in Figure 3. The bottom sediment was found to respond in an elastic manner to the ocean waves above. Decay rates computed from wave height and bottom motion measurements were found to be much greater than corresponding estimates based on bottom friction or percolation.

Marineland Test: Florida Atlantic Coast

The Marineland Test was conducted in December 1975 to determine, among other things, the transformation of waves across the continental shelf offshore of Marineland, Florida (see Figure 4). The test plan by Shemdin, et al. (1975) describes the scope of the experiment. Preliminary results on radar imaging of waves were

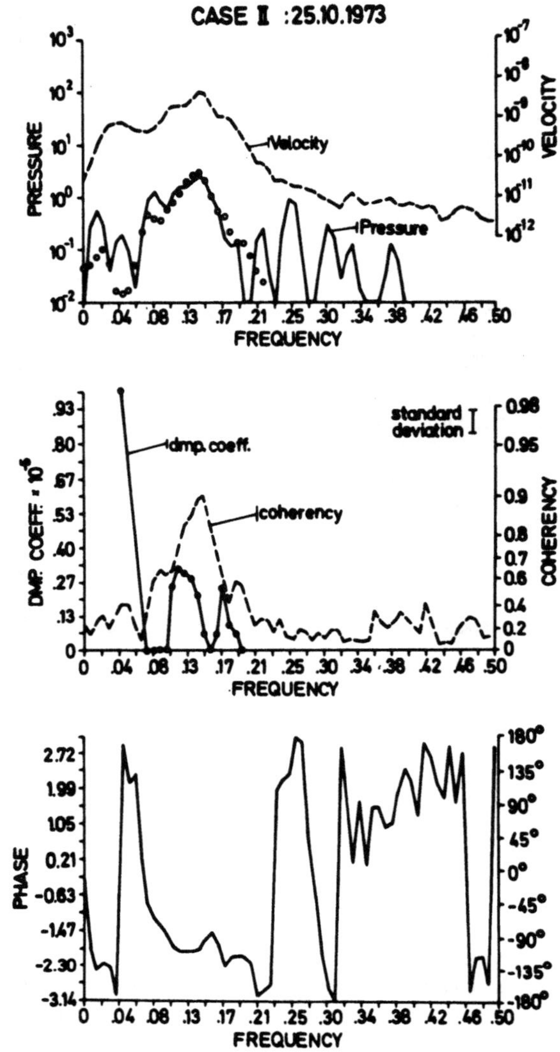

Figure 2. Bottom motion data for the JONSWAP area, after Rosenthal (1977): (a) --- bottom vertical velocity spectrum in m^2s^{-1}, o- bottom pressure power spectrum calculated from wave-rider data in m^2s, —— bottom-pressure power spectrum from pressure transducer in m^2s.
(b) --- coherency between bottom pressure and bottom vertical velocity, —— damping coefficient μ in s^{-1}
(c) phase angle between pressure and velocity.

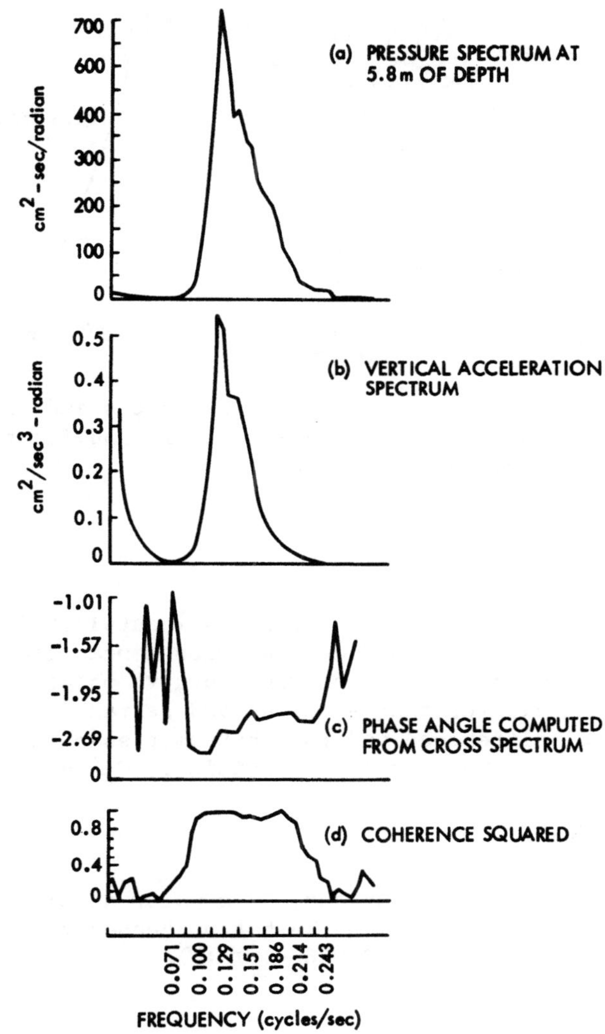

Figure 3. Bottom motion data for the Delta Region of the Gulf of Mexico, after Tubman and Suhayda (1976).

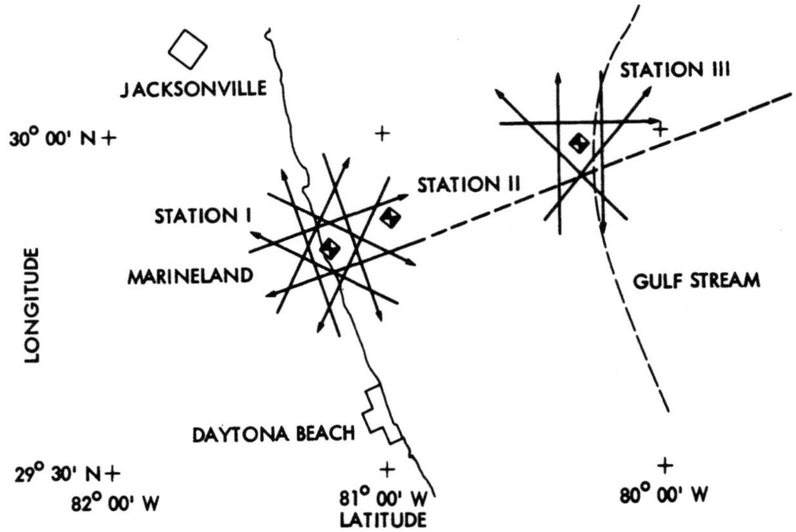

Figure 4. Geographic location, flight paths and surface truth stations (Stations I, II and II) for the Marineland Test site.

reported by Shemdin, et al. (1977). The continental shelf offshore of Marineland is covered with fine sand with a mean diameter less than 0.3 mm. A comparison between wave spectra obtained 100 km and 1.0 km offshore indicates that shoaling and refraction alone cannot explain the total difference and that bottom dissipation must be considered as well. The friction coefficient, C_f, defined by

$$\tau_b = C_f \, \rho \, U_b |U_b|, \tag{1}$$

where τ_b is the bottom shear stress, ρ is water density and U_b is orbital velocity just outside the bottom boundary layers, was calculated to be in the range 0.006 - 0.010.

Pamama City: Florida Gulf Coast

Wave spectra from two towers located 3.2 km and 17.7 km offshore of Panama City, Florida were analyzed by Hasselmann and Collins (1968). They calculated the friction coefficient, C_f, to be 0.015. A quasi-linear theory was developed in which the effect of the entire wave spectrum was taken into account when calculating dissipation at a given spectral component, in contrast to previous dissipation formulae in which the wave field is treated as a monochromatic wave. The bottom sediment in this area is fine sand and little, if any, mud is found because of its distance from the Mississippi river mouth.

Additional wave data from the Panama City towers were analyzed by the authors more recently using the Hasselmann and Collins (1968) technique. The friction coefficients were found to be somewhat higher, but still in the range 0.01-0.05.

Melkbosstrand: Cape Town, South Africa

Surprisingly high friction coefficients, C_f, were reported by Van Ieperen (1975) from measurement at Melkbosstrand, South Africa. Wave spectra from four wave gages located along a transect to shore (depth 10 - 44 m) were used to calculate the friction coefficients. The procedures outlined by Hasselmann and Collins (1968) and Bretschneider and Reid (1954) were both used. The friction coefficient, C_f, was found to be in the range 0.06 - 0.10. These values are one order of magnitude larger than calculated at Florida. Van Ieperen also used a linear velocity model and found it fit the data with less scatter. No explanation was provided for the high C_f values.

3. MECHANISMS AFFECTING WAVE TRANSFORMATION IN SHALLOW WATER

Linear Mechanisms

Percolation. This mechanism was investigated first by Putnam (1949) for an isotropic and infinitely deep sand layer. The wave pressure field induces flow in the sand below which is dependent on the sand permeability. The energy dissipation in waves can be computed from the cross-correlation between wave-induced pressure above the bed and vertical velocity in the bed. Definitions are given in Figure 5 for a finite-depth sand layer and monochromatic shallow-water waves.

It can be shown

$$\frac{\partial E}{\partial t} = \overline{pw}\Big|_{z=-h}, \qquad (2)$$

where E is the wave energy per unit area

$$E = \frac{1}{2}\rho g a^2, \qquad (3)$$

a is wave amplitude, g the gravitational acceleration, ρ the

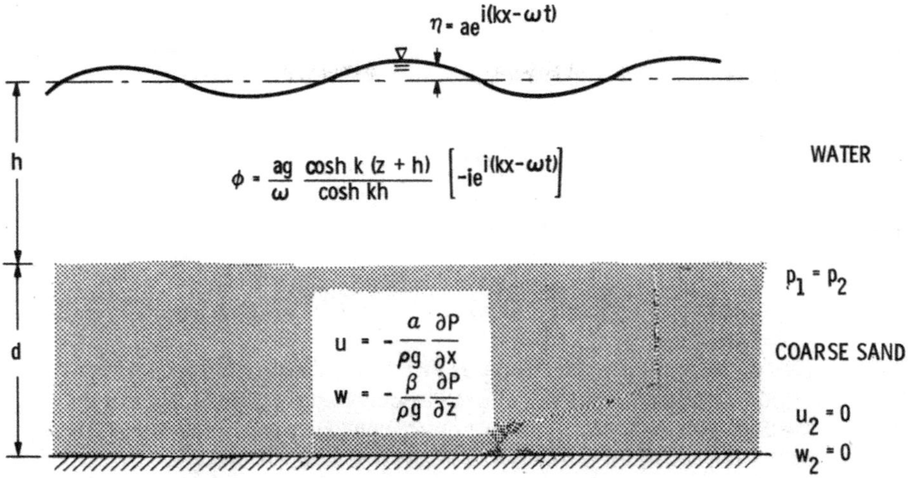

Figure 5. Schematic definition of percolation mechanism boundary value problem.

pressure and w the vertical velocity. The pressure field in the sand layer is governed by Laplace's equation

$$\frac{\partial^2 p}{\partial x^2} + \frac{\beta}{\alpha}\frac{\partial^2 p}{\partial z^2} = 0 \; , \tag{4}$$

where α and β are horizontal and vertical percolation coefficients defined in Figure 5. The solution that satisfies Eq. (4) and the boundary conditions, shown schematically in Figure 5, is (cf. Sleath, 1970)

$$p = ga \frac{\cosh\sqrt{\frac{\alpha}{\beta}}\, k\,(z+d)}{\cosh\, kh\, \cosh\sqrt{\frac{\alpha}{\beta}}\, kd}\, e^{i(kx\, -\, wt)} \; . \tag{5}$$

The rate of energy dissipation, computed from Eq. (2), is found to be

$$\frac{\partial E}{\partial t} = - Ek\sqrt{\alpha\beta}\, \frac{\tanh\sqrt{\frac{\alpha}{\beta}}\, kd}{\cosh^2\, kh} \; . \tag{6}$$

For an isotropic sand (α = β) and a deep sand layer (d > 0.3 wave-length) eq. (6) becomes

$$\frac{\partial E}{\partial t} = - Ek\alpha/\cosh^2 kh, \tag{7}$$

which is identical to Putnam's solution to within a constant.

Using laboratory experiments Sleath (1970) provided coefficients of permeability for a coarse sand (mean diameter, D_{50} = 1.3 mm with standard deviation = 0.46 mm) and a fine sand (mean diameter = 0.38 mm with standard deviation = 0.10 mm). The calculated α values for the coarse and fine sands are 1.15 and 0.124, respectively. For a deep sand layer the damping coefficient

$$\gamma_p = -\frac{1}{E}\frac{\partial E}{\partial t}$$

is found from Eq. (7) to be 2.55×10^{-4} for coarse sand and 0.69×10^{-4} for fine sand. In contrast, the measured bottom motion reported by Rosenthal (1977) yields a damping coefficient equal to 1.47×10^{-5}.

A preliminary examination of sediment conditions in the JONSWAP area (Richter and Figge, personal communication) revealed the existence of mean to coarse sand on the bottom with dispersed patches of fine sand. The latter is expected in an active ocean environment such as prevails in the JONSWAP area. The calculated damping coefficients from decay rates of waves were found to lie within the bounds $0.7 - 2.5 \times 10^{-4}$ calculated for coarse and fine sand, respectively. These estimates strongly suggest that percolation plays a primary role in wave decay in the JONSWAP area.

Wave-Bottom Interaction. When the bottom sediment is composed of soft material such as mud or decomposed organic matter, the bottom responds to the wave-induced pressure field in an elastic and viscous manner. The boundary value problem for a finite depth soft bottom is shown in Figure 6.

A detailed formulation and solution of the coupled two-layer problem was given by Hsiao and Shemdin (1977) (see also Gade, 1958). The conditions prevailing in the delta region of the Gulf of Mexico, reported by Tubman and Suhayda (1976), were compared to theoretical predictions. Both viscous and elastic aspects of bottom motion were considered. It was found that viscosity in the mud is primarily responsible for wave decay. The measured damping rate of waves reported is consistent with theory, using estimates of mud viscosity given for the area. The decay rate due to bottom motion is at least two orders of magnitude greater than estimated rates based on either friction or percolation mechanisms. The measured wave height at different depths is compared with theoretical estimates in Figure 7. For comparison, the predicted decay rate based on nonlinear bottom friction is also included; the friction mechanism is seen to be too small to explain the observed decay rate.

Nonlinear Mechanisms

Bottom Friction. This mechanism is perhaps the most explored of all bottom-interaction processes. Bottom friction is generally dominant when the sediment is composed of fine sand (D_m = 0.1 - 0.4 mm) where low permeability prohibits percolation and granular friction prevents viscous flow behaviour. The bottom friction coef-

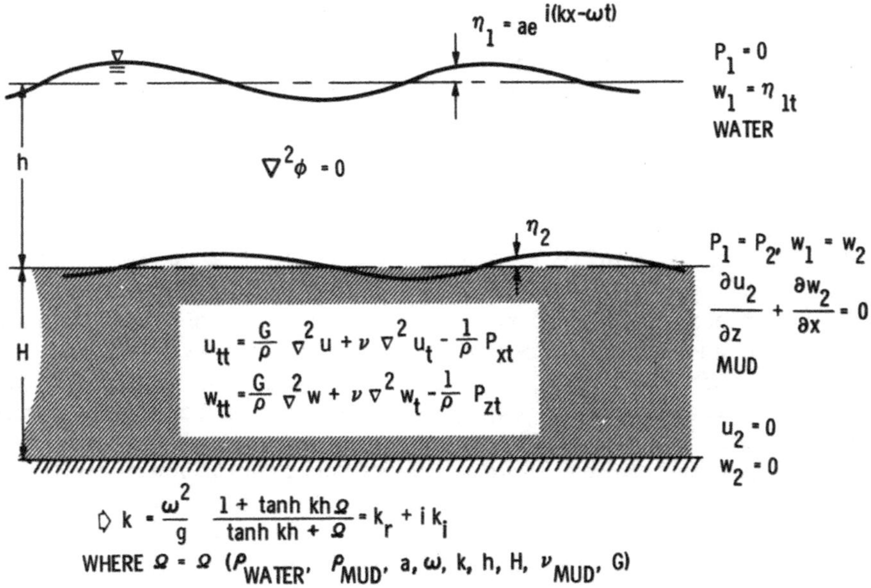

Figure 6. Schematic definition of bottom motion mechanism boundary value problem.

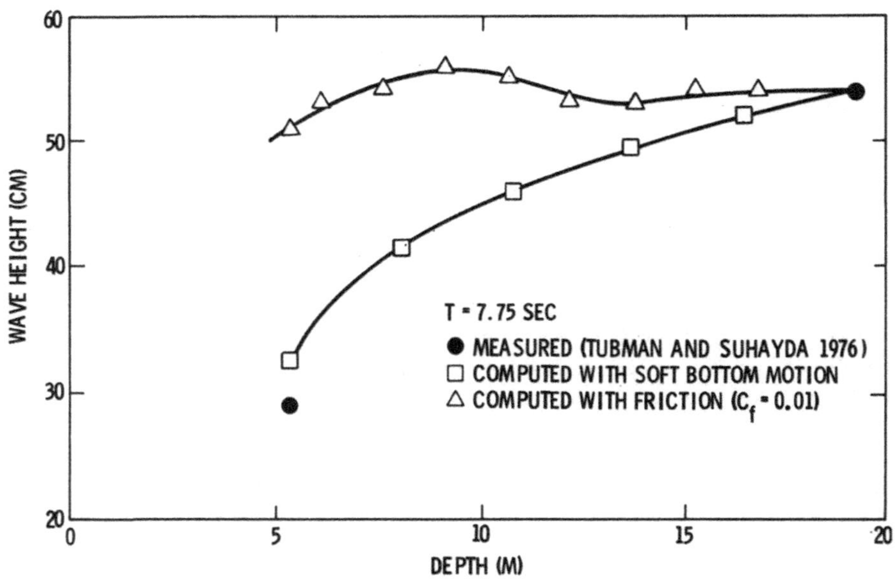

Figure 7. Comparison of measured wave height damping and predicted results using the wave-bottom interaction theory. Predicted wave height damping using bottom friction mechanism is also shown for comparison.

ficient depends on flow Reynolds number and bottom roughness height. Jonsson (1965) defines a Reynolds number, $R = U_b\, a_b/\nu$ based on the orbital velocity U_b above the bottom just outside the boundary layer, the maximum horizontal excursion amplitude, a_b, and kinematic viscosity ν.

The bed roughness factor is defined as a_b/k_o, where k_c is the roughness height. The latter is of the same order as a ripple height when ripples are present and of the order of a sand grain diameter when ripples are absent. On the basis of laboratory results Jonsson asserts that for different R and a_b/k_o values the friction coefficient, C_f, can vary throughout the range 0.007 - 0.5.

Bottom ripples were investigated by Tunstall and Inman (1975) in relation to their effect on the bottom shear stress. They concluded that ripples achieve equilibrium when $2.7 < a_b/k_o < 22$, at which state the bottom friction coefficient, C_f, lies in the range 0.045 - 0.25. The latter is one order of magnitude greater than values suggested by Hasselmann and Collins (1968) and Bretschneider and Reid (1954).

The mechanics of formation of ripples was investigated by Dingler (1974), who found their equilibrium configuration to be dependent on the near-bottom fluid flow and on the composition of the bed material. The ripple regime was found to be bounded at its lower limit by the onset of sand grain motion and at its upper limit by the development of sediment sheet flow at the water sediment interface. The ripples were found to disappear when the fluid stress far exceeded the gravity force on a sand grain. It is then presumed that k_o becomes one to two orders of magnitude smaller, a_b/k_o increases to 100 - 5000 and the friction coefficient drops to 0.007 - 0.01.

The above discussion clearly demonstrates the importance of ripples in determining the friction coefficient and suggests variability over two orders of magnitude. Such variability is not inconsistent with the wide range of observed decay rates at the sites discussed in (2).

Bottom Scattering. This mechanism was explored by Long (1973), who found it to be potentially important, but strongly dependent on the spectrum of bottom irregularities. In the JONSWAP area, measurements of the bottom irregularities by Richter et al. (1976) suggest that the mechanism is inadequate to explain the observed swell attenuation. The bottom scattering mechanism predicts increased wave energy of offshore propagating waves in the offshore direction. In contrast, other dissipation mechanisms, such as bottom friction or percolation, predict decreased wave energy of such waves in the offshore direction. A conclusive verification of this mechanism requires accurate measurements of swell in the onshore and offshore directions at a minimum of two stations.

Nonlinear Wave-Wave Energy Transfer. The nonlinear energy transfer due to third-order wave-wave interactions is given by a Boltzman integral

$$\frac{\partial F}{\partial t}(\vec{k}_4, h) = \int \pi\, \omega_4 \left(\frac{3g^2 D}{2\omega_1\omega_2\omega_3\omega_4}\right)^2 \cdot$$

$$[\omega_4 F_1 F_2 F_3 + \omega_3 F_1 F_2 F_4 - \omega_1 F_2 F_3 F_4 - \omega_2 F_1 F_3 F_4] \cdot \tag{8}$$

$$\cdot\ \delta(k_1 + k_2 - k_3 - k_4) \cdot \delta(\omega_1 + \omega_2 - \omega_3 - \omega_4) d\vec{k}_1\, d\vec{k}_2\, d\vec{k}_3\ ,$$

describing the rate of change of the spectrum at wave number $\vec{k}_4$ due to all interactions with other components $\vec{k}_1$, $\vec{k}_2$, $\vec{k}_3$ which satisfy the resonance conditions

$$\vec{k}_1 + \vec{k}_2 = \vec{k}_3 + \vec{k}_4$$

$$\omega_1 + \omega_2 = \omega_3 + \omega_4$$

The depth dependence of (8) enters through the nonlinear interaction coefficient D and the dispersion relationship, particularly in the frequency δ-function.

Herterich and Hasselmann (1977) have computed the transfer rate for a narrow-peak spectrum, for which the interactions are concentrated in a limited region around the peak wave number

$$\vec{k}_p \approx \vec{k}_1 \approx \vec{k}_2 \approx \vec{k}_3 \approx \vec{k}_4$$

using a modification of the approximate technique of Longuet-Higgins (1976) and Fox (1976). A straightforward application of the Longuet-Higgins - Fox method is not possible, because for finite h, the interaction coefficient D is discontinuous at $\underline{k}_1 = \underline{k}_2 = \underline{k}_3 = \underline{k}_4 = \underline{k}_p$.

Apart from this complication, the finite-depth case can be reduced to the infinite-depth case by scale transformations. The growth rate is changed by a factor

$$R = \hat{\alpha}\, f_D^{\,2} \tag{9}$$

in which the first component, $\hat{\alpha}$, represents the effect of the depth-dependent dispersion relation in the δ-function frequency resonance term in (8) and f_D denotes the change in the coupling coefficient D.

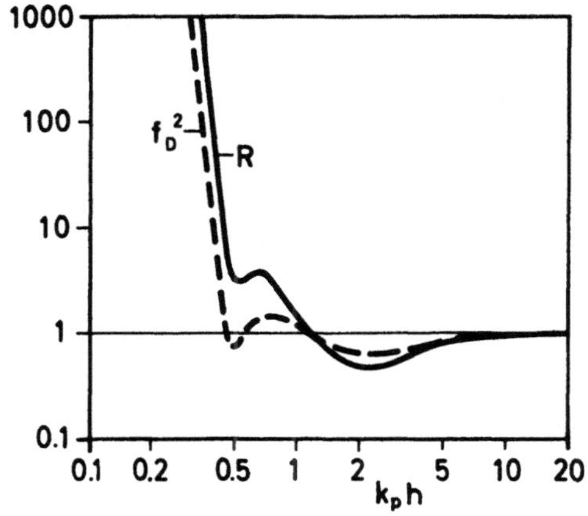

Figure 8. Depth dependence of the ratio R of the transfer rate in the finite and infinite-depth cases. R consists of the two factors f_D^2 and $\hat{\alpha}$ representing the contributions from the coupling strength and the resonance phase volume, respectively.

These factors are graphed in Figure 8 in terms of $k_p h$.

The results indicate that for $k_p h \lesssim O(1)$ the nonlinear wave-wave energy transfer is strongly dependent on $k_p h$ with a maximum near $k_p h \approx 0.6$. However, Figure 8 should be interpreted only qualitatively, as, in addition to the scale factors, the distribution of the nonlinear transfer itself is also slightly altered at finite depth through the discontinuity in the coupling coefficient.

2. INTERPRETATION OF DATA IN TERMS OF DISSIPATIVE MECHANISMS

The relative importance of the various dissipative mechanisms at a site is strongly dependent on the sediment mean grain size diameter. Percolation is most effective in coarse sand ($D_m \geq 0.5$ mm) (see Savage, 1953). In fine sand ($D_m < 0.4$ mm) bottom friction becomes more important than percolation. The friction coefficient is also critically dependent on the presence or absence of stable sand ripples on the bottom. Friction may even dominate over percolation in coarse sand if sand ripples are present. Finally, when the bottom is composed of silt, clay, or soft organic matter, bottom motion becomes dominant in dissipating wave energy. The relative strength of the three dissipative mechanisms is shown in Figure 9 for T = 7.75 sec as a function of kh. The high dissipation rate due to bottom motion corresponds to a very soft bottom layer. The decay rate due to friction increases with decreasing kh and, as shown in Figure 9, is highly dependent on the friction coefficient, which in turn is dependent on sand ripples.

The only published work known to us which discusses criteria for sand ripple formation is Dingler (1974), who proposed the ratio of fluid stress to sand grain gravitational stress as a relevant dimensionless number, namely

$$\theta = \frac{\rho\, U_b^2}{g(\rho_s - \rho\; D_m)} \quad ,$$

where ρ_s and ρ are sand and fluid densities, respectively. Dingler suggested that sediment ripples appear when the fluid stress overcomes the grain restoring force. The ripples become unstable when $\theta > 40$. At $\theta \approx 240$ the bed becomes flat again. Such sand bed regimes have their analogies in fluvial hydraulics. In Table 1 we have summarized wave and sediment parameters needed to evaluate θ for the different sites. The values of θ, together with D_m, are useful in identifying the most pertinent dissipation mechanism at a site.

In Table 1, the D_m values suggest that percolation is likely to be dominant in the JONSWAP area and that friction is likely to be dominant at the Marineland and Panama City sites. No sediment grain size information could be found for Melkbosstrand. We have speculated in Table 1 on the consequences of assuming coarse sand with D_m = 1.0 mm and fine sand with D_m = 0.3 mm. The fine silt and clay in the Delta Region site suggests that bottom motion is likely to be dominant there.

The θ parameter in the JONSWAP site is sufficiently small to suggest little granular motion, and consequently few ripples could be expected during the swell decay measurement period (although stronger rippling has been observed following severe storms). The estimated friction coefficient, C_f, from Jonsson (1965), is 0.005. The C_f value calculated from wave decay rates by Hasselmann et al. (1973) is 0.015, who also found the damping coefficient to be independent of tidal current, suggesting a linear rather than nonlinear dissipation mechanism. The difference between estimated and calculated C_f values and the linearity are both consistent with percolation rather than friction being the dominant mechanism for wave decay in this area.

The θ parameter for the Marineland site does not suggest the presence of ripples. The estimate C_f value from Jonsson (1965) is 0.005, which is of the same order as the calculated C_f value from wave decay rates and suggests friction as the dominant dissipation mechanism in the Marineland area.

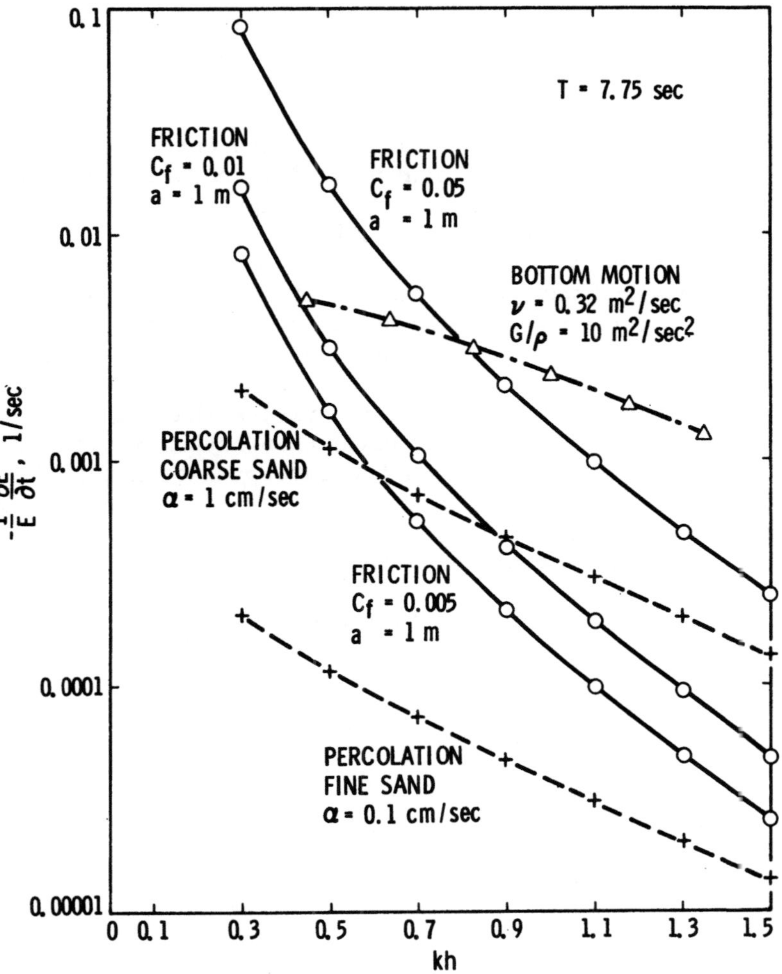

Figure 9. Damping coefficients associated with percolation, bottom motion and friction mechanism for different kh.

The interpretation of the results from the Panama City site is more difficult. There the θ parameter corresponds to values similar to those for the Marineland site, suggesting an estimated C_f value based on Jonsson (1965) in the proximity of 0.005. The C_f coefficient calculated from wave decay rates is 0.03, or six times as large as the estimated value. This can only be true if bottom ripples are present, which are not, however, predicted by the θ parameter. The estimated decay rate due to percolation is also an order of magnitude smaller than the observed decay. The dissipation by bottom motion was not determined, but an excessively soft bottom structure has not been reported. We conclude that a satisfying explanation of the Panama site data cannot be given in the framework of the present processes and parameterizations, and that further investigations are needed.

The interpretation of the Melkbosstrand data depends on the (unknown) sand structure. If coarse sand is assumed to be present, the θ parameter is sufficiently low to suggest the possible existence of bottom ripples. The estimated C_f, using Jonsson's procedure, is 0.05, which is of the same order as the value calculated from measured wave-decay rates. If, however, fine sand is assumed, the θ parameter is sufficiently high to suggest a flat bottom, for which the Jonsson technique gives $C_f \simeq 0.005$, or one order of magnitude smaller than the value calculated from wave-decay rates. Since there are no major rivers that inject silt and clay into the site, it may be assumed that bottom motion is not an important mechanism. This leaves the interpretation that the Melkbosstrand site is covered with coarse sand and that sand ripples must be present in order to arrive at the high friction coefficients (0.06 - 0.10) reported by Van Ieperen (1975).

Finally, it is clear that bottom motion governs wave energy dissipation in the Delta Region of the Gulf of Mexico.

A summary of the relevant significance of different mechanisms at the different sites is shown in Table 2.

5. INCORPORATION OF PROCESSES IN WAVE-PREDICTION MODELS

The radiative transfer equation

$$\frac{\partial F}{\partial t} + \dot{x}_i \frac{\partial F}{\partial x_i} + \dot{k}_i \frac{\partial F}{\partial k_i} = S \tag{11}$$

for the spectrum $F(\vec{x},\vec{k};t)$ is used where

$$\dot{x}_i = \frac{\partial \sigma}{\partial k_i} \quad (\vec{x},\vec{k};t) = V_i \ , \tag{12}$$

$$\dot{k}_i = -\frac{\partial \sigma}{\partial x_i} (\vec{x},\vec{k};t) \ . \tag{13}$$

Here V_i is the group velocity, σ is wave frequency and S is the source function. The latter is further divided into the various linear and nonlinear components

$$S = S_p + S_b + S_f + S_n + S_s \ . \tag{14}$$

TABLE 1. Summary of Wave and Sediment Parameters.

Data Set	JONSWAP, North Sea	Marineland, Florida Atlantic Coast	Panama City, Florida Gulf Coast	Melkbos-strand, Capetown, So. Africa	Delta Region, Gulf of Mexico
Dominant wave period T (sec)	10.0	8.0	12.0	12.0	8.0
Depth h(m) mean / range	15 (10-20)	20 (10-30)	25 (20-30)	27 (10-45)	12 (4.5-19.2)
kh mean / range	0.85 (0.68-1.0)	1.4 (0.89-2.0)	0.95 (0.83-1.06)	0.99 (0.55-1.41)	0.99 (0.56-1.37)
H_s (m) mean / range	0.5 (0.5)	1.5 (1.5)	2.5 (2.0-3.0)	2.6 (1.2-4.0)	0.42 (0.30-0.55)
U_b (m/s) mean / range	0.17 (0.13-0.22)	0.40 (0.16-0.58)	0.60 (0.41-0.84)	0.70 (0.16-1.8)	0.14 (0.12-0.20)
D_m (mm)	1.00	0.20 (0.15-0.25)	0.30 (0.2-0.4)	Unknown? coarse 1.0 fine 0.2	(0.001-0.004)
θ mean / range	2.0 (1-3)	48 (7-101)	72 (33-141)	23 (4-50) 147 (31-351)	very large
Bottom Texture	no ripples	no ripples	no ripples	ripples no ripples	not applicable
$(U_b/\sigma k_o)$ mean	270	2.5×10^3	3.8×10^3	20 1.4×10^3	
Estimated C_f from Jonsson	0.005	0.005	0.004	0.05 0.005	not applicable
C reported mean / range	0.015	0.008 (0.006-0.01)	0.03 (0.015-0.04)	0.08 (0.06-0.1)	not applicable

TABLE 2.

Relative Importance of Dissipation Mechanisms at Various Sites.

	JONSWAP, North Sea	Marine-land, Florida Atlantic Coast	Panama City, Florida Gulf Coast	Melkbos-strand, Cape Town, South Africa	Delta Region, Gulf of Mexico
Percolation	major	minor	minor	moderate	minor
Bottom Motion	minor	minor	minor	minor	major
Friction	moderate, no ripples	major, no ripples	major, if ripples present	major, if ripples present	minor
Dominant Mechanism	Perco-lation	Friction	Friction	Friction	Bottom Motion

The first three refer to energy dissipation by percolation, bottom motion and friction, respectively. The last two, S_n and S_s, refer to energy transfer by nonlinear wave-wave interaction and bottom scattering, respectively.

Equation (11) applies in the absence of currents (in the presence of currents, the energy spectrum should be replaced by the action spectrum F/σ', where σ' is the intrinsic wave frequency in the reference frame moving with the currents). If the wave field is independent of time and the depth contours are straight and parallel, equation (11) reduces to

$$\frac{\partial}{\partial x}(V \cos \theta F) - \frac{\partial}{\partial k_x}\left(\frac{\partial \sigma}{\partial x} F\right) = S \qquad (15)$$

where θ is the angle between the wave normal and the x-axis normal to the depth contours, and V is the magnitude of the group velocity.

In terms of the spectral density $E(\sigma,\theta)$ with respect to frequency σ and propagation direction θ,

$$E(\sigma,\theta)d\sigma d\theta = F(\vec{k})d\vec{k} \ , \tag{16}$$

or $E = \frac{2\pi k}{V} F$,

Equation (15) takes the form

$$\frac{\partial}{\partial x}(V \cos \theta E) + \frac{\partial}{\partial \theta}(\dot{\theta}E) = \frac{2\pi k}{V} S \tag{17}$$

where $\dot{\theta} = \frac{-dk}{dx} \frac{V\sin\theta}{k}$ (18)

or $$\frac{\partial E}{\partial x} = \frac{1}{k}\frac{dk}{dx}\left(E + \tan\theta \frac{\partial E}{\partial \theta}\right) - \frac{1}{V}\frac{dV}{dx}E + \frac{2\pi k}{V^2 \cos\theta} S \tag{19}$$

The first and second terms on the right-hand side of Eq. (19) represent wave refraction and shoaling, respectively. The last term represents the net energy transfer and dissipation. Each of the dissipation mechanisms can be incorporated in Eq. (19) by specifying the form of the individual contributions to S in Eq. (14). Thus the source function for percolation is given by

$$S_p = - \frac{VE}{2\pi} \sqrt{\alpha\beta} \frac{\tanh\sqrt{\frac{\alpha}{\beta}}\, kd}{\cosh^2 kh} \ . \tag{20}$$

Similarly, S_b has been calculated from the solution of the coupled wave-soft-bottom interaction, and expressions for the dissipation rate S_f are given in Hasselmann and Collins (1968) or Bretschneider and Reid (1954).

The energy transfer rate S_s due to bottom scattering has been evaluated by Long (1973), but was not included in the calculations shown below. Detailed nonlinear wave-wave transfer calculations for shallow water waves have not yet been made, but deep-water calculations were transformed to approximate shallow-water transfer rates using the relations of Herterich and Hasselmann (1977).

As example, the one-dimensional radiative transfer equation (19) was applied to compute the change of the wave spectrum from deep to shallow water at the Marineland site. Here friction is primarily responsible for wave-energy dissipation, so that only S_f and S_n are included in the source function computation.

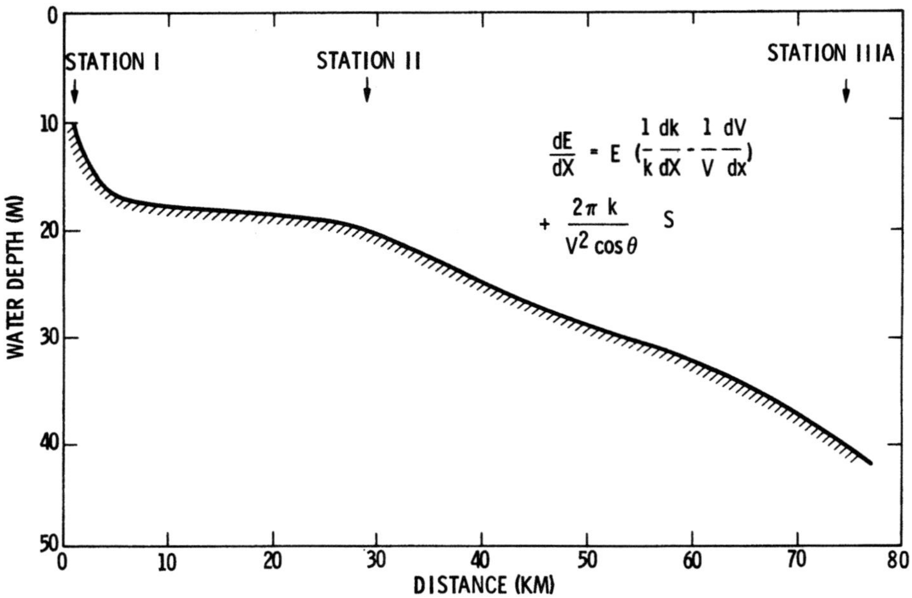

Figure 10. Marineland shelf profile.

The bottom profile over the shelf is shown in Figure 10. The bottom in the along-shore direction is reasonable uniform beyond the 10 m depth. Figure 11 shows the measured spectrum at 29 m depth and the observed and computed spectrum at 10 m depth, together with the relevant source function. The net change is decomposed into contributions from refraction and shoaling, friction, and nonlinear transfer. All transfer rates in Figure 11 and the following Figure 12 represent the relative contributions of the mechanism in question to the total change of the spectrum between the initial and final states shown. (The nonlinear transfer rates given in these examples are in fact too large, as they were based on earlier erroneous calculations of the nonlinear transfer rate before the complications arising from the discontinuity of the coupling coefficient at finite depth were discovered. However, nonlinear transfer rates of the relative magnitude indicated or even larger can in fact occur for swell of higher energy.)

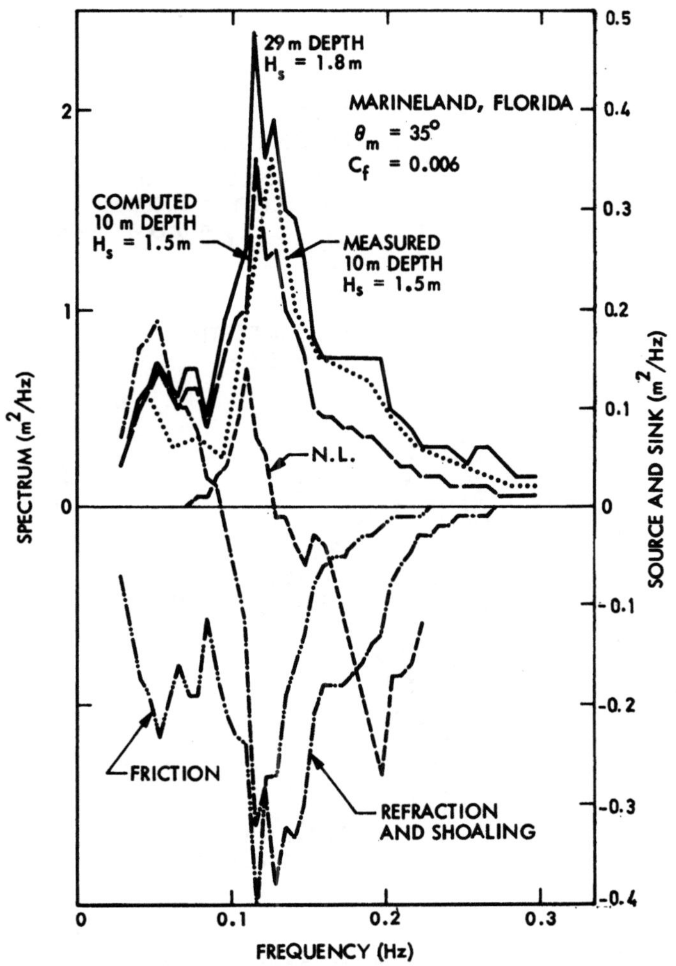

Figure 11. Computed and measured wave transformation at the Marineland site. The spectrum at 10 m depth is computed from the measured spectrum at 29 m depth as initial value. The source and sink scale refers to the total contribution from 29 to 10 m depth.

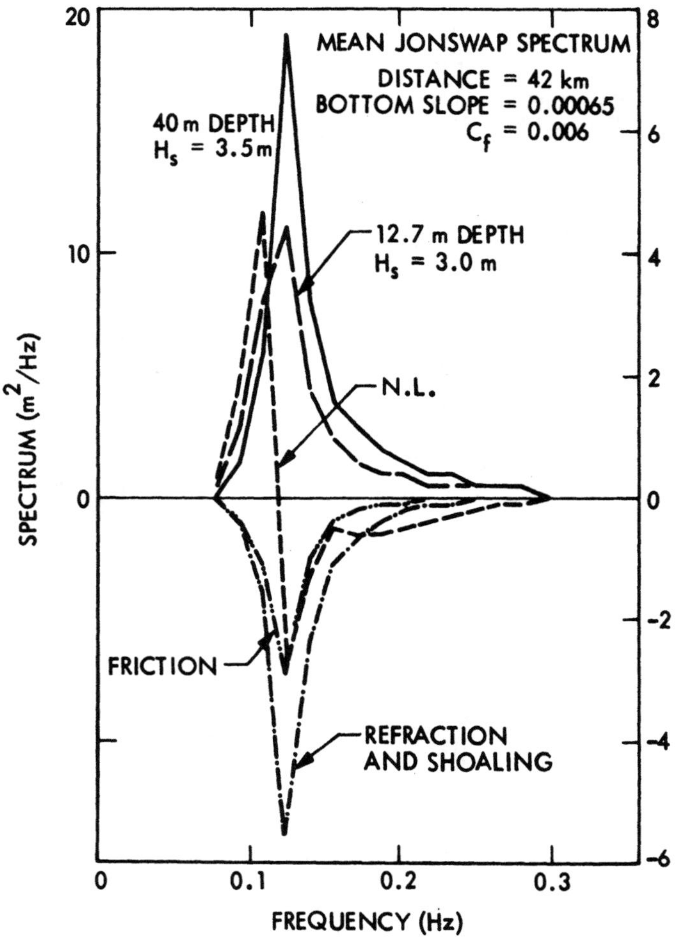

Figure 12. Computed transformation in shallow water using the mean JONSWAP spectrum in deep water as input.

6. SUMMARY AND CONCLUSION

We have examined wave-energy dissipation rates in shallow water calculated from measured wave spectra at different distances from shore. A variety of processes have been suggested as the cause of wave attenuation in shallow water, and we conclude that most of these processes can indeed become important in particular situations. The nonlinear transfer by wave-wave scattering, which has been neglected in most shallow-water models, can also become important for high energy swell. The relative significance of friction, percolation and bottom motion depends strongly on the type of sediment, grain diameter and local currents. These are therefore important inputs for identifying the dissipation mechanism dominant at a site.

Our analysis suggests that bottom motion is primarily responsible for wave-energy dissipation in the Delta Region of the Gulf of Mexico, that friction is primarily responsible for wave-energy dissipation in Marineland, Panama City and Melkbosstrand, and that percolation is probably the dominant mechanism in the JONSWAP area of the North Sea.

ACKNOWLEDGEMENT

Financial support for this study was provided by ONR-Geography Program, Contract No. N00014-76-MP60029, and by NATO Special Grant No. SRG 14. The Marineland data were gathered under NASA-Office of Applications through the Seasat-A Project Surface Truth Program. The deep-water wave spectrum at Marineland was provided by W. McLeish of NOAA-AOML. The Panama City data were provided by J.E. Breeding, Jr. of Naval Coastal Systems Laboratory. K. Richter and K. Figge of the Deutsches Hydrographisches Institut provided data on sand properties in the German Bight and were helpful in discussions.

REFERENCES

Bretschneider, C.L. and R.O. Reid (1954), "Modification of Wave Height due to Bottom Friction, Percolation, and Refraction", Beach Erosion Board Tech. Memo. No. 45.

Dingler, J.R. (1974), "Wave-Formed Ripples in Nearshore Sands", Ph.D. Dissertation, University of California, San Diego.

Fox, M.J.H. (1976), "On the Nonlinear Transfer of Energy in the Peak of a Gravity-Wave Spectrum. II." Proc.Roy.Soc. London A, 348, 467-483.

Gade, H.G. (1958), "Effects of Nonrigid, Impermeable Bottom on Plane Surface Waves in Shallow Water", J.Mar. Res., 16, 61-82.

Hasselmann, K., and J.I. Collins (1968), "Spectral Dissipation of Finite-Depth Gravity Waves Due to Turbulent Bottom Friction," J. Mar. Res., 26, 1, 1-12.
Hasselmann, K., T.P. Barnett, E. Bouws, H. Carlson, D.E. Cartwright, K. Enke, J.A. Ewing, H. Gienapp, D.E. Hasselmann, P. Kruseman, A. Meerburg, P. Müller, D.J. Olbers, K. Richter, W. Sell, and H. Walden (1973), "Measurements of Wind-Wave Growth and Swell Decay during the Joint North Sea Wave Project (JONSWAP)", Deutsche Hydrogr. Zeits., Reihe A (8°), Nr. 12, Hamburg.
Herterich, K., and K. Hasselmann (1977), "On the Effect of Finite-Depth on the Nonlinear Energy Transfer in a Surface-Wave Spectrum", (to be published).
Hsiao, S.V., and O.H. Shemdin (1977), "Interaction of Ocean Waves with a Soft Bottom", submitted to J. Phys. Oceanogr.
Jonsson, I.G. (1965), "Friction Factor Diagrams for Oscillatory Boundary Layers", Basic Research Progress No. 10, Tech. Univ. of Denmark, Copenhagen.
Long, R.B. (1973), "Scattering of Surface Waves by an Irregular Bottom", J. Geophys. Res., 78, 33, 7861-7870.
Longuet-Higgins, M.S. (1976), "On the Nonlinear Transfer of Energy in the Peak of a Gravity-Wave Spectrum: A Simplified Model", Proc. Roy. Soc. London A 347, 311-328.
Putnam, J.A. (1949), "Loss of Wave Energy due to Percolation in a Permeable Sea Bottom", Trans.Amer.Geophys.Union, 30, 3, 349-356.
Richter, K., B. Schmalfeldt, and J. Siebert (1976), "Bottom Irregularities in the North Sea, Deutsche Hydrogr. Zeits., 29, 1 -10.
Rosenthal, W. (1977), submitted to J. Geophys. Res.
Savage, R. P. (1953), "Laboratory Study of Wave Energy Loss by Bottom Friction and Percolation," Beach Erosion Board Tech. Memo. No. 31.
Shemdin, O. H., Blue, J. E. and Dunne, J. A. (1975), "Seasat-A Surface Truth Program: Marineland Test Plan," Memo 622-6, JPL internal document.
Shemdin, O. H., et al. (1977), "Comparison of In situ and Remotely Sensed Ocean Waves off Marineland, Florida," J. Boundary Layer Meteo., in press.
Sleath, J. F. H. (1970), "Wave-induced Pressures in Beds of Sand," J. Hydr. Div., HY2, ASCE, 367-378.
Tubman, M. W. and Suhayda, J. N. (1976), "Wave Action and Bottom Movements in Fine Sediments," Proc. 15th Int's Conf. on Coastal Engineering, Honolulu, Hawaii.
Tunstall, E. B. and Inman, D. L. (1975), "Vortex Generation by Oscillatory Flow Over Rippled Surfaces," J. Geophys. Res., 80, 24, 3475-3484.
Van Ieperen, M. P. (1975), "The Bottom Friction of the Sea-Bed off Melkbosstrand, South Africa," Deutsche Hydrographische Zeitschrift, 2, 72-88.

DISCUSSION

F. Dobson:

Have you considered the possibility that some swell energy may escape to low frequencies via mechanisms such as surf beats or edge waves?

O. Shemdin:

Surf beats and edge waves occur commonly within the surf zone where strong nonlinear interactions take place. We have confined our computations to the area between the surf zone and the edge of the continental shelf. We assume that in this region weak nonlinear coupling and transfer prevails.

R. Dorrestein:

Did you also consider the possible effect of a current (e.g. tidal current) in your calculations on the various dissipation mechanisms? (The nonlinear bottom friction is obviously affected by a prevailing current but I would presume that percolation and bottom motion would be hardly influenced.)

O. Shemdin:

Both percolation and bottom motion are linear mechanisms and would not be affected significantly by tidal current. The nonlinear bottom friction (square law) mechanism will be affected by tidal current. In fact, the lack of correlation between swell decay and tidal current suggested the inapplicability of the friction mechanism in the JONSWAP area.

M. Coantic:

My question is related to the use of the steady-state Darcy's law to compute the percolation velocity field within the sand layer to the pressure field. I should expect, in the limit of high frequencies, a phase lag and an attenuation of the velocity field with respect to the forcing pressure field. What is the justification for neglecting the acceleration terms in the computation of percolation velocities?

O. Shemdin:

The constant of proportionality between velocity and pressure gradient in sand were defined based on steady state experiments by Sleath. The same constants were used in wave tank studies to determine the velocity field in sand under wave induced pressure. The unsteadiness did not show a different proportionality constant. As Dr. K. Hasselmann said, the constant depends on the separation of flow around a sand particle and should be almost instantaneous. Consequently, constant of proportionality derived from steady flow should be valid for sand flow due to wave action.

L. Cavaleri:

The linear theory of waves assumes a fixed bottom - I wonder if the percolation or one of the other effects you mentioned could be a possible explanation of the discrepancy between the pressure distribution we have found on the vertical, in a wave field and the corresponding theoretical values?

P. Shemdin:

The pressure and velocity fields in wave propagating over a sandy bottom or very soft bottom may differ significantly from that over a fixed bottom. Both bottom motion and percolation mechanisms may be responsible for your discrepancy. It would be useful to obtain information on bottom material in the area of your site and also on decay rate of waves as measured by at least two sensors.

STRONG INTERACTIONS IN WIND-WAVE FIELDS

O. M. Phillips

Department of Earth and Planetary Sciences
Johns Hopkins University
Baltimore, Maryland 21218

1. Introduction

The theory of weak resonant non-linear interactions among surface wave components has been developed greatly in recent years as several contributions to this symposium have testified. The time scale for these interactions is relatively long, characteristically of the order hundreds of wave periods, but nevertheless in certain circumstances they can be influential in spectral transfers of action and energy and in the progression of the spectral peak towards lower frequencies. At the same time, however, there are also strong interactions among wave groups that can result in rapid and strong modification of the wave characteristics, perhaps sporadically leading to wave breaking and the formation of whitecaps, perhaps more generally on a smaller scale. The theme of this paper is to explore the nature and consequences of some of these processes, particularly as they influence the turbulent fluxes across the sea surface.

Wind wave spectra, measured in the laboratory and in the field have shapes that, superficially at least, are often remarkably similar. In each case, they rise to a sharp maximum, decreasing towards high frequencies approximately as n^{-5}. In laboratory spectra one frequently observes a secondary spectral maximum at about twice the frequency of the primary peak while in the field, this is often absent or less distinct. Nevertheless, the spectral shapes are quite similar and it is tempting to assume that laboratory wind waves are a scale model of an idealized field situation with dynamical similarity between the two. However, as we will see, this is certainly not so; the dynamical balances involved are in some respects quite different and the apparent similarities may be more misleading than illuminating. In particular, the extrapolation of trends observed in the laboratory to field situations and vice versa, should be accompanied with considerable caution since the balance of dynamical processes involved in each situation may be very different.

Although the wind wave spectra may be quite similar except for scale, even visual observation of the sea surface is capable of recognizing important differences. At sea, with a wind speed of perhaps 10 m/sec, the eye can pick out groups of long wave crests which overtake and move through shorter waves; the dispersive nature of the wave field is readily apparent. Wave breaking tends to be sporadic, the density of whitecaps increasing as the wind speed does. Breaking tends to occur when individual wave crests run together and it continues for a time before subsiding. During breaking, the whitecaps may catch up with shorter, more slowly moving wavelets which are engulfed in the breaking zone — the water behind a whitecap, though flecked with foam, is generally almost devoid of small-scale gravity wave structure. In the laboratory, when the wind is blowing at a comparable speed, the waves generated at a short fetch are much shorter and small scale breaking (generally accompanied by very little air entrainment and bubble formation) occurs at practically every wave crest. The coherence of each individual wave crest is greater; small scale structure, often generated by turbulence in the breaking zone, appears as capillary waves propagating at a relatively large angle to the wind and moving more or less coherently, attached to the breaking crest. Freely propagating small scale waves are virtually absent — the breaking zones catch up with and absorb any such waves and there is little opportunity for regeneration before the next breaking event. Even to the eye, the whole wave structure, large and small scale, moves coherently for several wavelengths at least. The waves look remarkably non-dispersive.

These qualitative observations can be made much more precise and the contrast between the two situations results in differences in what one would expect of the wave dispersion relation, the interfacial exchange of heat and the detailed spectral shapes at high wave-numbers and frequencies.

2. Small Scale Breaking and the Dispersion Relation

Some of the differences in structure between wind waves in the laboratory and the field can be traced to differences in the dimensionless parameter $g/n_0 u_*$ in the two cases, where n_0 is the frequency of the spectral peak and u_* the friction velocity of the air flow over the waves. If the waves near the spectral peak satisfy the dispersion relation, as they are expected to do to a good approximation, then the speed $c_0 = g/n_0$ and $g/n_0u_* = c_0/u_*$. Under laboratory conditions, if u_* is sufficiently large to generate waves at all, c_0 is relatively small and the ratio is generally of order unity, while in the field it may be as large as 25 or 30. The ratio c_0/u_* is very nearly proportional to the 'wave age' c_0/U_{10} which has long been used to parametrize field data, but use of the friction velocity in the ratio enables a direct and simple interpretation of it to be made. It will be shown that the numerical value of c_0/u_* is approximately equal to the ratio of the maximum frequency of freely travelling waves in the wind direction (waves that approximately satisfy the usual dispersion relation) to the peak frequency of the spectrum.

The tangential stress on the water surface must be supported, just below the interface, entirely by molecular viscosity so that when the wind blows a thin layer of high vorticity is generated.

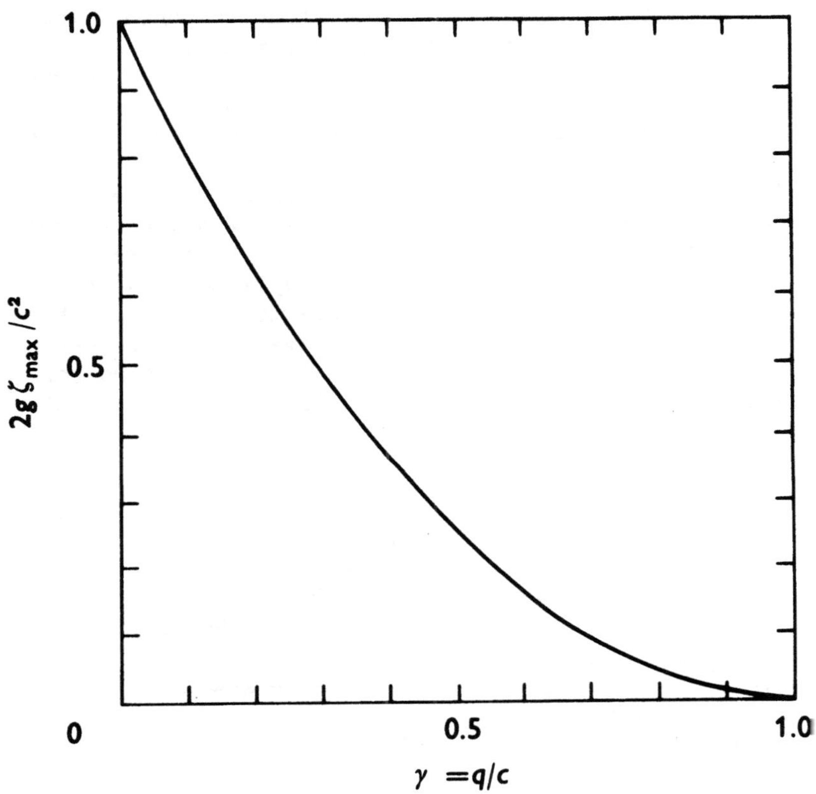

Figure 1. The wave height at the point of incipient breaking as a function of q/c, where q is the magnitude of the mean surface drift and c the phase velocity of the wave.

The mean drift q of the surface has been measured by Wu (1975) and Banner and Phillips (1974) among others; it is generally of the order 3% of the wind speed or approximately $0.6u_*$. This thin surface layer is strained by the wave motions and Banner and Phillips showed that the compression of the surface drift layer at short wave crests produces a non-linear amplification of the surface drift which can be sufficient to increase the speed of fluid elements near the wave crests to the phase speed of the wave itself. At this point the fluid elements at the wave crest are at rest relative to the moving wave profile; they may either fall forward so that the wave will break, or slip to the rear. This condition has been called the point of incipient breaking; with short waves whose phase speed is of the order u_* it is attained at a wave amplitude substantially less than the Stokes limiting form as Figure 1 illustrates. The effect is clearly a strone one, particularly for those wave components travelling in the direction of the wind at a speed only slightly larger than $q \sim 0.6u_*$; continued energy supply to these components either from the wind or from interaction with longer waves will lead to small scale breaking even at a small wave slope. For those components whose phase speed is equal to q, a wave of even infinitesimal amplitude will be at the point of incipient breaking. Microscale breaking can be recognized by lines, dimples or small steps on the surface, accompanied often by capillary waves where the surface is turned under. The mean streamlines in the motion are shown schematically in Figure 2, but the motion in the falling 'roller' is necessarily unsteady and superimposed on the mean flow is a trailing turbulent wake. Breaking on this scale is often observed to generate capillary waves of two types — very short capillaries whose phase speed matches the speed of the breaking (gravity) wave and which travel ahead, as well as longer capillary wave groups, generated probably by the unsteadiness of the motion in the breaking zone, trailing behind at angles to the primary direction of propagation.

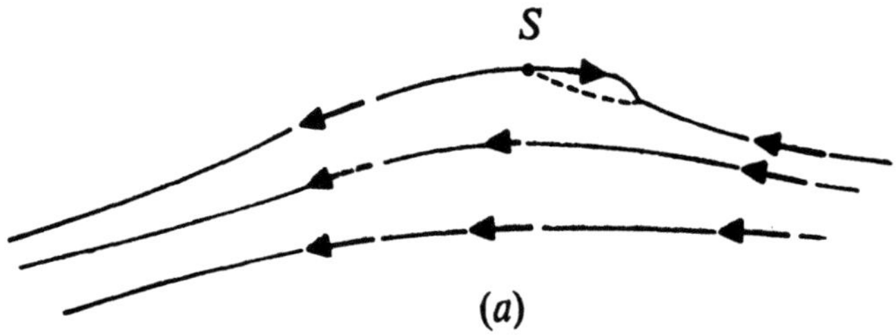

Figure 2. Mean streamlines in microscale breaking. The free surface in the breaking region is unsteady; the surface layer is turned under and mixed in the turbulent trailing wake.

The occurrence of these structures limits very effectively the minimum phase speed and consequently the maximum frequency of freely travelling gravity waves provided of course that the surface drift is at least comparable with the minimum phase velocity of free infinitesimal waves (about 20 cm/sec). Even if q is somewhat less than this, the amplification at wave crests will cause short gravity waves to break at a small but not infinitesimal amplitude; as Figure 1 shows the maximum energy density even of those wavelets for which c/q = 2 is only about 4% of the Stokes limiting form.

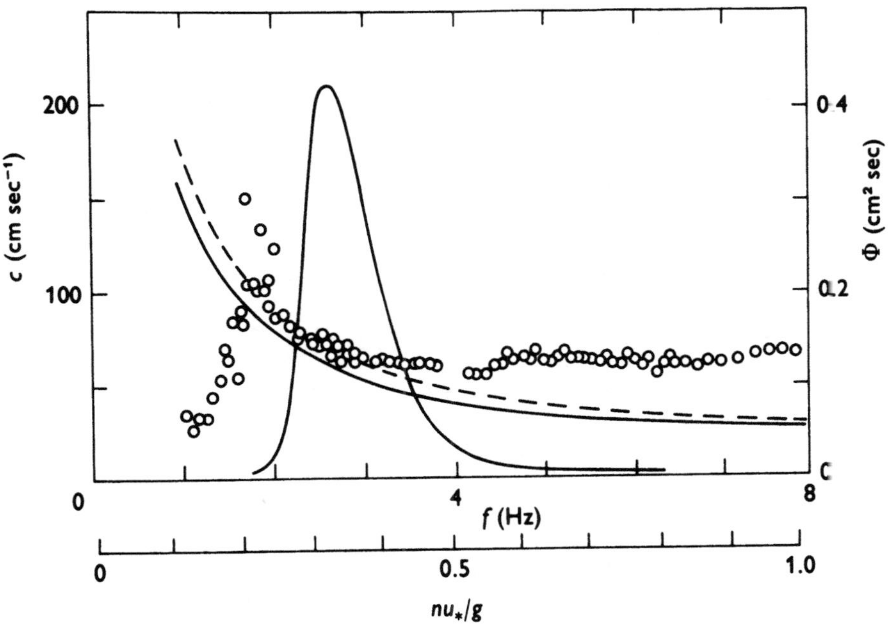

Figure 3. The variation in phase speed of spectral components in a laboratory wind-wave tunnel, measured by Ramamonjiarisoa (1974). The spectral shape is indicated by the continuous curve with a peak at about 2.6 Hz. The theoretical phase speed $c = g/2\pi f$ is also shown, the broken line immediately above this curve indicating the mean longitudinal phase speed if the energy density is distributed directionally as $\cos^2\theta$. Near the spectral peak, the measured phase speeds lie close to the dispersion relation for freely travelling components, while when $nu_*/g>0.4$, c is nearly constant, indicating that spectral components at these frequencies represent largely the distortions of lower frequency components near the spectral peak, rather than freely travelling waves. The small values of c at low frequencies, where the spectral density is very small, probably represent disturbances associated with groups of the more energetic waves and travelling with the appropriate group velocity.

In effect, then, the slowest moving gravity waves are those whose phase speed in the direction of the wind is approximately u_*; the ratio of their speed to that of components at the spectral peak is u_*/c_o and the ratio of their frequencies is the inverse of this.

Consequently, the parameter c_o/u_* can be interpreted directly as the ratio of the highest frequency of freely travelling waves to that of the spectral peak — it defines the range of frequencies in the wave field over which the dispersion relation for infinitesimal waves may be satisfied even approximately. This is not to say that higher frequency components may not be observed in the wave field, but they will be associated either with higher harmonics in the distorted lower frequency waves or with short waves travelling at high angles to the wind. In the laboratory, where c_o/u_* is characteristically of order unity, virtually all frequencies above that of the spectral peak represent 'bound' components of this kind — they would be expected to propagate essentially at the speed of the primary wave. This expectation is borne out by recent but already classic measurements by Ramamonjiarisoa and Coantic (1976) shown in Figure 3. In the field, where $c_o/u_* \gg 1$, the range of frequencies of freely travelling waves may be considerably greater but even in these circumstances, if there is relatively dense breaking on a larger scale (with whitecapping and air entrainment) shorter, slower freely travelling waves may be erased by the breaking zones and only partially regenerated before the next breaking event catches up with them.

These effects cast into question the applicability of weak resonant interaction calculations to the spectral evolution, particularly in the laboratory but also in the field in some circumstances. The conclusions of Fox (1976) concerning energy transfer near the peak of a narrow spectrum are likely not to be affected, but the pumping mechanism described by Webb elsewhere in this book, involving interactions between long waves and groups of short, freely travelling waves, is certainly not pertinent to laboratory conditions and possibly not generally to field situations. When the short wave components are non-dispersive, the weak resonant interaction theory degenerates; in fact every triad of collinear wave-numbers satisfies the second order resonance conditions (rather than the selected tetrads satisfying the third order conditions). The interactions are consequently in one sense much stronger though not yet explored in detail.

As Banner and Melville (1976) have shown, wave breaking, even on a small scale, is accompanied by air flow separation. This is in fact a kinematic consequence of the continuity of velocity across the air-water interface; a sketch of the mean streamlines in air and water is given in Figure 4. Since the water in the breaking region is moving forward faster than the

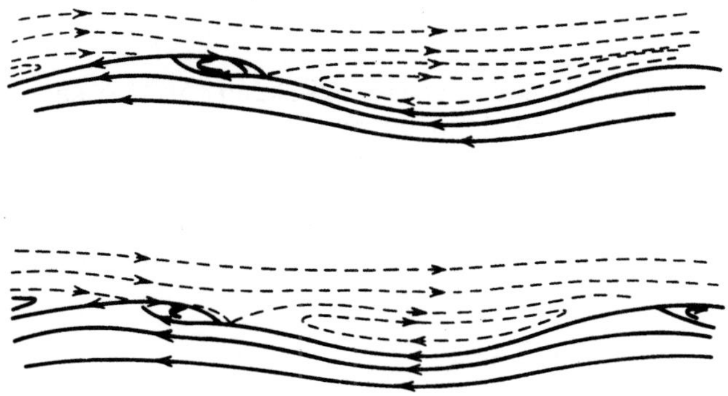

Figure 4. A schematic diagram of the streamlines of mean flow in the vicinity of breaking waves, as given by Banner and Melville (1976). In a frame of reference moving with the waves, the direction of motion of the surface in the breaking region is reversed as the water tumbles forward, and the stagnation point in the water near the wave crest is accompanied by one in the air flow. The top diagram shows the mean streamlines when a breaking wave follows an unbroken one; the bottom is for two successive breakers.

wave profile, so is the adjacent air layer, while ahead of the breaking region the water (and air) are approaching the crest. Relative to the wave profile, then, the air flow at the surface must change direction at a mean stagnation point. At the same time, the occurrence of small-scale breaking results in the surface layer of the water being turned under. If latent or sensible heat is being lost from the surface, the diffusive layers in both air and water are disrupted by the breaking; the cold water film at the surface is mixed by the turbulent wake trailing the breaking zone. It is interesting to speculate whether the discontinuous behavior observed by Street (and described elsewhere in this book) in the heat transfer coefficient is associated with the onset of small scale breaking. At very low wind speeds the transfer rate follows that of a smooth flat plate, but as the wind speed increases there appears to be an abrupt transition to higher values. Whether or not this transition is the result of the onset of micro-scale breaking is still uncertain, but be that as it may, the occurrence of this type of flow in the water and air should leave some imprint on the transfer not only of heat and water vapor, but momentum also. Preliminary measurements of Banner and Melville (1976) indicate that the influence on momentum transfer is very strong.

3. 'Asymptotic' Spectra

The occurrence of small scale breaking, with near discontinuities in slope at the dimples and forward faces of breaking regions, has an influence on the spectral shapes at very high wave numbers and frequencies. The occurrence of isolated irregular line segments across which the slope changes by an amount Δf is associated with an asymptotic form of the two-dimensional wave number spectrum

$$\Psi(\underline{k}) \to (2\pi)^{-1}\nu\overline{R(\Delta f)^2}\, k^{-3} \tag{1}$$

where ν is the number density of points along the breaking lines where the line is perpendicular to the line-number $\underline{k}$ and $\bar{R}$ is the mean radius of curvature of the line segments at these points. This contrasts with the usual saturation range result for lower wave-numbers in which $\Psi(\underline{k}) \propto k^{-4}$. It can also be shown that the asymptotic form of the frequency spectrum of slope at a fixed point is

$$\Phi_s(n) \to (2\pi)^{-1}\, \nu'\,\overline{(\Delta f)^2}\, n^{-2} \tag{2}$$

where ν' is the average number of breaking wavelets passing the point per unit time. Again, the saturation range result gives a dependence as n^{-1}. In the laboratory when $c_o/u_* \sim 1$, virtually every wave crest is breaking so that $\nu' \sim n_o$, the frequency of the spectral peak, while in the field the frequency of small scale breaking wavelets is of the order g/u_*. Components in this frequency range are non-dispersive and propagate with the speed of the breaking wavelet, which is proportional to u_* so that the frequency spectrum of the surface displacement is $n^{-2}u_*^2$ times that of the surface slope, or

$$\Phi(n) \sim u_* g\overline{(\Delta f)^2}\, n^{-4} \tag{3}$$

which, when $c_o/u_* \sim 1$ can be written

$$\Phi(n) \sim g^2 n_o^{-1}\overline{(\Delta f)^2}\, n^{-4} \tag{4}$$

These asymptotic spectral shapes are associated simply with the lines of slope discontinuity but there may in addition be high frequency contributions from capillary waves travelling at a large angle to the wind. It is not at all obvious *a priori* what the relative contributions will be but there are some observational indications that the above spectral forms can be observed both in the laboratory and in the field. Figure 5 shows some preliminary measurements obtained in collaboration with Dr. N. Huang and

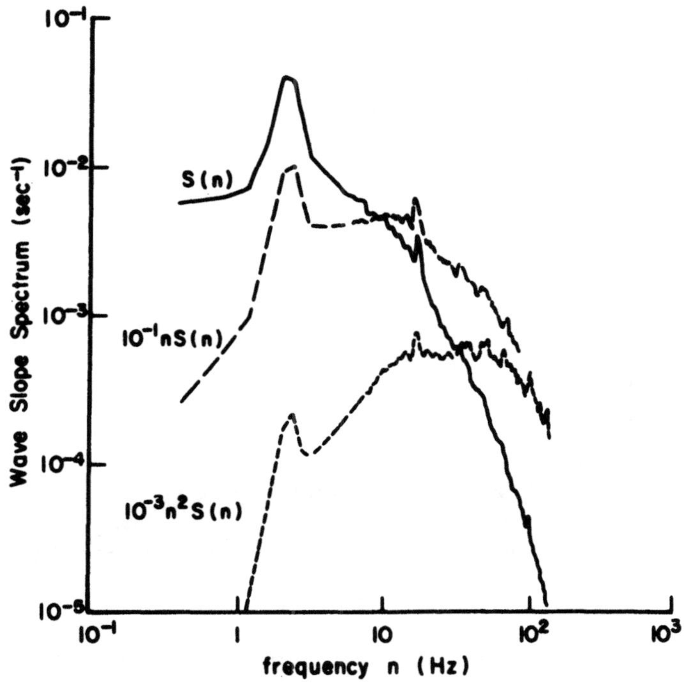

Figure 5. Preliminary measurements of high frequency slope spectra measured in the NASA-Wallops Island wind wave facility. The broken lines show the spectra weighted by n and n^2; evidently an n^{-1} range suggestive of saturation is followed by a range in which $\Phi_s(n) \propto n^{-2}$ as in Equation (2).

Dr. S. Long at the NASA wind-wave facility on Wallops Island, Virginia. The wind speed was approximately 11 m/sec, and almost all of the wave crests could be observed to be breaking. At lower wind speeds not every wave crest passing the observation point was breaking and the spectral level decreased. Finally, Figure 6 shows some field measurements taken and kindly supplied to me by Dr. H. Mitsuyasu. The frequency spectrum of surface displacement, weighted by n^{-4}, shows a definite plateau as Equation (4) would suggest. It remains to be seen, however, whether this form can be observed consistently. It can hardly be expected under light wind conditions when the friction velocity

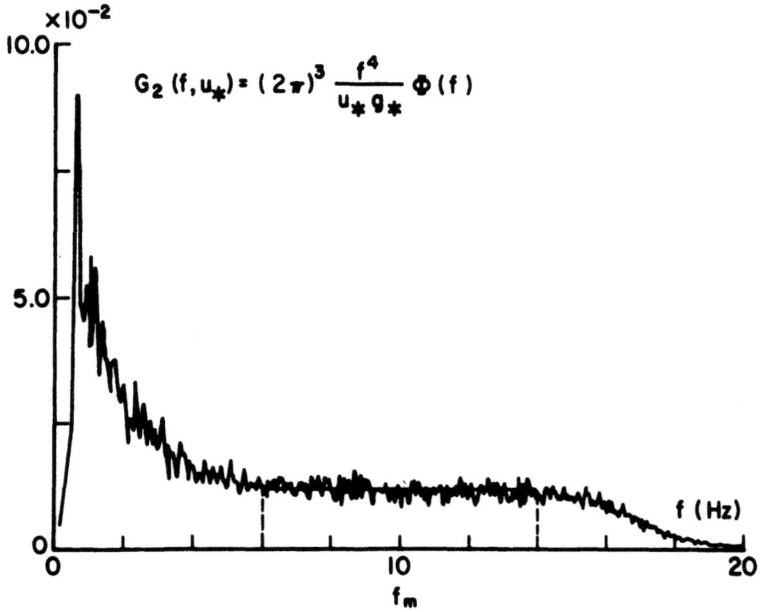

Figure 6. A mean square acceleration spectrum measured in the field by Dr. H. Mitsuyasu.

is substantially less than 20 cm/sec, since then the speed of surface drift may be considerably less than the minimum phase speed of gravity-capillary waves and insufficient to induce micro-scale breaking.

This presentation has concentrated on the influence of the ratio c_o/u_* upon the wave dynamics both in the laboratory and the field. This is, of course, not the only parameter of importance; the dimensionless fetch $k_oX = Xn_o^2/g$ is a better index of the importance of weak interactions in the wave dynamics and their contribution to such effects as the overshoot phenomenon and the advance of the spectral front. But that is another story, which is discussed elsewhere.

ACKNOWLEDGMENTS

The preparation of this paper and some of the research described herein was supported by the National Science Foundation under Grant Number ATM76-04050-A1.

References

Banner, M.L. and Melville, W.K. (1976). On the separation of air flow over water waves. J.Fluid Mech., 77, 825-42.

Banner, M.L. and Phillips, O.M. (1974). On the incipient breaking of small-scale waves. J. Fluid Mech., 65, 647-56.

Fox, M.J.H. (1976). On the nonlinear transfer of energy in the peak of a gravity wave spectrum - II. Proc. Roy. Soc. A 348, 467-83.

Phillips, O.M. (1977). The Dynamics of the Upper Ocean, Second Edition. Cambridge Univ. Press.

Ramamonjiarisoa, A. and Coantic, M. (1976). Loi éxperimentale de dispersion des vagues produites par le vent sur une faible longuer d'action. C.R. Acad. Sci., Paris, B, 282, 111-4.

Wu, J. (1975). Wind-induced drift currents. J. Fluid Mech., 68, 49-70.

DISCUSSION

W. Pierson:

Would you not expect a portion of the frequency spectrum in the open sea at high frequencies to behave like $u_* \omega^{-4}$

How do you go back to wave numbers?

O. Phillips:

Yes, indeed. To calculate the frequency spectrum one, in fact, calculates the two-dimensional wave-number spectrum first and infers the frequency spectrum from this, using now the fact that small scale structures (with their harmonics) move coherrently.

H. Mitsuyasu:

According to your theoretical inferences, the form of the high-frequency wave spectrum is proportional to f^{-4} both in the laboratory and for ocean waves, but the constants of proportionality are different. However, according to our observations, even the constants of proportionality are the same at the same wind speed. How can you explain this fact?

O. Phillips:

The spectral densities in these "asymptotic" ranges are dependent on the number of small scale breaking events per unit time at a point, and this is of order g/u_* both in the laboratory and the field. In the laboratory this is of the same order as the dominant wave frequency; in the field much larger than the dominant wave frequency, but when $c/u_* \sim 1$ the two coefficients are essentially the same.

A. Ramamonjiarisoa:

During experiments conducted in Mediterranean Sea we found a dispersion relation very similar to that found in the laboratory. The "Mistral" blew with almost constant direction and velocity. So the situation is a little confusing as far as arguments about the c/u_* influence on the physical mechanisms involved are concerned. During these field experiments the wave age was of order unity.

O. Phillips:

The same kind of suppression of freely travelling high frequency waves can occur even when c_0/u_* is large, if there is frequent large scale breaking with whitecaps. The breaking zones essentially catch up with shorter, free waves and erase them; if they cannot be regenerated to their saturation level before the next breaking event, their spectral density may be small. I should expect that when c_0/u_* is large, the dispersion relation may or may not be close to the linear form for an extended range depending on whether the density of whitecaps is small or relatively large.

The main point, however, is that in the laboratory, with $c_0/u_* \sim 1$, we will certainly not have the usual dispersion relation at high frequencies.

WIND-GENERATED WATER SURFACE WAVES: THE LABORATORY EVIDENCE

Erich J. Plate

Professor of Civil Engineering
University of Karlsruhe, West Germany

Summary: In the laboratory, fetch limited pattern of water surface waves at strong winds are seen to develop in three different ways associated with different and consecutive regions: a linear region, in which waves grow exponentially without length change, a saturation region in which waves lengthen (number of waves is not preserved) while maintaining their shape, and a breaking region. The generation process of the first ripples and some characteristics of the waves in the saturation zone are discussed in this paper.

1. REGIONS OF WAVE DEVELOPMENT

It is well known that for a given point of the water surface the wind wave pattern observed there depends on the wind speed, fetch length, and wind duration. Only when the wind duration is large enough do we obtain fetch limited waves whose statistical characteristics do not change with time. These are the maximum possible waves for a given wind intensity and fetch - in the sense that the individual waves follow some random distributions for their properties, but that the parameters of these distributions are stationary. The minimum duration necessary for a fetch limited case to occur after a step change in wind velocity equals fetch divided by wind velocity - so that at a fetch of about 300 km even a very strong wind of 40 m/s must have persisted unchanged for at least 2 hrs.

Because of the variability of the natural wind, truly fetch limited conditions do not often occur on the open sea, but since extreme waves are the result of these conditions, it is of in-

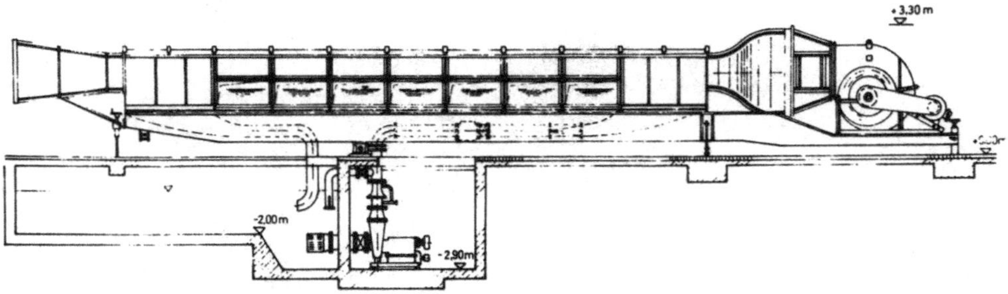

Technical Data:

Dimensions:		Slope Adjustment:		Max. Waterdepth:	Max. Wind - Velocity:	Max. Water - Velocity:
Total Length	24,5m	In Flow - Direction	4%	0,40m	20 m/s at 39 KW	0.3 m/s (0,40m Waterdepth)
Test Section	10,5m	Opposite Flow - Direction	1%		Motorpower	at 14,5 KW Motorpower
Width	1,8m					
Height	1,2m					

Fig. 1: Wind - Wave - Facility

terest to the engineer designing off-shore and similar structures to establish what these waves might be. One way of doing this is by extrapolating laboratory conditions.

It is an experience of hydraulic engineers which has been tested in numerous applications that hydraulic models represent the scaled down reality if certain scaling laws are kept, and it is my personal belief that this experience is also verifiable in the case of wind-generated waves. And just as in the case of other hydraulics models, wind wave information obtained in a laboratory should serve two purposes. The one is to repeat the workings of nature on a reduced scale, so that measurements taken, - for example on the action of wind waves on structures - , can be scaled up to yield prototype information (for a discussion see Plate and Nath, 1969). The second purpose is to reveal the physics of the processes of wave generation by wind under controlled conditions. This paper is concerned with this second purpose.

In the laboratory, the wind is generated by a fan whose steadiness can easily be controlled, and the water surface is that of a tank, such as in the wind wave tunnel of the University of Karlsruhe shown in Fig. 1. This tunnel differs from others of its kind by its large width (1,8 m) as compared to its length (12,5 m), which is considered important to avoid channeling of the waves. It is carefully designed to make sure that the air flow is a boundary layer flow with zero or near zero pressure gradient (which implies that the ambient wind u_∞ outside of the boundary layer is independent of x), and that the air channel has a sufficiently large height to prevent pressure feed back from the wind

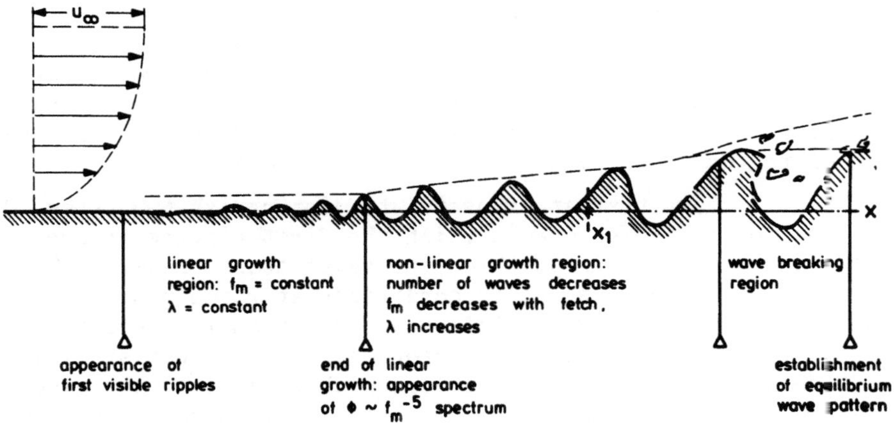

Fig. 2: Development of wind waves with fetch

waves (no blockage). In such a facility, wind-generated waves can be observed at leisure, and a surprising wealth of details is found which often conflicts with presently held theoretical views.

A striking observation for fetch limited data is that for high enough winds there exist four different regions of flow development which are schematically shown in Fig. 2. Region 1, at shortest fetches, is the region of wave generation with clearly visible, distinctly two dimensional waves established at the end of the region. Region 2 is the region of linear instability. The wave pattern is two dimensional, but not necessarily consisting of long crested waves. Its most significant characteristic is that the dominant frequency in the spectrum does not change with fetch while the amplitude of the waves corresponding to this frequency is increasing exponentially. Region 3 is the region of saturation growth: the waves increase in amplitude but at a slower rate, while their frequency decreases with fetch. The final region, region 4 might be called the mixing region in which air and water become mixed. It begins when the waves start to show white caps, and in its most developed form it has all the appearances of a layer with continuous transition of an air-water mixture from clear air to a dense mixture of air, water droplets, and mist near the water surface - much as in the mixing region of density stratified fluids with small interfacial Richardson numbers. For air-water flows region 4 has not been well observed in the laboratory because wind speeds and/or fetches have to be very large for it to become fully developed. I therefore shall not discuss this region here. However, in the initial stages, it overlaps the saturation region and therefore most of the laboratory data can be described through the processes in the latter.

Any of the fetch limited wave pattern is an asymptotic pattern also for the case of duration limited wave development. After a gradual change of wind velocity from zero to some large value of u_∞ one observes for example for fetch x_1 of Fig. 2 that the temporal development of the wave pattern goes through all the stages of region 1 to 3 until it becomes stationary.

Fig. 2 shows that it is useful to distinguish two types of stability limit: a lower one at which the first ripples become visible, and a higher one at which the waves start breaking. The lower one seems to correspond to conditions at which wind field and waves become linearly coupled: waves occasionally seen at shorter fetches than the one corresponding to the lower stability limit have wave heights which decrease with time, i.e. the waves are damped. The upper limit, on the other hand, corresponds to a breakdown of the stable waves: it is the stability limit investigated by Jeffreys (1926). In the remainder of the paper, the characteristics of the regions will be demonstrated by means of recent experimental data.

2. WAVE GENERATION REGION

The experimental evidence (Plate et al. 1969) does not support the concept, that the initial ripples are caused by resonance between the water surface and a pattern of turbulence eddies which is convected by the air flow across it (Phillips,1966). It is most likely that the first water waves are reactions of the water surface to the generation of spontaneous turbulence "bursts". There exist a number of experimental papers (for example, Morrison et al., 1971) on the burst phenomenon which show that the average length L of the burst for a smooth surface is related to the shear velocity u_* of the air flow and the viscosity ν through the relation

$$\frac{u_* L}{\nu} = \text{constant} \qquad (1)$$

with the constant equal to about 630.

There do not exist direct measurements of the wave length λ of these very small first ripples. Plate et al.(1969) have made some measurements on the growth of waves in region 2, in which the frequency f_m which corresponds to the peak in the spectrum of the waves was determined as function of shear velocity u_*. These data have been plotted in the form u_* vs f_m^2, with the results shown in Fig. 3. A straight line through the origin yields the relation $u_* = 0{,}3\ f^2$, and if it is assumed that the deep water dispersion relation is valid for the first waves, then this result corresponds to the relation

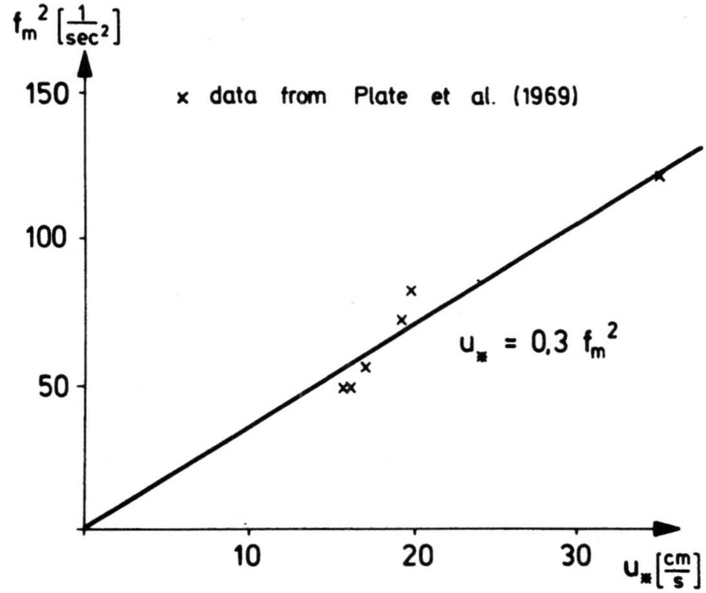

Fig. 3: Relation between frequency f_m of first ripples and shear velocity u_*

$$\frac{u_* \lambda}{\nu} = 300 \qquad (2)$$

which in functional form, if not in magnitude, corresponds to the result of Eq. 1.

The conjecture that first ripples are caused by bursts has been discussed previously by G. Hardtke (oral communication) for the case of generation of sediment ripples. His experiments had yielded a constant of about 600. Yalin (1977) reported a value of about 2000 for sediment waves and showed that Eq. 1 could be derived from dimensional analysis, without recourse to hypotheses on the turbulence structure. It must also be mentioned that Prof. Engelund and his students (see Larnaes, 1976) have found that for the temporal development of instabilities a small perturbation of the drift current yielded growth rates in surprisingly good agreement with observations. In view of the well known theoretical result that two dimensional waves of the Tollmien-Schlichting kind are more unstable than three dimensional disturbances, such a theory might explain the puzzling fact that the first waves are strikingly two dimensional and do not occur, as one might have expected, as random and short crested waves.

3. LINEAR REGION

In the linear region, waves grow exponentially without changing their length or frequency from the value at the lower stability limit. The pattern of these waves is basically that of a monochromatic, two dimensional wave train with crests aligned perpendicular to the wind. However, due to slight phase shifts between adjacent wave groups and due to slight differences in amplitudes of individual waves the pattern can be represented through a continous spectrum such as shown in Fig. 4 with a peak developing with fetch at the frequency of the two dimensional wave. The change of these peaks with fetch is exponential and corresponds in magnitude roughly to the growth rate predicted by the theory of Miles (Phillips, 1966) however, systematic deviations exist (Plate et al., 1969) which may require some modifications to Miles' original theory. I shall not go into the numerous papers which contain such possible modifications.

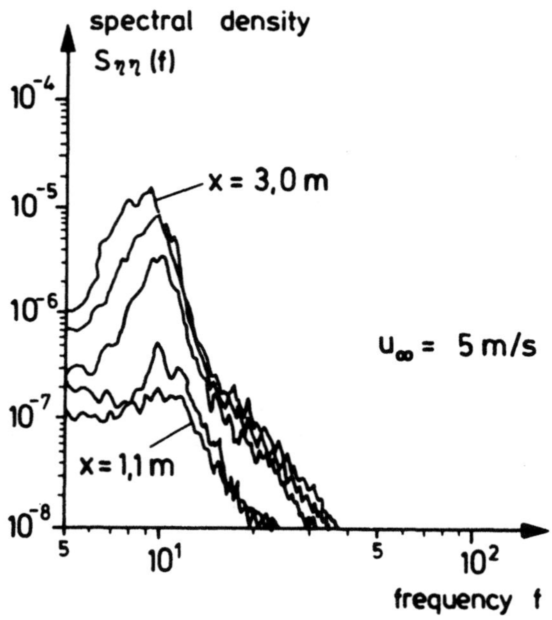

Fig. 4: Frequency spectra for short fetches (linear region)

What seems to me most important is that the linear region is always short: and the larger the wind, the shorter it is. I believe that this holds in the field as well as in the laboratory, although it is conceivable that in the case of wind acting on swell from a previous storm an energy feed back as envisioned by Miles might again occur.

4. SATURATION REGION

The region of greatest extent and theoretically perhaps the least explored is the saturation region in which all the characteristics change with fetch, without much evidence of wave breaking. This implies that the number of waves changes (decreases) with fetch. It is very difficult to detect the change of the wave length by observation alone, because the wave trains travelling along the surface consist of wave groups. But if one considers average wave parameters, these changes are clearly evident.

Consider the shape and size of the largest waves. We took a wave train of roughly 400 waves, i.e. a length of record consisting of $400 \cdot P_m$ where $P_m = 1/f_m$ is the period corresponding to the peak of the wave spectrum. From these records we obtained the highest third of the waves by the method explained in Chang et al. (1971) and determined the average wave form η_{AV} as function of fetch. Fig. 5 shows the results, and in dimensionless form it is seen in Fig. 6. The wave is obviously non-sinusoidal and strongly skewed, but similar in shape at all fetches. The skew extends into the probability distribution of the water surface elevation η, as seen in Fig. 7, and has also been observed in the field (Kinsman, see Phillips, 1966). More important is that the wave height H does not follow a Rayleigh but a normal distribution, as seen in Fig. 8 - even though a plot of the mean wave height H_m against the average of the third highest waves $H^{(1/3)}$ is well represented by the relationship

$$H^{(1/3)} = 1,598 \cdot H_m \tag{3}$$

corresponding to the Rayleigh distribution, as shown in Fig. 9. This graph has led Hess et al. (1969) and others to conclude that wave heights are Rayleigh distributed. Also, a linear relation exists between wave height and standard deviation σ_η of the water surface elevation. Since we also find that wave length λ and σ_η are linearly related according to Fig. 10, we can conclude that:

a. the wave train at a point in the saturation range is essentially made up of a sequence of skewed waves, which are

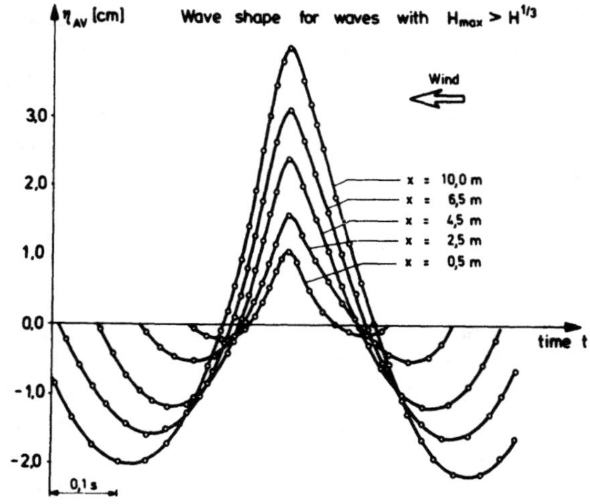

Fig. 5: Average wave shape for u_∞ = 15 m/s

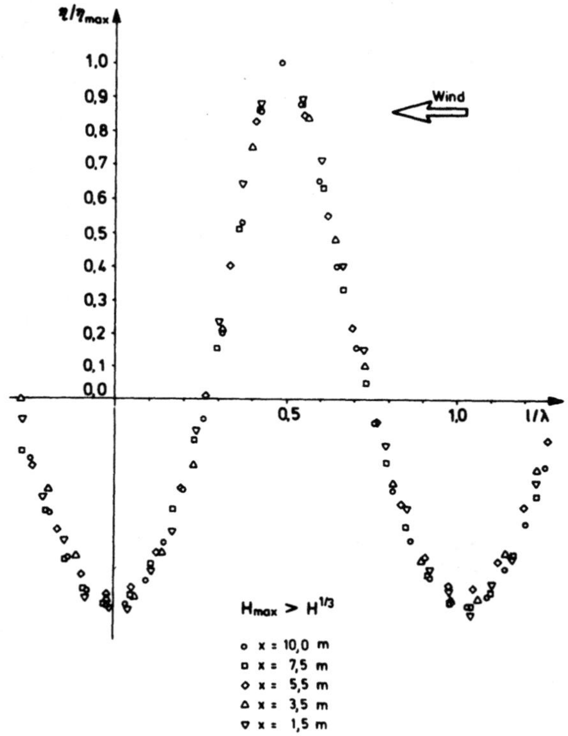

Fig. 6: Dimensionless wave shapes for u_∞ = 10 m/s

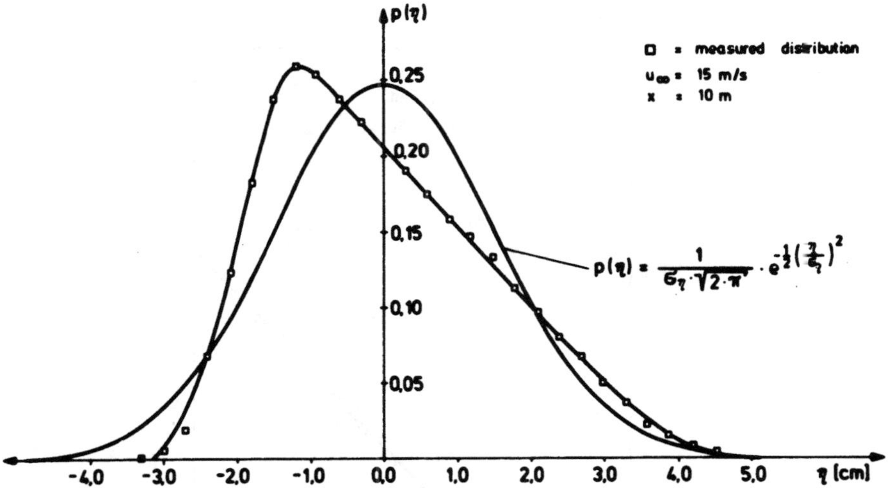

Fig. 7: Probability density distribution for water surface elevation η

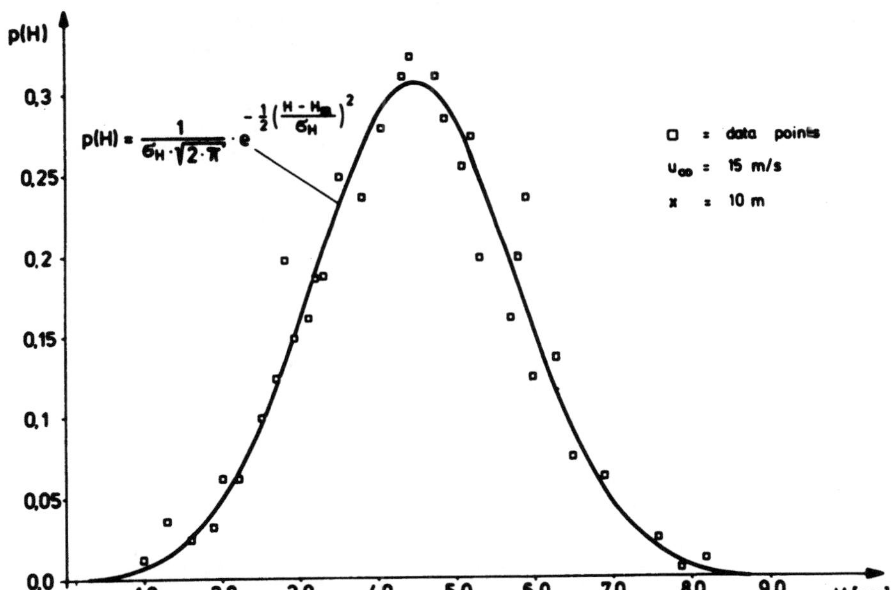

Fig. 8: Wave height distribution

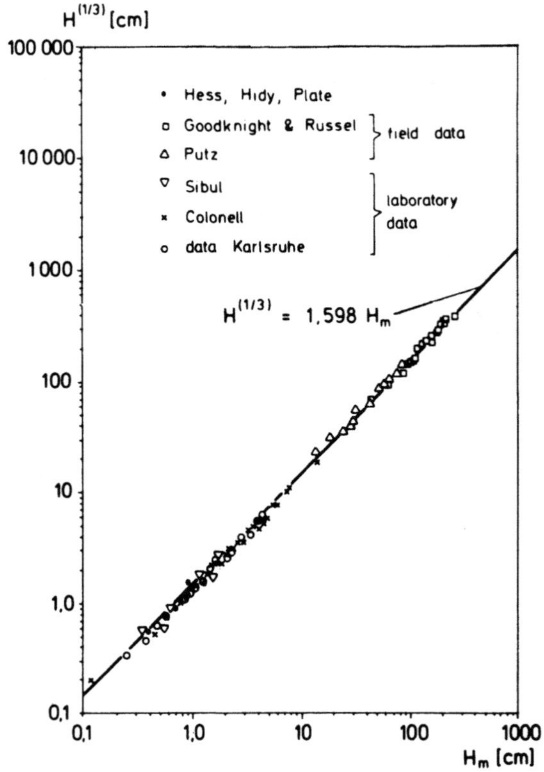

Fig. 9: Average height $H^{(1/3)}$ of highest third of 400 waves function of wave height H_m

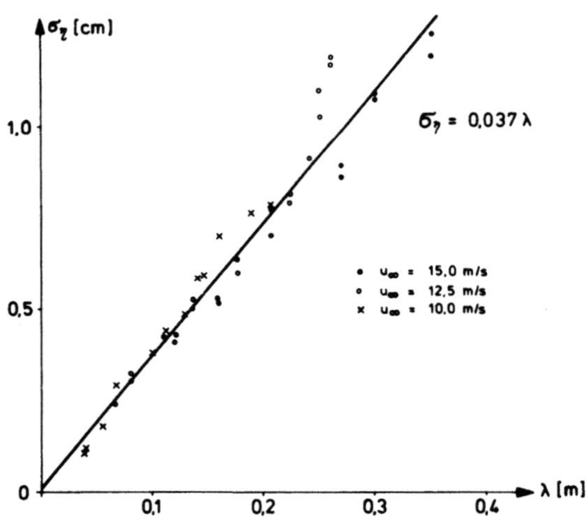

Fig. 10: Relationship between wavelength and standard deviation σ_η

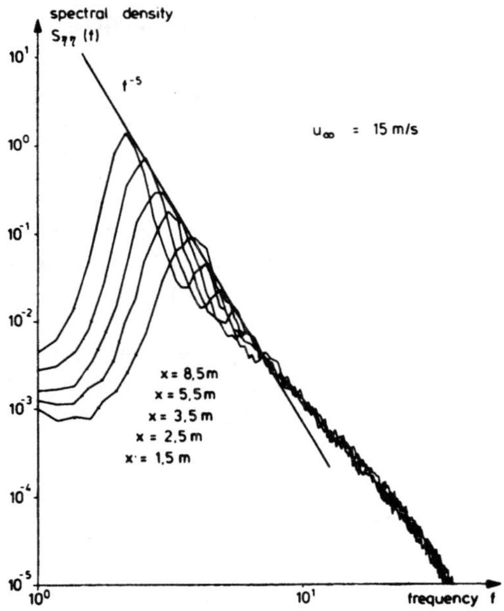

Fig. 11: Frequency spectra of water surface elevation

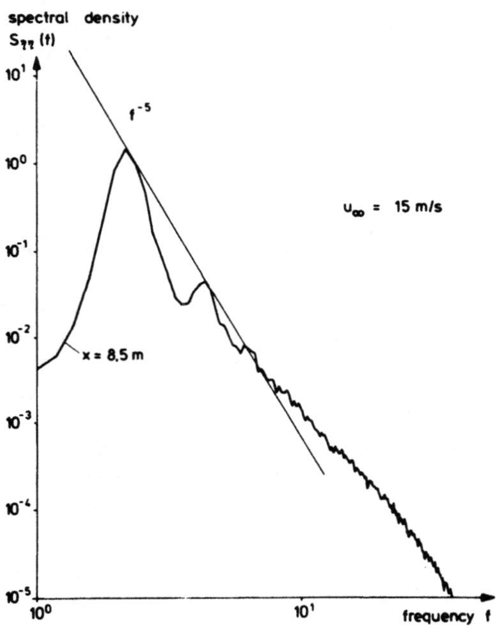

Fig. 12: Frequency spectrum of water surface elevation: example

slightly modulated by small random components;

b. the waves making up the wave train are geometrically similar with a constant height to length ratio of about 0,05; (although it appears possible that long waves may have slightly decreasing ratios of height to length ratios, see Plate, 1972)

c. all parts of the wave train travel with the same progression speed c which corresponds approximately to the phase speed of the large waves (see also Ramamonjiarisoa, 1971).

These observations imply that spectra must also have a similarity shape, and if this is so, then the peak density ϕ_m of the spectrum must depend on the frequency of the peak f_m according to the Phillips law

$$\phi_m(f_m) \sim f_m^{-5} \tag{4}$$

as shown by Plate et al. (1969) and as is seen in Fig. 11. Furthermore, due to the skewness of the waves the spectra must have secondary maxima at the higher harmonics of the peak frequency, as is indeed observed - as an example see Fig. 12. Because these secondary peaks fall roughly on a -5 curve in a double logarithmic representation(for reasons see Plate, 1972), a -5 subrange seems to appear in the individual spectra. Also, these secondary peaks are responsible for the so-called "over-shoot phenomenon".

There does not seem to exist a theory at this time which can fully explain and predict the wave growth in the saturation region.

5. COMPARISON WITH FIELD SITUATIONS

Although I am convinced from numerous observations that the zones observed in the laboratory occur under ideal conditions (i.e. for steady winds of long duration) also on the surface of lakes and oceans, there are a number of factors which complicate the wave pattern in nature. First of all the largest waves are so large that on their backs additional waves might form, which then are travelling independently of the large waves. However, it seems that this would cause a "spectral gap" to appear in the wave spectrum, which at strong winds is not observed. During such situations it is also unlikely that strong interactions occur of wave trains from different directions; the experimental evidence (for example, Mitsuyasu et al., 1975) points toward a narrow beam of wave directions. Thirdly it is possible that an equilibrium condition can be reached in nature (large wave-age parameter c/u_*) at which the energy fed into the wave by the wind exactly equals dissipation, so that no change in wave pattern takes place. How these conditions might come about is not known, there does not exist, to my knowledge, supporting evidence for the existence of such regions. And finally, the waves shapes might differ from those of the laboratory. Since it is my belief that the shape of the wave might be an important factor in the wave load on structures, we have set up a field experiment at the Forschungsplattform Nordsee near Helgoland to obtain data which will be analysed exactly like the laboratory data were.

Acknowledgement: The figures were in part taken from an unpublished report by Bächlin. The work on which this paper is based has been sponsored by the German Research Foundation through its support of the Sonderforschungsbereich 80.

Literature:

Chang, P.C., E.J. Plate and G.M. Hidy, 1971: "Turbulent air flow over the dominant component of wind generated waves." Journal Fluid Mechanics, Vol. 47, pp. 183-208.

Hess, G.D., Hidy, G.M. and Plate, E.J., 1969: "Comparison between wind waves at sea and in the laboratory." Journal of Marine Research, Vol. 27, pp. 216-225.

Jeffreys 1926 - see Phillips (1966) below

Larnaes, G. 1976: "Formation of wind waves" Institute of Hydrodynamics and Hydraulic Engineering, Technical University of Denmark, Series Paper No. 9, Lyngby.

Mitsuyasu, H., Tasai, F., Suhara, T., Mizuno, S., Ohkusu, M., Honda, T. and Rikiishi, K., 1975: "Observations of the directional spectrum of ocean waves using a clover leaf buoy", Journal of Physical Oceanography Vol. 5 p. 750-760.

Morrison, W.R.B., Bullock, K.J. and Kronauer, R.E., 1971: "Experimental evidence of waves in the sublayer", Journal Fluid Mech., Vol. 47, pp. 639.

Phillips, O.M. 1966: "The dynamics of the upper ocean", Cambridge University Press.

Plate, E.J. and Nath, 1969: "Modelling of structures subjected to wind generated waves", Proceedings ASCE, Vol. 95, Waterways and Harbors Division, pp. 491-511.

Plate, E.J., Chang, P.C. and Hidy, G.M., 1969: "Experiments on the generation of small water waves by wind", Journal of Fluid Mechanics, Vol. 35, pp. 625-656.

Plate, E.J., 1972: "The role of the dominant wave in the spectrum of wind generated water surface waves", Proceedings 9th Symposium on Naval Hydrodynamics, Paris.

Ramamonjiarisoa, A., 1971: "Mesures en soufflerie de certaines propriétés statistiques de vagues de vent", Cahiers océanographiques No. 3(12), S. 957-979.

Yalin, M.S., 1977: "On the determination of ripple length", Proc. ASCE, Vol. 1o3, Journal Hydraulics Division, pp. 439-442.

DISCUSSION

O. M. Phillips:

The decrease in dominant wave frequency as the field develops must, I think, be accompanied by the disappearance of individual wave crests as they move downwind. For in a steady wave field, if crests are conserved, the same number must pass a given point per unit time. So the frequency must be constant, hence the wavelength, but it isn't! The fact is that the waves must necessarily occur as group, with intervals between them in which crests disappear, or run together.

E. J. Plate:

The individual waves of the dominant type are exposed to the wind over distances of a few wave lengths before they disappear due to modulation of the wave group. I believe that this distance is sufficient for them to be deformed by the wind, to lengthen, and perhaps to increase in speed. There are theoretical models (Lake and Yuen, TRW Report No. 26062 - 6014 Ru-00 May 76, Deardorff, J. Oceanographic Soc. Japan, 1967) which are based on this assumption. Whether it is true should be verified experimentally.

M. S. Longuet-Higgins:

(1) In your first zone of wave growth where the waves are initiated, it may be possible to compare your critical "Reynolds number" with the theoretical calculations of R. Lock (c. 1950) for the stability of a laminar boundary layer in both air and water. Have you tried this?

(2) A theory for the nonlinear distribution of surface elevation, generating the Gaussian distribution and including higher-order terms in the Gram-Charlier series, was given by Longuet-Higgins (J. Fluid Mech., 1962). It was compared with Kinsman's results with fair agreement. Have you compared your observed skewness and kurtosis with that theory?

(3) The Rayleigh distribution of crest-to-trough wave heights was shown originally to hold for narrow spectra only, but it has since been applied (incorrectly) to broad spectra. However, a theoretical distribution for crest-to-trough heights for a spectrum of arbitrary width (defined in terms of the higher moments of the spectrum) has been given by French workers (A. Cavarié) and does show a tendency towards a Gaussian form at large spectral widths. It would be interesting to compare this with your measurements.

(4) The increase in both the amplitude and wavelength of the perturbations suggests that we might appropriately resolve the waves into normal modes with complex wave numbers (growing with horizontal distance). The possibility of modes of increasing wavelength may also be appropriate. Incidentally, the linear solution to the Cauchy-Poisson problem (see Lamb's Hydrodynamics, Chapter 9) displays a wave form with crest-to-crest separation increasing with x (but decreasing with t).

E. J. Plate:

(1) Some years ago we investigated the applicability of Lock's model to the problem of wave generation, but it is very difficult to untangle his results. On the other hand, the "generation by burst" seems to be an attractive alternative to the generation model of Phillips. It is difficult however to visualize a mechanism by which the burst-generated standing wave becomes periodic in space and two-dimensional.

(2) No. I only intended to show that the older results of Figure 9 do not prove what they seem to prove: the validity of the Rayleigh distribution for wave heights.

(3) We shall analyze the data further and shall base the analysis on the work referenced by you.

(4) The Cauchy-Poisson problem applies to "free waves' and I expect it to be related to the problem treated by Mitsuyasu in this conference. For forced waves, the approach recommended by you seems to correspond to the one used by Lake and Yuen (1976).

H. Mitsuyasu:

(1) When we analyze the statistical distribution of the wave surface we must take care to include the contribution of the high frequency components. On one side, higher moments are much affected by the errors introduced in the high frequency side of the spectrum. On the other side, the measured distribution is also affected by the high frequency component overlapping on the dominant waves.

(2) It is true that the similarity of the spectrum is widely satisfied in various conditions in the laboratory tank, but situations are slightly different in the ocean.

(3) The same thing can be said on the relation between the mean wave height and dominant wave length. That is, if the dimensionless fetch becomes large, the relation must be slightly different.

Question: You have shown that: $u_* = 0.3 f_m^2$. On the other hand, you have shown that spectral peak does change irrespective of the wind speed. This seems to be contradictory.

E. J. Plate:

It is not my intention to say that laboratory and field are exactly equal, and I agree with the comments. As far as the question is concerned, the data in Fig. 3 refer to experiments in which u_* was varied, while Fig. 4 is for constant u_* and thus f_m.

K. Hasselmann:

Did you investigate whether your laboratory data scales in the same way as the field data? Mitsuyasu's and the Stanford laboratory data for example, do, and one can then conclude that the stable shape of the spectrum and the shift of the peak towards lower frequencies can be explained by the same nonlinear interaction mechanisms as in the ocean. Both Mitsuyasu and the Stanford group demonstrated this directly by calculation of the nonlinear transfer rates for their spectra.

E. J. Plate:

Our attempts at parametrization of the spectrum peak, phase velocity, etc. led to graphs which could not be described by two-parameter curves, although we are roughly in agreement with the cited pieces of work. But I am not sure that their agreement proves your theory for the laboratory because there are other theories which also yield the "fetch graphs" as solutions (i.e. Deardorff, 67). I think that in order to be complete it must explain the waves of finite amplitude and constant shape that we observe. I would suggest that we agree on a set of "critical experiments" to be conducted in the laboratory, and I would be willing to have them done in our laboratory.

LABORATORY STUDIES ON WIND-WAVE GENERATION, AMPLIFICATION AND EVOLUTION

A.RAMAMONJIARISOA , S.BALDY and I.CHOI

INSTITUT DE MECANIQUE STATISTIQUE DE LA TURBULENCE
I2 AVENUE DU GENERAL LECLERC I3003 MARSEILLE FRANCE

INTRODUCTION

We report here on some works done by the IMST research group on the physical processes of wave generation by wind.Brevity dictates that this paper will deal mainly with the experimental data and we shall not go into many details of the analytical studies.

On the basis of various theories,we first carried out general considerations(Ramamonjiarisoa,I974;Coantic et al. I976) which led us to conclude that some interesting aspects of the principal mechanisms involved can be,at least partially,reproduced in the two IMST wind-water facilities.

We then turned our attention to some specific questions.

ON THE STATISTICAL PROPERTIES OF WIND-GENERATED GRAVITY WAVES

These properties were first analyzed.After some relatively classical studies about the water surface deflection spectrum,distribution function etc,much effort was devoted to the analysis of wind wave propagation.Indeed,it seemed to be a matter of some importance to know whether or not wind waves propagate according to the same dispersion relation that governs small amplitude linear waves,the following facts being generally observed(Ramamonjiarisoa and Coantic,I976):

-From simple visual observations,the wind wave profile is strongly asymmetrical;

-The water surface deflection spectrum exhibits almost systematically a secondary peak at twice the frequency of the spectral maximum.

-The water surface deflection distribution function is asymmetrical, and a marked departure from the normal distribution is observed, especially at high surface deflection levels.

Figure 1 illustrates the first experimental results on the dispersion relation, as obtained by estimating the cross correlation or the cross-spectral density function of two capacitance wave gauge signals, the gauges being spaced along the centerline of the large wind-water facility.

The following features of the experimental propagation curve are noticeable:

-Around the frequency of the spectral maximum, the observed celerities are close to the theoretical values (the experimental results being corrected for drift current and mass transport velocity, the theoretical curve can be associated with both linear and non-linear analyses).

-The lowest celerity in the low frequency range is close to half the celerity corresponding to the spectral maximum frequency. This suggests that the field is made up of groups of waves. It may be inferred that the lowest frequency corresponds to the interference phenomenon, so that the lowest celerity will be of the order of the group velocity.

-The most striking departures from the theoretical values appear in the high frequency range: there, the celerity is almost constant and close to the dominant wave celerity.

For the reasons mentioned above, it was expected to find some discrepancy between the experimental values and the linear prediction of celerity. However, the difference observed is so large, especially in the high frequency range, that it became necessary to carry out further experiments to obtain supplementary verification.

The new experiments were conducted at a higher wind velocity (10 m/s instead of 5 m/s or 8 m/s), and therefore at a higher velocity of the dominant wave, so that a wider high frequency range could be obtained and, if the previous observations were correct, still larger discrepancies between the experimental values and the linear prediction for celerities should appear. This is readily apparent in figure 2a, which shows phase measurement results for wave gauge spacings of $\Delta X = .03$ m and $\Delta X = .10$ m.

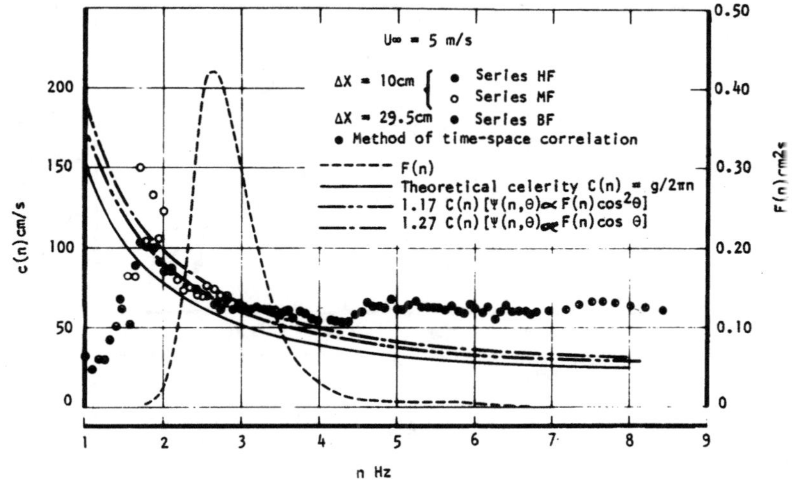

Fig. 1. Experimental phase celerity of harmonic components of a laboratory wind wave field, following Ramamonjiarisoa (1974). The distinction between three frequency ranges (Low, mean and high) refers to signal conditioning operations made to obtain accurate celerity estimates.

In the average, with a good approximation, the phase can be considered proportional to frequency. Thus, the phase celerity is nearly constant. By calculating the slopes of the mean straight lines, one finds:

c=1.10 m/s for ΔX=.03 m and c=1.05 m/s for ΔX=.10 m

The theoretical value of the dominant wave celerity is .92 m/s. The difference, of the order of .30 m/s could be attributed to the drift current, which would be of the order of 3% of the wind velocity or .30 m/s.

Figure 2b shows the coherence between the two wave gauge signals. One notes the following features:

–The coherence is a maximum at the dominant frequency. It decreases rapidly with decreasing frequency and more slowly with increasing frequency.

–Two other maxima occur at, or near, the second and third harmonics of the dominant frequency.

Some attempts have been made by various authors, in order to explain the anomalous feature of the observed dispersion relation in terms of, especially, weak non-linear interactions and angular spread

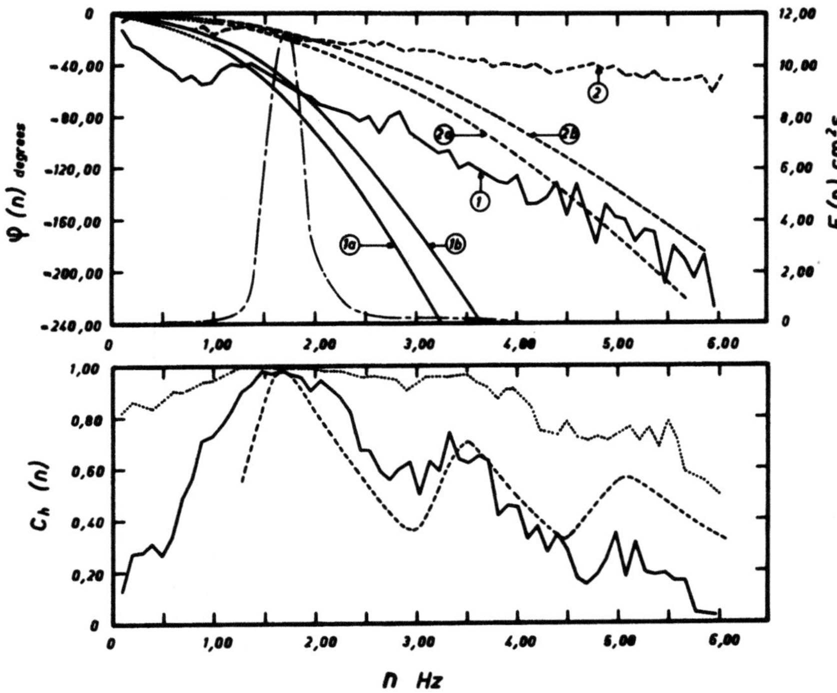

Fig. 2a (upper) and 2b (lower). Phase $\varphi(n)$ and Coherency $C_h(n)$ between two wave gauge signals. Curves (1) and ——— correspond to gauges spacing $\Delta X = .10$ m ; (2) and ... to $\Delta X = .03$ m . Curves (1a) and (2a) refer to one-dimensional linear wave field: (1b) and (2b) to a two-dimensional field with a spectral function $\Psi(n,\theta) \propto F(n) \cos\theta$. The dashed line on the lower figure represents the asymptotic form (large ΔX) of the coherency function for a two-dimensional $[\Psi(n,\theta) \propto F(n) \cos^s\theta/2]$ non-linear "Stokes" wave group.

effects (Huang and Tung, 1977; Longuet-Higgins, 1976). On the other hand, on the basis of both field and laboratory experimental data, Mollo-Christensen and Ramamonjiarisoa (1977) considered the possibility that the wave field consists of waves that are periodic in the phase of the fundamental component, as for Stokes waves. In which case, the modulational instability of such waves could be responsible, in particular, for the wave number variation with respect to the fetch or to the wind velocity.

For the moment, as any of the proposed explanations seems to be entirely satisfactory, we would suggest that a non-linear process, not yet well understood, acts so that, for a given wind velocity, longer and longer waves are generated in the wave tank when the fetch

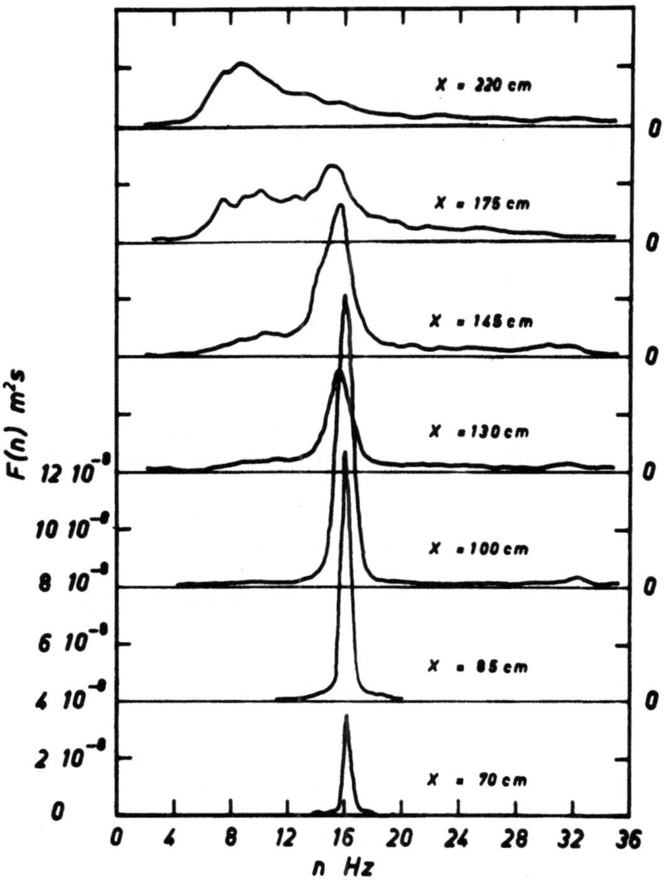

Fig. 3. Fetch (X) evolution of the spectral density function of the first capillarity-gravity waves. The zero energy level for each fetch value is given on the right.

increases. The understanding of the physical mechanisms of the generation of the shortest waves is therefore of interest.

ON THE GENERATION OF THE CAPILLARITY-GRAVITY WAVES

The corresponding experiments were conducted in the small I.M.S.T. wind-water facility. They involved very delicate measurements, the vertical thickness of the air and water flows to be explored being of the order of, at most, 1 cm and the first waves having heights and lengths of the order of 1 mm and 1 cm respectively. After much preliminary work to find a compromise between scientific interest and technical constraints, it was decided to perform the experiments at a wind velocity U of 5 m/s. No general forced circulation of the water was introduced. A 1 m length rigid plate was placed at the entrance section of the water tank, insuring that the air motion established a fully developed turbulent boundary layer before it joined the water surface. As apparent on photograph 1, under such conditions three different aspects of the water surface can be distinguished at the lower fetch values:

-Adjacent to the rigid plate is a relatively smooth zone, with no visible or "measurable" wave motion, but with very rapid drift current development as will be shown from experimental data.

-This smooth zone is followed by a second domain where the first visible waves are generated inside divergent "streaks". Their motion is almost two-dimensional, with a rather remarkable regularity.

-Finally the streaks meet each other and the wave motion becomes apparently random and three-dimensional.

As far as the characteristics of the first wavelots are concerned, figure 3 shows the fetch evolution of their spectral density function. At first, they have a frequency around 16 Hz; the spectral bandwidth is very narrow, supporting the visual observation regarding the strong regularity of the motion. Then the amplitude increases, although no significant change of the frequency is observed. after that, the dominant frequency decreases slightly as well as the corresponding energy. The energy content in the low frequency range increases. Finally, the peak around 16 Hz disappears, and another peak emerges around 8 Hz. The generation of subharmonic perturbations can sometimes be observed visually. The phenomenon is somewhat similar to the vortex paring, noted in two-dimensional shear flows. Possibly, because of the randomness in the propagation, an individual wave catches the wave in front of it, and then a longer wave is produced. The slight decrease of the frequency peak could be associated again with the modulational instability of a non-linear wave group, the waves being of finite amplitude. Figure 4 shows the propagation celerity of the wavelets. As in the case of well-developed gravity waves, the observed celerity is almost constant (of order .32 m/s) and higher than the theoretical values. However, actually, the departures can be explained in terms of the drift current effect, as already proved by others.

With regard to the origin of the first wavelets, the Phillips theory unfortunately cannot give an immediate explanation, the resonant condition being evidently not satisfied. Following Mc Leish and Putland , these wavelets could be due to the instability of the drift current which develops in the water as another effect of the air-water dynamical interactions. This instability was then investigated.

PHOTOGRAPH I

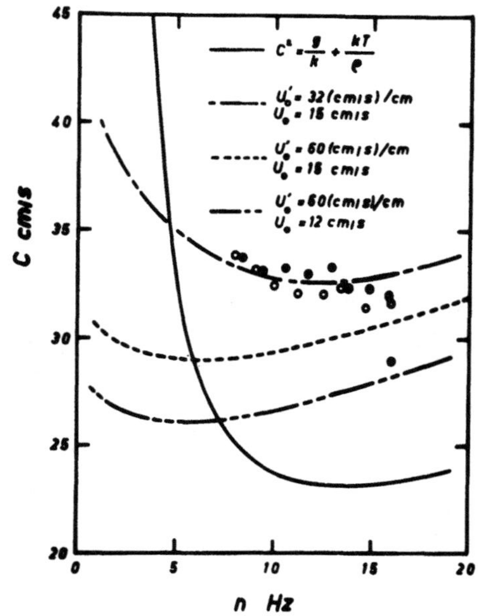

Fig. 4. Phase celerity of the harmonic components of the capilarity-gravity wave field. Circles represent measured values. Interupted lines refer to the expression :

$$C = U_0 - U_0'/2k + \left[(U_0'/2k)^2 + g/k + k\,T/\rho \right]^{1/2}$$

which takes into account the effect of drift current upon the phase celerity (U_0 water surface velocity; U_0' velocity gradient at the interface).

Figure 5 shows how the drift current, measured by hot film anemometry, evolves as a function of both depth and fetch. For obvious reasons, measurements were restricted to the first zone, as defined before of the water surface where no wave motion was present. A log-linear plot was used, showing that an exponential evolution of the velocity with depth provides a reasonable description.

When the fetch increases, first, both the surface velocity and the boundary layer H_1 thickness increase rapidly. Then, the surface velocity seems to keep an almost constant value ($\simeq$ 3% wind velocity), but the boundary layer thickness continues to increase.

A complete analysis of the stability problem, involving the Orr-Sommerfeld equation, is a rather large task. In order to determine ,the order of magnitude of the characteristics of the unstable modes, a first analysis ignoring viscosity effects was carried out: the stability of the coupled air and water motions is

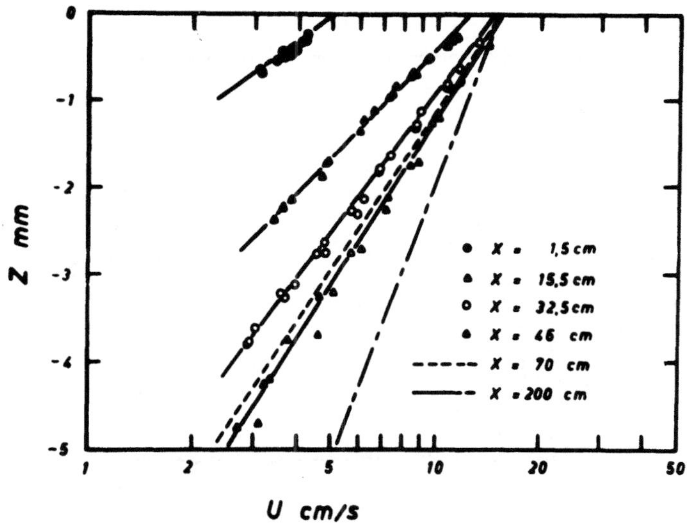

Fig. 5. Drift current distribution as a function of fetch(X) and depth (z). At X=70 cm , and X=200 cm , no measurement being possible, the distributions were estimated on the basis of CHOI (1977) analysis.

then governed by a sixth order determinant (Choi, 1977), which reduces to third order when applied to the water flow only. Figure 6 displays the neutral stability curves which correspond to this flow. The stability parameters are U_o and H_1. Figure 6 also shows the phase celerity of the unstable modes. The previous experimental data have made it possible to give an order of magnitude of U_o and H_1 near the fetch where the first wavelets are generated, $U_o \simeq .15$ m/s ; $H_1 \simeq .43$ cm). For these values, the water flow is stable at all wave numbers. Slightly different stability parameters could lead to the instability of the observed wavelength, but a large discrepancy would remain between the measured and the analytical phase celerities: the former is of the order of $2U_o$ while the latter is smaller than U_o.

Finally, the first wavelets seem not to be related to Tollmien-Schlichting waves associated with the drift current instability. In the light of the experimental data, this result does not seem surprising because the Tollmien-Schlichting waves will normally propagate at a velocity within the velocity range present in the flow. Actually, the wavelets propagate at velocities well away from this range. Some preliminary theoretical analysis (Choi, 1977) indicates that the drift current could influence the water surface response function to external excitations such as turbulent pressure fluctu-

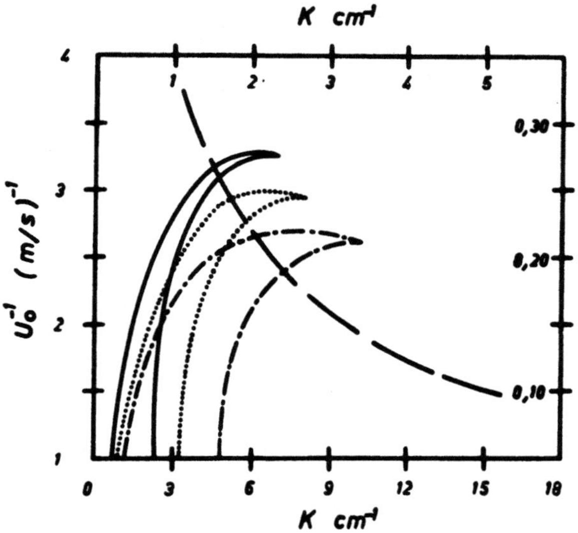

Fig. 6. Neutral stability curves(—— H_1=.87cm; ...H_1=.61cm; H_1=.43cm) and, phase celerity (—— ——) of the unstable modes associated with the water drift current. The left and lower scales refer to stability curves; the right and upper to phase celerity.

ations in the air. A resonance machanism taking into account this effect could be responsible for the generation of the first wavelets.

ON THE ENERGY TRANSFER FROM WIND TO GRAVITY WAVES

Experimental data have been obtained concerning the mean wind flow structure between the crest and the trough of a regular wave generated by an electrohydraulic system. A wave follower (Baldy, Ramamonjiarisoa and Coantic, 1977) servomechanism equipped with hot wire anemometers was used. After some preliminary studies concerning, in particular, the effects of the initial amplitude and of the wind velocity upon wave amplification, the wind flow structure experiment was carried out under the following conditions:

Wind velocity U = 5 m/s ; initial wave amplitude: 5.30 cm ; wave frequency: I Hz .

For such conditions, the wave is effectively amplified with a rate roughly comparable to that predicted by Miles' theory (Baldy, 1977). Figure 7 shows typical samples of the time evolution of the water surface deflection, the hot wires probe position the longitudinal and vertical components of the wind velocity turbulent fluctuations. Six hundred samples of that type gave, after ensemble

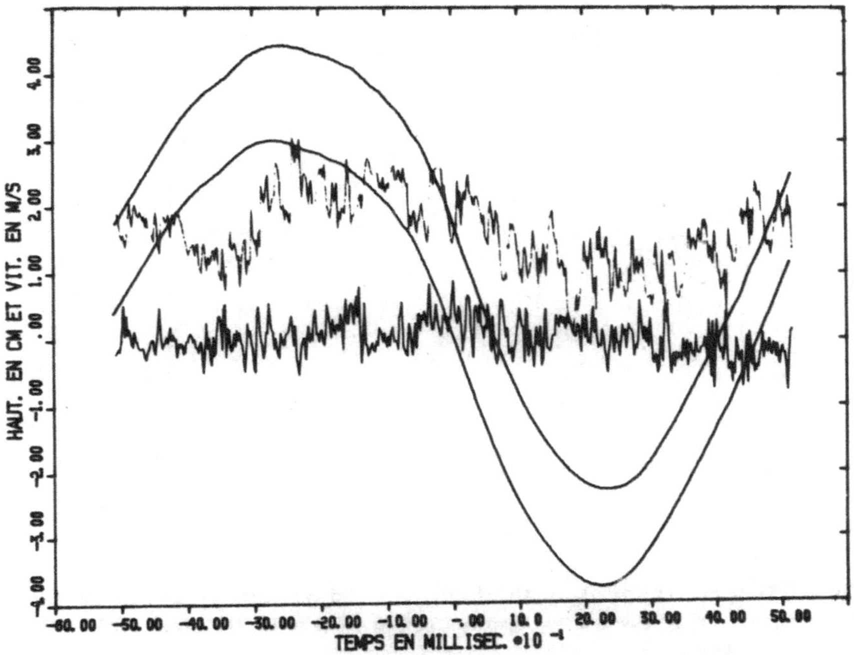

Fig. 7. Typical samples of time-evolution of the surface deflection level (lower regular line), the hot wires position (upper regular line), the vertical component (lower fluctuating line) and the horizontal component (upper fluctuating line) of the air velocity turbulent fluctuations. Ordinates represent wave height and h hot wires position in cm. The surface deflection zero down crossing is taken as the time origin.

averaging operations, all the requested mean evolutions and, especially, the wave-induced velocity components $\tilde{u}$ and $\tilde{w}$. Figure 8a gives the detailed form of $\tilde{u}$, the probes being maintained 1 cm above the moving water surface. The non-linear character of the coupling between wave and air motions is clearly proved. Figure 8b summarizes the results on $\tilde{u}$ for various probe levels, z, above the water surface.

At z=10cm ($\simeq$ 1/10 wavelength), the wave-induced perturbation seems already strongly attenuated. The velocity maximum is shifted with respect to the wave crest and the phase lag depends upon the probe level. Finally, combinations of the $\tilde{u}$ and $\tilde{w}$ components measurements lead to the velocity field shown on figure 9, as viewed in a frame of reference moving with the wave profile.

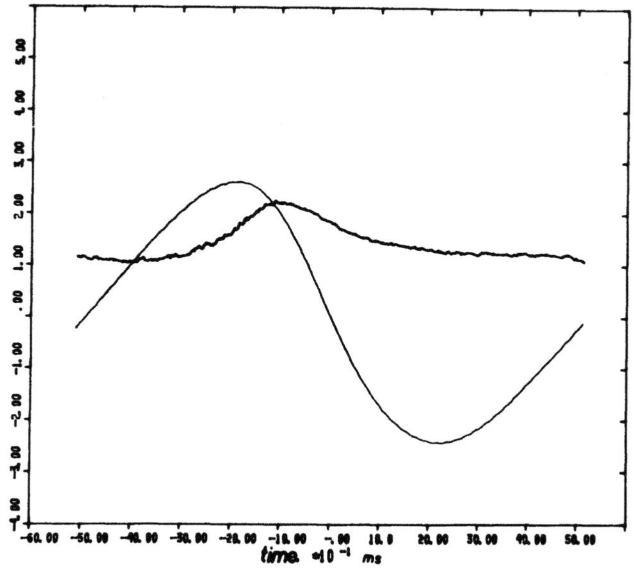

Fig. 8a. The horizontal component $\tilde{u}$ of the wave-induced velocity field in the air flow at z=1cm above the instantaneous water surface level. Wind direction is from the right to the left.

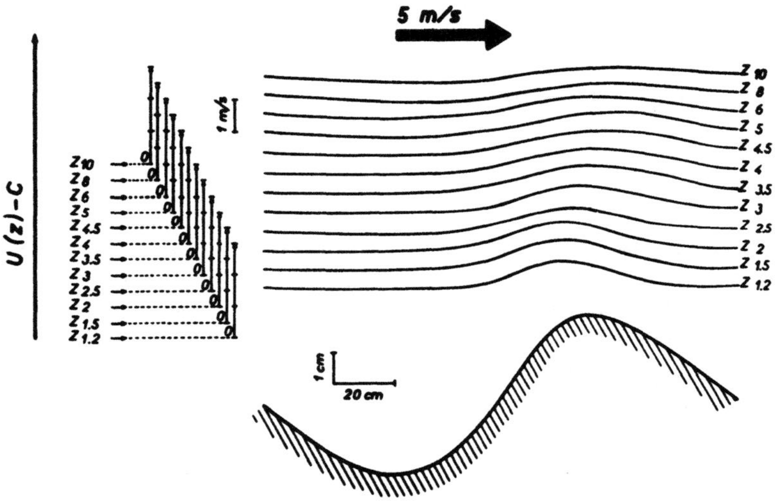

Fig. 8b. Evolution with respect to altitude z of the horizontal component $\tilde{u}$ of the wave-induced velocity field in the air flow. z_n means z=n cm above the instantaneous water surface level. The corresponding origin of velocity is given on the left.

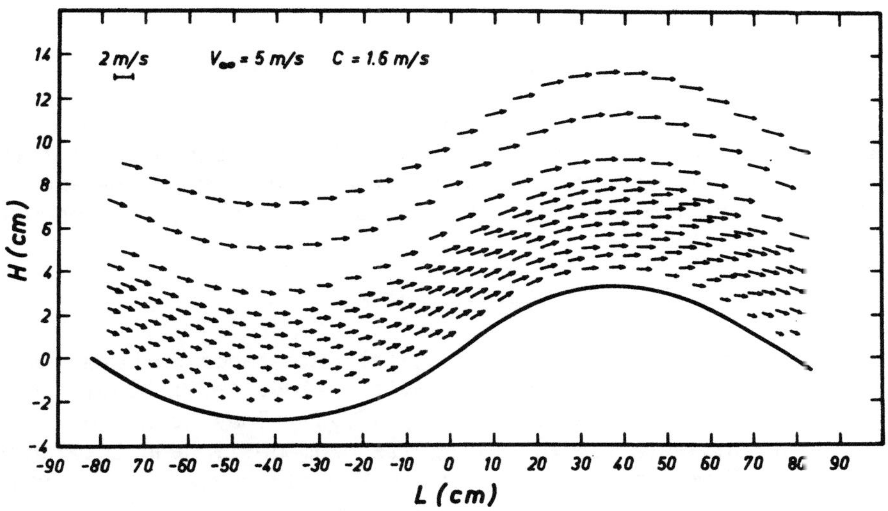

Fig.9. Wave-induced velocity field as viewed in a frame of reference moving with the wave profile.

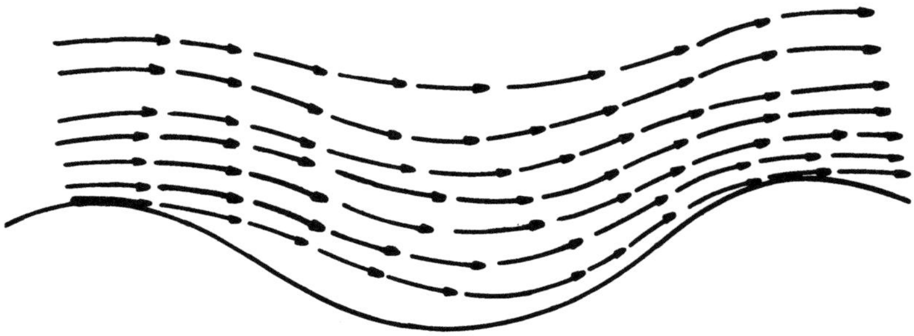

Fig. 10. Streamlines configuration suggested by the velocity field shown on figure 9.

The velocity field seems quite consistent with the type of streamline configuration shown in figure 10 : compensating for the waves, the streamlines are pressed together near the crests and spread apart near the troughs. No flow separation seems to occur.

CONCLUDING REMARKS

Further analyses of the experimental data reported here have been carried out (Ramamonjiarisoa, 1974; Baldy, 1977; Choi, 1977; Mollo-Christensen and Ramamonjiarisoa, 1977). They have led us to think that a wind-wave field, during its propagation and interaction with the wind flow, could be more appropriately regarded as a random field of non-linear wave groups rather than a superposition of free waves. Then, the main mechanism responsible for the evolution of the wind waves spectra should be related to the phenomena analyzed earlier by Whitham (1974) and more recently by Chu and Mei (1971), Huen and Lake (1975) and others. This point of view provides, in particular, a partial explanation of the anomalous observed dispersion relation. On the other hand, the scale modification of the wave field could be explained in terms of the modulational instability of these non-linear wave groups.

ACKNOWLEDGMENTS

The work presented here constitutes part of studies performed at the I.M.S.T. on the small scale air-sea interaction processes. It received financial support from, essentially, the "Centre National pour l'Exploitation des Océans" (C.N.E.X.O., contract N° 75/1234), the "Centre National de la Recherche Scientifique" (C.N.R.S.) and the "Institut Français du Pétrole" (I.F.P.). The authors would like to take this opportunity to express their thanks to these organisations.

REFERENCES

Ramamonjiarisoa, A., (1974): Contribution à l'étude de la structure statistique et des mécanisms de génération des vagues de vent. Thèse de Doctorat d'Etat, Université de Provence.

Coantic, M., Ramamonjiarisoa, A., Mestayer, P.G., and Peri, G. : Structure of wind and waves and energy transfer mechanisms in a simulated air-sea interfacial zone. Communication AMS Conference on Ocean-Atmoshere interaction, Seattle. Abstract, Bull. AMS, st., p. 141.

Ramamonjiarisoa, A., and Coantic, M. (1976): Loi expérimentale de dispersion des vagues produites par le vent sur une faible longueur d'action. Comptes rendus Acad. Sc., Paris, t.283, B.111-114.

Huang, N.E., and Tung, C.C. (1977): The influence of the directional Energy distribution on the Non-Linear Dispersion Relation in a Random Gravity Wave Field. Jour. of Phys. Oceano. Vol. 7, N°3; pp. 403-414.

Longuet-Higgins, M.S. (1976): Some effects of finite steepness on the generation of waves by wind. Submitted to "Deep-Sea Research".

Mollo-Christersen, E.J., and Ramamonjiarisoa A. (1977): Occurence of Stokes wave groups in random wave fields. Submitted to "Journal of Geophysical Research".

McLeish, W.L., and Putland, G.E., (1975): The initial water circulation and waves induced by an airflow. NOAA Techn. Report. ERL 316-AOML 16, Boulder, Colorado.

Choi, I. (1977): Contribution à l'étude des mécanismes physiques de la génération des ondes de capillarité-gravité à une interface air-eau. Thèse de Doctorat de 3e Cycle, Université d'Aix-Marseille II.

Baldy, S., Ramamonjiarisoa, A., and Coantic, M. (1977): Description and characteristics of a "wave follower" system for energy exchange studies in the vicinity of an air-water interface. To appear in "Review of Scientific Instrument".

Baldy (1977): Etude statistique de mécansimes de transfert dynamique à travers une interface air-eau. Thèse de Doctorat-Ingénieur, Université d'Aix-Marseille II.

W hitham, G.B. (1974): Linear and Non-linear waves. John Wiley and Sons, N.Y. 636 pp.

Chu, U.H. and Mei C.C. (1971): The non-linear evolution of Stokes waves in deep water. J. Fluids Mech., 47, 337-351.

Yuen, H.C., and Lake, B.D. (1975): Non-linear deep water waves: theory and experiment, Physics of Fluids, 18, 956-960.

DISCUSSION

W. Pierson:

The cross spectra vary much more rapidly at high frequency. Are the estimation techniques sufficient to resolve this? Did you try other angular spreading functions such as $A + B(\cos\theta)^2$ or even bimodal shapes?

A. Ramamonjiarisoa:

Cross spectra were estimated by Fast Fourier Transform techniques. Particular attention has been given to effect of spacing on cross spectra especially for high frequencies. Appropriate adjustment of this spacing allowed estimate of the cross spectra for such frequencies, as accurately as for low frequencies.

On the basis of various experimental results, in particular, those obtained by Mobarek (1965) in a wind-wave tank, an angular spreading function on the form $A \cos^s(\theta/2)$, s varying with frequency, was used in a more detailed analysis. As far as we know there is not yet experimental evidence on the existence of bimodal distribution.

M. S. Longuet-Higgins:

In your paper with M. Coantic you considered the effect on the apparent phase-speeds of having a rather smooth distribution of energy with regard to direction at fixed frequency. I believe the spreading factors were $\cos^2\theta$ or $\cos^4\theta$. However, if the distribution were more bimodal, corresponding to waves with phase-speed c such that $c/c_o = \cos\theta$ then it might be possible to explain the observations as due partly to free waves generated by strong nonlinear interactions near the wave crests. Have you measured the directional distributuion with sufficient resolution to determine whether or not it was bimodal?

A. Ramamonjiarisoa:

The possible existence of a bimodal distribution constitutes, I think, a fundamental question. As far as I know, there is no strong experimental evidence for that. As you showed in a recent article, a bimodal distribution seems to be consistent with some basic physical mechanisms of wind waves generation and development. I would like to mention that Mobarek in 1965 measured directional spectrum of wind waves in conditions resembling closely to those of our experiments. The observed distribution was unimodal as far as frequencies around the peak frequency are concerned. No information was given for frequencies above twice the peak frequency. For my point of view, the question remains largely open.

D. Hasselmann:

Did you try to do a second order approximation as Prof. Mitsuyasu did, which I think might explain the data, especially the fast decay of coherence.

A. Ramamonjiarisoa:

Prof. Mitsuyasu reported anomalous dispersion relation curves very similar to those I presented here and indeed it could be of interest to try to explain the anomaly by second order effects. However, he did experiments in the decay area - and our experiments concern the generating area. For our case, it seems that strong nonlinear processes occured; breaking of the highest waves was clearly observed.

E. J. Plate:

The phase velocity calculated with the space-time correlation method shows an increase in phase velocity at low frequencies, whereas the phase speed calculated from the phase spectrum does not. Can you account for this discrepancy?

A. Ramamonjiarisoa:

It is the cross-spectral method which gave in the low frequency domain some points which seem to show an increase of phase speed. Many other data have been obtained in this domain. In view of the statistical errors in estimates associated in particular with the wave gauge spacing influence I believe that the space-time correlation method, which allows appropriate signal conditioning processes before calculating celerity, gave the correct evolution, namely a decrease of phase velocity.

G. E. Mattingly:

(a) Is the stability analysis you refer to done using a complex phase velocity and a real wave number?

(b) I would suggest that it is much more realistic to compare experimental data to stability analyses done using complex wave numbers, complex phase velocities and real frequencies. There is evidence available from stability calculations in homogenous shear flow that the differences are important and that the complex wave number model is the proper choice.

A. Ramamonjiarisoa:

(a) Yes.

(b) Perturbations of interest here being space-dependent instead of time-dependent, this point of view is largely justified. In the first analysis presented, calculations have been made in the time domain and if necessary results are translated into the space domain using classical time-space equivalence by means of the group velocity. Analysis in the space domain could be specially of interest if it can lead to the conclusion that some unstable modes can move faster than the flow.

M. A. Donelan:

This point arose in conversation with John Wills over "un verre de Ricard hier soir". Does your result on the wave celerity above the peak of the spectrum (viz. the waves with frequencies above the peak have an apparent celerity about the same as the peak rather than the free-wave celerity) really mean that the short waves are "bound" or all move at sharp angles to the wind? Or is another interpretation of your space-time correlation measurements possible? What I have in mind is this. The auto-correlation function of waves persists for tens of group periods (as inferred by Plate in the text "Stochastic Hydraulics") not just a few wave periods (as implied by Phillips in his book "The Dynamics of the Upper Ocean" 1966, Camb. Univ. Press). Furthermore the intra-group phase is not entirely random, so that the dominant waves can be regarded as an amplitude - modulated and, much more slowly, phase-modulated carrier. If this is so the space-time correlation method would tend to bias the result in favour of those higher frequencies which are bound to the carrier and against the free ones which would tend to have a much more rapidly varying phase.

A. Ramamonjiarisoa:

The possibility of high frequency components moving at sharp angles to the wind was already discussed (comments of Prof. Longuet-Higgins and Dr. Pierson). Interpretation in terms of modulation process has been attempted by Prof. Mollo-Christensen and myself in a joint paper submitted recently for publication. There exist indeed other relatively strong arguments in favour of that type of interpretation. I believe however that the randomness of the modulation constitutes an important character to be taken into account.

WAVE-PRESSURE CORRELATION MEASUREMENTS OVER GROWING SEA WAVES WITH A WAVE FOLLOWER AND FIXED-HEIGHT PRESSURE SENSORS

F.W. DOBSON AND J.A. ELLIOTT

BEDFORD INSTITUTE OF OCEANOGRAPHY
DARTMOUTH, N.S., CANADA

ABSTRACT

A wave follower has been used to measure wave-pressure correlations a few cm above growing wind waves, and the results compared with simultaneous correlations from fixed-height sensors extrapolated down to the surface, assuming a unidirectional wind wave field. In conditions where swell introduced significant contributions to the correlations, agreement was poor, and estimation of wave-supported energy and momentum fluxes from either set of data was not possible. In conditions when no significant swell components were present and the directional characteristics of the wave field were well-measured, the two techniques gave results which agreed within experimental error, after an unlooked-for vertical acceleration sensitivity of the wave follower pressure sensor was allowed for. Estimates of the total wave-supported momentum flux τ_w are difficult to make because the $\tau_w(f)$ spectrum does not close before the data become unreliable, at a ratio f/f_m of frequency to wave spectrum peak frequency of about two.

INTRODUCTION

This paper will present wave-pressure correlations measured with the Bedford Institute of Oceanography wave follower in two field experiments, the 1973 Joint North Sea Wave Project (JONSWAP 73), and the 1974 Bight of Abaco pressure experiment. Results will be presented as comparisons of the cross spectra of wave elevation and air pressure fluctuations, recorded by the wave follower and by fixed-height sensors.

THE BIO WAVE FOLLOWER

The BIO wave follower (Harvey and Dobson, 1976) is an electro-hydraulic servomechanism which senses the instantaneous position of the water surface and drives a piston topped with a sensor package to maintain pressure and wave height sensors at a fixed distance above the waves. The BIO device (see Figure 3) is a direct descendant of one devised by Blair Kinsman and designed and built by Peep and Flower (1969) at Chesapeake Bay Institute, The Johns Hopkins University.

The wave follower is mounted on an undersea tower, and its height can be adjusted to allow for tidal variations in water level. Signals from the wave follower itself and from fixed-height sensors on an adjacent radio mast are telemetered via cables to the control station on board ship, where they are monitored for possible faults and recorded in FM form on analogue magnetic tape for subsequent digitization and analysis.

THE JONSWAP 73 RESULTS

In the fall of 1973 the BIO wave follower was taken to the JONSWAP site in the North Sea, and added to the array at Station 4, about 6 km offshore from the island of Sylt. The array of sensors included, in addition to the wave follower, two fixed-height sensors of the Elliott (1972a) design, a resistance-wire wave staff, and a Kaijo-Denki PAT-III three-component sonic anemometer.

After a first look through the data set, it was found that the wave-pressure correlations from the fixed-height instruments in general showed low coherences. The phase differences measured by the wave follower, where only a small sensor separation (1.5 cm) is involved produced wave-pressure phase differences which differed by up to 20° from those measured by the fixed sensor-wave staff combination, where the sensor separation was about 1 m. In an attempt to explain the differences a model was proposed by K. Hasselmann, whereby the effects of the large observed swell spectrum were included by extrapolating the swell spectrum to sea frequencies (Figure 1). The model allowed limited variations of the swell and sea direction and of the extrapolated swell amplitude, using cross spectra from the two wave sensors to determine the allowed limits of direction variation. The main hope for the model was that the effects of swell and sea could be superimposed in a simple way. The difficulty was that strong evidence exists suggesting the presence of two or more separate swell trains travelling through the area at the time.

Using two different forms of the model (fixed and variable swell direction) and the sensor separation vectors of the array, the observed $X_{\zeta p}(f)$ spectra were corrected by removing from them the effects of swell at sea frequencies ($X_{\zeta p}(f)$ is the wave-pressure cross spectrum). The resulting wave pressure phase angles are given in Figure 2 (the wave follower phases are not corrected for vertical motion in the air pressure gradient). The wave follower phases are essentially 180°, but are on the damping side; the fixed sensor phases are not inconsistent with earlier work by Elliott (1972b), but are brought closer to the wave follower results by the model corrections.

The principal conclusions drawn from the work are:

1. The wave spectra for all the data, with or without the wave follower in operation, showed the presence of strong swell components;

2. The coherences between the fixed-height sensors and the sea were generally low, making it impossible to make useful estimates of wave growth rates by downward extrapolation of the wave-pressure quadrature spectra;

3. For the one run where wave follower pressure data were available, attempts at allowing for the effects of swell at sea frequencies failed to account for discrepancies between the wave follower and fixed-height sensor results, in particular for discrepancies in the observed wave-pressure phase.

4. The best estimates of wave-pressure phase from the wave follower data indicate essentially zero wave growth for $U_5/C>1$; this is in disagreement with earlier work.

5. A partial explanation for the many internal inconsistencies in the data set is the presence of other, northward-travelling swell components, which masked the wave-generating pressure field but cannot be allowed for in the model, principally because not enough information is available on their period, amplitude, and direction.

THE BIGHT OF ABACO EXPERIMENT

The Bight of Abaco experiment took place in November-December 1974; the rationale and general history of the experiment have been described in a companion paper in this volume (Snyder *et al*, 1977). The arrangement of sensors used is shown in Figure 3.

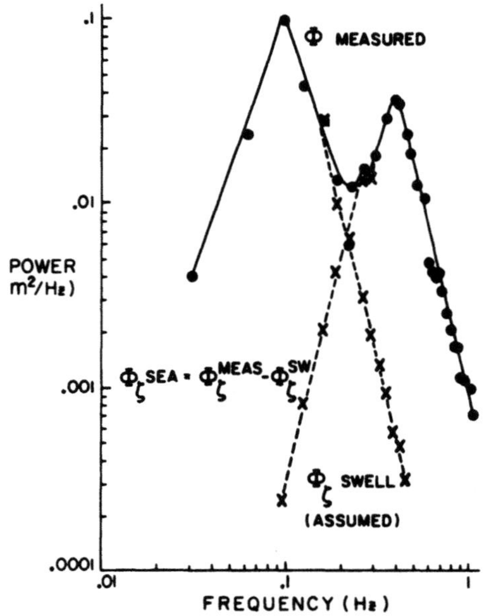

Figure 1
Wave staff power spectrum for JONSWAP 73 wave follower tracking run. Dotted lines are extrapolations of swell and sea spectra assumed in model calculations.

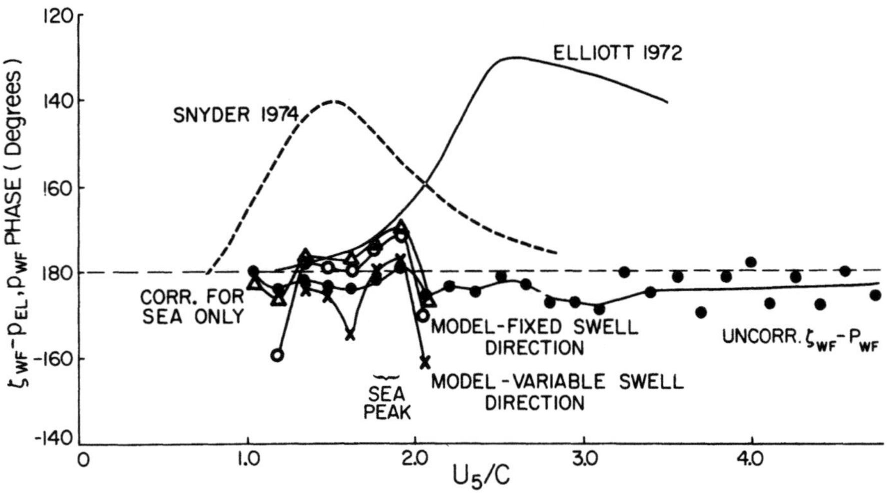

Figure 2
Wave pressure phase differences: ● as measured by wave follower, not corrected for vertical movement, Δ as measured by lower fixed-height pressure sensor, corrected only for sensor separation due to the sea; o,x fixed-height measurements corrected for swell, using respectively the fixed swell-direction and the variable swell-direction models.

The BIO data have been analysed in terms of a unidirectional model of the wave-pressure correlations, that is, by making the implicit assumption that the observed correlations are caused by wave, wind , and pressure fields which have delta-function directional distributions.

RESULTS

The preliminary BIO results are presented (Table 1) in two groups:

a. Runs with the wave follower operating in a tracking mode The tracking height was set at various levels from 5 to 10 cm from run to run. The wind speeds ranged from 4-5.5 m/sec, mostly from the North; the wave directional spectra were double-lobed at the spectral peak, presumably due to the nonuniform fetch distribution at the site. The spectra tended towards a single-peaked distribution in the wind direction at $f/f_m \simeq 2$, where f_m is the wave spectrum peak frequency. The wave follower tracking runs were in general not ideal cases for application of the unidirectional analysis.

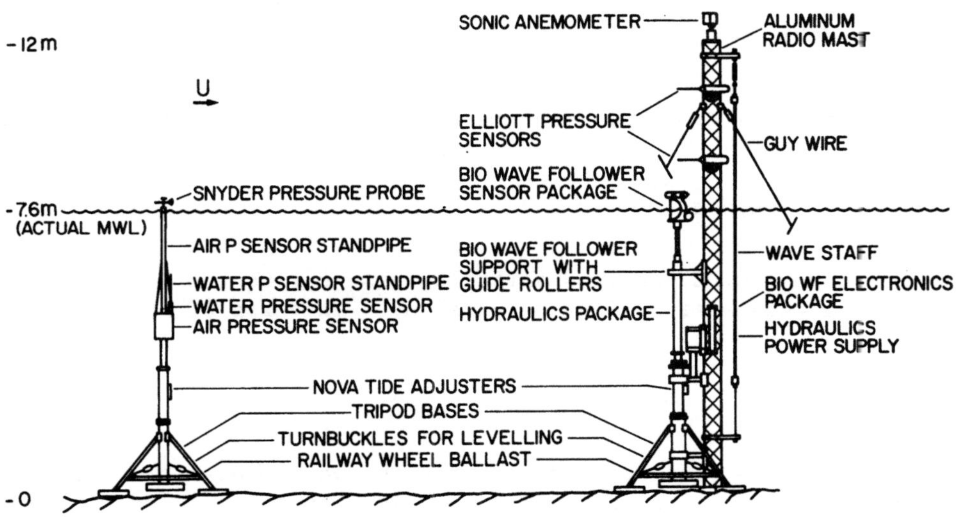

Figure 3
Arrangement of sensors for Bight of Abaco experiment: profile view. Four Nova pressure probes, of which one is shown here, were arranged in an array (see Snyder *et al*, 1977).

b. The Runs 15 to 20, which are for fixed-height sensors only. These runs are considered to form the most representative group for unidirectional modelling since they are high wind speed cases (6 to 7 m/s) for which the fetch is relatively uniform and the wave field is single lobed and travelling downwind.

The fixed-height data are from the two Elliott pressure instruments, vertically separated and positioned within 0.3 m of a wave staff or other wave gauge. The distance between the two pressure sensors and between the lower pressure sensor and the mean water level was typically 1/2 to 1 meter. Because of the horizontal and vertical separation of the sensors, the pressure-wave height cross correlations must be corrected for downwave separation and must be extrapolated to the sea surface.

The wave-pressure correlations are presented in spectral form, in terms of the dimensionless pressure profile parameter γ. The wave-induced pressure

$$p_w(t) = \rho_a g \gamma(\mu,\lambda) \zeta(t)$$

where ρ_a is air density, g is gravitational acceleration, $\zeta(t)$ is instantaneous water elevation, and

$$\mu = \vec{k} \cdot \vec{U}/\omega, \text{ and } \lambda = kz$$

where k is wave number, U is wind velocity, ω is radian frequency, and z is measurement height. μ is equivalent to U/c in former publications, and λ allows a parameterization of the pressure field height variation. From measurements with the fixed-height sensors in Runs 15-20

$$\gamma(\mu,\lambda) = \gamma(\mu,o)e^{-d\lambda}$$

where d = 0.86 $\pm$ 0.02, in substantial agreement with the Elliott (1972b) data when analysed in the same way. In terms of measured quantities,

$$\gamma(f) = X_{\zeta p}(f)/\rho_a g G_{\zeta^2}(f)$$

where $X_{\zeta p}(f)$ and $G_{\zeta^2}(f)$ are respectively the wave pressure cross spectrum and the wave power spectrum obtained using the unidirectional model.

Arg(γ), the wave-pressure phase shift (not shown) shows the largest deviations from 180°-the potential flow prediction - at the lowest μ values for which data are reliable; a mean curve

TABLE I
BIGHT OF ABACO EXPERIMENT - RUN DATA SUMMARY

Run/[1] Type	Day (Dec. 74)	Time (L)	Run Length min	U_5 m/s	Wind[2] Direction degrees	u_* m/s	Stability[3] z/L	Current Speed m/s	Current[2] Direction m/s	$H_{1/3}$ m	Peak Wave Period s	Peak Wave[2] Direction degrees
15/F	14	1315	42	6.51	039	.220	.078	-	-	.304	2.4	065
16/F	14	1424	44	6.02	050	.168	.030	.072	049	.306	2.4	065
17/F	14	1521	42	6.66	048	.210	.068	-	-	.320	2.4	070
18/F	14	1651	44	6.19	050	.208	.048	-	-	.363	2.6	070
19/F	14	1835	44	6.22	054	.191	.039	-	-	.254	2.6	070
20/F	14	1943	13.5	7.02	061	.224	.027	-	-	.420	2.6	075
4/WF	10	1020	48	4.66	350	.173	-.13	.12	340	.247	2.0	230/030
7/WF	10	1745	44	4.21	005	.173	-.18	-	-	.161	1.6	290/100
33/WF	18	1110	40	5.12	343	.217	-.29	-	-	.257	2.2	330
34/WF	18	1236	45	4.26	004	.165	-.62	-	-	.233	1.9	325/065

[1] F = fixed-height pressure sensors only
WF = all pressure sensors, including wave follower.

[2] All directions <u>from</u>. Two values indicate a two-lobed directional distribution.

[3] L = Monin-Obukhov length

faired through the rather scattered data points passes through 125° at $\mu \simeq 1.2$, and 140° at $\mu \simeq 3$. At higher μ values the data become scattered and unreliable.

γ, the pressure profile parameter (Figure 4) has a negative real part (γ_r) and a positive imaginary part (γ_i, the growth rate). No data values are given for frequencies less than the wave spectrum peak frequency. The γ_i curves lie close to a curve derived from the Elliott (1972b) data. The Elliott curve is not identical with that in Elliott (1972b: Figure 10); to arrive at the curve the extrapolation technique used for the present data was applied to Elliott's original data. The original curve has a slope 30% less than the one shown. The Miles curve shown is derived from Conte and Miles (1959) for a value 10^{-2} of the Miles "profile parameter"

$$\Omega = gz_o\kappa^2/u_*^2$$

The values of Ω estimated from sonic anemometer measurements for the runs presented varied from 3×10^{-5} to 2×10^{-2}. Therefore a differenct but experimentally reasonable choice of Ω would produce a curve fitting the observed data. The estimates of γ_r from the wave follower data are in some sense constrained to agree with the fixed-sensor data, since a fit to the fixed sensor wave-

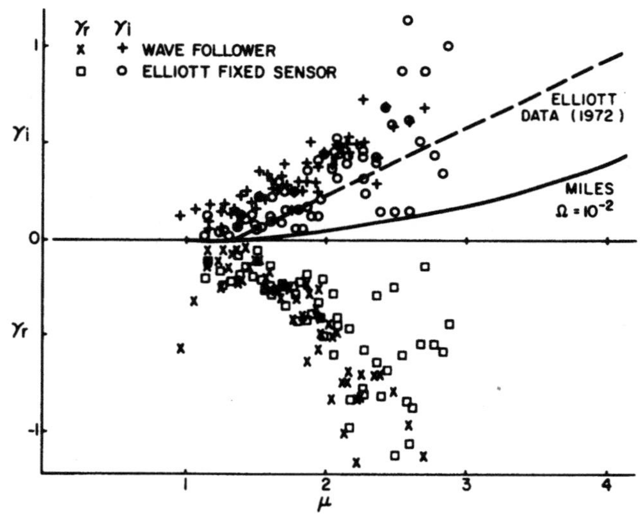

Figure 4a

Wave-coherent pressure parameter $\gamma = \gamma_r + i\gamma_i$ (see text) vs μ, for wave follower tracking runs (BOA 4,7,33,34). x, +, γ_r and γ_i respectively from wave follower data; □ , o, γ_r and γ_i respectively from lower fixed-height sensor data.

pressure cospectra was used to determine the vertical acceleration sensitivity of the wave follower pressure sensor. The γ_i estimates from the two sensors are independent estimates of the same process, and can be seen to agree with one another. The γ_i agreement strongly suggests that downward extrapolation of fixed-height pressure sensor data is adequate for obtaining estimates of the energy input to the waves via pressure fluctuations, provided that coherence levels are high enough.

Dobson (1971), reported values for the growth rate parameter which could be parameterized by

$$\gamma_i = C(\mu-1)$$

where $C \sim 1.0$; Elliott (1972b) found $C \sim 0.2$, and Snyder (1974b) found $C \sim 0.1$. The present data indicate $C \sim 0.25$. Dobson failed to allow in his computation of μ for the presence of significant currents running against the wind; when these are allowed for, his estimate of $C \sim 0.8$. The reason for the residual discrepancy remains a mystery. For similar μ values Dobson's $Q_{\zeta p}(f)$ spectra are similar to the Bight of Abaco data, but his $G_{\zeta^2}(f)$ spectra are smaller.

Estimates of the spectrum of wave-supported momentum flux

$$\tau_w(f) = \omega^2 Q_{\zeta p}(f)/g$$

indicate, when plotted against the ratio f/f_m of frequency to wave spectral peak frequency, that all frequencies above f_m contribute about equally to the wave-supported stress, up to $f \sim 2.5\ f_m$ where the data become unreliable. The conclusion is that this data set does not allow a direct estimate of the total wave-supported wind stress.

One way to estimate the total wave-supported momentum flux is to use a linear extrapolation for γ_i and assume that the high-frequency wave spectrum varies as f^{-5};

$$\overline{\tau}_w = \int_0^\infty \tau_w(f)df = 4\pi^2\rho_a \int_0^\infty f^2\gamma_i(f)\ G_{\zeta^2}(f)df.$$

A linear extrapolation of $\gamma_i(f)$ gives $\tau_w(f)$ an f^{-2} behavior. Allowing the observed average $\tau_w(f)/\tau_o$ spectrum to fall off as f^{-2} and integrating, and using an average value (from the sonic anemometer data) for the measured τ_o gives for Run 20 an estimated ratio of wave-supported to total momentum flux of almost 1, and the extrapolated $\tau_w(f)/\tau_o$ spectrum reaches negligible values at $f/f_m \sim 5$. Integrating the spectra out to $f/f_m \simeq 2.5$ gives, for the wave follower runs, partial τ_w/τ_o ratios of about 0.3.

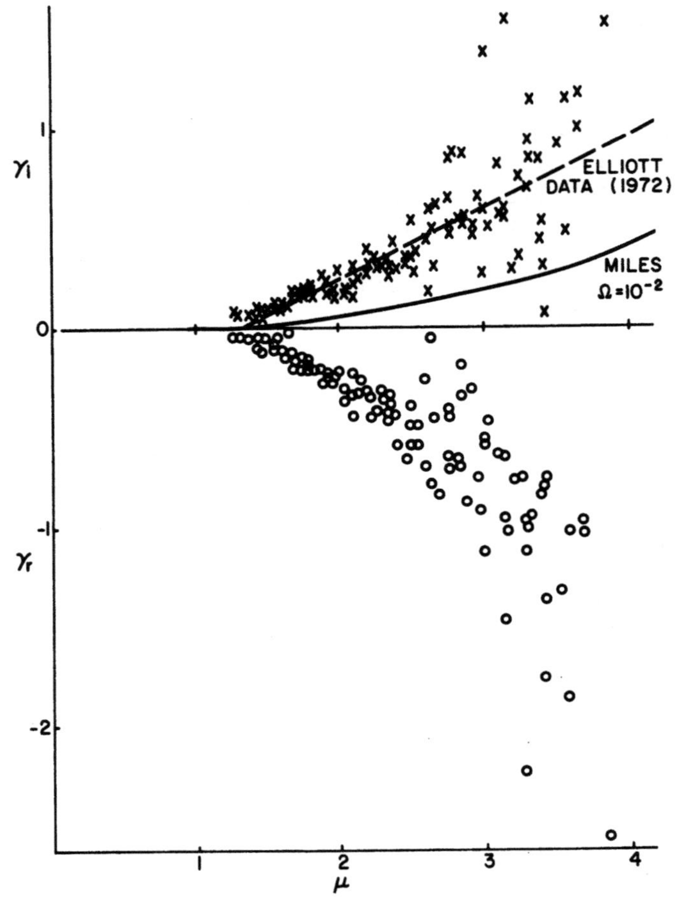

Figure 4b

As for Figure 4a, but from fixed-height data only (Runs 15-20 BOA); o γ_r; x γ_i. The two lines are ___ the prediction of Conte and Miles (1959) theory for $\Omega = 10^{-2}$; - - - data from Elliott (1972b) (see text).

CONCLUSIONS

1. The fixed height sensor and the wave follower results agree on γ_i; therefore fixed-height results may be extrapolated downwards to obtain $Q_{\zeta p}$ and hence γ and τ_w, provided that the fixed-height sensors can be placed low enough to obtain high coherences with the wave signal. Our linear approximation of the growth rate curve is

$$\gamma_i = 0.25\ (\mu - 1).$$

2. When an unlooked-for sensitivity to vertical acceleration of the WF pressure sensor is allowed for, we agree on γ_r.

3. The spectra of wave-supported momentum flux do not close by $f/f_m = 2$. An extrapolation to allow integration under the spectra indicates τ_w/τ_o is nearly one with significant contributions up to $f/f_m = 5$.

4. The results in general agree with those of Elliott (1972b) reworked using a different vertical extrapolation scheme.

5. The presence of upwind-travelling wave components can effectively mask the wave-pressure correlations from a growing sea. If swell components are present, they can prevent the accurate determination of wave-pressure correlations by fixed-height sensors at a single point, either by preventing the placement of the sensors at low enough heights, or by containing energy over a sufficiently wide direction range to contaminate the wave-pressure correlations produced by the sea.

6. Because of 5, it is necessary when making wave-pressure correlation measurements to obtain high-resolution directional spectra of the wave energy and the wave-pressure correlation fields, in the frequency range of the sea.

REFERENCES

Conte, S.D. and Miles, J.W., 1959. On the numerical integration of the Orr-Sommerfeld equation. J. Soc. Indust. Appl. Math. 7(4), pp. 361-366.

Dobson, F.W., 1971. Measurements of atmospheric pressure on wind-generated sea waves. J. Fluid Mech. 48(1), pp. 91-127.

Elliott, J.A., 1972a. Instrumentation for measuring static pressure fluctuations within the atmospheric boundary layer. Boundary-Layer Meteorol. 2, pp. 476-495.

Elliott, J.A., 1972b. Microscale pressure fluctuations near waves being generated by the wind. J. Fluid Mech. 54(3), pp. 427-448.

Harvey, D.R. and Dobson, F.W., 1976. The Bedford Institute wave follower. Bedford Inst. of Oceanography Int. Rep. BI-R-76-13. 67 pp.

Peep, M. and Flower, R.J., 1969. The Chesapeake Bay Institute wave follower. The Johns Hopkins Univ. Tech. Rep. 73, 53 pp.

Snyder, R.L., Long, R.B., Irish, J., Hunley, D.G., Pflaum, N.C., 1974a. An instrument to measure atmospheric pressure fluctuations above surface gravity waves. J. Mar. Res. 32(3), pp. 485-496.

Snyder, R.L., 1974b. A field study of wave-induced pressure fluctuations above surface gravity waves. J. Mar. Res. 32(3), pp. 497-531.

Dobson, F.W., Elliott, J.A., Long, R.B., Snyder, R.L., 1977 The Bight of Abaco pressure experiment. Contribution to NATO Symposium on Turbulent Fluxes Through the Sea Surface, Wave Dynamics and Prediction, Marseilles, September 1977. In press.

DISCUSSION

See following paper.

THE BIGHT OF ABACO PRESSURE EXPERIMENT

R.L. SNYDER,[1] R.B. LONG,[2] F.W. DOBSON,[3] AND J.A. ELLIOTT[3]

1. NOVA UNIVERSITY, DANIA, FLORIDA, USA
2. NOAA/AOML/SAIL, MIAMI, FLORIDA, USA
3. BEDFORD INSTITUTE OF OCEANOGRAPHY, DARTMOUTH, N.S., CANADA

In November-December of 1974, we conducted a joint experiment to investigate microscale air pressure fluctuations above surface gravity waves. The primary objective of this experiment was to study the vertical structure of the wave-induced component of these fluctuations and to estimate the corresponding differential and integral momentum transfer to the wave field. Secondary objectives included the resolution of differences between previous independent field measurements of wave-induced pressure by Dobson (1971), Elliott (1972), and Snyder (1974), and the construction of a multidimensional model of the turbulent component of the air pressure field. We have, to one extent or another, succeeded in achieving all of these objectives and in due course will publish details in the literature. In this brief report, we wish to convey only the flavour of the investigation along with several somewhat preliminary findings.

The evolution of the wave field is controlled by the radiative transfer equation (Hasselman, 1960)

$$\frac{\partial}{\partial t} E_{\zeta^2} + \vec{V} \cdot \vec{\nabla} E_{\zeta^2} = \alpha + \beta E_{\zeta^2} + S_{NL} + S_D$$

Here $E_{\zeta^2}(\vec{x}, t, \omega, \theta)$ is the spectral density of the wave component with radial frequency ω and direction of travel θ, evaluated at horizontal position $\vec{x}$ and time t, $\vec{V}(\omega,\theta)$ is the group velocity,

$\alpha(\vec{x},t,\omega,\theta)$ and $\beta(\vec{x},t,\omega,\theta)$ are coefficients representing two sources of atmospheric input of spectral density corresponding to the theories of Phillips (1957) and Miles (1957), S_{NL} is the nonlinear source term (Hasselmann, 1962), and S_D represents dissipation. Both α and β are directly related to the microscale air pressure field at the mean surface, α to the three-dimensional spectral density of turbulent pressure F_{Q^2}

$$\alpha = \frac{2\pi}{\rho^2 g^4} \omega^5 F_{Q^2}(\vec{k},-\omega),$$

and β to the imaginary part γ_i of the dimensionless ratio between wave-induced air pressure and surface elevation defined by

$$\hat{P} = s \rho g \gamma \hat{\zeta}$$

In the above equations, Q is the turbulent component of the air pressure, $\hat{P}$ and $\hat{\zeta}$ are complex wave-coherent amplitudes, g is the acceleration of gravity, ρ is the density of water, s is the air/water density ratio, and $\vec{k}$ is defined by

$$\vec{k} = \frac{\omega^2}{g} (\cos\theta, \sin\theta)$$

The earlier measurements of wave-induced pressure by Dobson, Elliott, and Snyder had emphasized single aspects of the experimental task, while ignoring or dealing only partially with other aspects. Dobson's measurements from a floating buoy provided a direct estimate of momentum transfer, integrated over wave azimuth. Elliott's fixed vertical array measurements allowed a coarse extrapolation to the mean surface, also resulting in an estimate of integral momentum transfer. Neither experimental design provided the directional information necessary to properly estimate the differential transfer for a single wave component. Snyder's fixed horizontal array measurements allowed a resolution of the directional dependence of the wave-induced fields, but did not permit a proper extrapolation to the mean surface. The present effort was conceived as an attempt to combine in a single array experiment the separate features of these three earlier experiments in order to deal properly both with the extrapolation of fixed observations to the mean surface and with the resolution of the directional dependence of the wave-induced fields.

The experiment began with a two-week-long instrument intercalibration at Nova University's Ocean Sciences Center, during which instruments and calibration techniques were compared.

This intercalibration revealed that the response of the pressure standard used by Snyder *et al.* (1974) in previous calibrations of the response of the Nova air pressure instruments was slower than assumed, resulting in a systematic bias in Snyder's (1974) results at higher frequencies. Correction for this bias helps to resolve the discrepancy between his results and those of Elliott.

The month-long field experiment was conducted at the same Bight of Abaco site previously used by Snyder. Principal vehicle for the experiment was the University of Miami's 20 m laboratory vessel *Calanus*, which provided power for the instruments and a site for the acquisition of data. The field instruments were deployed in two areas northwest of the *Calanus* mooring. An existing tower 100 m to the northwest supported a three-component vertical array of cup anemometers, a wind vane, tide gauge, and, to monitor the mean surface current, an electromagnetic current meter (which unfortunately did not work). The main instrument array 200 m to the northwest contained seven air pressure sensors and six wave sensors surrounding an instrument mast in a three-legged star pattern. In the centre of the array was Dobson's wave follower-mounted pressure sensor (Dobson and Elliott, 1977). Stacked vertically on the instrument mast were two Elliott sensors. Four Nova instruments, mounted on remotely adjustable stands, formed the legs of the array. Each of the Nova instruments measured both air pressure and water pressure. A fifth wave sensor, a resistance wire wave staff, was attached to the instrument mast. The wave follower itself constituted the sixth wave sensor. Finally, a sonic anemometer was mounted at the top of the instrument mast.

Signals from the instruments were conducted to the *Calanus* by a series of cables lying on the bottom. These signals were acquired simultaneously on two data acquisition systems, a 12-channel FM system used by the Canadians, and a 40-channel digital system used by the Americans. The latter system acquired all signals except those of the sonic anemometer.

The field experiment lasted approximately one month, yielding 37 runs of 40 to 45 minutes duration. Seven of these runs contain wave follower information. Most of our preliminary conclusions are derived from an analysis of runs 14 to 20, during which the wind blew from northeast to east at 6 to 7 m/s (dimensionless fetch $\sim 2\times 10^3$), and during which the wave follower was not operating. During run 14, a second vessel, the 11 m ketch *Catspaw*, was placed immediately downwind of the main instrument array in order to backscatter waves.

It is not possible in the limited space available to adequately describe the conceptually straightforward yet operationally very complex analysis by which we have attempted to understand the data. Suffice to say we have proceeded along two parallel fronts: one (Bedford), a simplified analysis of a subset of the data very similar in spirit to the previous analyses of Dobson and Elliott; and the second (Nova), a very much more sophisticated and considerably less transparent analysis of the full data set which attempts to account for a number of effects in a fundamental way. It is the latter analysis which we will discuss briefly in this report.

The Nova analysis can be understood qualitatively in the following terms: we assume that the relevant experimental data for each run are the mean wind parameters, the instrument positions, and the matrix of cross spectra for all instrument pairs in the main array $G_{ij}(\omega)$. Analytical tasks are (1) to identify the set of physical unknowns which is responsible for the values which the matrix elements assume, (2) to work out the relationships between these unknowns and the matrix elements, and (3) to use these relationships to estimate the physical unknowns. The physical unknowns in turn form the basis for the physical interpretation of the results of the experiment.

As unknowns, we identify three run-dependent variables, the spectral density E_{ζ^2}, the effective mean current $\vec{D}(\omega,\theta)$ seen by a given wave component, and the spectral density of turbulent pressure F_{Q^2}. We also identify one global variable, the previously-defined dimensionless wave-induced air pressure ratio γ. γ is taken to be a function of two parameters, a dimensionless wind speed μ

$$\mu \equiv \frac{\vec{k} \cdot \vec{U}_5}{\omega}$$

where U_5 is the mean wind velocity at 5 m elevation, and a dimensionless height γ

$$\gamma \equiv kz$$

where z is the height above the mean surface. We note that this parameterization is certainly not complete. At least two parameters are required to fix the mean wind profile. Therefore γ should be a function of at least three parameters. Such a parameterization is compatible with our analytical scheme, but we doubt that the data are sufficient to resolve a dependence on a third parameter. Rather we shall interpret our results as being characteristic of some mean value of all implicit parameters.

To implement the Nova analysis, we expand the physical unknowns in series containing unknown coefficients and then fix these coefficients by a series of linear, bilinear, and more highly linear least squares analyses. In this manner we estimate the variables E_{ζ^2} and $\vec{D}$ from an iterative computation involving wave sensor/wave sensor elements. γ is in turn estimated from wave sensor/air pressure sensor elements. Two computational schemes are employed, one linear and the other bilinear. In the bilinear case, the phase of the resulting estimate for γ is independent of λ. These schemes are based on the representations

$$\gamma(\mu,\lambda) = \sum_{r,s}^{R,S} \Gamma_{rs}\, \mu^r L_s(\lambda)$$

and

$$\gamma(\mu,\lambda) = \sum_{q}^{Q} \Gamma_q\, \mu^q \cdot \sum_{r,s}^{R,S} \Lambda_{rs} \mu^r L_s(\lambda)$$

respectively. Here the coefficients Γ_{rs} and Γ_q are complex and the coefficients Λ_{rs} real. These coefficients are typically constrained to match potential theory for $\mu = 0$. Two sets of basis functions to represent the λ dependence are employed. The first of these is the set

$$L_s(\lambda) \equiv \lambda^s\, e^{-\lambda}$$

The second is a set of empirical orthogonal eigenfunctions recently computed for the Miles theory by R.B. Long, who finds that over the customary range of wind profile parameters, a linear combination of the first three eigenfunctions does an extremely good job of representing the Miles solutions.

A third iterative analysis is used to estimate both the upwind travelling component of E_{ζ^2} and γ for negative μ by fitting wave sensor/air pressure sensor and air pressure sensor/air pressure sensor elements (backscatter runs). Finally, F_{Q^2} is estimated from air pressure sensor/air pressure sensor elements. This analysis assumes that the turbulence is 'frozen' in the air flow.

Application of the above analyses to runs 14 to 20 results in the following preliminary conclusions:

(1) In the range $1 < \mu < 3$ and $1 < \lambda < 3$, the vertical dependence of the wave-induced air pressure is approximately exponential and slightly less steep than the potential theory result

$$|\gamma| \sim e^{-\lambda}$$

The phase of γ is independent of λ. These conslusions are consistent with the previous findings of Elliott (1972) and Snyder (1974). The Nova analysis yields a somewhat larger effective vertical decay (closer to potential theory) than does Elliott's analysis or the Bedford analysis of the present data (Dobson and Elliott, 1977).

(2) For $1 < \mu < 3$, this analysis yields an estimate for the imaginary part of γ, extrapolated to the mean surface, which is in good agreement with both Elliott and Snyder (corrected for the response of his pressure standard) and with the Bedford analysis of the present data. These estimates are approximated by the simple relation

$$\gamma_i = 0.2 \begin{pmatrix} +0.1 \\ -0.05 \end{pmatrix} (\mu-1) \tag{1}$$

(3) Over the same range, this analysis yields an estimate for the real part of γ which is in reasonable agreement with Snyder (corrected) but is somewhat larger than both Elliott's estimate and the Bedford estimate for the present data. In all cases excepting Elliott, Arg γ has its maximum departure from 180° near $\mu = 1$.

(4) For $3 < \mu < 5$, it appears that γ_i continues to increase more or less linearly with μ, although there is some indication that the increase may be faster than linear.

(5) The Nova analysis of limited backscatter observations does not support earlier conclusions by Dobson (1971) and Snyder (1974) that for $\mu < 0$ the wave field is strongly damped. This analysis suggests that the phase of the wave-induced pressure is essentially 180°. Further experimental investigation of waves travelling against the wind is warranted.

(6) The observed wave-induced air pressure field is comparable with, but not identical to, the predictions of Miles' 1957 theory.

(7) Based on an extrapolation of (1) to higher μ, with $\gamma_i \equiv 0$ for $\mu < 1$, we estimate that, for a JONSWAP spectrum (Hasselmann *et al.*, 1973), the total momentum transfer is an inverse function of frequency roughly equal to 100% of the wind stress at dimensionless fetch of 10^2 and 10% of this stress at dimensionless fetch of 10^5.

(8) For small to moderate dimensionless fetch, momentum transfer to the spectral peak is a significant but not dominant fraction of the total momentum transfer to the wave field. This fraction is an inverse function of dimensionless fetch. For the present experiment, transfer at frequencies less than twice the peak frequency is estimated to be roughly one-third of the total transfer to the wave field.

(9) An accurate determination of the total momentum transfer to the wave field will be possible only after reliable field observations of wave-induced pressure have been made for large $\mu(3 < \mu < 10)$. Such observations could be obtained from a Bight-like experiment designed for small dimensionless fetch.

(10) An understanding of the momentum and energy balance for fully developed seas will be possible only after reliable field observations of wave-induced pressure have been made for small $\mu(0 < \mu < 1)$. Such observations will probably require the development of open ocean instrumentation for use at large dimensionless fetch.

(11) The turbulent component of the air pressure field close to the mean surface is reasonably consistent with the 'frozen turbulence' hypothesis.

(12) A two-dimensional turbulent pressure spectrum can be estimated from the cross spectral data.

REFERENCES

Dobson, F.W. (1971) Measurements of atmospheric pressure on wind-generated sea waves. *Journal of Fluid Mechanics, 48:* 91-127.

Dobson, F.W. and J.A. Elliott (1977) Wave-pressure correlation measurements over growing sea waves with a wave follower and fixed-height pressure sensors. Proceedings, NATO Symposium on Fluxes through the Air/Sea Interface and Wave Prediction, at Ile de Bendor, France, September 1977.

Elliott, J.A. (1972) Microscale pressure fluctuations near waves being generated by wind. *Journal of Fluid Mechanics, 54:* 427-448.

Hasselmann, K. (1960) Grundgleichungen der Seegangsvoraussage. *Schiffstechnik, 7:* 191-195.

Hasselmann, K. (1962) On the non-linear energy transfer in a gravity-wave spectrum. I. General theory. *Journal of Fluid Mechanics, 12:* 481-500.

Hasselmann, K., T.P. Barnett, E. Bouws, H. Carlson, D.E. Cartwright, K. Enke, J.A. Ewing, H. Gienapp, D.E. Hasselmann, P. Krusemann, A. Meerburg, P. Müller, D.J. Olbers, K. Richter, W. Sell, and H. Walden (1973) Measurement of wind-wave growth and swell decay during the Joint North Sea Wave Project (JONSWAP). Erganzungsheft zur Deutschen Hydrographischen Zeitschrift, Reihe A (8°), Nr. 12.

Miles, J.W. (1957) On the generation of surface waves by shear flows. *Journal of Fluid Mechanics, 2:* 417-445.

Snyder, R.L. (1974) A field study of wave-induced pressure fluctuations above surface gravity waves. *Journal of Marine Research, 32:* 497-531.

Snyder, R.L., R.B. Long, J. Irish, D.G. Hunley and N.C. Pflaum (1974) An instrument to measure atmospheric pressure fluctuations above surface gravity waves. *Journal of Marine Research, 32:* 485-496.

DISCUSSION

joint discussion of papers 3.1.3

F.W. DOBSON AND J.A. ELLIOTT
BEDFORD INSTITUTE OF OCEANOGRAPHY
DARTMOUTH, N.S., CANADA

WAVE-PRESSURE CORRELATION MEASUREMENTS OVER GROWING SEA WAVES WITH A WAVE FOLLOWER AND FIXED-HEIGHT PRESSURE SENSORS

and 3.1.4

R.L. SNYDER, R.B. LONG, F.W. DOBSON, and J.A. ELLIOTT

THE BIGHT OF ABACO PRESSURE EXPERIMENT

O. M. Phillips:

These two papers, by Dobson, Elliott, Long and Snyder and by Dobson describe very difficult measurements and their successful outcome is a considerable tribute to the skill of the authors.

When they started out on this, I thought it was an impossible task, but I am happy to have been wrong.

E. J. Plate:

We have learned to expect that in the ocean momentum is transferred through the high frequency waves. Your data seem to show that momentum is transferred more or less uniformly from frequencies near f_m to at least $2f_m$. Could you estimate the value of f_m (smaller than the one observed by you?) for which we may expect our old notion to be true?

F. W. Dobson:

The relevant information is in Snyder's figure showing the momentum-transfer integrand plotted against f/f_m for various values of dimensionless fetch. You see there that at large dimensionless fetch, the integrand, though small, has its peak at the highest values of f/f_m. Your "notion" is most true when the momentum transfer is most widely separated from f_m and so the answer is that this occurs at large dimensionless fetch.

K. Hasselmann:

(1) It is encouraging to see that you obtain similar ratios for the wave induced to total momentum flux across the air-sea interface by direct measurement as was inferred indirectly from the JONSWAP measurements: up to 100 % for small fetches, decreasing to 10 - 20 % for large fetches.

(2) Is it possible that some of the scatter in the atmospheric wave input data may be due to the fact that the data cannot be pacified in terms of the single parameter $k \cdot U/\omega$. This would be the case, for example, if the generating mechanisms are nonlinear, as suggested by recent investigations on the role of whitecaps in flow separation over the wave crests. It should be possible to estimate the degree of nonlinearity from the pressure-wave height coherence. It is possible that the relatively low coherences typically observed in the wind-wave peak as compared with the swell peaks (even when there is a pronounced wind-wave induced peak observed in the pressure spectrum) may be an expression of this nonlinearity?

F. W. Dobson:

Elliott, in his 1972 J. F. M. paper, attempts to quantify this by removing from his pressure power spectra the wave induced part; he finds a substantial residual, not found over land, but incoherent with the waves. The waves therefore are contributing a significant component to the field of turbulent pressure fluctuations in the general frequency range of the wave spectrum. The mechanism for this contribution is not understood.

R. L. Snyder:

As Fred Dobson has indicated, there is some evidence for a wave-turbulence whose effect would be to reduce the coherence between the surface elevation and air pressure. There is, however, another mechanism which in many experimental situations will dominate. For non-dimensional fetches of order 10^3 to 10^4, the peak of the spectrum corresponds to $\mu \sim 1$ to 2, where γ is a minimum. Thus the pressure signal for downwind travelling waves can easily be dominated by the pressure signal for waves of the same frequency travelling in other directions, in particular, waves travelling against the wind. This latter signal is incoherent with the dominant part of the wave signal, resulting from waves travelling downwind.

Thus the overall coherence between surface elevation and air pressure is low. In my experience, a well defined wave-induced peak in a generation situation is indicative of waves travelling in directions other than the wind direction. In a situation where these waves are absent, the wave induced peak is not that well defined.

V. Cardone:

The Phillips linear growth rate parameter, α, remains a part of all contemporary discrete and hybrid-parametric wave prediction models. It is usually simply prescribed on the basis of net wave growth and incorporates Priestley's (1964) form for the shape of the atmospheric pressure spectrum and a wind speed power law relation for the intensity of the fluctuations (e.g. u^4 or u^6). How does your measured spectrum of the incoherent atmospheric turbulent pressure fluctuations over the sea surface compare to Priestley's? Also, do your experimental conditions extend over a broad enough range of wind speeds to reveal the variation of the root mean square pressure intensity with wind speed?

R. L. Snyder:

Priestley's (1964) measurements yielded downwind and crosswind coherences, both of which fell off as exp $(-\mu_i|\xi_i|)$, where μ_1 and μ_2 are comparable coefficients and ξ_1 and ξ_2 are the downwind and crosswind components of instrument displacement. Priestley assumed that where the displacement was neither downwind nor crosswind, the coherence would go as the product of these two exponentials. The Bight of Abaco data do not support this picture. We find that the coherence is largely independent of downwind displacement, consistent with the frozen turbulence assumption. We explain Priestley's downwind coherences in terms of statistical variations in mean wind direction. Unfortunately, this kind of model confines $F_Q^2(k,\omega)$ to the dispersion curve $\omega = k \cdot U$ (where it is infinite).

Elsewhere, F_Q^2 is zero. Thus we can not presently estimate the α term in the radiative transfer equation. As indicated in our paper, it may be possible to make such an estimate by relaxing the delta function dependence of F_Q^2 in some appropriate way. My expectation would be that such a computation would produce smaller α's than does Priestley's model.

We can probably learn something about the wind speed dependence of the r.m.s. turbulent pressure, but as yet have not looked at this particular aspect of the data.

OBSERVATIONS OF NONLINEAR PROCESSES IN SURFACE WAVE GENERATION

Craig E. Dorman and Erik Mollo-Christensen

Department of Meteorology,
Massachusetts Institute of Technology
Cambridge, Massachusetts, U.S.A.

Abstract

Arguments are presented that suggest the possible existence of an air-sea interaction soliton , that is a wave group of permanent type envelope, of a significant half-life, and which consist of coupled disturbances in the air and water velocity fields. It is argued that instabilities are modulationally unstable, and thus, to a linear approximation, favor the generation of groups. The phase between wind and sea surface disturbances for a disturbance of permanent type is shown to be consistent with observations. Other experimental evidence is cited in support of the interaction soliton. The possible usefulness of such a description of wind generation is also discussed.

Introduction

Recent work on non-linear wave propagation suggest that surface waves may tend to occur in solitons or group solitons. Group solitons will tend to form in deep water; they are capable of attaining envelopes of permanent shape (see Chu and Mei, and also Yuen and Lake, 1975). It has also been demonstrated both theoretically and experimentally that group solitons are capable of surviving collisions (Yuen and Lake, loc. cit.). Mollo-Christensen and Ramamonjiarisoa, 1977, have constructed models of random fields and compared the results with observations. In the present paper, we shall examine the possibility that there exists an air-sea interaction soliton, give heuristic reasons for suggesting the possibility, and discuss the experimental consequences and compare with observations.

Modulational Instability of Instability Waves

A simple argument that suggests that wind waves may tend to be generated in groups rather than in periodic wave trains is as follows. The instability theory of wave generation by wind, as was presented by Miles, 1957, predicts that the wave energy will increase exponentially with time as long as the wave amplitude is small, so that linear theory is approximately valid. The growth rate depends on velocity profile parameters and wavelength, among others. The choice of a real wavenumber and posing the problem of temporal growth is one of convenience. Gaster, 1962, has shown that the problem of finding the growth rate of waves sinusoidal in time is solved by setting the spatial growth rate equal to the temporal growth rate multiplied by the wave group velocity. We shall now argue that the result of instability theory is also valid for a wave train where the initial amplitude varies slowly in space and time, and that the subsequent history of a disturbance of a certain initial wavenumber and appropriate frequency can be written in terms of energy E:

$$E(x,t) = E_0 \quad f(x-c_g\ t) \quad e^{bt} \tag{1}$$

f(x) is the initial energy distribution, E_0 is a parameter that scales the initial energy level, and b is the growth rate. C_g is the group velocity. One then finds for the growth rate of the group disturbance:

$$DE/Dt = \partial E/\partial t + c_g\ \partial E/\partial x = b\ E_0\ \ f(x-c_g t)\ e^{bt} \tag{2}$$

This shows that the growth rate is the largest where the initial amplitude is the largest, and that therefore modulational maxima will be amplified the most by the instability. This rather obvious conclusion says that waves formed through a process of instability will tend to be modulationally unstable. The energy is supplied from the mean flow, in contrast to the modulational instability of freely propagating neutrally stable waves, which get the energy involved in modulational instability from neighbors in the wave field. This suggests that modulational instability of instability waves, such as wind waves in the process of generation, may grow faster than the modulational instability fluctuations of neutrally propagating waves, such as free surface waves in the absence of wind. To the above reasoning must be added some cautionary notes. First, the reasoning requires that there be a frequency band of unstable waves, so that the initial state can be described as a fourier integral over wavenumbers, some of which are unstable. Furthermore, the reasoning only will hold for small amplitude, and as soon as finite amplitude effects become important, the initial modulational instability may be converted to modulational stability, tending to eliminate bumps in amplitude. Here, we can only quote the results of Whitham, 1965, (see Whitham, 1974) which show that finite amplitude Stokes waves in deep water are modulationally unstable, so that waves that are only slightly affected by wind may still possibly be modulationally unstable.

The process of modulational instability of waves being generated by an instability process is clearly deserving of further more careful theoretical work. But we shall not wait for the completion of further theory to suggest that there may exist an air sea interaction group soliton, and proceed to describe some of its properties and look for them in observational data.

Description of Interaction Group Solution

Although there is much work to be done before one can be assured that there is an interaction group soliton, it will still be useful to venture a tentative description that can be used as a guide for the design of experiments, and as support in the formulation of questions to be answered by analysis of by observation.

First take a look at the time series of velocity and wave observations observed during a passage of what may be an interaction soliton. The observations were made by Dorman, 1971. Figure 1 shows 5 second wind profiles, water current profiles, isotach of horizontal wind velocity, wind velocity signal at a height of 2.4 m, filtered through a bandpass filter with low and high frequency limits set at 0.1 and 0.2 Hz, respectively, the filter output being squared to obtain the record shown. The two bottom traces represent bandpass filtered and squared wave gauge signals, showing low and high frequency content of the wave signal as indicated. The third trace from the bottom is the unfiltered wave gauge signal. This record is a two minute portion of a ten minute record, the ten minute record being one of 56 ten minute records taken at one hour intervals during a period when the wind increased gradually from 8 m/s to 13 m/s. The "event" is characterized by a burst of high frequency wave energy. We see such a burst at the most once per ten minute period, and in many ten minute periods no bursts occurred.

Note that the wake of the burst arrives first, then the burst, and then the up wind portion of the burst. First there seems to be a tendency for an inflection to occur in the wind profile, the isotachs show one or two undulations, just before the high frequency wave burst occurs. This is associated with a large signal in low frequency wind and irregular wave profiles.

Taking thirty second average wave spectra, in this case for

another burst than the one shown in Figure 1, one sees the sudden increase in high frequency wave energy during the burst, in time period 16, for example, as shown in Figure 2. During what we believe is a typical burst, there is also an almost linear variation of the relative phase between wind and the waves with frequency, as shown in Figure 3, where we show thirty second coherence and phase between wind signal and wave gauge output. The third record is typical of a burst. Note that the coherence is high during the bursts, which also means that one can attribute significance to the phase observation. We shall show that a phase that varies linearly with frequency is characteristic of an interacting pattern of permanent form. Yet another important feature of an interaction pattern of permanent type, such as a joint group soliton, is the phase relationship between variables.

Phase Relation between Variables

Consider, as an example of the relationship between variables, the phase relation between stress and sea surface deflection. For a stress field that moves with and is locked to the surface deflection field, the relationship can be written:

$$\text{Stress} = \tau(x,z,t) = B(z) \quad \eta(x,t+\Delta(z))$$

η is the sea surface deflection, $\Delta(z)$ is the relative delay of the latter with respect to the stress. $B(z)$ is a function of z describing the vertical variation of Reynolds stress.

Now consider the cross-spectral density of the two variables. Calculate the cross spectral density by first taking the time cross correlation for a relative time delay of τ, and then take the Fourier transform, to obtain successively,

$$\begin{aligned}
&\iint \eta(x,t)\, \tau(x,z,t+T)\, e^{i\omega T}\, dt\, dT \\
&= B(z) \iint \eta(x,t)\, \eta(x,t+\Delta(z)+T)\, e^{i\omega(T+t)}\, e^{-i\omega t}\, dt\, d(t+T) \\
&= B(z)\, Y^*(x,\omega)\, Y(x,\omega)\, e^{-i\omega \Delta(z)} \qquad (3) \\
&= Coh(\omega)\, |Y|^2\, e^{i\theta(\omega,z)}
\end{aligned}$$

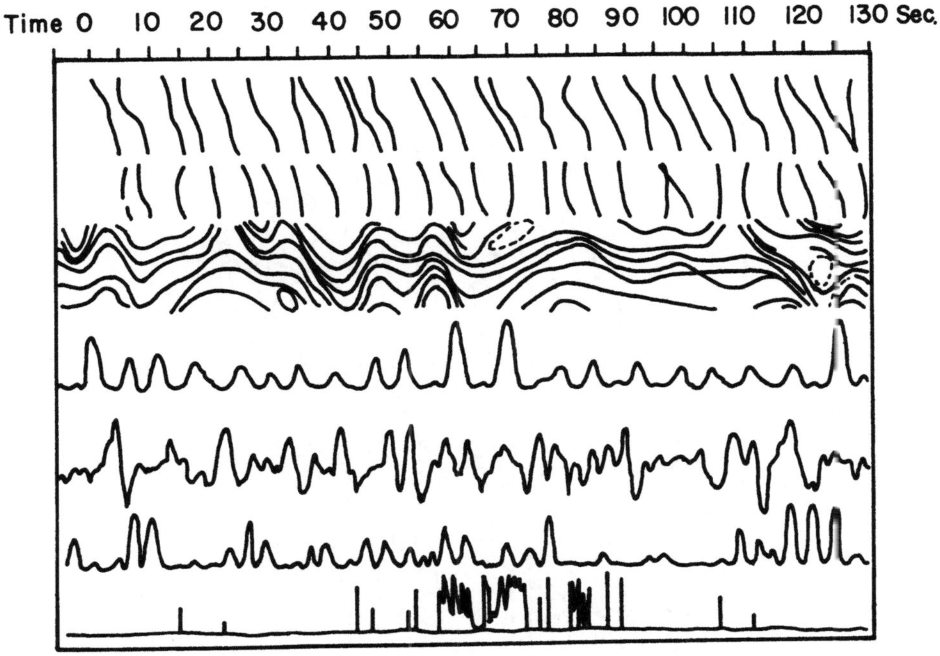

Fig.1. Two minute segment of ten minute observation, selected because it contains an "active" event in high frequency waves.

Upper traces: Five-second average wind profiles

Second traces: Five second average current profiles

Third traces : Five second average isotachs of horizontal wind speed

Fourth trace: Wind at 2.4 m , Bandpass filtered and squared. Bandpass 0.1-0.2 hz

Fifth trace: Wave gauge signal

Lower two traces: Wave gauge signal filtered and squared. Bandpass 0.1-0.2 hz and 5- 20 hz, respectively

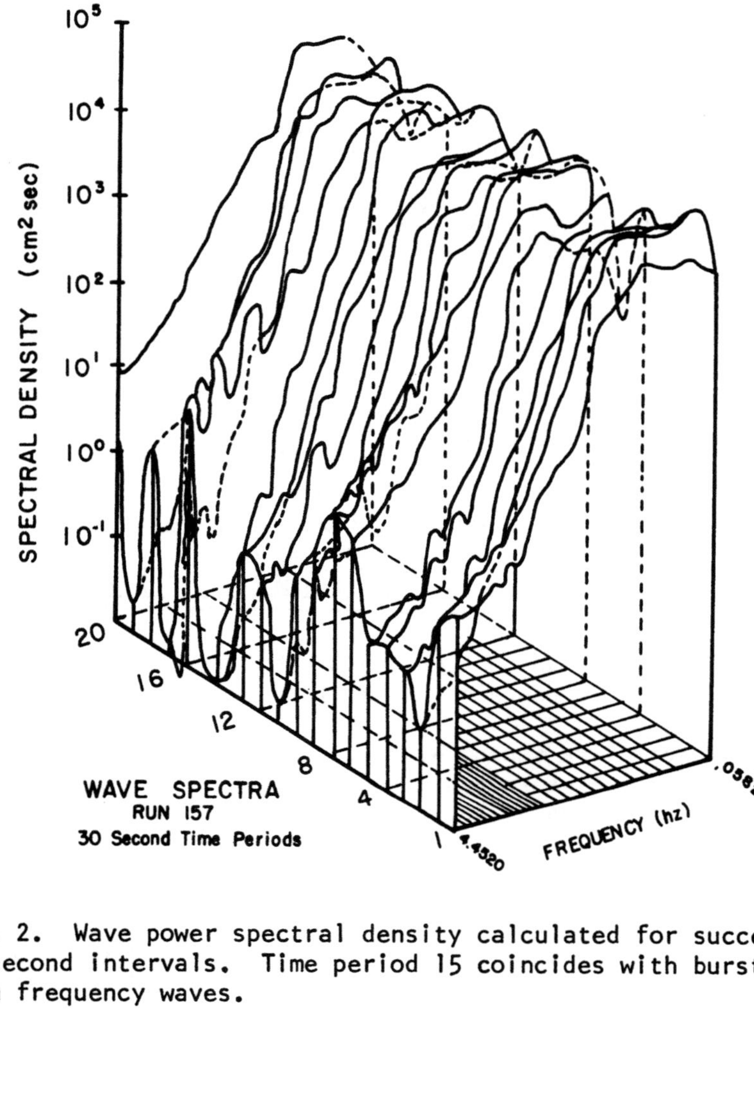

Fig. 2. Wave power spectral density calculated for successive 30-second intervals. Time period 15 coincides with burst of high frequency waves.

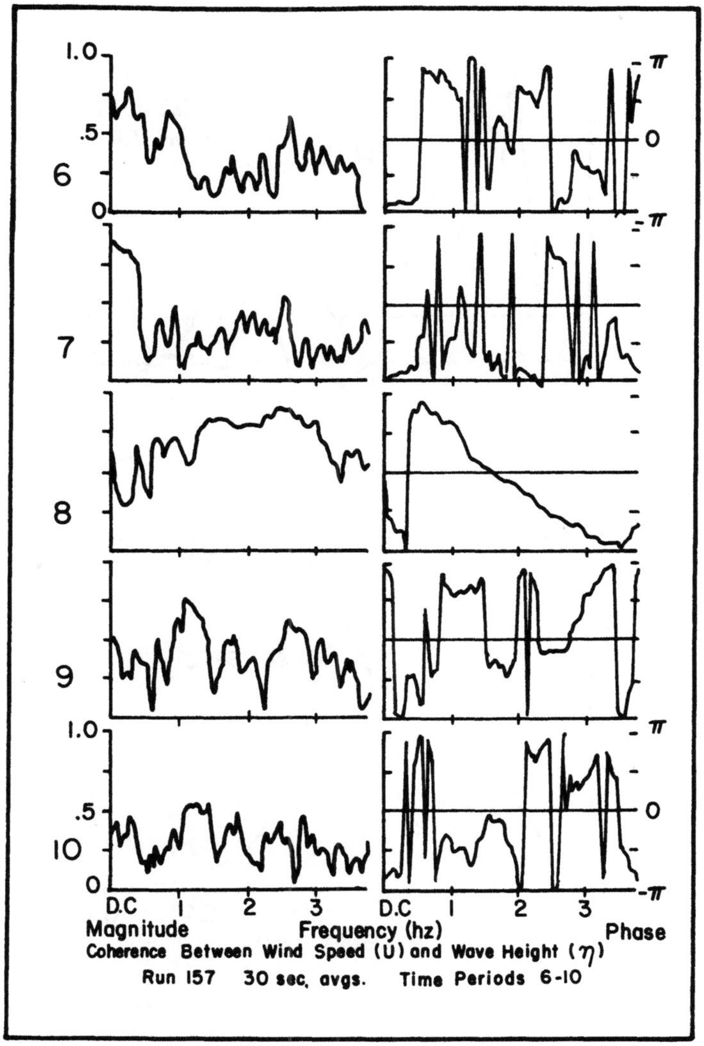
1.0
.5
0
6
7
8
9
10
1.0
.5
0
π
0
-π
π
0
-π
D.C 1 2 3
D.C 1 2 3
Magnitude
Frequency (hz)
Phase
Coherence Between Wind Speed (U) and Wave Height (η)
Run 157 30 sec. avgs. Time Periods 6-10

Fig. 3

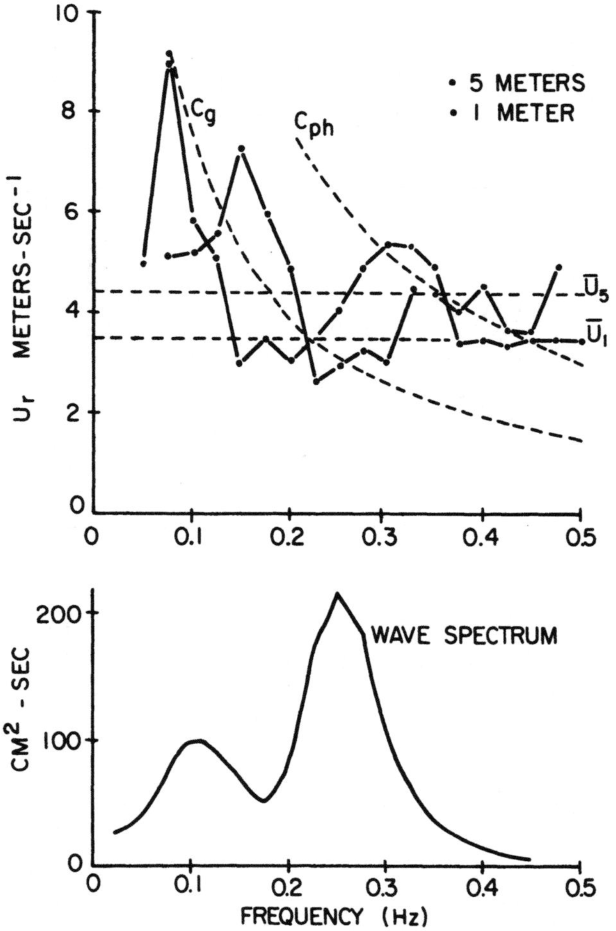

Fig. 4. Horizontal downwind correlation speed of wind fluctuations vs. frequency. Free surface wave phase and group velocities shown as dashed line. Lower plot wave spectral density during measurements.

Y is the Fourier transform of η, θ is the phase angle, Coh(ω,) is the coherence. Comparing expressions, one finds for the phase angle: $\theta(\omega, z) = -\omega\,\Delta(z)$. This shows that the phase varies linearly with frequency. The phase variation shown in Figure 3 is consistent with the existence of an interaction pattern of permanent form such as an interaction group soliton. In order to lend further support to the existence of a group soliton rather than a solitary wave kind of interaction, we show Figure 4, taken from Ruggles, 196 .

Propagation Speeds

Figure 4 shows the propagation speed of wind fluctuations observed using two buoys with hot film sensors, separated in the wind direction. For comparison, the phase velocity of surface waves according to linear theory is shown, and the corresponding group velocity is also shown, as well as the power spectral density of the waves.

We were surprised at the time the result was obtained, that the gusts did not seem to propagate at linear wave phase velocity, at least for frequencies above the spectral maximum. We thought that our measurements were insufficiently precise to give reliable results, and interpreted them to suggest that the propagation speed of gusts for low frequencies was equal to group velocity, and that, for frequencies above the spectral maximum, the propagation speed matched wave phase velocity.

Now, from the measurements of Ramamonjiarisoa, 1974, it appears that the phase velocity of the waves are very different from linear theory. Mollo-Christensen and Ramamonjiarisoa, 1977, have interpreted the observations as explainable in terms of the wave field consisting of group solitons. This may provide a better interpretation of the observations of Ruggles, 1969, shown in Figure 4.

Discussion and Conclusions

While the experimental evidence is not conclusive, it suggests the possible existence of an air sea interaction group soliton. This supports further the explanation of the observed anaomalous celerity by Ramamonjiarisoa, and the explanation offered by Mollo-Christensen and Ramamonjiarisoa.

One may question the usefulness of describing air sea interaction and wavefields in terms of "particles" such as group solitons and interaction group solitons. The advantages of the particle description is similar to the advantage in such usage in particle physics, namely that an energy containing structure in a field can be usefully given a name if it has a permanent existence or occurs under certain invariant conditions. A group soliton can be considered the result of a multiplicity of nonlinear interactions that has finally resulted in a semipermanent disturbance that has the property of being able to survive collisions with other particles or whose collisions with other particles result in the generation of other particles under rather simple rules. Thus, rather than having to describe the processes of creation and interactions in detail every time one needs to describe a field, one can define the important particle parameters, their dependence upon field parameters and their population density. By perhaps far-fetched analogy, an electron beam is defined in terms of particle energy and number flux.

This description is much simpler than trying to define an electron beam in terms of a solution to a nonlinear partial differential equation. In air-sea interaction, we do not have a quantum which defines a universal scale. Therefore, we cannot have the same simplicity as in particle physics, but we may still be able to emulate the physicists in developing a particle description of air sea interaction and wave fields, and if useful and successful, such descriptions may also be found useful in other turbulent flows.

References

Chu, V.H. and C.C. Mei, 1971. The non-linear evolution of Stokes waves in deep water. J. Fluid Mech., 47, 337-351.

Dorman, C.E., 1971. The Relationship between Microscales and Wind Wave Development. Report 71-6. Department of Meteorology, Massachusetts Institute of Technology.

Gaster, M., 1962. A note on the relation between temporally-increasing and spatially-increasing disturbances on hydrodynamic stability. J. Fluid Mech., 14, 222-224.

Miles, J.W., 1957. On the generation of surface waves by shear flows. J. Fluid Mech., 3, 185-204.

Mollo-Christensen, E. and A. Ramamonjiarisoa, 1977. Occurrence of Stokes wave groups in random wave fields.

Ramamonjiarisoa, A., 1974. Sur l'evolution des spectres d'energie des vagues de vent a fetch court. Fifth Liege Colloquium on Ocean Hydrodynamics.

Ruggles, K.W., 1969. Observation of the wind field in the first ten meters above the ocean. Ph.D. Thesis, MIT Department of Meteorology.

Whitman, G.B., 1965. Nonlinear dispersion of water waves. J. Fluid Mech., 27, 399-412.

Whitman, G.B., 1974. Linear and Nonlinear Waves. John Wiley & Sons.

This work was sponsored by the Office of Naval Research, under Contract N00014-76-C-0413.

DISCUSSION

W. Rosenthal:

What would you suggest might be observed at sea surface when an "event" occurs?

E. Mollo-Christensen:

We have no algorithm which will identify a "generating wave group"; the presence of many wavelets and a breaking crest may be a possible indication.

E. J. Plate:

As a comment: it is interesting to note that the "interaction solitons" are quite isolated events, indicative of a strong intermittency - in contrast to the situation observed in the laboratory. I would like to know, though, if the wind part of the "interaction soliton" is generated by the waves, or vice versa. Can this be inferred from the phase shift between air and water pattern?

E. Mollo-Christensen:

It is a coupled event. The boundary layer is modified by wave generation, of course.

O. M. Phillips:

The wind field is largely given by the large scale meteorological situation, of course; beyond that, I suppose everything is coupled.

ENERGY AND MOMENTUM FLUX TO NONRESONANT FORCED WAVES

Dieter Hasselmann

Meteorological Institute, University Hamburg
Bundesallee 55, 2000 Hamburg 13

1. INTRODUCTION

This paper is a result of the author's struggle to understand the momentum balance in Longuet-Higgins' (1969) fluctuating horizontal stress mechanism. A surface stress oscillating in phase with a gravity wave will cause the wave to grow. If no other external forces than the fluctuating stress act on the surface then there is no average transfer of momentum across the surface and yet clearly the momentum of the wave is increasing. After some time it became clear that we have to be careful in defining what a fluctuating stress means. We can distinguish two limiting cases:

Case A: The fluctuating stress is purely viscous, and all viscosity effects in the fluid motion are taken care of.

Case B: The fluctuating stress is purely inviscid, and the fluid is also inviscid.

1.1 Case A:

For case A the problem can be analysed easily if we restrict the discussion to extremely small waves, so that $k\zeta \ll R^{-1/2}$, $R = c/(\nu k)$ (ν is the dynamic viscosity of water, c and k are the phasespeed and wavenumber of the surface wave.) The result of the calculations is as must be expected: The wave energy growth is balanced by the work done by the external tangential stress and the momentum growth is balanced by a decay of mean flow momentum. A second order negative mean current is set up by the viscous forces in order to balance the growth of wavemomentum (D. Hasselmann, 1974). This result, both surprising and yet obvious, is purely academic, because the waveheight is restricted to such small values and also because nature is probably much closer to the limiting case B, than to the purely viscous case A.

1.2 Case B:

What is a purely inviscid fluctuating horizontal stress? In inviscid motion we can only allow for normal surface stress p and so we are considering the horizontal component of the normal stress, i.e. $p \cdot \zeta_{,x}$. So on first sight it seems that we are led to the classical problem of wave instability due to atmospheric pressure. Indeed, if we consider the fluid motion as an entirely linear process, we may decompose the motion into linear waves and consider the pressure forcing of each component; in this case there is nothing left for Longuet-Higgins' mechanism. Yet, intuitively, it is obvious that this cannot be the whole story. It is certainly possible that for instance small short waves riding on long waves receive more momentum from the atmosphere on the crest of a long wave than in its trough. In some limit it must be possible to describe such a process in terms of a fluctuating stress and Longuet-Higgins' mechanism must give a valid asymptotic description. Thus it becomes clear that Longuet-Higgins' mechanism for Case B is closely linked to nonlinear wavedynamics and the energy and momentum transfer to forced waves. But then we might as well first consider the general problem of transfer to nonresonant waves and establish the connection to Longuet-Higgins' mechanism later on.

2. THE TRANSFER TO FORCED WAVES

2.1 The Basic Mechanism

The opinion that the transfer to forced waves is a negligible effect and smaller than the transfer to linear waves by a factor $\rho(\text{air})/\rho(\text{water}) \approx 10^{-3}$ was firmly entrenched in the author's mind and seems to have been wideheld. Since this opinion is wrong it is difficult to explain how it developed in the first place, but it has to do with an incorrect application of the correct statement that linear waves will grow only if forced in resonance. At the same time, believing the false conclusion, it always appeared as a puzzle why the atmosphere or indeed any external forcing agens should bother about the dynamics of ocean waves. Only after really analyzing the puzzle did it emerge that the referred to apparently straightforward conclusion was wrong. Let us consider a model example:
Let two linear waves, (Z_1,k_1,ω_1) interact $(\sigma^2=k\,g)$ and thus produce two forced waves, (Z_3^D,k_3,ω_3) and (Z_4^D,k_4,ω_4) with

$$k_3 = k_1+k_2, \quad \omega_3 = \sigma_1+\sigma_2 \quad \text{and} \tag{1a}$$

$$k_4 = k_2-k_1, \quad \omega_4 = \sigma_2-\sigma_1 \tag{1b}$$

and suppose that $\sigma_2 > \sigma_1$.
Now suppose we had constructed a machine that would generate surface pressure in a way we would prescribe. Then let this machine generate a pressure amplitude (p_3,k_3,ω_3). If then p_3 is imaginary relative to Z_3 energy and momentum are being supplied to Z_3 at the rate

$$J_M = k_3p_3Z_3^{*D} = k_3p_3G^{(+)}Z_1^*Z_2^* = k_3A^{(+)} \quad (2)$$

$$J_E = \omega_3p_3Z_3^{*D} = \omega_3A^{(+)} \quad (3)$$

$G^{(+)}$ is a kinematical factor, such that

$$Z_3{}^D = G^{(+)}Z_1Z_2 \quad (4)$$

Since a surface gravity wave field is a conservative system the energy and momentum transferred to the system must show up as growth of energy and momentum.

But certainly Z_3 is not growing, because Z_3 is not being forced in resonance, $\omega_3 \neq \sigma_3$. The governing equation for Z_3 is

$$\ddot{Z}_3 + \sigma_3{}^2Z_3 = -\frac{\sigma_3{}^2}{\rho g}p_3 + \sigma_3{}^2D\cdot Z_1Z_2 \quad (5)$$

and the r.h.s. is oscillating as $\exp(-i\omega_3t)$ so that Z_3 is certainly stationary, and Z_3 can be written as a sum

$$Z_3 = Z_3{}^P + Z_3{}^D \quad (6)$$

and we note that Z_3^P is imaginary relative to Z_3^D. This is of course the clue to the energy and momentum balance.

The two linear waves are not exactly linear but are governed by

$$\ddot{Z}_1 + \sigma_1{}^2Z_1 = \sigma_1{}^2D\cdot Z_3Z_2^* \quad (7)$$

and

$$\ddot{Z}_2 + \sigma_2{}^2Z_2 = \sigma_2{}^2D\cdot Z_3Z_1^* \quad (8)$$

If there was no atmospheric forcing the system (5), (7), (8) would remain stationary, the only significant nonlinear effect being a slight real shift in the resonance frequency. Now, however, the small imaginary component Z_3^P gives rise to a small imaginary shift in the resonance frequency and both Z_1 and Z_2 become unstable.

It is a simple matter to establish the balance and we obtain

$$\dot{M}_1 = k_1 \cdot A^{(+)}$$
$$\dot{M}_2 = k_2 \cdot A^{(+)} \qquad (9)$$
$$\dot{E}_1 = \sigma_1 \cdot A^{(+)}$$
$$\dot{E}_2 = \sigma_2 \cdot A^{(+)} \qquad (10)$$

for the momentum increase $\dot{M}_i$ and energy increase $\dot{E}_i$.
Thus

$$\dot{E} = \dot{E}_1 + \dot{E}_2 = J_E$$
$$\dot{M} = \dot{M}_1 + \dot{M}_2 = J_M \qquad (11)$$

This derivation is not quite exact, the equations of motion are slightly more complicated than (5), (7), (8), a more detailed derivation is to appear elsewhere (D.Hasselmann, 1977, hereafter referred to as I).

If we now measured the growth and the input fluxes we would find positive input of energy and momentum due to $\overline{p \cdot \zeta_{,x}}$ and $-\overline{p \cdot \zeta}$ at the frequency ω_3 and wavenumber k_3 and would find wave growth at (k_1,σ_1) and (k_2,σ_2).

The mechanism works equally well if we tune our machine to generate pressure p_4 at (k_4, ω_4). However, if we now transfer positive energy and momentum to the wave Z_4, then according to the change of sign in (1b), and in order to satisfy the energy and momentum balance, we now have

$$\dot{M}_1 = -k_1 \cdot A^{(-)}$$
$$\dot{M}_2 = k_2 \cdot A^{(-)}$$
$$\dot{E}_1 = -\sigma_1 \cdot A^{(-)}$$
$$\dot{E}_2 = \sigma_2 \cdot A^{(-)} \qquad (12)$$

and

$$J_E = \omega_4 \cdot A^{(-)}$$
$$J_E = k_4 \cdot A^{(-)}$$
$$A^{(-)} = p_4 \cdot Z_1 \cdot Z_2^* \cdot G^{(-)} \qquad (13)$$

with $G^{(-)}$ again a kinematical factor

$$Z_4^D = G^{(-)} Z_2 Z_1^* \qquad (14)$$

Forcing at ω_3 increases the action of the wave system, whereas forcing at ω_4 leaves the action unchanged. Up to this point statistic has played no rôle in the entire argument. It is straightforward to introduce statistics.

The total transfer to forced waves we can in principle measure by evaluating the quadspectra $\langle p(k,\omega)\ Z\ (-k,-\omega)\rangle$, provided we have both good time and spatial measurements of p and ζ. If we want to know how this input results in wavegrowth, we have to measure the higher order spectra

$$\langle p(k,\omega)\, Z(k_1,\omega_1)\, Z(k_2,\omega_2)\rangle = B(k_1,\omega_1;k_2,\omega_2) \tag{15}$$

for all pairs

$$k_1 + k_2 + k = 0$$
$$\omega_1 + \omega_2 + \omega = 0$$
$$\omega_i = \pm\, \sigma_i \tag{16}$$

This bispectrum then has to be multiplied with the corresponding coupling functions to yield the growth.

The total growth at (k_1, σ_1) would then arise from two sources:

a) The linear input at (k_1,σ_1)
b) The second order input on a curve in (k,ω) space given by

$$(k,\omega) = (k_1,\sigma_1) \pm (k_c,\sigma_c) \tag{17}$$

with k_c running through all points occupied by the linear spectrum.

Obviously it is possible to generalize these results to include all input to higher order waves. A third order wave produced by the product of $Z_1Z_2Z_3$ may then receive energy and momentum and this will cause growth at the wavenumbers and frequencies (k_i,σ_i), $i = 1..3$.

It has already been demonstrated in the symposium that a simple second order approximation for an interaction of a long wave Z_1 with a short wave Z_2 yields only a poor approximation if $\varepsilon=k_2Z_1 > 1$.

Second order treatment yields

$$Z_3(k_2+k_1,\sigma_2+\sigma_1) = k_2Z_1Z_2(1+0(k_1/k_2))$$
$$Z_4(k_2-k_1,\sigma_2-\sigma_1) = -Z_3\cdot(1+0(k_1/k_2)) \tag{18}$$

whereas the correct treatment yields

$$Z(k_2+nk_1,\sigma_2+n\sigma_1) = \tilde{Z}_2\cdot J_n(2k_2Z_1)(1+0(k_1/k_2)) \tag{19}$$

Thus, if $2k_2Z_1 = m$ the energy is spread out to roughly m bands on both sides of the resonance curve. This result will be useful for speculative comments later on.

3. LONGUET-HIGGINS' MECHANISM (CASE B)

3.1 Weak Interaction Case

A long wave Z_1 and a short wave Z_2 will on interaction yield the second order waves Z_3 and Z_4 given by (18), let us assume $k_2Z_1 < 1$.

Assume we had tuned our pressure generator to yield $p_3 = +p_4$, and we have chosen the phase imaginary relative to Z_3. Thus, the total pressure is , $p_3 = ip_o$,

$$p = 2ip_o e^{i(k_2x-\sigma_2t)} \cdot \cos(k_1x-\sigma_1t) \tag{20}$$

If we now perform a twoscale measurement of the atmospheric input, treating the low frequency wave as a mean field, we neither resolve the bands Z_2, Z_3, Z_4, nor the wavenumber modulation. All we see is a strong modulation of the product $\tau = \langle p\ Z_{2,x}\rangle$ in addition to a negligible modulation of $\langle Z_2Z_2\rangle$ of the order $O(k_1/k_2)$. The modulation of τ is in phase with the long wave and we are therefore not surprised to also discover that the long wave is slowly growing,

$$\dot{M}_1 = k_1\overline{\tau Z_1} \quad , \quad \dot{E}_1 = \sigma_1/k_1 \cdot \dot{M}_1 \tag{21}$$

The puzzle with the momentum balance still remains in this two scaling set up. It is only when we go to a finer resolution, so that we no longer treat the long wave as a mean field that we discover that there are actually sidebands, and that the sideband Z_3 is receiving more momentum than the sideband Z_4. Thus we see that Case B is not really a pure case of a fluctuating stress, but that there is some mean input, this mean input balancing the low frequency growth. The result (21) is then identical to that obtained from (9) through (14).

3.2 Some Speculative Comments on the Strong Interaction Case

We now go to the case $k_2Z_1 > 1$, say $(k_2Z_1) = 3.5$ and using (19) we see that the third sidebands on each side of the resonance band still have considerable energy, let us call these amplitudes $Z(3)$ and $Z(-3)$. We now again tune our pressure generator so that $p(3)$ is imagninary relative to $Z(3)$, and $p(3) = p(-3)$.
Further note $Z(3) = -Z(-3)\{1+O(k_1/k_2)\}$.

This is now a strong interaction situation and while the details have not been worked out, the situation seems to be fairly obvious: The input flux at high frequencies has a nonvanishing mean and is obtained as

$$J_E = 6\sigma_1\ p(3)Z_3^* = \frac{\sigma_1}{k_1} J_M \tag{22}$$

In (19) we have introduced $\tilde{Z}_2$ instead of Z_2, and $\tilde{Z}_2$ is to have the same value as Z_2, but is defined as

$$\tilde{Z}_2 = \{\sum_{n=-\infty}^{+\infty} Z^2(n)\}^{1/2} \tag{23}$$

so that $\tilde{Z}_2{}^2$ is the total variance of the high frequency eigenfunction. Thus

$$J_E = c_1 J_M = 6\sigma_1 p_3 Z_2 J_3(2k_2 Z_1) \tag{24}$$

It would be very surprising if not all of this flux went into long wave growth. This is what we shall assume. According to Longuet-Higgins (1969) then the momentum growth should equal

$$\dot{M}_1 = (k_1 Z_1)\ \delta\tau \tag{25}$$

where $\delta\tau$ is the amplitude of the stress fluctuating in phase with the long wave

$$\tau = J_M + \delta\tau \cos(k_1 x-\sigma_1 t) + \delta\tau_i \sin(k_1 x-\sigma_1 t) \tag{26}$$

In order to evaluate $\delta\tau$ we calculate the fluctuating component of $p\cdot\zeta,_x$. With

$$p = 2ip_o e^{i(k_2 x-\sigma_2 t)}\ J_3(2k_2 Z_1)\cos(3(k_1 x-\sigma_1 t)) \tag{27}$$

$$p(3) = ip_o$$

and

$$\zeta = Z_2 e^{i(k_2 x-\sigma_2 t)} \sum_{n=-\infty}^{+\infty} J_n(2k_2 Z_1) e^{in(k_1 x-\sigma_1 t)} \tag{28}$$

and making use of

$$J_{n-1}(\varepsilon) + J_{n+1}(\varepsilon) = \frac{2n}{\varepsilon} J_n(\varepsilon) \tag{29}$$

we can in fact establish that (25) is identical to (24).

The analysis works equally well for any pressure forcing with the imaginary components p(n) satisfying

$$p(n) = (-)^{n+1} p(-n) \tag{30}$$

and such forcing always transfers the input to low frequencies.

A forcing, which has the same symmetry as the high frequency wave,

$$p(n) = (-)^n \, p(-n) \tag{31}$$

yields mean fluxes larger than those given in (24) by a factor (k_2/k_1) for the momentum flux and a factor (σ_2/σ_1) for the energy flux, and this flux, we may confidently assume, will lead to growth at high frequencies. Such a pressure distribution will also lead to stress fluctuations, these are out of phase with the long wave, corresponding to $\delta\tau_i$ in (26), and are related to the real part of the pressure amplitude.

4. MODELS

4.1 Waves of Similar Wavelengths

Crude models for the bispectral transfer for frequencies ranging from a peakfrequency σ_p to a cutoff frequency $\sigma_c \approx 5\sigma_p$ will be presented in(I).Each model included second order interactions only and is poor for long wave-short wave interactions, in which range it certainly overpredicts. These calculations yielded energy and momentum transfer in the order of 30% of the transfer due to the underlying linear model. The models are complementary to those considered by Garrett and Smith (1976). The treatment presented by Garrett and Smith does well at widely separated wavenumbers, the bispectral treatment is best for closely neighboured wavenumbers.

4.2 One Long Wave and a Spectrum of Very Short Waves

We can in this case apply the considerations of section 3.2. Let us describe the pressure response of the atmosphere as a simple linear response

$$p(k,\omega) = (T_r + iT)(k,\omega) \cdot Z(k,\omega) \tag{32}$$

we shall be concerned with T only. (T may still depend on the statistical parameteres of the wavefield, for instance on the spectrum.) We concentrate again on one high frequency wave as given by (19), but keep in mind that a whole spectrum of such waves is to exist. We wish to estimate how much transfer to the long wave might occur, if we assume that the flux enters only into the short waves. The ratio $\sigma_2/\sigma_1 \approx 7$ shall be assumed.

We use T(n) for $T(k_2+nk_1, \sigma_2+n\sigma_1)$ and assuming a value of $2k_1Z_1=.1$ we have $2k_2Z_1 \approx 5$.

The results (19), (30) and (31) state:

If $T(n) = T(-n)$ then all of the flux is used for short wave growth. If $T(n) = -T(-n)$ then all of the flux is used for long wave growth. Thus we have to split $T(n)$ into its symmetric and antisymmetric part.

The value $2k_2Z_1 \approx 5$ means that approximately the sidebands $-5 \leq n \leq 5$ are occupied.

We assume that $T(k,\omega)$ approximately varies as

$$T(k,\omega) = (U - \omega/k)\cdot\omega \tag{33}$$

and assume $U \gg \omega/k$. The wavenumber k_2+nk_1 remains approximately constant, but $\sigma_2+n\sigma_1$ varies from about $2\sigma_1$ to $12\sigma_1$. The even and odd part of the transfer function is then

$$T_s(n) \approx U \cdot \sigma_2$$

$$T_a(n) \approx U n \sigma_1 \tag{34}$$

Thus the antisymmetric part is on average about half as large as the symmetric part. This would then by the considerations of 3.2 imply that roughly

$$\dot{M}_1 \approx .35 \frac{k_1}{k_2} J_M$$

$$\dot{E}_1 \approx .35 \frac{\sigma_1}{\sigma_2} J_E \quad , \quad J_E \approx \frac{\sigma_2}{k_2} J_m \tag{35}$$

The factor .35 in (35) arises from

$$.35 = \bar{n} \cdot \frac{\sigma_1}{\sigma_2} = \frac{1}{2}(2k_2Z_1) \cdot \frac{\sigma_1}{\sigma_2} = (k_1Z_1) \cdot \frac{\sigma_2}{\sigma_1}$$

Thus we could also write

$$\dot{E}_1 = (k_1Z_1) \cdot J_E$$

$$\dot{M}_1 = (k_1Z_1)\cdot\frac{\sigma_1}{\sigma_2} J_M \tag{36}$$

expressing the low frequency growth rates roughly in terms of the total high frequency input.

Of course these estimates give orders of magnitudes at best, so that we may as well say that about 1% of the total momentum flux to short waves, and roughly 7% of the total energy flux may in this way be supplied to the long waves.

If instead of (33) and (34) we had imposed the limitation that the modulus of the fluctuating stress $\delta\tau$ should not exceed the mean stress,

$$\delta\tau \leq J_M$$

we arrive at

$$\begin{aligned} \dot{M}_1 &\leq (k_1 Z_1)\, J_m \\ \dot{E}_1 &\leq \frac{\sigma_2}{\sigma_1}(k_1 Z_1)\, J_E \end{aligned} \tag{37}$$

These estimates still assume that the major energy and momentum input is due to the symmetric part of the transfer function. If the entire input is due to the antisymmetric part, we of course have

$$\begin{aligned} \dot{M}_1 &= J_M \\ \dot{E}_1 &= J_E \end{aligned} \tag{38}$$

To the author, the estimates (36) seem to be the most reasonable.

4.3 Comparison of the Two Models

It would be very nice to have a composite model which allows for the strong interactions at high frequencies and yet contains many long waves. Such a model is not in sight.

Whether model 4.1 or model 4.2 gives a better description of the transfer to forced waves really depends essentially on the linear transfer function $T(k,\omega)$. If the transfer function is large at low or intermediate frequencies, then model 4.1 should be quite adequate. This would imply a large linear input at low frequencies,but also a non-neglible bispectral input.

If the transfer function becomes very large only at high frequencies then clearly something similar to model 4.2 is needed.

5. EDDY VISCOUS STRESSES

It is not obvious whether the analysis allows an extension to include truly turbulent effects, as opposed to small scale wave motion. The interplay of viscous effects and wave motion is likely to produce turbulent motion, which may then be essential for the transfer of high frequency wave momentum to currents.

A rigorous analysis of such a mechanism is exceedingly difficult, because of the enormous differences in scale between the length responsible for viscous effects, typical waveheights characterizing a length for nonlinear effects, and finally typical wavelengths. Nevertheless, a model description in which we assume that in a balanced situation a field of short waves receives momentum from the atmosphere and passes on this momentum to currents can be studied. The passing on to currents is then described by an eddy viscous mean stress, which has such a small vertical gradient that the current can be assumed to be stationary.

If we then introduce a long wave and assume that the only additional input mechanism triggered by the long wave is a fluctuating stress, we have the eddy viscous analogue of Longuet-Higgins' (Case A)mechanism. We must then again expect that the current now receives less momentum than in the absence of the long wave and this in fact turns out to be the case for this model (D.Hasselmann, 1974). This result is in agreement with the analysis of Case B, the only difference being, that in Case B the short waves grow less rapidly, whereas in this model they give less momentum to the current.

6. CONCLUSIONS AND SUMMARY

The effectiveness of the fluctuating stress mechanism is difficult to estimate, but it appears most likely that the total momentum transfer to long waves is in the order of a few percent (1-5%) of the total stress.

The difficulties in obtaining an estimate have two reasons:

a) The pressure transfer fuction T (k,ω) is only poorly known, not to mention a possible dependence on the spectrum.
b) At the moment no tool is available to properly treat the strong interaction of short waves with long waves in a statistical wavefield.

It has been shown that at large wavenumbers k_2 the spectral band around the dispersion curve has the width

$$\Delta\sigma = 2(k_2 Z_1)\sigma_1$$

so that the spectral area in which momentum transfer can occur is considerably enlarged. This result is in agreement with and to be expected from WKBJ-theory.

The rôle of the symmectric and antisymmetric part of the transfer function $T(k,\omega)$ is such that the symmetric part is responsible for growth of high frequency waves, while the antisymmetric part transfers the input to low frequencies.

Whether or not an additional input mechanism, such as the one presented here, is needed to close the JONSWAP balance has been discussed in (I), and is still an open question.

There is mainly one experimental clue which might indicate that the air flow over waves is strongly nonlinear. The pressure spectra over waves show a very pronounced increase in the wave energy containing frequency range, as compared to pressure spectra over a flat surface, the spectra being larger by a factor of typically 5 (Elliott, 1972). Thus, one should expect a large coherence between waves and pressure at least for a wavespectrum which is close to unidirectional. The low coherence, in the order of .5, can probably mainly be attributed to the directional spread. The other tempting explanation is that the atmosphere reacts strongly nonlinearly, so that pressure waves, not proportional to the free ocean waves, but proportional to sidebands are produced.

REFERENCES

Elliott, J.A. 1972:Microscale pressure fluctuations near waves being generated by wind. J.Fluid Mech. 54, 427

Garrett,C. and Smith, J. 1976:On the interaction between long and short surface waves. J.Phy. Oceanogr. 6, 925

Hasselmann, D.E. 1974:The momentum balance between wind, gravity waves and currents. Berichte aus dem Sonderforschungsbereich Meeresforschung SFB 94;Universität Hamburg (unpublished)

Hasselmann, D.E. 1977: Wind wave generation by energy and momentum flux to the forced components of a wavefield. (Submitted to J.Fluid Mech.)

Longuet-Higgins, M.S. 1969: On the action of a variable stress at the surface of water waves. Physics of Fluids 12, 737

OPEN SEA SIMULTANEOUS OBSERVATIONS OF AIR AND WATER MOTIONS DURING ACTIVE AIR-SEA INTERACTIONS

J. P. Giovanangeli (+*), M. Revault D'Allonnes (+),
A. Ramamonjiarisoa (*)
(+) Laboratoire D'Oceanographie Physique. Museum National D'Histoire Naturelle. 43 Rue Cuvier,
(*) Institut de Mecanique Statistique de la Turbulence
12 Av du General Leclerc, Marseille

INTRODUCTION

During active air-sea interactions the mechanical energy of air enters the sea across the surface to appear as energies of wind waves, of drift current and of turbulence in water. Although the subject of many investigations, the physical mechanims involved during energy transfer from wind to sea, as well as in interactions between the various forms of water motions (waves, current, turbulence), are not yet properly understood.

This paper presents preliminary results of a joint experiment conducted in open sea in October, 1976, by the "Laboratoire d'Océanographie Physique (L.O.P.) du Muséum National d'Histoire Naturelle - Paris" and the "Institut de Mécanique Statistique de la Turbulence (I.M.S.T.) - Marseille".

The main objective was to try to give first estimates of the part of energy which enters the waves and that one which enters the drift current. A rather sophisticated experimental set-up has been operated on-board the C.N.E.X.O."Bouée Laboratoire BORHA II" moored on three points in the Mediteranean Sea at 42° N, 5° E (figure 1). These experiments had made it possible to record simultaneously the principal characteristics of the interface , the air and the water motions, during an observational period when an abrupt increase of wind speed occured.

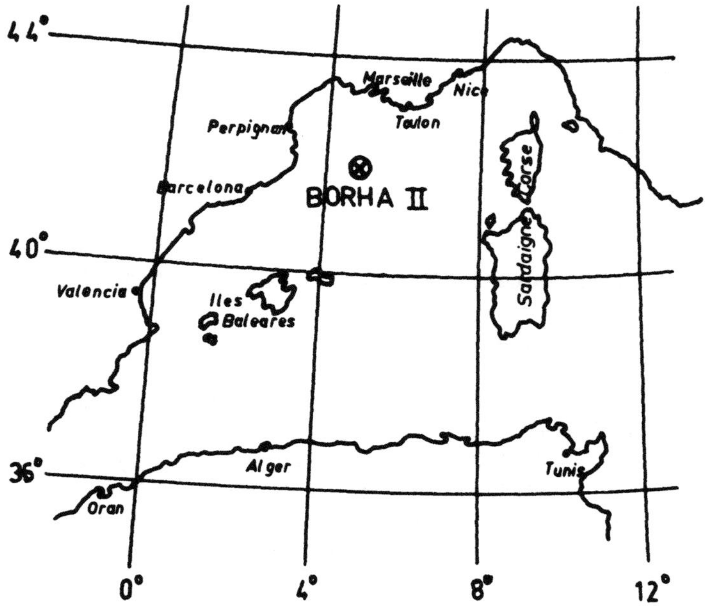

Fig. 1. Site of experiment.

EXPERIMENTAL ARRANGEMENT AND METHODS

A number of probes were disposed above the sea-surface as shown in figure 2. The wind speed vertical distribution, $U(z)$, was measured between the mean sea level and an elevation of about 13 meters using six "3 cup" anemometers. Six thermistors shielded from solar radiation and located close to the cup anemometers provided the vertical temperature profile. At $z \simeq 11.5$ meters an X hot wires anemometer, fixed on a vane, recorded the small-scale turbulent fluctuations. The probes were disposed about 9 meters away from the vertical buoy center line.

Three wave gauges gave the instantaneous sea surface deflection levels. The sea surface temperature was measured by means of 2 thermistors fitted under a small buoy. In water, the vertical distribution of mean current was determined by means of two current meters, one fixed at 6.5 m depth and the other beeing moved between the surface and $z = -20$ m (Figure 3). A "Conductivity-Temperature-

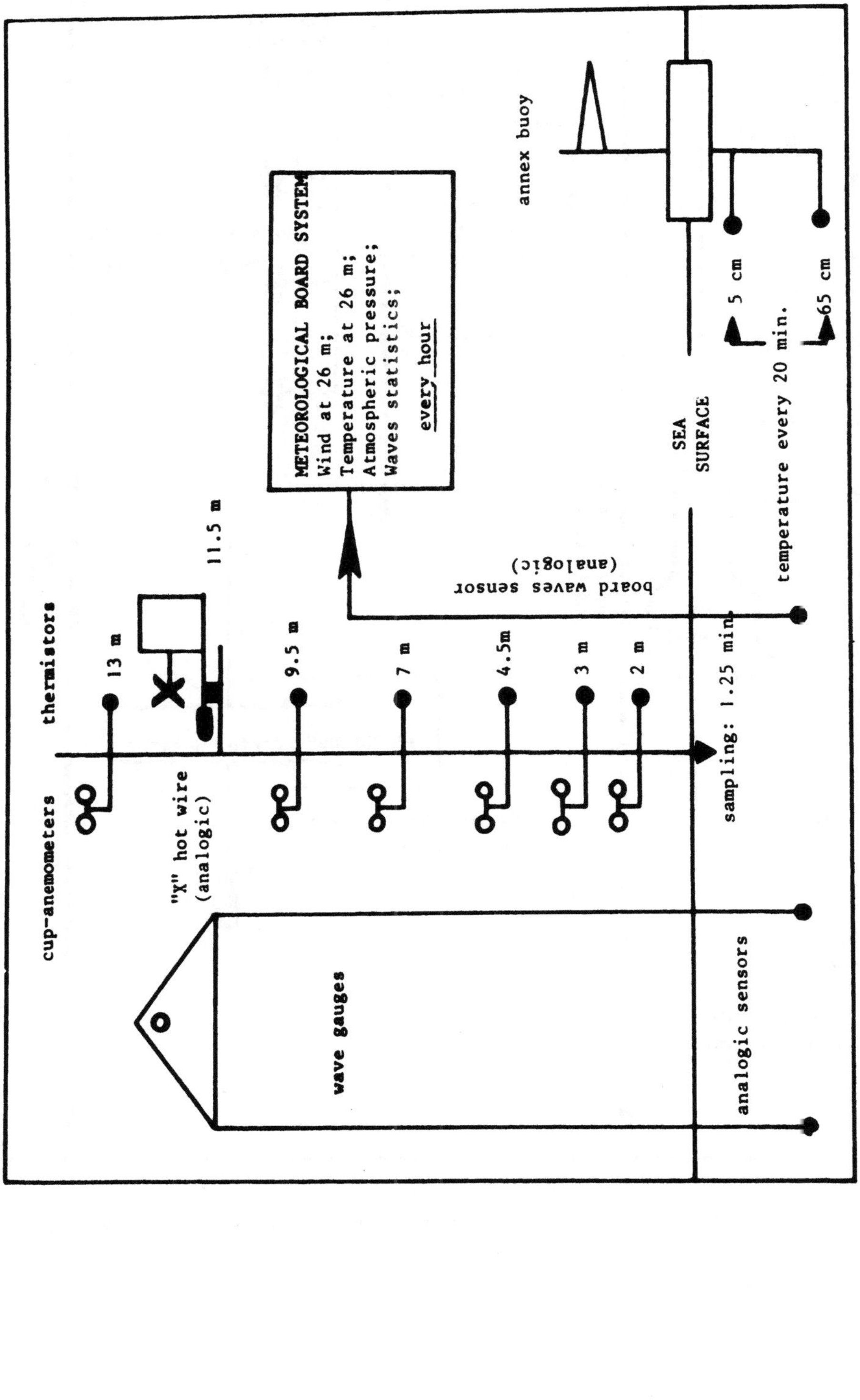

Fig. 2. General lay out of experimental set up in air.

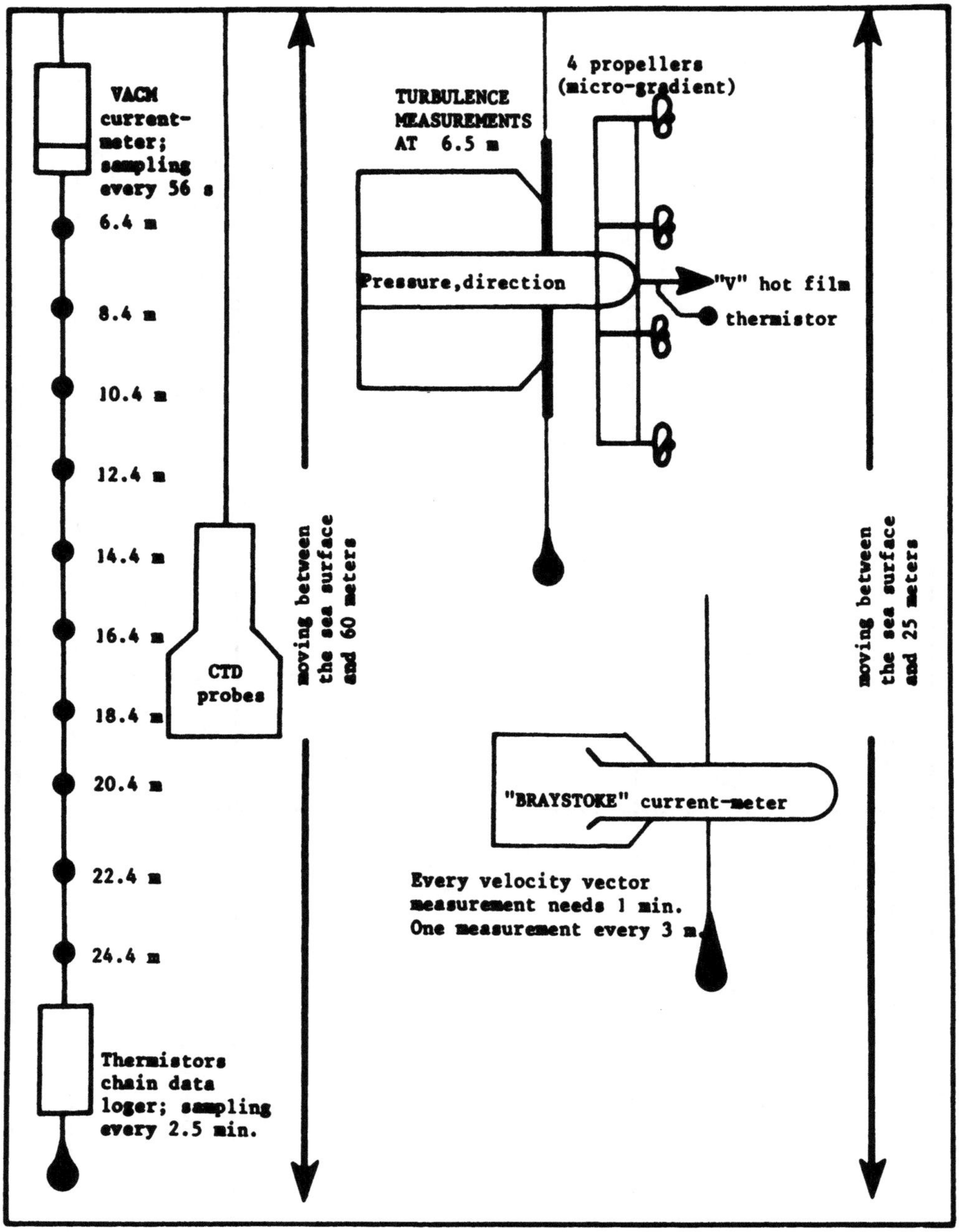

Fig. 3. General lay-out of experimental set-up in water.

Depth" instrument and 10 thermistors measured the vertical temperature and density distributions. Small scale measurements were performed using a complete system, including four fast response propellers, four thermistors, direction and pressure sensors and a double wedge hot films probe, mounted on an aligning vane. Hot films and hot wires probes were calibrated in a wind-water facility before and after experiments. The output voltages from cup anemometers, air and water thermistors and fixed current meter were digitized and tape-recorded (the sampling intervals are given in Figures 2 and 3). Fluctuating output voltages from hot films, hot wires, wave gauges, vane direction sensor, micro propellers and pressure and direction sensors in water were recorded on two Analog-tape instruments. The records were later processed in the laboratory, using an Analog/Digital Converter and a numerical system (Hewlett-Packard model 2100A).

Different sampling frequencies were adopted to match the frequency domains of interest in the various parameters. Furthermore, using data from the meteorological equipment of the buoy, evolutions of wind, interface motion and current were studied within a rather large range of time scales (From 10^{-3} sec to about 10 days).

Prior to the analysis on the energy partition between waves and current, it is necessary, first, to estimate the air friction velocity $u_* = (\tau/\rho_a)^{1/2}$ (where τ is the total stress and ρ_a the air density) which represents the total rate momentum transfer from the wind to sea. This has been done by means of the well-known "Profile" and "Inertio-Dissipative" methods. Earlier experiments had led us to conclude that the air flow is distorted by the presence of the buoy. However, two different methods have been set up to correct the experimental data from this parasitical effect. The first was an analytical method based on the properties of the inviscid fluid flow around a solid cylinder. The second was an experimental method which called on the determination, within a wind tunnel simulation of the atmospheric boundary layer, of the flow structure around a model of the buoy (Ruimy, 1977). Then, the undisturbed wind profiles were deduced from the disturbed ones so that the use of the "Profile" method could be considered.

Momemtum flux determination by the "Inertio-Dissipative" method was based on the X wires anemometer data. One has to recall that these probes were fitted on a vane so that it was first necessary to estimate the dynamical response of this device, which was done during a laboratory study (Dupé, 1976).

As far as the interface motion is concerned, the spectrum $F(n)$ of the surface deflection level was determined by a FFT technique (Resch and Bourguel, 1976). Then, by integration, the total variance $\overline{\eta^2}$ is estimated. The phase celerity $C(n)$ has been measured by calculating the space-time correlation (Favre and al., 1964) (Ramamonjiarisoa, 1974) and the cross-spectral density functions for the

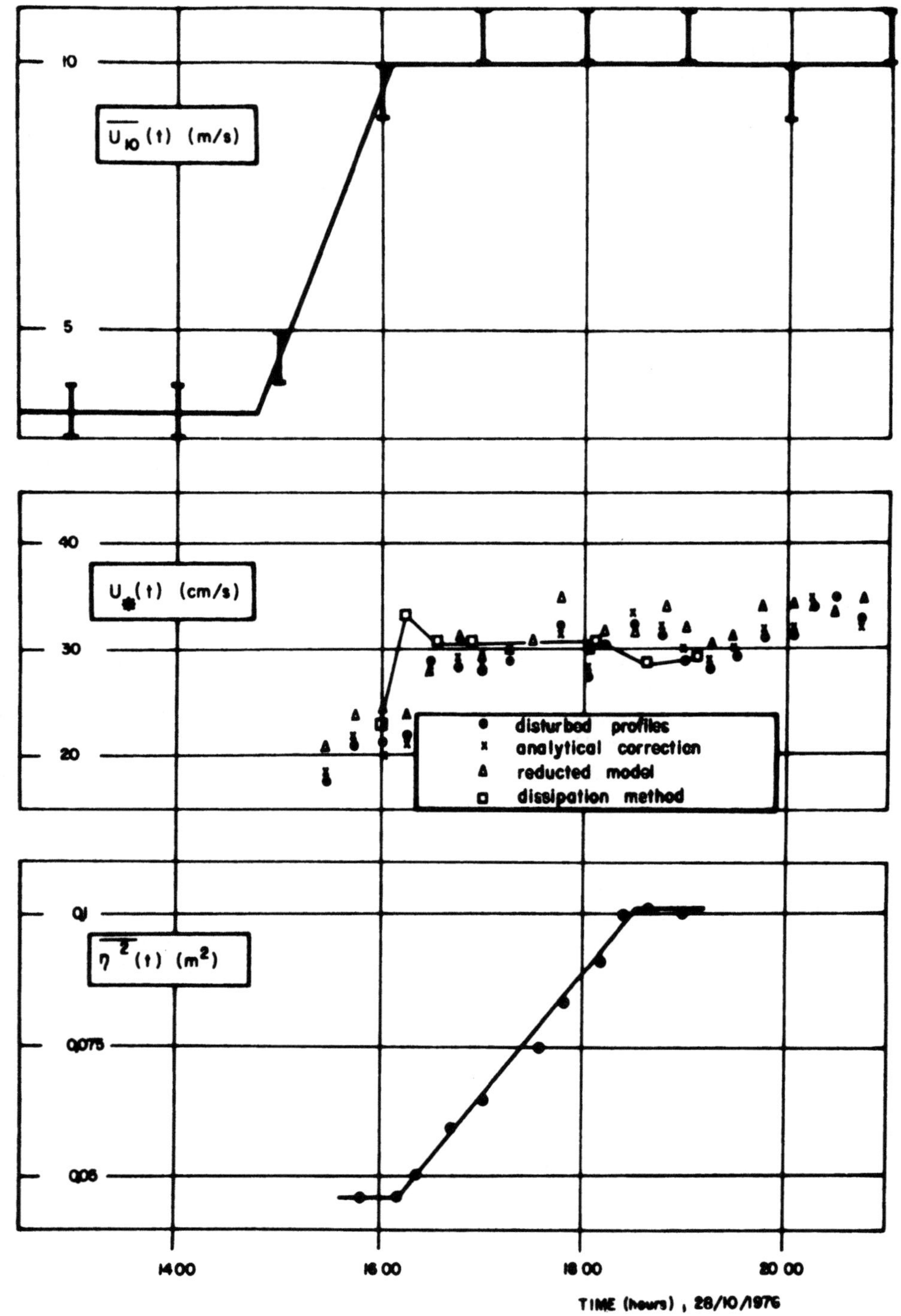

Fig. 4. Evolution with time of U_{10}, u^* and $\overline{\eta^2}$.

signals of two wave gauges spaced about 3 m along the mean wind direction.

OBSERVATIONAL CONDITIONS AND FIRST EXPERIMENTAL DATA

The meteorological conditions at the observational site are parparticulary appropriate to study the early stage of wave and current generation by wind. Indeed, an abrupt increase of the wind velocity, starting from a low value, often occures. Thus, during a period of fifteen days, this situation happened two or three times. However all the instruments of the experimental set up were simultaneously available only once (28 October 1976), during a period of about ten hours.

Figure 4 shows the simultaneous time evolutions of the wind velocity U_{10} at z=10 m (averaged over 10 minutes), the friction velocity u* measured as afore-said, and the wave field total energy $\overline{\eta^2}$, determined by integrating the spectra shown in fig. 5. In approximately two hours, U_{10} increased about from 3 m/s to 10 m/s, and kept this latter value, while u* varied about from 19 cm/s to 31 cm/s. After U_{10} and u* had reached their nearly constant values, a wave growth with a nearly constant rate was clearly observed. During this period, there was no significant change in the dominant wave frequency ($n_o \simeq 0.19$ Hz), the corresponding wave age and celerity were respectively close to 0.75 and to $g/2\pi n_o$ (linear theory).

Figure 6 displays the evolution of the mean drift current velocity at various depths. As for the wave field a significant evolution was observed only after U_{10} and u* reached their nearly constant values. For instance, for a two meter depth, the velocity increased from 15 to 32 cm/s (3.2 % of U_{10}). For larger depths, the velocity increase occured with a time lag depending on depth. At 20 m no significant water entrainment appeared.

Figure 7 summarizes results on time evolution of the dynamical mixed layer depth, z_o, which seems proportional to $t^{1/2}$. The mean current kinetic energy can then be estimated

$$E_c(t) = \frac{1}{2}\rho_w \int_o^{z_o(t)} v^2 \, dz$$

This estimate is also shown on fig. 7. E_c is approximately proportional to time.

DISCUSSION

The observational conditions, as well as the experimental data reported above, suggest that we were concerned with the early stage of wave and current generation by wind. These data allow to estimate the rate of growth of the wave field total energy $E_v = \frac{1}{2}\rho_w g \overline{\eta^2}$ and that of the mean current kinetic energy E_c.

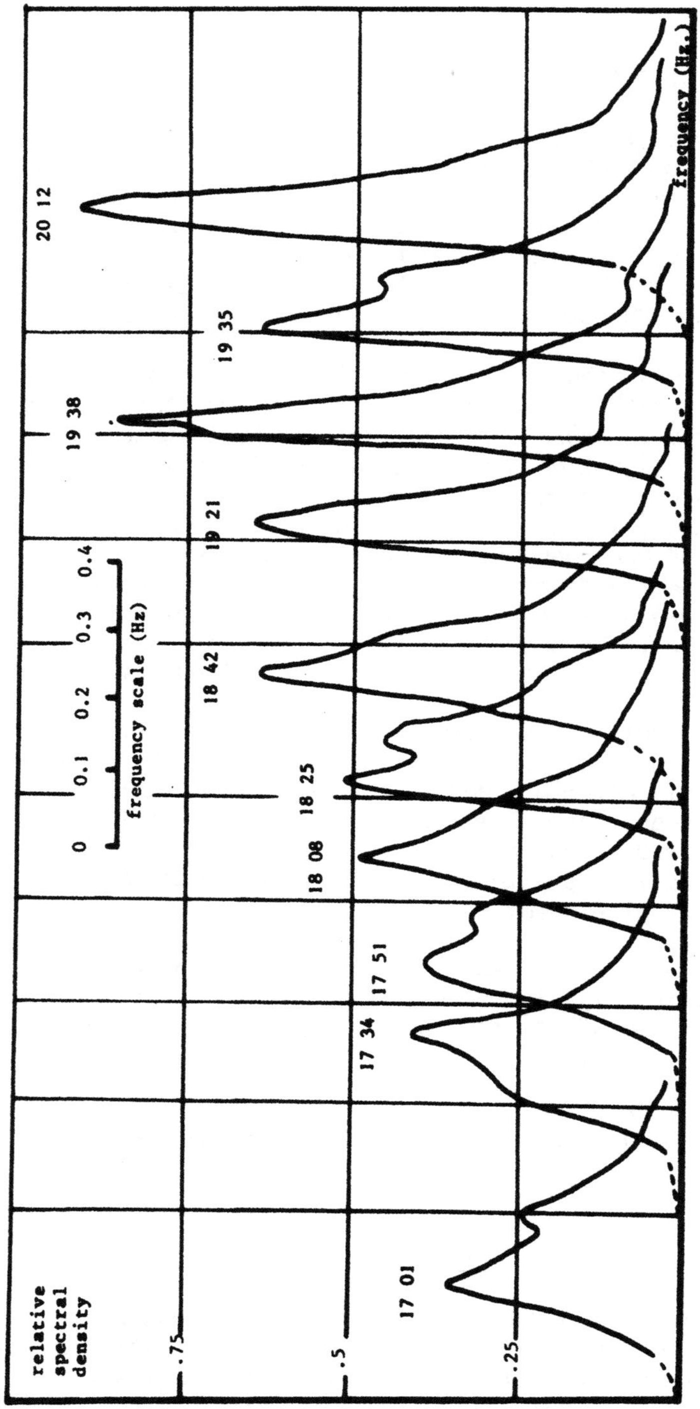

Fig. 5. Evolution with time of wave spectra .

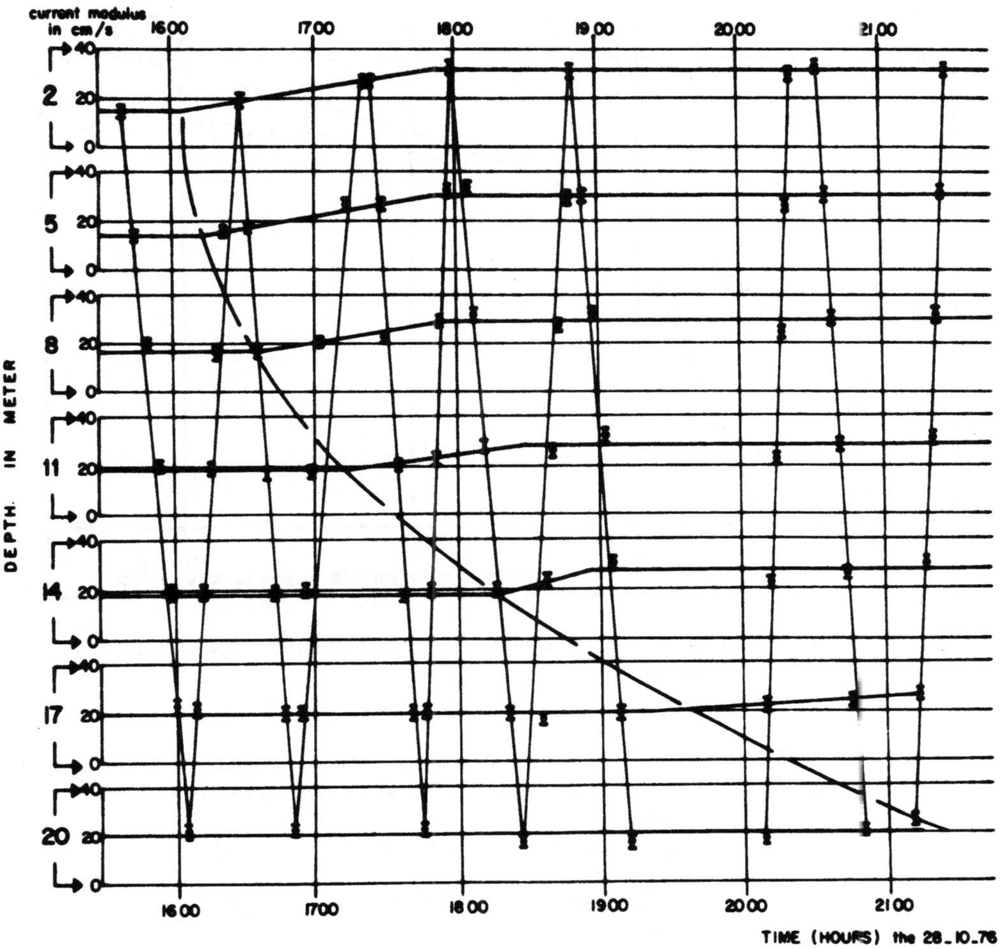

Fig. 6. Variation with time of mean drift current velocity with respect to depth.

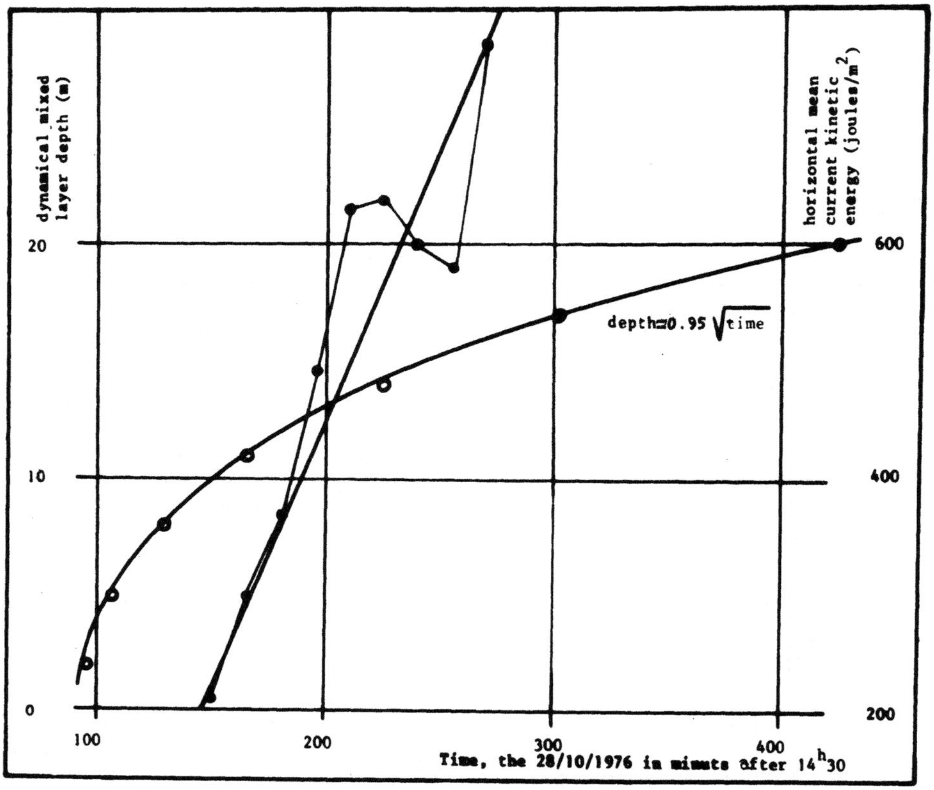

Fig. 7. Mean current kinetic energy and dynamical mixed layer depth versus time.

During the period of 5 p.m. to 7 p.m. where no significant change of the air friction velocity was observed, one obtains

$$\frac{\partial E_v}{\partial t} = 0.019 \ \mathrm{J/m^2/sec.} \qquad (1)$$

$$\frac{\partial E_c}{\partial t} = 0.113 \ \mathrm{J/m^2/sec.} \qquad (2)$$

The most important question is obviously to attempt to relate these rates to the total momentum transfer rate $\rho_a u^{*2}$. At present, no definitive general answer can be given because a variety of physical phenomena such as turbulence-wave-current interactions, dissipation and so on, should be considered simultaneously.

The experimental conditions were such that a first approach seems possible, as far as the wave field is concerned, on the basis of Deardorff (1967) analysis. The equation

$$\frac{\partial \overline{\eta^2}}{\partial t} \simeq \frac{\partial \overline{\tilde{\eta}^2}}{\partial t} = \frac{c_o}{\rho_w g} \overline{\tilde{p}_\eta \frac{\partial \tilde{\eta}}{\partial x}}$$

is taken as governing the energy growth of the wave field. It can be derived starting from the general radiative transfer equation

$$\frac{\partial}{\partial t} F(n,\theta,\underline{x},t) + \underline{c}_g \cdot \nabla F(n,\theta,\underline{x},t) = S(n,\theta,\underline{x},t)$$

and assuming, especially, that : a) the wave field is two dimensional; b) the total energy $\overline{\eta^2}$ is almost entirely associated with the dominant wave (symbol $\sim$)$\left(\overline{\eta^2} = \int_0^\infty F(n)\,dn \simeq \int_{n_0-\Delta n}^{n_0+\Delta n} F(n)\,dn = \overline{\tilde{\eta}^2}\right)$; c) the only one source term is the work done by the pressure force, and the different dissipative mechanisms can be neglected.

One should mention that these assumptions, although not generally valid, are adopted here in first approximation because of the experimental conditions.

The form drag $\overline{\tilde{p}_\eta \frac{\partial \tilde{\eta}}{\partial x}}$,supposed to be responsible of the growth of the energy which remains in the wave field, is written as a fraction χ_v of the total stress $\rho_a u^{*2}$.

$$\overline{\tilde{p}_\eta \frac{\partial \tilde{\eta}}{\partial x}} = \chi_v \rho_a u^{*2}$$

Then

$$\frac{\partial E_v}{\partial t} = \chi_v \frac{c_0}{2} \rho_a u^{*2}$$

A simple calculation, using the present experimental data gives $\chi_v = 5\ \%$. This value compares favourably with the order of magnitude found by Hasselmann and al. (1973) (3 % to 10 %) and that given by Toba (1972) analysis which predicts

$$\chi_v = 0.15 \exp(-1.9\, c_0/U_{10})$$

so that for $c_0/U_{10} = 0.75$, $\chi_v = 4\ \%$.

CONCLUDING REMARKS

This paper reports on work performed by L.O.P. and I.M.S.T. during a joint field observation on the dynamical energy exchange between atmosphere and ocean. It concerns mainly the observational conditions, experimental arrangement and the methods which have been used to obtain the various statistical parameters of interest. First estimates have been made concerning the growths of the wind wave field total energy and the mean current kinetic energy. The time rate of change observed in the wave field energy should approximately correspond to 5 % of the total stress working with the group celerity. At the present stage of analysis, such a relatively simple estimate cannot be given as far as the drift current is concerned. Indeed, study of various phenomena such as wave-turbulence, wave-

wave and wave-current interactions are first necessary. The main informations needed for that beeing available from ours experiments, this study will constitute our near future principal activity.

ACKNOWLEDGMENTS

The work presented received financial support of, essentially, "Centre National pour l'Exploitation des Océans" and "Centre National de la Recherche Scientifique" (A.T.P. "Océanographie Physique 73"). The authors would like to thank here.

REFERENCES

Deardorff J.W. - "Aerodynamic theory of wave growth with constant wave steepness". Journal of the Ocean. Soc. of Japan. Vol. 23 n° 6;p.p. 278-297. (1967).

Dupé M. - "Réponse d'un anémomètre à fils chauds en X monté sur une girouette à faible inertie". L.O.P. Rapport de D.E.A.(1976).

Favre A., Gaviglio J. and Fohr J.P. - "Répartition spectrale de corrélations spatiotemporelles de vitesse en couche limite turbulente". Proceedings of the Eleventh International Congress of Applied Mechanics. Munich(1964). Springer Verlag. Berlin (1966). Ed. Görtler.

K. Hasselmann, T.P. Barnett, E. Bouws, H. Carlson, D.E. Cartwright, K. Enke, J.A. Ewing, H. Gienapp, D.E. Hasselmann, P. Kruseman, A. Meerburg, P. Müller, D.J. Olbers, K. Richter, W. Sell, H. Walden. - "Measurements of Wind-Wave Growth and Swell Decay during the Joint North Sea Wave Project(JONSWAP) Deutsches Hydrographisches Institut. Hamburg (1973).

Ramamonjiarisoa A. - "Contribution à l'étude de la structure statistique et des mécanismes de génération des vagues de vent". Thèse de Doctorat d'Etat. Marseille (1974).

F.J. Resch, M. Bourguel - "Analyse spectrale par méthodes numériques". Rapport I.M.S.T. (1976).

Ruimy D. - "Etude en similitude de la déformation du profil moyen de vent au voisinage de la Bouée Laboratoire BORHA II". Thèse de Docteur-Ingénieur. Paris (1977).

Y. Toba - "Local balance in the air-sea boundary processes". I-On the growth process of wind-waves. Journal of Ocean. Soc. of Japan. Vol. 28, N° 3. (1972).

DISCUSSION

E. J. Plate:

The constancy of shear stress u_*^2 during the growth of the waves is very surprising and against expectation. It should increase, at least for small waves. Would you care to comment on the meaning of this result?

J.P. Giovanangeli:

The constancy of total shear stress is observed for values of u_* and determined by two different methods. However, we should remark that the waves were not very small ($H_{1/3} \simeq 1$ meter).

E.L. Mollo-Christensen:

Since the observations took place during a cold-front passage, you may want to mention whether the waves were locally generated or radiated from elsewhere.

J.P. Giovanangeli:

This problem has been considered since we have not been able to get data on space evolution of wind and wave field. However, we can suppose that we have observed local processes. Indeed, if the waves were generated away, dispersion relation suggests that we should observe an increase of the maximum peak frequency.

FIELD OBSERVATIONS OF AIR FLOW ABOVE THE WAVES

Lutz Haase, Manfred Gruenewald, Dieter E. Hasselmann

Meteorologisches Institut der Universitaet Hamburg
Bundesstrasse 55, 2000 Hamburg 13

BUNDESSTRASSE 55, 2000 HAMBURG 13

The flow of the air above the waves has been of considerable interest for a long time (e.g.Motzfeld 1937), for at least two reasons. The first is in connection with wave generation, and need not be detailed in this type of symposium. The second reason is : if the air flow is modulated by the wavy surface, the mean wind profile may be distorted. It would then not be possible any more to use common profile theory to deduce the stress at the sea surface from wind profile measurements.

Although the relevance of the problem has been recognized for a long time, experimental evidence for the influence of waves on wind profiles measured in open sea conditions has been marginal until recently. This is not surprising since it will be shown in the following that the influence of waves on the windprofile is confined to fairly low heights,which are not easily accessible.

For illustration Fig.1 (Dittmer 1977) shows two average profiles from GATE, for wave heights below 25 cm and between 25 and 75 cm. The full lines are KEYPS profiles, fitted through the upper 4 levels. There is evidently a more pronounced deformation of the profile with increasing wave height. Unfortunately, for the GATE expedition up to now only standard WMO estimated wave heights are available. Therefore a more complete data set has been used in the following.

During JONSWAP 1973 we have measured wind profiles with a surface following buoy. This allows us to measure the wind profile near the surface even in the presence of higher waves. By aid of 3 accelerometers and a gyro at the same buoy we also have direc-

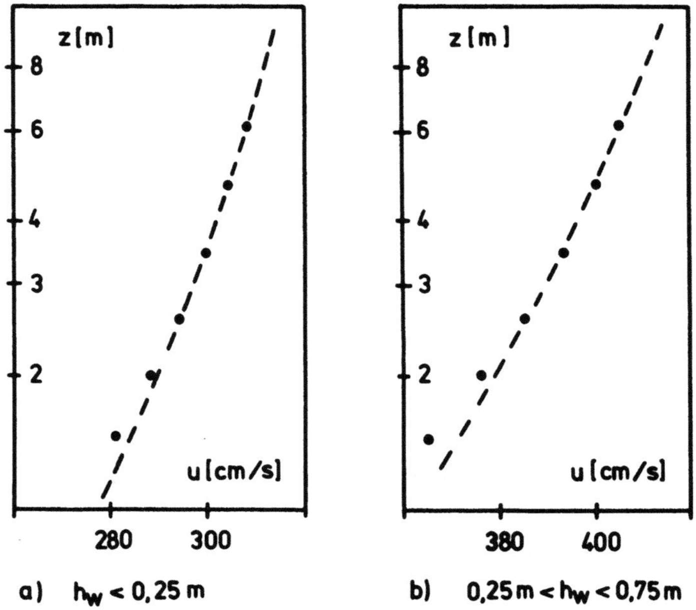

Fig.1: Average wind profiles observed at a surface following buoy during GATE, grouped according to the estimated wave height h_w. The full line is the KEYPS diabatic flux profile relationship fitting the upper 4 levels best through selection of surface roughness, while the curvature of the profile is given from the observed stability (from Dittmer 1977).

tional wave spectra of quality comparable to those obtained from customary pitch and roll buoys. We have used the wave energy integrated over the total spectrum - that includes wind waves and swell - in this paper. For purpose of better illustration this was converted to a representative wave height, say $H_{1/3}$.

We also measured other variables. Specifically, we considered the influence of density stratification on the profiles. It is well known that in non-neutral conditions the profiles are curved. Neglect of these buoyancy influences on the profiles may easily mask any influence of the waves on the wind profile. This seems to be the explanation for the contradicting results by Benilov et al (1976). We therefore have essentially reduced all the profiles to neutral conditions.

It would be nice to determine the strength of wave influence on the profiles in terms of the relevant wave parameters. Unfortu-

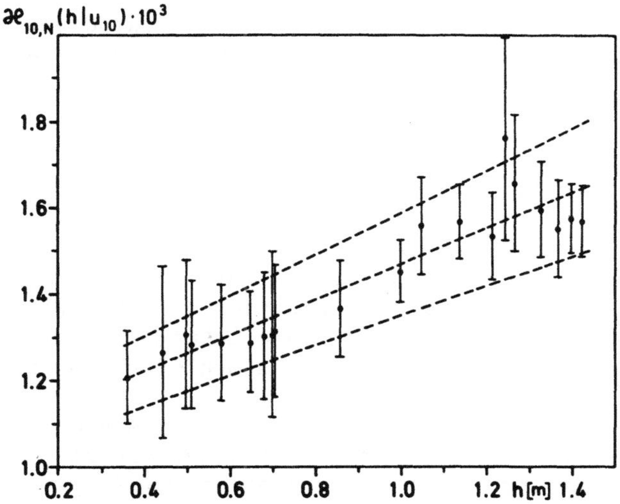

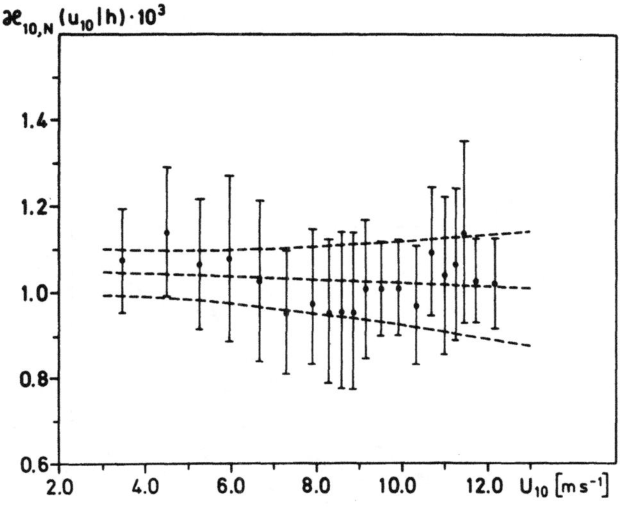

Fig.2: Regression analysis of the dimensionless profile slope κ versus wave energy and wind speed. Upper part gives variation of κ with wave energy independent of wind speed variations, lower part gives variation of κ with mean wind speed independent of wave energy variations (from Gruenewald, Kruegermeyer and Dunckel 1977).

nately, even with our reasonably good profile data it is not really possible to do so. The reason is that with higher waves the number of levels and the height range over which the undisturbed profile could be determined becomes too small and the gradients become fairly uncertain. Therefore, we have used an average slope over all levels including those influenced by waves. As a nondimensional profile slope we used

$$\kappa = \left(\frac{K}{\overline{U}_{10}} \cdot \frac{\partial \overline{U}}{\partial \ln z}\right)^2$$

where K is von Karman's constant. This would be the common definition of the neutral drag coefficient. But since the deformation of of the profile inhibits the determination of the stress from the profile unless further precautions are taken, we cannot say that this is a drag coefficient,it is a measure of the profile slope only.

This immediately poses a problem. Some researchers claim that there is a dependence of the drag coefficient on wind speed. Thus, there are at least two relevant parameters: wind speed and wave energy. Unfortunately, wind and wave energy are somewhat correlated. In order to separate different factors a multiple regression analysis was performed. The outcome is shown in Fig.2 (Gruenewald, Kruegermeyer and Dunckel, 1977). We see that the profile slope depends on wave energy, but not on mean wind speed.

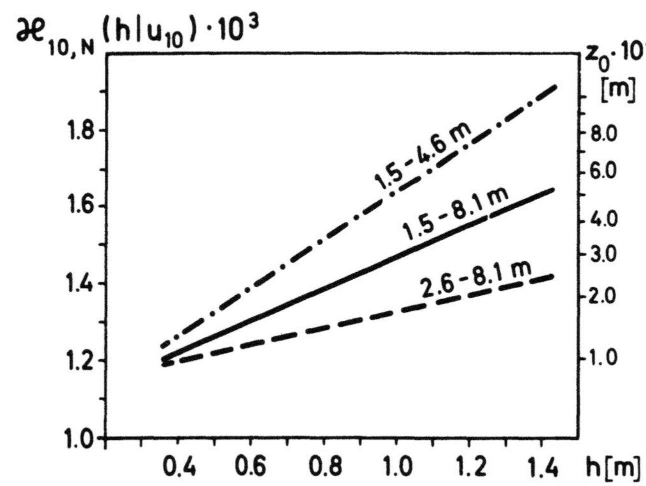

Fig.3: Variation of dimensionless profile slope parameter κ with wave energy, independent of wind speed variations. Different height ranges of the profile are used as indicated at the lines. (Kruegermeyer,Gruenewald and Dunckel,1977).

In order to delineate the height range in which the wave influence is felt, the same type of analysis was done for different height intervals as indicated in Fig.3. This figure shows clearly that the wave-influence on the profile is confined mainly to the lower heights which are comparable to the wave height. Incidentially, this also shows what hazards one runs into if one determines the stress from profiles without taking due precautions. That may also - so to say posthumous - clarify some of the old controversies regarding the dependence of the profile derived drag coefficient on wind speed. Possibly, a good deal of the inferred variation of the drag coefficient with wind speed is due to the profile deformation by waves. If anything, the profile slope does not depend on wind speed, but on wave energy, which of course is correlated to wind speed. From the present data alone it is not possible to conclude if the momentum flux,say at 10m, would depend on wave energy.

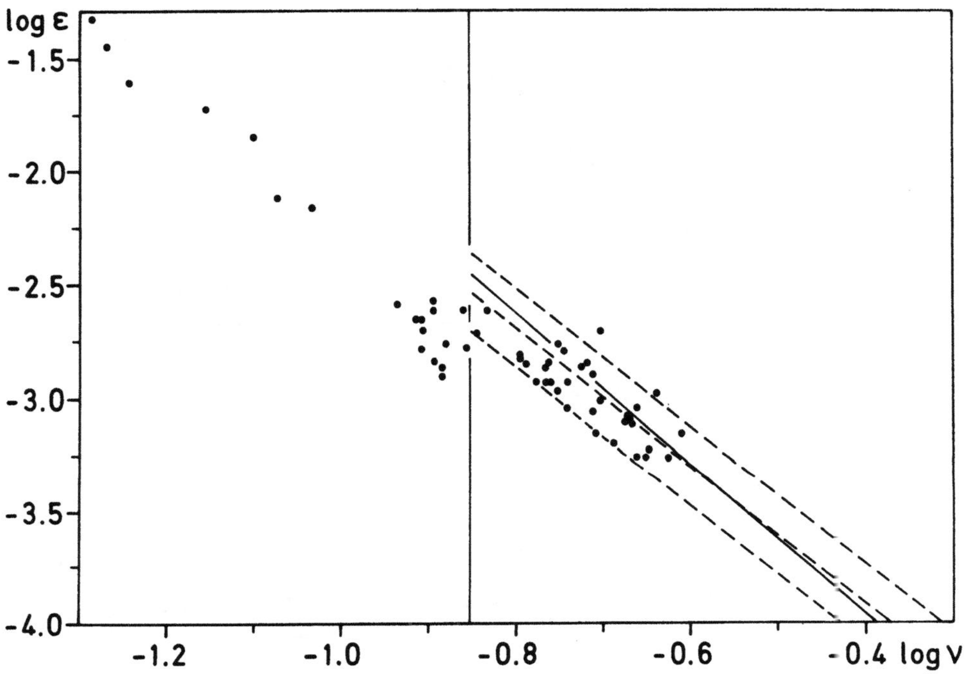

Fig.4: Wave spectra parameter during JONSWAP 1973, measured at the surface following buoy of the Meteorologisches Institut, Universitaet Hamburg. Abscissa is dimensionless peak frequency, ordinate dimensionless wave energy. Dashed lines are best fit plus minus standard deviation from Hasselmann et al (1976), full line their quasi-equilibrium relation.

The preceeding analysis is unsatisfactory in the sense that the dimensionless profile slope is related to wind speed and wave energy, both of which carry dimensions. A dimensionless quantity should depend only on other dimensionless quantities. Therefore, the analysis was repeated with respect to the dimensionless parameters found to be useful in studies of wave generation (Hasselmann, Ross, Mueller, Sell, 1976). These are the dimensionless wave energy ε and the dimensionless peak frequency ν of the wave spectrum.

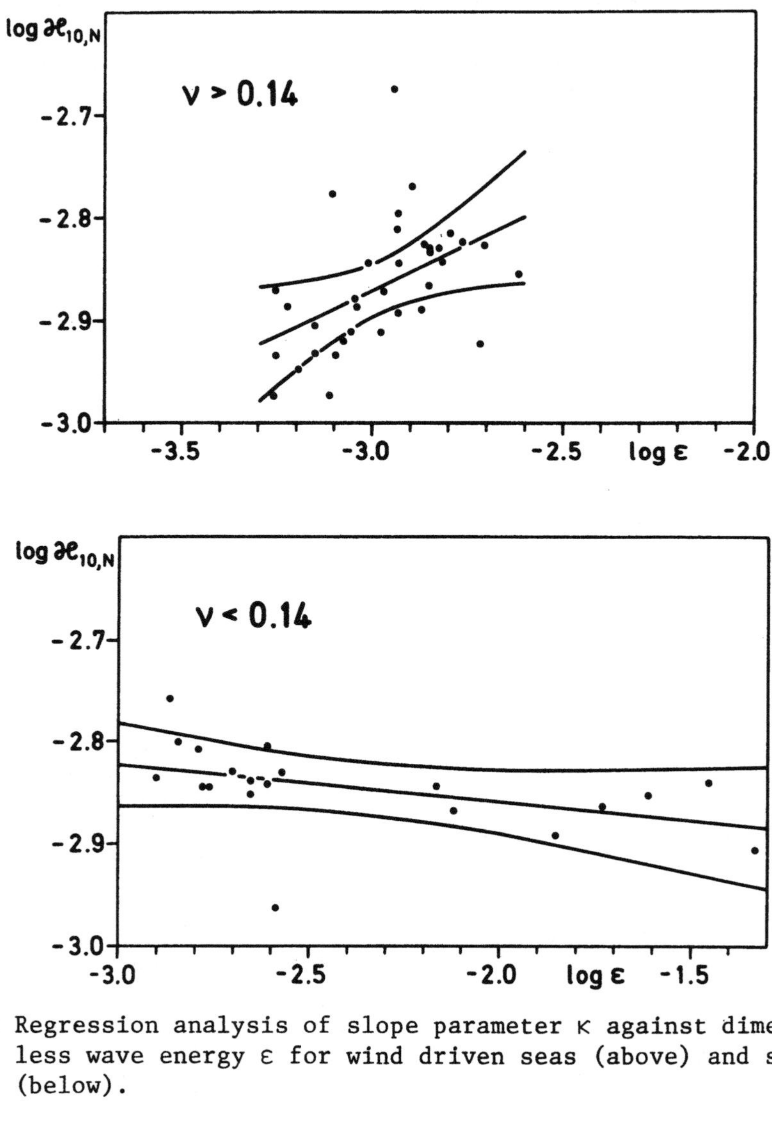

Fig.5: Regression analysis of slope parameter κ against dimensionless wave energy ε for wind driven seas (above) and swell (below).

A plot of the wave spectra used in this study in terms of ε and ν is given in Fig.4. The full line is the average relationship given by Hasselmann et al (1976) for wave generation cases. ν=0.14 devides wind driven seas and swell. The regression analysis was then done separately for wind driven seas and swell, Fig.5. It is remarkable that the dimensionless parameter ε yields a much poorer correlation with the dimensionless profile slope than the simple wave energy, Fig.2. It is argued that this indicates that the profile deformation is a kinematic rather than a dynamic effect.

We have compared the observations with theory, and made some model calculations. The model assumes a logarithmic profile above the undisturbed surface. For the flow above the wavy suface it is assumed that the flow is bent to follow the surface undulations similar to Brooke Benjamin's (1959) analysis. This is a first approximation, which has been found to give a reasonable description of the wind flow above the waves from observations of the Atlantic Trade Wind Experiment (Wucknitz 1976). Let the windfield be given by

$$\hat{U}(z,x,t) = U(z) + u(z,x,t) + u_t(z,x,t)$$

where the turbulent fluctuating velocity u_t need not be treated in detail and is ommitted in the following, since we wish to compare the calculation with mean profiles. The deformation of the windfield is described by introducing a curvilinear coordinate system

$$\eta = z - \sum_{k>0} A(k)\cos(kx - \sigma t)\, e^{-k\eta}$$

$$\xi = x - \sum_{k>0} A(k)\sin(kx - \sigma t)\, e^{-k\eta}$$

This is somewhat different from the Brooke Benjamin formulation, but has the advantage to transform the sea surface $z = \zeta$ exactly onto $\eta = 0$. $A(k)$ is a wave amplitude, k wave number, σ frequency.

It is assumed that a logarithmic profile exists in this coordinate system:

$$U(\eta) = \frac{u_*}{K} \ln \frac{\eta}{\eta_0}$$

where u_* is friction velocity, η_0 roughness length. For a surface following buoy an instrument level with height d above the mean water line of the buoy has a height

$$z(t) = \zeta(t) + d = d + \eta(d,t)$$

defining the time dependence of $\eta(d,t)$ For a wave field $A(k)$ we use the stream function

$$\Psi(\eta) = \int_o^\eta U(\eta)\,d\eta - \sum_{k>o} \int_o^\eta c(k)A(k)\cos(kx-\sigma t)\,k e^{-k\eta} d\eta$$

corresponding to the simplest approximation $\Psi(\eta) = \int(U-c)(\eta)d\eta$ to the solution of the Rayleigh equation if only one wave is present. The horizontal velocities derived from this stream function have to be averaged for fixed instrument levels d over at least the longest wave length contained in the wave field. Keeping up to second order terms and changing to a continuous wave spectrum F(k) one obtains

$$\overline{\hat{U}(d)} = \overline{U(d)} + \int[U(\eta) - c(k)]k^2F(k)e^{-k\eta}(2e^{-k\eta}-1)dk +$$
$$+2\frac{\partial U}{\partial \eta}\int kF(k)e^{-k\eta}(1-e^{-k\eta})dk + \frac{1}{2}\frac{\partial^2 U}{\partial \eta^2}\int F(k)(1-e^{-k\eta})^2 dk$$

The results of the model calculation show a deviation of the profile from the logarithmic profile, which depends on wave height, Fig.6. The spectrum used for these calculations is of the equilibrium range type

$$F(\omega) = \alpha g^2 \omega^{-5}, \quad \alpha = 10^{-2}$$

(Phillips 1969,pp 111). This is not a very realistic spectrum, but since this first approximation reproduces very nearly the observed profile deformation, and as it seems to be nearly impossible to experimentally determine the deviation from the logarithmic profile with more accuracy and detail, it was decided that a more detailed calculation would not give more insight into the dominant processes.

In order to avoid misinterpretation of the calculations we emphasize that we have assumed that $\Psi(\eta)$,as given above,completely describes the wave induced flow. We definitely do not claim that this is an adequate description of the flow to second order, but the point is that the qualitative departure of the measured mean horizontal velocity from a logarithmic profile can be explained in this crude model.

There are essentially 3 effects which produce a deviation from the undisturbed profile:

(1) The average horizontal wind velocity along a streamline is decreased compared to the undisturbed flow at the mean height, due to the nonlinear variation of the streamline with x and z.

(2) The buoy follows the streamline at the surface. That is, instruments at higher levels at a fixed distance to the surface do not follow a streamline. Due to nonlinearity there is a net effect of the variation around the streamline even in a linear velocity profile.

(3) Since the profile is assumed to be logarithmic, even a linear

variation around the streamline would produce a net reduction of the velocity due to the increase of the gradient towards the surface. For a surface following buoy, test calculations show this to be the main effect.

We may note that the dip motions of the buoy also imply a motion through the profile, which gives a reduction of the wind speed. This effect is not included in the model calculations, since the measured dip motions are fairly small and have been neglected in

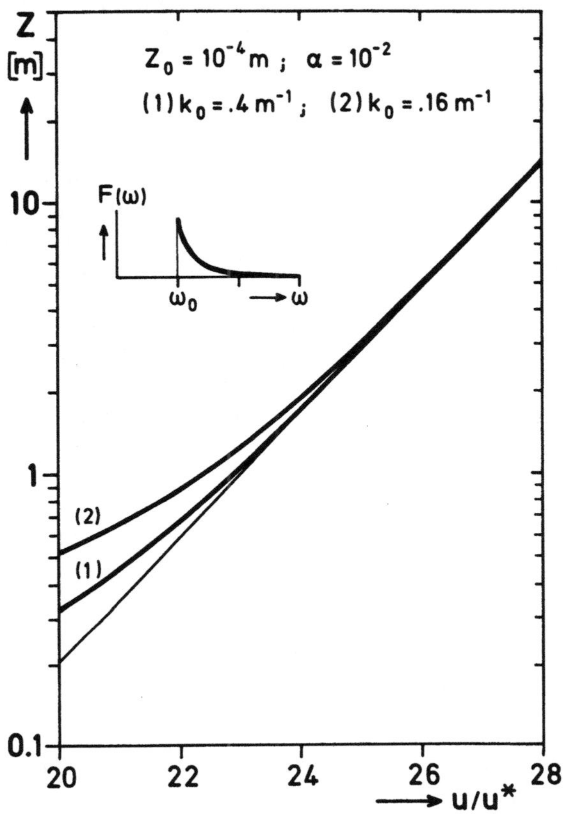

Fig.6: Results of model calculation. For better comparision, the calculations are given for typical parameters as encountered at sea. The assumed wave spectrum is indicated in the inlet. The two thick lines correspond to calculations with different wave spectra ($h_W \sim 0.25$m and 0.8m), with parameters as indicated. Thin line is the logarithmic profile.

the calculation. Buoy dip motions would produce an effect similar to the effect (3) mentioned above. A rough estimate shows that for our buoy the influence of the buoy dip motions would be smaller by a factor of 2 to 5.

CONCLUSIONS

The profile deformation observed seems to be an influence of the waves on the wind field. The influence is mainly of kinematic origin. That is to say, the profile deformation is to be expected also if the wind field and the wave field are in phase and no momentum is transferred to the waves. The observed deformation probably has little to do with the dynamic response of the wind to wave induced stresses. To extract any information on wave generation one would have to go into considerable more effort in data analysis.

Most of the older profile measurements, which have been used to determine drag coefficients, are obsolete, since it will be nearly impossible to correct older observations for the profile deformation. We have seen in Fig.3 that the effect on the slope and hence the inferred drag coefficients is quite considerable.

It is also noted that the considerable change of the mean gradient will have implications on the energy balance of turbulence near to the surface, since the production term contains the gradient of mean wind speed. Obviously, when discussing balance equations near to the surface, some care must be taken in selecting the 'best' coordinate system.

REFERENCES

Benilov, A.Yu., A.L.Gumbatov, M.M.Zaslavskiy and S.A.Kitaygorodskiy, 1976: Interpretation of measurements of average wind velocity in the marine atmospheric surface layer. Izv., Atmos.Oceanic Phys. 12, No.10, 1011-1019, engl.ed.

Brooke Benjamin,T.,1959: Shearing flow over a wavy boundary. J.Fluid Mech.6, 161-205

Dittmer,K., 1977: The hydrodynamic roughness of the sea surface at low wind speeds. "Meteor" Forschungs-Erg. B 12, 10-15

Dunckel,M., L.Hasse, L.Kruegermeyer,D.Schriever, J.Wucknitz, 1974: Turbulent fluxes of momentum, heat and water vapor in the atmospheric surface layer at sea during ATEX. Bound.Lay.Met.6, 81-106.

Gent, P.R., and P.A.Taylor, 1976: A numerical model of the air flow above water waves. J.Fluid Mech.77, part 1, 105-128

Hasselmann,K., D.B.Ross,P.Mueller, W.Sell, 1976: A parametric wave prediction model. J.Phys.Oceano. 6, No.2, 200-228

Kruegermeyer,L.,M.Gruenewald,M.Dunckel, 1977: The influence of the significant wave height on the wind profile. Submitted to: Bound.Lay.Met.

Motzfeld, H., 1937: Die turbulente Strömung an welligen Wänden. Z.Angew.Math.Mech.17, 193-212

Phillips, O.M.,1966: The dynamics of the upper ocean. Cambridge Univ.Press, 261 pp

Wucknitz, J.,1975: Determination of turbulent fluxes of momentum and sensible heat from fluctuation measurements and the structure of wind field above waves at the tropical Atlantic during ATEX. "Meteor" Forschungs-Erg. B 11, 25-50.

DISCUSSION

H. Mitsuyasu:

(1) In many cases wind profile deviate to the higher wind side from the extrapolated logarithmic distribution. But the data you have shown shows the shift of reverse direction (low wind-side shift). Have you obtained data which show a high wind-side shift of the profile?

(2) What was the wave period when you measured the wind profile? Wind profile may depend on whether or not the wind is separated, and wind separation may be controlled by the u/c (wind speed/wave speed).

L. Hasse:

(1) It is agreed that a single short-time profile may show deviations to either side, since we are dealing with a turbulent medium anyway. On the average, the wave induced deviation of the profile, measured by a surface-following buoy, seems to be always a decrease in wind speed at the lowest heights in agreement with the theory. In the literature, deviations to both sides have been reported but with non-surface following instrumentation. Also the influence of stability of the density stratification on the profiles was obviously neglected, i.e. the deviation of the

diabatic profile from the neutral logarithmic profile due to buoyancy was erroneously ascribed to wave influences.

(2) All measurements were with windspeed below 15 m/s. Since the paper is based on 95 wave spectra from three weeks it is not possible to give an average wave period. Different conditions of wave generation and swell were encountered. An impression of the conditions is given in the figures.

O. Phillips:

Are the wave-induced distortions sufficiently well accounted for by simple theory to make this useful as a basis for correction in stress measurements by the profile method?

L. Hasse:

To obtain an estimate of the reliability of the correction, we should repeat the calculations with the actual measured wave spectra. This has not been done yet. In practice, I would rather advocate the use of profiles at heights where the profiles are not affected by waves.

MODELS OF TURBULENT AIRFLOW ABOVE IDEALISED WATER WAVES

P. A. Taylor, K. J. Richards and R. A. Nunes

Department of Oceanography
University of Southampton, U.K.

INTRODUCTION

In a series of papers Gent and Taylor (1976, 1977) and Gent (1977) have developed and applied a numerical model of deep turbulent boundary-layer flow above idealised, monochromatic, two dimensional water waves. The present paper will discuss some of their results and describe some recent developments to the model to study flow over differing shapes of wave. The Gent and Taylor model is essentially a numerical solution of the averaged continuity and Navier Stokes equations with closure hypotheses to relate second order correlations (the Reynolds stresses) to mean gradients via an isotropic eddy viscosity, K, which depends upon the local value of mean turbulent kinetic energy, E, and a mixing-length $\ell(z,z_0)$ which is prescribed as a function of position. It is a relatively simple closure scheme with few parameters. More complex higher order closure schemes or slightly different simple assumptions may be found to give better results but they preferred to use rather basic, simple closure hypotheses for their numerical experiments. Slightly different hypotheses were used by Townsend (1972) in his model of airflow over water waves which was linear in the waveslope (ak). Moderately good agreement between Gent and Taylor's non-linear model and Townsend's results are reported for small values of ak (a is the amplitude and k the horizontal wavenumber).

Gent and Taylor (1976), hereafter referred to as GT1, and Gent (1977) concentrate their attention on relatively small amplitude waves, ak typically 0.157 ($\pi/20$). In addition to the effects of surface elevation and the surface velocities associated with the wave motion on the airflow they consider some cases where the local surface roughness of the water (z_0) is varied with position on the waves to try to model the effect of the increased slope of small capillary and short gravity waves near the crests of the larger, dominant waves. Results for these cases show generally good agreement (Figs. 11 and 12 of GT1) with Elliott's (1972) observations of the phase of the wave induced pressure perturbation and ζ, the fractional rate of energy transfer from the wind to the waves due to working of the pressure forces. Although there are some considerable differences in the pressure phase results for $1 < U_5/c < 3$ (where U_5 is the mean wind at 5 m and c is the phase speed of the waves) there is some scatter in the observations in this range and considerable disagreement between Elliott's results and those of Dobson (1970) and Snyder (1974). We note with particular interest the results of the Bight of Abaco experiment presented by Dobson, Elliott, Long and Snyder at this conference. We should, however, remark that detailed comparisons between Gent and Taylor's non-linear model results and Fourier analysed data from the sea are a little suspect if we argue the case for non-linearity very strongly.

In Gent and Taylor (1977), hereafter referred to as GT2, attention is concentrated on cases of larger amplitude waves, slopes up to about 0.6, and a discussion of flow separation. We will discuss some of their results in section 3. An important point to note from results presented in that paper is the considerable difference between results for situations typical of the sea and of those that may be modelled in wind tunnels.

LINEAR OR NON-LINEAR MODELS?

For small amplitude water waves with low values of wave slope, ak, there have been a series of models developed for the flow over water waves which linearise in terms of ak and assume that the flow perturbations are sinusoidal. Most notable are the essentially inviscid models of Miles (1957) and Brooke Benjamin (1959) and the turbulent flow models of Townsend (1972) and Long (1971). This is clearly a natural approach and matches the usual Fourier analysis of the wave field. There is, however, some question as to the limits on ak to which the theory should be applied. As an aside, it can perhaps be reported that Gent and Taylor's non-linear model evolved from a study of water flow over sand waves and airflow over hills for which the linear approach was felt to be inappropriate. Had they set out with only the wind wave problem in mind it is most likely they would have chosen a linear approach similar to that of Townsend.

From the results given in GT1 for cases where z_o is constant along the wave form, $z = z_o(x)$, we observe that there is certainly some divergence from linear behaviour for moderate wave slopes.

As a rough estimate they find results for ζ, the surface pressure and stress perturbations, and other quantities that diverge from linear behaviour (inferred from numerical solutions at very small wave slope) by about 25% at ak = 0.157. For ak > 0.3 the results are often totally different to those for small amplitude. Features such as variations of roughness length with position on the wave are, in our view, essentially non-linear features, as small amplitude modulations in z_o, say mean ± 20% will not produce any significant effect. As a general estimate we expect solutions obtained by a linear theory to hold within about 10% for ak < 0.05 but to have possible errors of 20% at ak = 0.1. In most cases linear theory tends to overestimate the flow perturbations due to the waves at these amplitudes. For many purposes linear theories may suffice but are clearly inadequate in considerations of the airflow above steep (ak > 0.2) and possibly breaking waves. There are, however, severe practical limitations to what can be achieved with the type of non-linear models developed by Gent and Taylor. Their work considers only the flow above an infinite train of monochromatic two dimensional waves. By using a frame of reference moving with the phase speed of these waves and ignoring changes in the wave field due to the airflow they are able to reduce the problem to a steady flow situation. Even in this highly idealised case numerical solutions take about 5 minutes of CDC7600 CPU time with a coarse 11 x 11 finite difference grid and up to 40 minutes for a 21 x 15 grid in order to adequately converge to a steady state. The computational effort required for three dimensional studies or for two dimensional cases with a spectrum of waves, when the ability to translate at the phase speed would be lost, is truly enormous. Improved numerical techniques and some relaxation of the convergence criteria would certainly help but it would still be a very major computational effort.

The treatment of the lower boundary condition on velocity requires some caution in the linear models. Townsend (1972) correctly matches the surface air and water velocities by assuming a profile near the surface which, for the simple case of waves normal to the wind, is of the form

$$U_a + u = \frac{\tau^{\frac{1}{2}}_{xz}}{\kappa} \ln\left(\frac{z - z_b}{z_o}\right) + ckz_b - c \tag{1}$$

where $U_a = \frac{u_o}{\kappa} \ln\left(\frac{z}{z_o} - c\right)$ is the velocity distribution averaged over a wavelength, u_o is an average friction velocity, u is the wave induced velocity perturbation and τ_{xz} is the local Reynolds stress, $-\overline{u'w'}$. With this form of profile the surface is to be assumed at $z = z_b + z_o$. In order to apply equation (1) at a fixed z, near the surface it is linearised to give

$$u = \frac{1}{2\kappa} \frac{(\tau_{xz})_o}{u_o} \ln \frac{z}{z_o} + ckz_b - u_o z_b/\kappa z \tag{2}$$

where $(\tau_{xz})_o$ is the stress fluctuation at the surface. To obtain this however the expression

$\ln \frac{z-z_b}{z}$ has been approximated by $-\frac{z_b}{z}$ and $\ln \frac{z-z_b}{z_o}$ has been approximated by $\ln \frac{z}{z_o}$ which would appear to require that the boundary condition be applied at a value $z >> z_b$. Townsend applies the lower boundary condition at $kz = 0.0183$ and the requirement $z >> z_b$ would presumably require $ak << 0.0183$, a rather lower value for wave slope than one might hope for. We should however comment that although the method is formally only applicable for these very small wave slopes we would expect the results to hold for a wider range, say $ak < 0.05$.

This problem with the lower boundary condition is a manifestation of the difficulty noted by Miles (1957) and Brooke Benjamin (1959) arising because the mean profile U(z) has a large gradient at the surface. We can avoid this difficulty in much the same way as was proposed by Brooke Benjamin by incorporating part of the effect of the surface undulations into the basic profile U and setting

$$U_a = \frac{u_o}{\kappa} \ln \frac{z^*+z_o}{z_o} - c \tag{3}$$

where $z^* = z - z_b e^{-kz^*}$, and the surface is $z^* = 0$. With a transformation from (x,z) to (x,z*) coordinates (see Taylor (1977) for details) we have obtained solutions to the linearised flow problem in cases with $c = 0$. So far these studies have been primarily intended for use in an investigation of flow above gravel waves and bedform stability problems (Richards (1978)) but we hope to extend them to the wind wave case shortly. In Fig. 1 we illustrate the comparison between the two dimensional finite difference and the linear model results for a pure sinusoidal sandwave with $L/z_o = 10{,}000$ by plotting profiles of shear stress and turbulent energy above the crest. Agreement between the two methods is very good for $ak = 0.003$ while the results for $ak = 0.157$ indicate considerable divergence from linear behaviour at this amplitude.

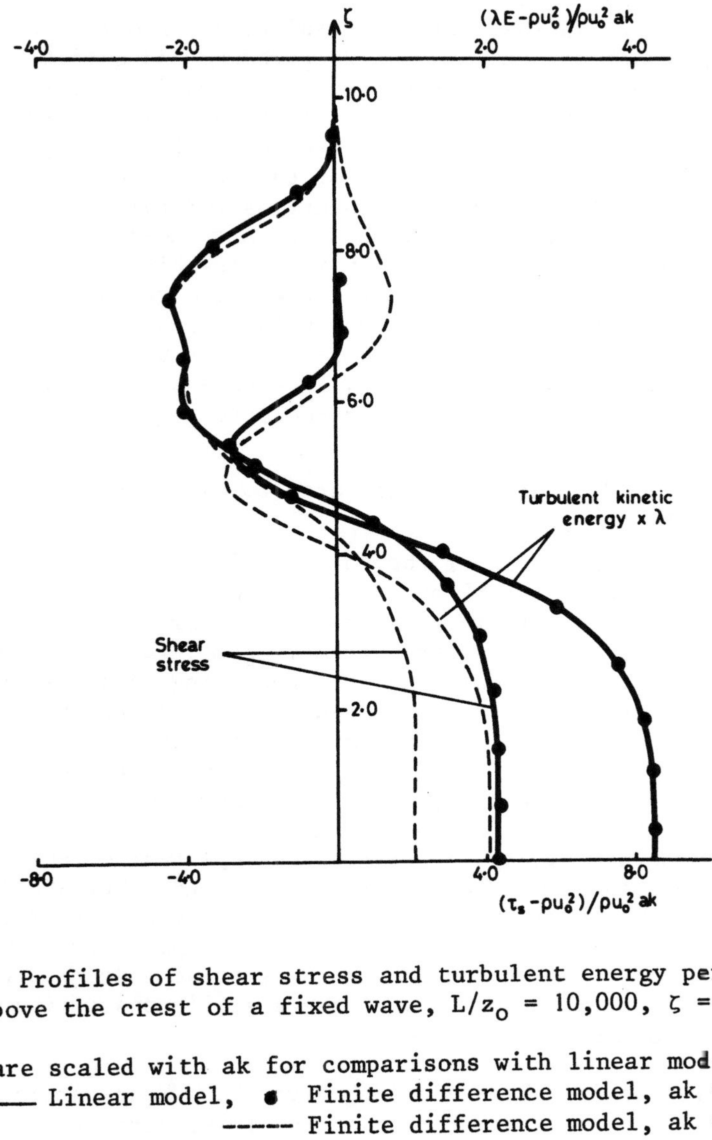

Fig. 1. Profiles of shear stress and turbulent energy perturbations above the crest of a fixed wave, $L/z_o = 10{,}000$, $\zeta = \ln \frac{z^*+z_o}{z_o}$

Values are scaled with ak for comparisons with linear model.
_____ Linear model, ● Finite difference model, ak = 0.003
----- Finite difference model, ak = 0.157

WAVES OF FINITE STEEPNESS AND OTHER FACTORS

Recent papers by Banner and Melville (1976), Longuet-Higgins (1977) and others have considered aspects of the airflow over relatively steep water waves. In GT2 results were presented for flow over waves with $R = -\ln(kz_o) = 8$ and $c = 8u_o$ for values of ak up to 0.628. In these studies, and those of GT1, a conformal mapping approach is used and the 2D waveform and surface velocities correspond to the Gerstner wave ($z_b = -a \cos k\xi$, $x = \xi + a \sin k\xi$) - see for example Kinsman (1965, p. 241). These, trochoidal, waves have sharper crests and flatter troughs than a pure sinusoid, with the sharpness of the crests increasing for increasing a. In the numerical solutions for this case there is a distinct change in the streamline pattern (viewed in a frame of reference moving with the waves) with increasking ak - see Fig. 8 of GT2. For ak = 0.157 the closed streamline region or 'cats eye' is relatively shallow and confined in the trough of the wave, while for ak = 0.314 something rather like 'separation' occurs in the flow at the crest of the wave. The pressure contribution to the horizontal stress on the lower boundary,

$$\int_o^L p \frac{dz_o}{dx} dx, \text{ which we will refer to as the form drag,}$$

increases from about 19% of the total (skin friction plus form drag) at ak = 0.157 to 56% at ak = 0.314 while the corresponding increase in drag coefficient expressed in terms of the mean wind at a height L above the water surface, $C_D(L)$, is from 1.90×10^{-3} to 2.85×10^{-3}. Our model then, is predicting quite significant increases in drag and changes in behaviour with increase in ak but not anywhere as large as the 40 to 50 fold increases in drag observed by Banner and Melville in their flume experiments on breaking and non breaking waves. This disagreement, plus a certain amount of prodding by various workers in this field, led us to ask whether simply increasing the amplitude of our Gerstner waves was enough or whether other factors, and other shapes of waves should be modelled. The range of possible factors and parameters, tabulated below, do not seem too great but coupled with the parameters R, ak and c/u_o already present do give us a vast parameter space to explore - and we are still considering monochromatic two dimensional waves.

The ranges for ak, R and c/u_* given have been partially studied for the cases underlined in the lower part of the table in GT1, GT2 and Gent (1977).

We are presently engaged in developing a model to incorporate the ranges of possibilities for shape, z_o variations and

Table 1. 'Parameters' for simple monochromatic, two dimensional wave trains

Parameter	Range of possible cases
ak	0 → 0.3 (for realistic application), perhaps only 0 → 0.1 for longer waves
R	5 to 12 (wavelengths 5 cm to 50 m with z_o = 0.005 cm)
c/u_o	3 to 40 and -3 to -40 for waves against the wind
Shape	Pure sinusoidal, Gerstner, Other steep symmetric waves, Asymmetric waves
z_o variations	z_o constant, slight modulation, large modulation to represent wave breaking, viscous sublayer simulation
Surface water velocities	Gerstner, Banner and Phillips (1974) windrift, other

surface water velocities indicated. A brief description and some preliminary results are given in the next section.

A brief investigation of the effects of surface drift velocities was included in the results given in GT2. They find quite significant changes in the flow pattern (Fig. 17 of their paper) and surface pressure distributions (Fig. 16 of their paper) when surface drift velocities are included although the overall form drag and the surface shear stress predictions are only changed slightly. They conclude that surface wind drift may well be an important factor in determining some features of the airflow above the waves.

A 'NON ORTHOGONAL' MODEL

Although in principle the conformal mapping approach adopted in Gent and Taylor's work could be used for arbitrary wave shape the difficulties of obtaining the conformal mappings in this case led us to try a different approach in which a non orthogonal mapping is used to transform the region above the waves to a rectangle prior to attempting numerical solution. The mapping is simply

$$x^* = x \qquad z^* = z - z_b\, e^{-kz^*} \tag{4}$$

Full details of the transformed equations are given, for the case of flow above a rigid lower boundary, in Taylor (1977) which also gives some comparisons between results obtained with this and the conformal mapping methods. For the case of flow above water waves the same equations hold with modified lower boundary conditions. Accurate numerical solution of the equations in this case has caused considerable problems but we have recently tracked down the sources of discretisation error in the finite difference representation and can present some preliminary results.

Our first test was to run the case with uniform z_o, R = 8 and $c = 8u_o$ with ak = 0.157 used as a standard case in GT1 and GT2. Some results of this comparison are given in Fig. 2 showing comparisons between surface tangential shear stress and pressure distributions. Unfortunately the 21 x 15 non-orthogonal grid results have not completely converged (a longer computer run is required) in the results presented here but they are adequate to indicate good agreement between the two methods. There is a tendency for the results with the 11 x 11 grid to underestimate the surface shear stress near the crest but the results are really quite satisfactory with such a coarse grid.

Some preliminary results on the effects of changes in wave shape have been obtained using an 11 x 11 grid with R = 8, c = $8u_o$, and ak = 0.157 and are illustrated in Fig. 3. The wave forms used are

1) the Gerstner wave $z_b = - a \text{ Cos } k\xi \; ; \; x = \xi + a \text{ Sin } k\xi$

2) a Sin^4 wave $z_b = a(2 \text{ Sin}^4(kx/2) - 1)$

3) & 4) $z_b = -a \text{ Cos } (kx + c_3(1 - \text{Cos}(kx + c_2(1 - \text{Cos}(kx + c_1(1 - \text{Cos } (kx))))))$

with 3) $c_1 = -\pi/4, \; c_2 = \pi/4, \; c_3 = -\pi/10,$

and 4) $c_1 = -\pi/12, \; c_2 = -\pi/8, \; c_3 = -\pi/4$

In all cases the surface water velocities have been taken as $U = -c(1 - kz_b)$ and $W = U \, dz_b/dx$ relative to the moving frame of reference, where U and W are the horizontal and vertical velocity components.

This is consistent with the Gerstner wave solution but somewhat arbitrary in the other cases.

For the two symmetric cases and for case 3) the effects of changes in shape are rather less than we would have anticipated. Even in the strongly asymmetric case 4) there is no major change in behaviour although the surface pressure and stress distribution clearly reflect the asymmetry of the wave. We also find no

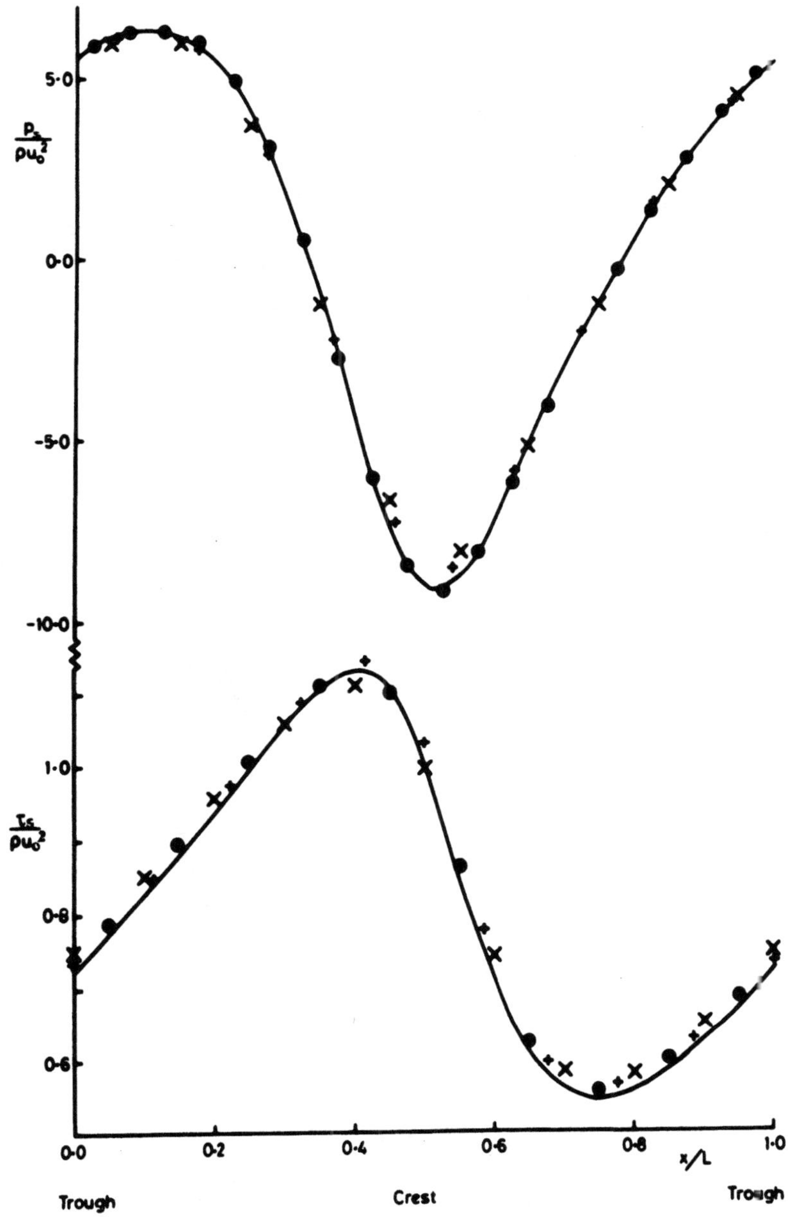

Fig. 2. A Comparison of surface shear stress and pressure results with non-orthogonal mapping and conformal mapping methods, R = 8, C = $8u_o$, ak = 0.157
——— Conformal mapping, 21 x 15 grid from GT2
\+ Conformal mapping, 11 x 11 grid from GT1
● Non-orthogonal mapping, 21 x 15 grid (not fully converged)
x Non-orthogonal mapping, 11 x 11 grid

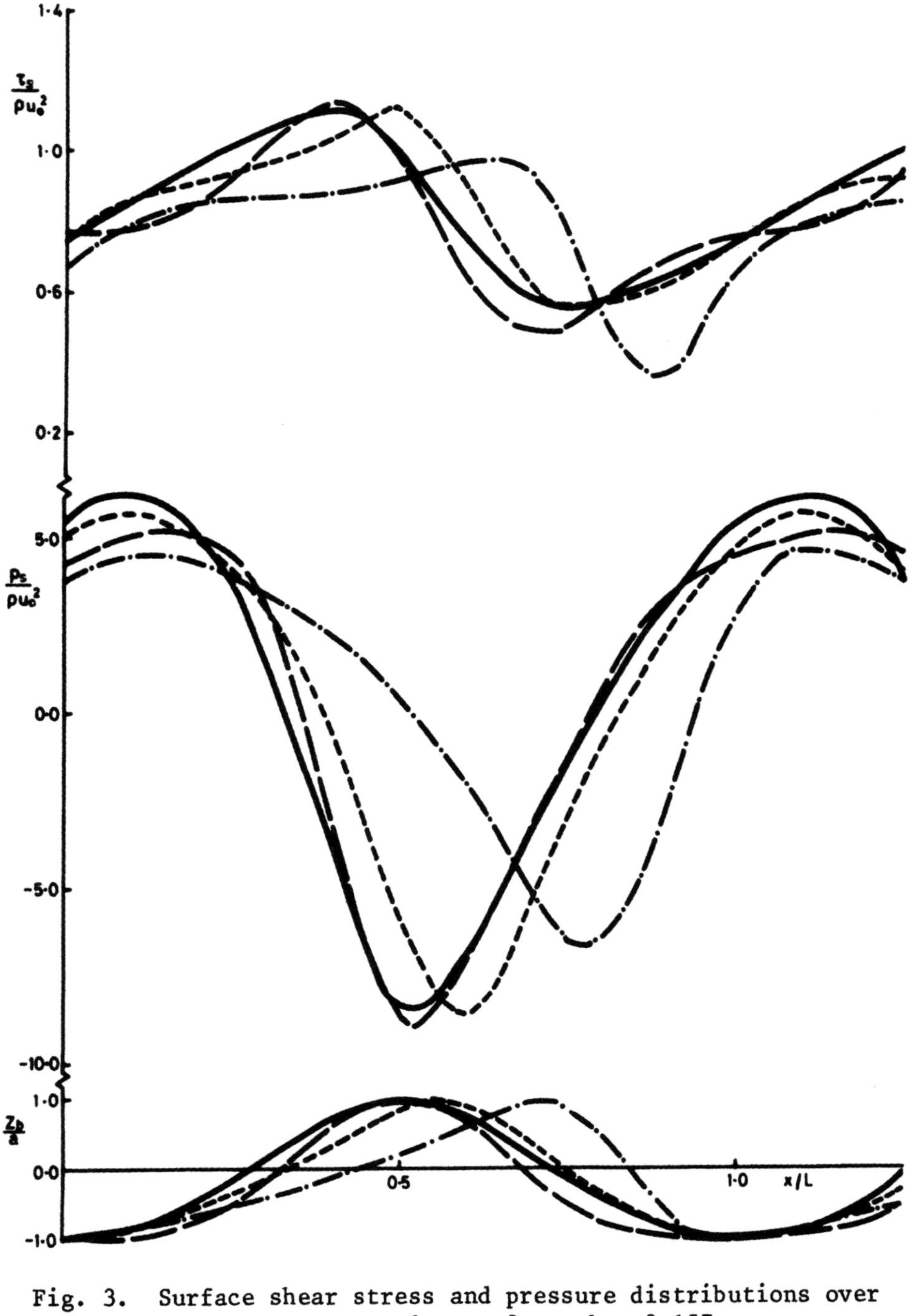

Fig. 3. Surface shear stress and pressure distributions over different wave shapes, R = 8, c = $8u_0$, ak = 0.157

——— Gerstner wave, — — — — — — Sin^4 wave,
- - - - - - Wave (3), —— - —— - —— Wave (4).

significant changes in the contributions from form drag to the total stress on the water. They are 21, 25, 21 and 27% respectively for the four cases.

Further studies are planned with this model and we may well find that shape is an important factor for slightly steeper waves or for different values of c/u_0. We also plan to incorporate changes in surface roughness and to investigate further the effects of wave breaking on the airflow above the waves.

ACKNOWLEDGEMENTS

This work has been supported under Natural Environment Research Council (U.K.) Grant GR3/3014.

REFERENCES

Banner, M. L. and Melville, W. K.; 1976, 'On the separation of airflow over water waves', J. Fluid Mech. 77, 825-842.

Banner, M. L. and Phillips, O. M.,; 1974, 'On the incipient breaking of small scale waves', J. Fluid Mech. 65, 647-656.

Benjamin, T. B.; 1959, 'Shearing flow over a wavy boundary', J. Fluid Mech. 6, 161-205.

Dobson, F. W.; 1971, 'Measurements of atmospheric pressure on wind generated sea waves', J. Fluid Mech. 48, 91-127.

Elliott, J. A.; 1972, 'Microscale pressure fluctuations near waves being generated by the wind', J. Fluid Mech. 54, 427-448.

Gent, P. R.; 1977, 'A numerical model of the airflow above water waves, Part 2', J. Fluid Mech. 82, 349-369.

Gent, P. R. and Taylor, P. A.; 1976, 'A numerical model of the airflow above water waves', J. Fluid Mech. 77, 105-128.

Gent, P. R. and Taylor, P. A.; 1977, 'A note on 'separation' over short wind waves', Boundary-Layer Meteorol. 11, 65-87.

Kinsman, B.; 1965, Wind waves, Prentice-Hall, New Jersey, 676 pp.

Long, R. B.; 1971, 'On generation of ocean waves by a turbulent wind', Ph.D. thesis, University of Miami, Florida.

Longuet-Higgins, M. S.; 1977, 'Some effects of finite steepness on the generation of waves by wind', To appear in Deep-Sea Research (Deacon supplement).

Miles, J. W.; 1957, 'On the generation of surface waves by shear flows', J. Fluid Mech. 3, 185-204.

Richards, K. J.; 1978, Ph.D. thesis, University of Southampton.

Snyder, R. L.; 1974, 'A field study of wave induced pressure fluctuations above surface gravity waves', J. Mar. Res. 32, 485-496.

Taylor, P. A.; 1977, 'Some numerical solutions of surface boundary-layer flow above gentle topography', Boundary-Layer Meteorol. 11, 439-465.

Townsend, A. A.; 1972, 'Flow in a deep turbulent boundary-layer over a surface distorted by water waves', J. Fluid Mech. 55, 719-735.

DISCUSSION

O. M. Phillips:

It might be pointed out that there have been several previous attempts to calculate the properties of turbulent air flow over waves, all notably unsuccessful. The critical element in these calculations is the careful treatment of the boundary conditions on the water surface and it is gratifying to see that the results seem reasonable and stable. The particular turbulent closure does not seem to be very critical; the boundary conditions are.

THE POTENTIAL IMPACT OF SEASAT-A ON THE STUDY OF THE PLANETARY BOUNDARY LAYER OVER THE OCEAN

WILLARD J. PIERSON

THE CITY COLLEGE OF NEW YORK
CONVENT AVE. AT 138TH ST., NEW YORK, N.Y. 10031

ABSTRACT

At an IUGG-IUTAM Symposium in 1966 at Kyoto, Japan, a brief paper by me made some general comments on our inadequate understanding and knowledge of the planetary boundary layer over the oceans.* The possibility that earth orbiting spacecraft could measure waves over the ocean and provide data from which very high quality synoptic surface weather maps and wind fields could be determined was described.

In 1978, an experimental spacecraft at an altitude of 800 km in a 108° retrograde orbit with a period of 100 minutes will test these concepts and provide global coverage of the vector wind field over the oceans, of water vapor and rainfall, and of wave conditions. Partial coverage of sea ice conditions in the arctic and antarctic will also be obtained. Three different kinds of radar, a passive microwave system at five frequencies, with two polarizations for each frequency, and a visible band and infrared band imaging system make up the remote sensors on this spacecraft.

Two of the radars, the altimeter and the scatterometer, were tested on SKYLAB, and an altimeter is currently on GEOS-3. Extensive tests of the scatterometer and the third radar (a synthetic aperture imaging radar) have been made from aircraft.

*Pierson (1967)

The present status of our research on the scatterometer is that the speed and direction of winds from about 4 M/S to 24 M/S (and probably even higher) can be determined and that above about 7 M/S the speed can be found to about ±0.5 M/S and the direction to about ±15°. Several hundred thousand vector winds will be determined each day.

SEASAT-A, and the spacecraft that will follow it, will initiate a new era in improved meteorological forecasts because of the improved initial value specification of the planetary boundary layer over the ocean. Fluxes of momentum, heat and H_2O across the air-sea interface can be determined without gaps due to missing ship reports to great accuracy by means of the bulk aerodynamic equations and other methods. The winds over the ocean will be known more accurately than ever before. Also information on radiation fluxes is potentially available.

INTRODUCTION

At first, it seems strange to want to measure such things as sea surface temperature, sea ice, waves, wind stress, winds, atmospheric water vapor content, cloud properties, precipitation, infrared fluxes in an infrared window, sea surface topography and the slope of the sea surface relative to the geoid with instruments 800 kilometers away from the ocean surface on a platform traveling at about 7.5 km/sec. But these are quantities that SEASAT-A has been designed to measure. If successfully launched, orbited and operated in May 1978, as SEASAT-1, these quantities will be measured by four continuously operated instruments over the ocean twice each day with coverage that varies from nearly 100% in 12 hours, through 100% in 72 hours, on to selected areas and then, finally, along the subsatellite track, depending upon which of the instruments is considered. The synthetic aperture radar can be operated only on an intermittent basis.

Once the measurements made by this spacecraft are properly calibrated and interpreted to obtain the quantities mentioned above, this one spacecraft could replace nearly all of the conventionally measured quantities at the ocean surface with the important exceptions of sea surface salinities, air-sea temperature differences, atmospheric surface pressures, dew points and wave periods. It appears possible to measure atmospheric surface pressures by means of another instrument on a future spacecraft. The air-sea temperature difference enters into the determination of the variation of wind with height and of sensible heat transfer. The inability to determine this quantity with a remote sensor may require its determination from conventional sources - or it might possibly be determined from the space data indirectly.

Many of the quantities measured, may well be measured more accurately than by the conventional techniques presently used over the ocean. In contrast to ship reports, data buoys (both operational and for scientific studies), instrumented platforms, and aircraft measurements, there will be no problems of cross calibration between different instruments that measure the same quantity, no problem of deployment and operation to study some different region of the ocean, and no problems of intermittent operation from one season to another, if the spacecraft system operates as designed. For the first time in Oceanography and Meteorology, many of the properties of the ocean, the planetary boundary layer over the ocean, and the atmosphere above the ocean can be studied in a truly global way with a truly global measuring system. Thus, it is not really strange to want to make these measurements from a rapidly moving platform 800 kilometers above the ocean surface.

SEASAT-A has been described in numerous publications scattered throughout the literature. Two that may be helpful are the reports by McCandless and Cardone (1976) and Pierson and Cardone (1977). Other references and descriptive material may be obtained from the SEASAT-A Program Office at NASA Headquarters in Washington, D.C.

The purpose of this paper is not to review all of the instruments on SEASAT. Rather an attempt will be made to relate the quantities that this spacecraft will measure to the study of fluxes through the sea surface and to wave dynamics.

FLUXES

In addition to the turbulent fluxes, there are other fluxes into and out of the ocean that must also be determined in order to obtain a complete understanding of the land-ocean-atmosphere system. Although the 100 meters of atmosphere over the ocean and the first 10 meters of ocean depth, in terms of turbulent fluxes, have proved to be a most difficult part of the total problem,

(Pierson (1967)), some of the other fluxes are also difficult to measure. The measurement of precipitation is in an even poorer stage of development than that of the various turbulent fluxes.

The emphasis of this symposium on turbulent fluxes is, in part, attributable to the fact that understanding them is proving to be a long and difficult problem. Nevertheless, they form only a subset of the full range of fluxes that enter into the study of the land-ocean-atmosphere system.

The turbulent fluxes have been studied in the microscale by means of rapidly responding sensors with a frequency range extending to twenty hertz, or higher. Fortunately, the valley in the spectra of the winds (when plotted on $f\ S(f)$ versus $\log f$ axes so as to preserve the total variance within frequency bands) at frequencies corresponding to periods of an hour to twenty minutes appears to be real so that records of that duration are useful and can be considered to be quasi-steady state.

The goal, of course, of the study of turbulent fluxes is to be able to parameterize them correctly by means of the bulk aerodynamic equations and temporally and spatially averaged values of the variables involved so as to be able to incorporate the effects of these fluxes in the appropriate hydrodynamic equations for the atmosphere and the oceans at the synoptic scale and the mesoscale.

The measurements to be made by SEASAT-A substitute area averages for time averages in order to obtain stable synoptic scale estimates of the quantities being measured. As will be illustrated later for winds and waves, these area averages of wave properties, wind stress, atmospheric water vapor and other quantities are more stable and more representative of the point at the center of the area sampled than a corresponding 20 to 60 minute average of a time history at that same point.

RADIATION FLUXES AND THE HEAT AND RADIATION BUDGET OF THE EARTH

The microwave radiation sensed by the passive microwave instrument on SEASAT is negligible in terms of the radiation budget of the earth.

The infrared imaging sensor is actually measuring the upward flux of infrared radiation from the sea surface in the absence of clouds and from cloud tops (more or less) when there are clouds, as attenuated and scattered on the way through the atmosphere to the spacecraft. A correction is thus involved if the radiation comes from the sea surface so as to obtain the sea surface temperature. A study of SKYLAB infrared measurements by Anding and Walker (1975) showed that the use of two infrared wavelengths to view the same point would provide a way to remove the effects of the atmosphere and determine a more accurate sea surface temperature since the two I R temperatures that are sensed are different. This feature will be incorporated on future TIROS spacecraft and might be incorporated on future SEASAT spacecraft.

The visible sensor on SEASAT actually, of course, senses the upward flux of certain wavelengths of light. The quantity sensed is a part of the total radiation budget for sunlight. It enters into the calculation of the albedo of the earth and the determination of the amount of sunlight entering the ocean and absorbed to heat the ocean.

Other sensors on other spacecraft are sensing radiation fluxes in various bands of the infrared and visible. What is disturbing to the writer is that in order to obtain such quantities as sea surface temperature, cloud cover, and so on, the real nature of these measurements as fluxes is not being used to study the day to day, week to week, month to month, seasonal and annual variations of the radiation budget of the earth to and from space, and back and forth in the land-ocean-atmosphere system. These fluxes, though not turbulent, are of extreme importance in the complete understanding of our planet and of changes in climate. Scientific studies in which radiation budgets and the albedo of the earth were obtained from space data have been published as a follow on of the references given by Pierson (1967), but continuously processed data stratified in a different way are really needed.

MOMENTUM FLUX, WIND STRESS, AND THE WIND PROFILE FOR THE SYNOPTIC SCALES

If agreement could be reached on the values of u_*, z_0 and k in equation (1), then for a neutrally stratified atmosphere the mean wind as a function of elevation for the first several hundred meters above the sea surface could be specified.

$$\bar{U}(z) = \frac{u_*}{K} \ln(z/z_0) \qquad (1)$$

Also the values of C_{10} for the drag coefficient would be known as in equation (2).

$$\tau/\rho = u_*^2 = C_{10}(\bar{U}(10))^2 \quad (2)$$

The extension to stable and unstable stratifications can then be made in terms of the Monin-Oboukhov length and the Richardson number (obtainable from the air-sea temperature difference) so as to describe the variation of wind with height for the first several hundred meters above the sea surface. These log and log linear profiles can be merged with the gradient wind and the thermal wind to continue the wind profile to the top of the friction layer. Once the friction velocity and z_0, or the equivalent in terms of the vector wind at a known elevation above the sea surface and the air-sea temperature difference are known, it follows that the wind shear, the turbulent wind spectra, the flux of momentum, and any other desired quantities can be computed.

Pierson (1964) illustrated the differences in the winds at 20 m and at 5 m starting with the same wind at 10 m and assuming each of the many published forms for the drag coefficient at 10 m. The differences can amount to several meters per second at the other elevations.

Cardone (1969) closed the problem for a neutrally stratified atmosphere by finding an emperical function relating z_0 to u_* as given by equation (3)

$$z_0 = 0.648/u_* + 4.28 \times 10^{-5} u_*^2 - 4.43 \times 10^{-3} \quad (3)$$

When combined with air-sea temperature difference effects via the Richardson number and the log linear profiles that result, this theory then yields a continuous variation of C_{10} as a function of u_* and the Richardson number. This result emphasizes the difficulty of obtaining wind profile measurements for truly neutral stability and helps to explain part of the scatter in plots of C_{10} versus wind speed at 10 meters.

The planetary boundary layer theory developed by Cardone (1969) has been applied in studies by Elsberry et al. (1974), Kaitala (1974) and Isozaki and Uji (1975). It provides a consistent way to process the data from conventional ship reports and data buoy reports so as to account for variations in the heights of the anemometers used to record the wind.

In a review of the available published data on this subject, Garratt (1977) has reached a number of important conclusions. One is that the scatter in the various published results on drag coef-fiecients as a function of wind speed and of the systematic differ-ences between data sets is "mainly due to calibration uncertain-ties associated with sensor (and electronics) performance in the field". He finds that k = 0.41 ± 0.025 and recommends that

$$z_0 = \alpha u_*^2/g \qquad (4)$$

with $\alpha = 1.44 \times 10^{-2}$ be used to close equation (1).

This requires a change in the relationship between z_0 and u_* used by Cardone (1969). It is believed, however, that z_0 should first decrease and then increase as a function of u_*. If found to be more nearly correct, the effect of this change will have to be incorporated in the interpretation of radar backscatter versus vector wind data.

The major difficulty in applying any boundary layer theory over the ocean to the synoptic scale and to various problems in the mesoscale is the sparsity and poor quality of the convention-ally obtained data on the average wind speed and direction at some known elevation above the sea surface. Apart from a few data buoys scattered over the oceans (it would help our SEASAT program to know the locations of all of them), these reports come from ships. Their total number is about three or four thousand with many more for the northern hemisphere, compared to the southern hemisphere. Trenberth and Neale (1977) have described numerical weather forecasting procedures for New Zealand. Many of the diffi-culties encountered can be attributed to sparse data.

The study of radar backscatter data and winds by means of SKYLAB data, (Cardone, Young, Pierson et al. (1976)) showed that the winds over the ocean were very poorly determined by ship reports and that the accuracy of the report was a function of the type of ship making the report. Weather ship reports were the best, ships with anemometers at a known height were next, and ships without anemometers were the poorest.

The most obvious reason for the large errors in the reports of wind speed and direction by ships is that only a two minute average is obtained. Most of the microscale fluctuations are still present in a two minute average, and so the reported wind, even with a perfectly calibrated and exposed anemometer, is not representative of the desired synoptic scale wind. The concentration of reports along the shipping lanes and the vast distances over which the reports have to be interpolated produce additional sources of error in the specification of the wind field over the ocean.

Information on the winds over the ocean will be obtained by means of two different instruments on SEASAT. Both instruments depend upon effects of the wind on the sea surface. The sea surface is in effect the transducer for the remote sensing anemometer, and the electromagnetic radiation reaching the spacecraft from an area of the sea surface is the communications link between the sea surface and the spacecraft.

The scanning multichannel microwave radiometer (SMMR) scans a conical scan at an incident angle near 50°. It measures the passive microwave radiation at five different frequencies for each of two polarizations and records the appropriate data for a series of elliptically spaced areas (or cells) along the scan of varying size depending on the frequency involved. Certain combinations of the data contain information on the increase in the microwave radiation from the sea surface over that from a smooth surface with the same absolute temperature and emissivity. The increased passive microwave radiation is a function of the increased roughness and increased foaminess (caused by breaking waves) of the sea surface. These two effects are in turn related to the speed of the wind over the ocean. Data from the passive microwave system provide only the magnitude of the vector wind over a swath to one side of the spacecraft from the subsatellite track to 650 km to one side.

The SEASAT-A scatterometer (SASS) has four antennas that point at $\pm 45^{o}$ and $\pm 135^{o}$ to the direction of movement of the spacecraft. Range cells are produced by separating the doppler frequencies along the beam so that fifteen areas approximately 40 km by 20 km are sampled. The cells from the fore and aft beams produce nearly overlapping pairs of cells such that the incident angles are almost equal and such that the pointing directions of the radar beams are almost 90° apart.

The information obtained in the form of normalized radar backscattering cross section measurements will be related to the winds over the ocean by means of backscatter data obtained from a radar on an aircraft operated by the NASA Langley AAFE program.* The radar backscattering cross section is a function of wind speed, V, wind direction relative to the pointing direction of the radar beam, χ, and incident angle, θ, as in equations (5) and (6).

$$\sigma_{VV} = \sigma_{VV}{}^{o}(V,\chi,\theta) \tag{5}$$

$$\sigma_{HH} = \sigma_{HH}{}^{o}(V,\chi,0) \tag{6}$$

For this program, for which the earlier results have been described by Jones, Schroeder and Mitchell (1977), the wind speed has been kept fixed except for the microscale turbulent fluctuations which have spatial dimensions larger than the size of the area of the ocean illuminated by the pencil beam radar on the aircraft. Then by flying in a circle, the values of $\sigma_{VV}{}^{o}$ and $\sigma_{HH}{}^{o}$ are found as functions of χ for various fixed θ's. Other data for $\chi = 0$, 90° and 180° for various θ's are also obtained.

All wind speeds are referred to that wind speed that would have been measured at an elevation of 20 meters above the sea surface in a neutral atmosphere, and thus the effects of atmospheric stability and varying anemometer height have been calibrated out. Because of equation (1), the wind speeds used are fully the equiva- of u_* and of the wind stress on the sea surface in the bulk aerodynamic sense. The logarithmic and log-linear wind profiles all reduce to the same form close to the sea surface in a given wind stress. Also the effect of air sea temperature differences is much weaker for high winds.

Data on σ^{o}_{VV} and σ^{o}_{HH} for wind speed ranging from 4.6 M/S to 20 M/S and for $\theta = 19^{o}$ to 20°, 30°, 39° to 40°, 50°, 58° and 65° have been obtained for χ varying in 10° steps around the full circle. For this range of wind speeds, aspect angles and incident angles, the values of σ^{o}_{HH} and σ^{o}_{VV} range over nearly six orders of magnitude.

For θ near 30° or 40°, the backscatter varies as approximately the square of the wind speed.

---

*Advanced Applications Flight Experiments.

As pointed out by Pierson and Moore (1972), the backscatter measurements made from a spacecraft will be very stable from the point of view of representing the effect of the mean vector wind over the area contributing to a given measurement. From Taylor's hypothesis, which is conservative, the length scales of the horizontal components of the turbulence can be related to the periods by advecting the gusts at the speed of the mean wind. The times required for the air traveling at a given velocity to travel 20 or 40 km is a measure of the periods in the turbulence that will be filtered out by averaging the backscatter over the illuminated area of the cell.

TABLE 1 Equivalent Anemometer Averaging Times for Various Mean Wind Speeds

Wind Speed (M/S)	5	10	15	20	25
20 Km Dimension	1h 7m	33 m	22 m	17 m	13 m
40 Km Dimension	2h14m	1h 6 m	44 m	34 m	26 m

This effect is illustrated by Table 1, If the instantaneous wind could be measured along a line 20 Km long in the upwind-downwind direction and averaged, the average would be the equivalent of a 1h7m time average at a point, if it turned out to equal 5 m/s. For 25 m/s it would be equal to a 13 minute average. Since the wind will not always blow directly across the shortest distance, the actual averaging time will be between the upper value and the lower value tabulated for each wind speed.

The above values are certainly conservative. What is really needed is the effective number of independent points in a sample so that the sampling variability of the mean can be found. Certainly, there must be a two-dimensional decorrelation distance involved such that different portions of the radar cell are effectively independent. The equivalent anemometer averaging time must therefore be even longer than the smaller of the two values tabulated and perhaps may even be longer than the larger of the two values.

The most intriguing feature of this result is that the dimensions of these cells are still small compared to synoptic scale variations in the horizontal wind and that the horizontal gradients of the synoptic scale wind over the dimensions of the cell can be neglected.

Running centered averages of the instantaneous wind at a point of the form

$$\bar{U}(t) = \frac{1}{T}\int_{-\frac{T}{2}+t}^{\frac{T}{2}+t} U(\tau)\, d\tau \qquad (7)$$

would show trends and other variations over several hours, yet these equivalent anemometer times are representative of a much more stable wind for the area involved. The benefit of obtaining an area average at an instant of time with the instruments on SEASAT has a distinct advantage compared to time averages at a point. The two minute averages of the winds reported by ships and the 10 or 20 minute averages obtained by data buoys are definitely at a disadvantage compared to the data to be obtained by SEASAT.

The reason for this striking difference between area averages and time averages has to do with the synoptic scale changes in the winds from hour to hour as the cyclones and anticyclones form, develop, move and dissipate over the globe. This contribution of the time variation of the synoptic scale wind at a point is removed when an instantaneous average over an area is obtained.

It is expected that for winds above the 7 M/S, the wind speed will be determined to within ± 0.5 M/S and that the wind direction, will be known to within ± 15°, for incident angles of 30° to 50°, according to the results of Pierson, Cardone and Greenwood (1974). Based on a worst case analysis at every stage, the radar scatterometer on SEASAT has been designed to provide winds within ± 2 M/S or ± 10%, whichever is worse (Grantham et al. (1975)).

There is another important point to be made concerning this instrument. For the paired cells on one side of the spacecraft, the data obtained can be considered to be of the form

$$\sigma^o_1 = \sigma^o_1\ (V, \chi, \theta) \qquad (8)$$

and

$$\sigma^o_2 = \sigma^o_2\ (V, \chi + 90^\circ, \theta) \qquad (9)$$

for a fixed θ and for any polarization. The angle, χ, is the direction of the wind relative to the pointing direction of the radar beams. The problem is to find those values of V and χ that produce the measured values of σ^o. For perfect data, there is always one true solution for V and χ, but there can be as many as three more incorrect solutions. The incorrect solutions must be eliminated on the basis of other synoptic meteorological data.

The errors in the determination of the vector wind (or the vector wind stress, or vector u_*) with this instrument are caused by sampling variability in the estimates of the received backscattered power P_R which has to be separated from instrument noise. The estimate of the received power which is computed from an estimate of the signal plus the noise, $\hat{P}_{SN}$, minus an estimate of the noise, $\hat{N}$, as in

$$\hat{P}_R = \hat{P}_{SN} - \hat{N} \tag{10}$$

is a normally distributed random variable with an expected value given by $\hat{P}_R$ and a standard deviation given by

$$\text{STANDARD DEVIATION} = \left(\frac{(P_R + N)^2 + \frac{1}{2} N^2}{\beta_c \, \tau_{SN}} \right) \tag{11}$$

where the noise is either measured or computed from theory. β_c is the band width of the doppler band pass filter for each cell, and τ_{SN} is the averaging time. Typical values of $\beta_c \, \tau_{SN}$ range from 1000 to 2500, or so, as a function of incident angle, but for low values of P_R the noise can be 10 to 30 times stronger than the estimate of P_R. The sampling variability of the estimates of P_R is the major factor in determining the accuracy of the winds measured by this system.

A second factor of lesser importance is the errors in the measurement of the attitude of the spacecraft (roll, pitch, and yaw) as these errors propagate through the radar equation. Both effects can be treated theoretically.

When operational, SEASAT will provide several hundred thousand vector winds each day for two swaths on each side of the spacecraft track. For 12 hours, there will be lenticularly spaced gaps at the equator extending north and south to about 45° and about 10° wide at the equator. To compensate for this, successive swaths will overlap poleward of 45° and it will be possible to determine the changes in the wind after 100 minutes for some areas.

The payoff of all of the scientific studies of the turbulent flux of momentum from the atmosphere to the ocean is thus a global system for measuring $\bar{U}$ and its direction at any elevation above the sea surface. The quantity, $\bar{U}$, will thus be known to an accuracy never before achieved for the bulk aerodynamic equations in which it is employed as in equations (12), (13) and (14) where the subscript, a, represents the anemometer height.

$$u_*^2 = \frac{\tau}{\rho} = C_a \bar{U}_a^2 \qquad (12)$$

$$E = k(\bar{e}_s - \bar{e}_a)\, \bar{U}_a \qquad (13)$$

$$Q_c = h_a \rho C_p (\theta_s - \theta_a)\, \bar{U}_a \qquad (14)$$

H_2O FLUX

The fluxes of H_2O are in the form of precipitation from the air to the ocean and evaporation from the ocean to the air. According to Wilheit*, the Scanning Multichannel Microwave Radiometer (SMMR) on SEASAT will make it possible to determine precipitation rates from 0 to 5 mm/hr from one combination of passive microwave signals and for rates greater than 5 mm/hr from another combination. The theory has been tested from other spacecraft data (Wilheit et al. (1975, 1977)) and only needs modification to be applicable to the SEASAT scanning system and frequencies. It is desirable to know the height of the freezing level in order to get more accurate results, which have been shown to be correct within a factor of two. An atlas by Rao et al. (1976) demonstrates the kind of results that can be obtained as computed from passive microwave data obtained by the Electrically Scanning Microwave Radiometer (ESMR) on the NIMBUS 5 spacecraft. The integrated zonal values are similar to those of Jacobs (1951). The equatorial Pacific shows a marked change from January 1973, an El Niño year, to January 1974, a non El Niño year.

---

*Personal Communication

These efforts represent the first steps in determining rainfall rates over the oceans of the world. They are possible because the passive microwave temperatures of water drops and clouds are higher than the passive microwave temperatures of the sea surface, where for certain microwave frequencies the emissivity of the ocean surface is about 0.4 to 0.5. Many numerical weather forecasting systems predict precipitation, and the data from SEASAT will provide a start toward verifying these models.

The increase in the microwave radiation from clouds will also be used to correct the backscatter measurements for attenuation. Rainfall, however, masks the backscatter from the sea surface to such an extent that the winds cannot be found where it is raining.

The complete H_2O budget of the oceans also requires measurements of evaporation. The budget is balanced by including the effect of river runoff.

To compute evaporation from equation (13), the value of $\bar{U}$ is known. The temperature used to compute the saturation mixing ratio, e_s, for the air immediately above the ocean surface is based upon the ocean surface temperature.

The sea surface temperature will be remotely sensed by both the infrared imaging system in the absence of clouds and by the passive microwave system (SMMR) through clouds. The measurement of surface temperatures through clouds will be a first for SEASAT. The only remaining unknown term in equation (13), if the constant, is assumed to be known, is the value of e_a. This symbol represents the mixing ratio for the air at a known height above the surface so as to provide an estimate of the gradient of e.

The quantity, e_a, will not be remotely sensed by SEASAT. Nevertheless, it appears to be possible to compute the evaporation from the sea surface. By treating E, U and e_s as known, e_a can be computed if desired.

The SMMR on SEASAT will measure the total water vapor and the total liquid water in the air column above each point on the sea surface scanned by the instrument. The greatest amount of water vapor is found near the sea surface. If the air trajectories for these atmospheric columns are all over water, then an increase in total water vapor plus liquid water along the trajectory must have been produced by evaporation from the sea surface.

If for a cloud free area, the water vapor in a one square centimeter atmospheric column increases along the air trajectory from 2 cm of water vapor per square cm to 2.5 cm of water vapor per square cm over a distance of 100 km and if the effective wind is ten meters per second over this distance, then 0.5 cm of water must have evaporated from that area of the ocean in 1000 seconds. The value of E would then be 1.8 mm/hr. The liquid water and the water vapor measured by SEASAT can be considered to be an integrated effect of evaporation in the absence of precipitation, and, as such, gradients in these quantities can be used to determine evaporation. These effects should be particularly pronounced as the air leaves the center of a subtropical high and travels toward the equator as the trade winds. The trade wind model of Malkus (1962) could be used.

Although many doubts about the bulk aerodynamic method have been expressed in the past by various authors (see Neumann and Pierson (1966), pages 412 to 416 for a summary), improved versions of them based on data of better quality may make it possible to study the H_2O budget of the ocean and to obtain fields of evaporation, precipitation, and evaporation minus precipitation for weekly, monthly seasonal and annual values for all the oceans, thus continuing and improving upon our knowledge of an important scientific problem studied by Jacobs (1951) for the Northern Hemisphere a quarter of a century ago and continued by Privett (1960) for the Southern Hemisphere.

SENSIBLE HEAT FLUX

The sensible heat flux given by equation (14) requires the air sea temperature difference. The air temperature at 10 or 20 meters above the sea surface is not presently available from spacecraft data. It is possible to obtain fields of the air temperature and of the air sea temperature difference from conventional data. The more accurate wind fields would then improve on the accuracy of the calculation of Q_c.

It may be possible to relate the air temperature to the surface wind direction relative to the gradient wind if the wind directions are obtained with enough accuracy by SEASAT. Areas where the air temperature is well known conventionally could then be used to determine the relationship, and it could then be used for data sparse areas.

The accuracy of the air temperature field can at least be checked for consistency with the calculated evaporation as described above. For no clouds, the air temperature cannot be colder than that value that would produce saturation from the value of e_a solved for in equation (13).

WAVES

The flux of momentum from the air to the ocean generates waves. More accurate winds over the ocean will make it possible to specify (or hindcast) waves more accurately. Also more accurate winds will make better wind forecasts possible and, hence, better wave forecasts.

SEASAT will, hopefully, measure wave properties with two instruments. They are the altimeter and the synthetic aperture radar. Doubt has been expressed by the writer, (Pierson (1976)), about the ability of the synthetic aperture radar to image high waves, and even if an image is obtained the interpretation of the image in terms of estimating a vector wave number spectrum leaves many unanswered questions.

If the waves are imaged, and if the image can be converted to a vector wave number spectrum, then the change from a wave record as a function of time to an essentially instantaneous area image (or measurement) provides the same kind of advantage for waves as the area average does for winds. In deep water, an area 10 km on a side should be essentially stationary from a statistical point of view. For typical wave number resolutions, an image of this size should yield about 500 degrees of freedom per spectral estimate. The sampling variability of the spectral estimates would be about ± 11% at the 90% confidence interval. For a twenty minute long ocean wave time history, for a useful spectral resolution, only about 50 degrees of freedom are obtained, and the 90% confidence interval typically spans a factor of two. Other dimensions, say 5 km by 20 km with the long axis parallel to a coast line could be used to study waves as a function of fetch. With this higher number of degrees of freedom it would be possible to test the argument given by Pierson (1977) that the estimates of γ in the JONSWAP spectrum (Hasselmann et al. (1976)) are biassed too high.

The altimeter will measure the significant wave height at intervals along the subsatellite track. This has already been accomplished on an intermittent basis by the altimeter on GEOS-3, and recently a series of papers by Walsh, to be published in the near future, has eliminated many of the problems associated with the analysis of the GEOS-3 altimeter data.

For GEOS-3, there were timing errors in the 12.5 nanosecond gates so that they were not actually 12.5 nanoseconds apart, as shown in Hofmeister et al. (1976) and there were slight constant biases in the measurements of the power in each gate as shown by Walsh (1977). If corrections are made for these effects, it can then be shown that the return altimeter wave form is the convolution of three probability density functions with a step function.

These probability density functions are

$$f_1(t) = (2\pi)^{-\frac{1}{2}} \sigma_P^{-1} \exp\left(- (t/\sigma_P)^2/2\right) \tag{15}$$

$$f_2(t) = (2\pi)^{-\frac{1}{2}} \sigma_J^{-1} \exp\left(- (t/\sigma_J)^2/2\right) \tag{16}$$

and

$$f_3(t) = (2\pi)^{-\frac{1}{2}} \sigma_W^{-1} \left(1 + \frac{\lambda}{6} \left(\frac{t^3}{\sigma_W^3} - \frac{3t}{\sigma_W}\right)\right) \exp\left(- (t/\sigma_W)^2/2\right) \tag{17}$$

where t, σ_P, σ_J and σ_W are in nanoseconds. (The times in equation (17) are in nanoseconds and are obtained by dividing the variance of the wave elevations in meters by 0.15 since a one nanosecond change in round trip time equals a 15 cm change in range).

The first probability density function (pdf) represents the effect of the transmitted pulse and accounts for the fact that it is not a perfectly square pulse. The second pdf accounts for the jitter, which is usually normally distributed, of the tracking circuitry. The third function is the pdf of wave elevations as derived by Longuit-Higgins (1963).

The convolution of the first two functions is again normally distributed with a variance given by equation (18).

$$\sigma_{PJ}^2 = \sigma_P^2 + \sigma_J^2 \tag{18}$$

and the pdf is

$$f_4(t) = (2\pi)^{-\frac{1}{2}} \sigma_{PJ}^{-1} \exp\left(- (t/\sigma_{PJ})^2/2\right) \tag{19}$$

Feldman (1977) has shown that the convolution of $f_3(t)$ and $f_4(t)$ yields the pdf given by equation (20).

$$f_5(t) = (2\pi)^{-\frac{1}{2}} \sigma_I^{-1}\left(1 + \frac{\lambda^*}{6} \left(\frac{t^3}{\sigma_I^3} - \frac{3t}{\sigma_I}\right)\right) \exp\left(-(t/\sigma_I)^2/2\right) \tag{20}$$

where

$$\lambda^* = \lambda\sigma_W^3/\sigma_I^3 \tag{21}$$

and where

$$\sigma_I^2 = \sigma_W^2 + \sigma_{PJ}^2 \tag{22}$$

Although λ^* is reduced in proportion to the cube of the ratio of σ_W to σ_I, the third moments (skewness) of both equation (17) and equation (20) are the same and are equal to equation (23).

$$E(t^3) = \lambda\, \sigma_W^3 = \frac{\lambda\sigma_W^3\, \sigma_I^3}{\sigma_I^3} = \lambda^* \sigma_I^3 \tag{23}$$

For GEOS 3, σ_{PJ} is often larger than σ_W, and λ^* is quite a bit smaller than λ.

The return pulse form is obtained by a convolution with a step function (Feldman (1977)), and the result is obtained by changing.

$$1 + \frac{\lambda}{6}\left(\frac{t^3}{\sigma_I^3} - \frac{3t}{\sigma_I}\right) \tag{24}$$

to

$$1 - \frac{\lambda}{6}\left(\frac{t^3}{\sigma_I^3} - \frac{3t}{\sigma_I}\right) \tag{25}$$

and integrating from $-\infty$ to t^*.

The expected value of the altimeter pulse shape in non-dimensional form is then

$$P(\tau) = \text{erf}\,\tau - \frac{\lambda^*}{(2\pi)^{\frac{1}{2}}6} (1-\tau^2) \exp(-\tau^2/2) \tag{26}$$

where

$$\tau = t^*/\sigma_I. \tag{27}$$

For GEOS 3 data, Walsh (1977) has determined the value of σ_P σ_J, σ_W, and $\lambda^*\sigma_I^3 = \lambda\sigma_W^3$. Thus, a radar altimeter can provide a measurement of wave height and of the skewness of the pdf of wave elevation. This skewness can be used to discriminate between sea and swell as it is expected to be large for wind waves and small for swell when both have the same significant height.

For the SEASAT altimeter, both σ_P and σ_J will be much smaller because the pulse will be 3 nanoseconds (compared to 12.5) and the jitter will be reduced by better tracking because of a much higher pulse repetition frequency. The PRF will be ambiguous because the round trip time of a pulse will be longer than time between pulses, but not redundant because the Rayleigh fading is each return pulse will be independent of the other pulses.

The SEASAT wave height measurements can be used to verify ocean wave forecasting models on a day to day basis and, thus, contribute to their improvement.

CONCLUDING REMARKS

The impact of SEASAT on the computation of the various fluxes across the air sea interface, many of which are turbulent fluxes will be that many of these fluxes can be determined globally with greater accuracy than ever before, since the wind field will be known very accurately. The flux of momentum that generates waves and the wind stress and, hence, ocean currents will be more accurately known. The H_2O fluxes, E and P, may perhaps be found. Finally, a number of the quantities that enter into the combined heat-radiation budget are at least, in part, available in the data. Changes in these latter quantities are important in medium and long range forecasting and in the study of short term climate changes. Perhaps, at some future symposium on these subjects, the contribution of the data from SEASAT will be shown to have been of great value in obtaining results on a global scale about the various interactions of the ocean and the atmosphere.

ACKNOWLEDGEMENTS

The research on which this paper is based has been supported by the Jet Propulsion Laboratory and the Wallops Flight Center both of the National Aeronautics and Space Administration under Contracts NAS 9-13642 and NAS 6-2439.

REFERENCES

Anding, C. and P. Walker (1975): Use of Skylab EREP Data in a Sea-Surface Temperature Experiment. NASA CR-144479.

Cardone, V. J. (1969): Specification of the Wind Field Distribution in the Marine Boundary Layer for Wave Forecasting Rep. TR 69-1, Geophys. Science Lab., New York Univ.

Cardone, V. J., J. D. Young, W. J. Pierson, R. K. Moore, et al. (1976): The Measurement of the Winds Near the Ocean Surface with a Radiometer Scatterometer on SKYLAB. A Joint Meteorological, Oceanographic, and Sensor Evaluation Program for Experiment S193 on SKYLAB. NASA CR-147487.

Elsberry, R. F., N.A.S. Pearson, and L. G. Corgnati Jr. (1974): A Quasi-imperical Model of the Hurricane Boundary Layer. J. Geophys. Res., Vol. 79, pp. 3033-3040.

Feldman, G. (1977): Average Return Pulse Form for the GEOS-3 Radar Altimeter as a Function of Ocean Wave Conditions. Master's Thesis The City College of New York.

Garrett, J. R. (1977): Review of Drag Coefficients over Oceans and Continents. Monthly Weather Rev., Vol. 105, No. 7, pp. 915-929.

Grantham, W. L., E. M. Bracalente, W. L. Jones, et. al. (1975): An Operational Satellite Scatterometer for Wind Vector Measurements over the Ocean. NASA TM X-72672 Langley Research Center, Hampton, Virginia.

Hasselmann, K., D. B. Ross, P. Muller and W. Sell, (1976): A Parametric Wave Prediction Model. J. Phys. Oceanogr., Vol. 6, pp. 200-228.

Hofmeister, E. L., R. N. Keeney, T. W. Godbey and R. J. Berg (1976): Data User's Handbook and Design Error Analysis GEOS-C Radar Altimeter Volume 1, Prepared under Contract Number NASA 62619 and APL 372165. General Electric Co., Utica, New York.

Isozaki, I. and T. Uji (1974): Numerical Model of Marine Surface Winds and its Application to the Prediction of Ocean Wind Waves. Papers in Meteorology and Geophysics,

Jacobs, W. C. (1951): The Energy Exchange between the Sea and the Atmosphere and Some of its Consequences. Bull. Scripps Inst. of Oceanography, Univ. of Calif., Vol. 6, pp. 27-122.

Jones, W. L., L. C. Schroeder, and J. L. Mitchell (1977): Aircraft Measurements of the Microwave Scattering Signature of the Ocean. IEEE Transactions on Antennas and Propagation. IEEE Journal of Oceanic Engineering, Special Joint Issue on Radio Oceanography, Volume OE-2, pp. 52-60.

Kaitala, J. E. (1974): Heating Functions and Moisture Source Terms in the FNWC Primative Equation Models Printed Lecture Notes, Naval Post Graduate School, Monterey, California.

Longuit-Higgins, M. S. (1963): The Effect of Non-linearities on Statistical Distributions in the Theory of Sea Waves. J. Fluid Mech., Vol. 17, 3, pp. 459-480.

Malkus, J. S. (1962): Large Scale Interactions. The Sea, Vol. 1, pp. 88-294. Edited by M. N. Hill, Interscience, New York-London.

McCandless, S. W. and V. J. Cardone (1976): SEASAT-A Oceanographic Data Systems and Users. IAF XXVII Congress, 76-061 Session 33.

Neumann, G. and W. J. Pierson (1966): Principles of Physical Oceanography. Prentice Hall 545 pp.

Pierson, W. J. (1964): The Interpretation of Wave Spectrums in Terms of the Wind Profile instead of the Wind Measured at a Constant Height J. Geophys. Res., 69, 5191-5203.

Pierson, W. J. (1967): Importance of the Atmospheric Boundary Layer over the Oceans in Synoptic Scale Meteorology in Boundary Layers and Turbulence. The Physics of Fluids Supplement 1976 American Institute of Physics Paper S203.

Pierson, W. J. (1976): The Theory and Applications of Ocean Wave Measuring Systems at and below the Sea Surface, on the Land, from Aircraft, and from Spacecraft. NASA CR-2642.

Pierson, W. J. (1977): Comments on "A Parametric Wave Prediction Model". J. Phys. Oceanogr. Vol. 7, No. 1, pp. 127-134.

Pierson, W. J. and R. K. Moore (1972): The Extrapolation of Laboratory and Aircraft Radar Sea Return Data to Spacecraft Altitudes in Proc. Fourth Annual Earth Resources Review MSC Houston, Texas (NASA).

Pierson, W. J. V. J. Cardone, and J. A. Greenwood (1974): The Applications of SEASAT-A to Meteorology. University Institute of Oceanography, prepared for the SPOC group of NOAA/NESS.

Pierson, W. J. and V. J. Cardone (1977): Ocean Surface Winds from SEASAT-A OTC paper 2762. Offshore Technology Conference, Houston, Texas May 2-5 1977.

Privett, D. W. (1960): The Exchange of Energy Between the Atmosphere and the Oceans of the Southern Hemisphere. Geophysical Memoir No. 104 M.O. 631d LONDON; Her Majesty's Stationery Office.

Rao, M. S. V., W. V. Abbott III and J. S. Theon (1976): Satellite-Derived Global Oceanic Rainfall Atlas (1973 and 1974). NASA SP-410 Scientific and Technical Information Office NASA Washington, D.C.

Trenberth, K. E. and A. A. Neale (1977): Numerical Weather Prediction in New Zealand. Monthly Wea. Rev., Vol. 105, No. 7, pp. 817-825.

Walsh, E. J. (1977): Extraction of Ocean Wave Heights from GEOS-3 Altimeter Data NASA Wallops Flight Center (Unpublished manuscript).

Wilheit, T. T. MSV Rao, T. C. Chang, E. G. Rogers and J. S. Theon (1975): A Satellite Technique for Quantitatively Mapping Rainfall Rates over the Oceans NASA/GSFC TMX-70904 also (1977). J. App. Meteor., Vol. 16, No. 5, May pp. 551-560.

DISCUSSION

S. M. Lazanoff :

How do you differentiate between swell height and wind wave height from the altimeter in a confused sea?

W. J. Pierson:

The skewness of the sea surface, as measured by the parameter σw^3 will be part way between sea values and swell values.

K. Katsaros:

Will your surface wind data be good enough so you can obtain divergence and convergence and thereby obtain airtrajectories? There is considerable interest in that from surface transfer people.

W. J. Pierson:

Hopefully!

ON THE SPECTRAL ENERGY BALANCE AND NUMERICAL PREDICTION OF OCEAN WAVES

Klaus Hasselmann

Max-Planck-Institut für Meteorologie
Bundesstrasse 55, 2000 Hamburg 13, FRG

1. THE LIMITING CASES OF DECOUPLED AND STRONGLY COUPLED WAVE TRAJECTORIES

In contrast to earlier, largely empirical methods of wave prediction, modern wave models normally start from a common, physically accepted basis: the energy-balance or wave-transport equation. However, considerable differences still exist between different models regarding the structure of the source function in this equation. Although systematic comparisions have not been made, different models may be expected to yield significantly different predictions when applied to non-uniform wind fields exhibiting the complex space-time structures typical of many generation situations.

To some part these discrepancies may be attributed to inadequate knowledge on the form of the source function at the time that the models were constructed. Parallel with the development of numerical wave prediction models, considerable progress has been made in the last years towards a clarification of the structure of the spectral energy balance, and these findings have not yet been fully incorporated into wave-prediction models. This applies particularly to the non-linear energy transfer, which was found to play an important role in the evolution of a growing wind-sea spectrum. Unfortunately, these effects are not easily included in the numerical framework of existing wave models and therefore require a general reassessment of the basic approach to numerical wave prediction.

In the past, numerical techniques have been strongly influenced by the assumption of decoupled wave trajectories. Considerable care is given to an accurate representation of the independent propagation of wave groups along their ray characteristics - ideally by the use of characteristic coordinates. Wave growth is usually represented by a linear source function (i.e. a combined Phillips-Miles mechanism) and wave dissipation - although recognized as inherently non-linear - is again normally represented as a ray-decoupled process, in which the dissipation for a particular wave component is expressed as a non-linear function depending on the spectrum only at that wavenumber. Numerical wave prediction in such a ray-decoupled model is largely a book-keeping task of keeping track of the evolution of a large number of wave groups of different frequency, direction, and starting positions as they propagate independently along their individual rays. The problem is analogous to particle propagation in a thin gas or plasma in which the mean free path is large compared with the macroscopic length scale of the system.

This approach appears appropriate for swell propagation, where the advective terms in the transport equation dominate, and the source functions, if present, can probably be regarded as linear. However, the theoretical and experimental evidence (Mitsuyasu, 1966, 1968, 1969, 1973; Hasselmann et al., 1973, 1976; Wu et al., 1977) suggests that for a typical wind sea the more appropriate gas dynamical analogy may be the opposite, continuum-mechanical limit, in which the mean free path is small compared with the macroscale of the system. Indeed, computations of the Boltzmann scattering integral for a growing wind sea indicate that, for typical wind-field scales, the non-linear transfer rates are an order of magnitude larger than the transfer rates associated with advection. Accordingly, the spectrum adjusts very rapidly to a local quasi-equilibrium in which the wind input, the non-linear transfer, and the dissipation approximately balance (except for a narrow frequency band around the peak). The shape of the spectrum itself is largely governed by the non-linear transfer, which generates a sharp, self-stabilised peak with a steep forward face and a more gradual decay on the high-frequency side. The observed migration of the peak towards lower frequencies is also a consequence of the non-linear transfer. According to these calculations, the "life history" of a wave group is so short, due to its non-linear interaction with other wave groups, that there is little point, numerically, in keeping track of its evolution along a ray over distances of the scale of the wind field. In the gas dynamical analogy one is similarly not interested in following individual particles, but only in the governing equations for the mean properties of the gas (pressure, temperature, density, and mean velocity). Because of the rapid establishment of a quasi-equilibrium Boltzmann distribution, these parameters in fact completely describe the statistical properties of the gas. In the same way, the existence of a quasi-equilibrium wind-sea spectrum also enables the sea state to be characterised by a few variables (essentially two scale parameters), and it can be argued that the appropriate approach to wave prediction under these conditions should similarly be the derivation of appropriate transport equations for these characteristic variables (Hasselmann et al., 1976).

Unfortunately, ocean waves occur not only as swell and growing wind seas, but also in various intermediate states. One therefore faces the difficulty of constructing wave models which operate not only in the two relative simple limiting cases of very small or large mean free wave paths, but also in the more complex transition regimes. Although first attempts to treat this problem are presented at this meeting, the question cannot yet be regarded as satisfactorily resolved. The difficulties are perhaps less physical than numerical; they arise primarily because it is not feasible to carry out repeated integrations of the complex Boltzmann scattering integral describing the non-linear transfer in the framework of a numerical wave model. Before addressing this question, however, it may be useful to provide first some orientation on the regions of applicability of the small and large mean free path approximations in terms of appropriate non-dimensional scale parameters.

2. LIMITS OF THE QUASI-EQUILIBRIUM WAVE MODEL

The structure of the quasi-equilibrium energy balance of a growing wind sea, as inferred from fetch-limited field and tank experiments, is indicated schematically in Fig. 1 (Hasselmann et al., 1973, 1976). Shown are the various terms of the one-dimensional energy-balance equation obtained by integrating the transport equation for the two-dimensional surface variance spectrum $F(f,\theta)$ (without refraction)

$$\frac{\partial F}{\partial t} + v_i \frac{\partial F}{\partial x_i} = S^{(2)} = S^{(2)}_{in} + S^{(2)}_{nl} + S^{(2)}_{ds} \tag{2.1}$$

over the propagation direction θ. Here f = frequency, v_i = group velocity, and the one-dimensional spectra and source function in Fig. 1 are defined as $E = \int F d\theta$ and $S = \int S^{(2)} d\theta$, respectively. The net source function S can be decomposed into the input source function S_{in}, the non-linear transfer S_{nl} due to conservative wave-wave scattering, and the residual dissipative source function S_{ds}, which includes also any energy redistribution associated with non-conservative non-linear processes such as wave breaking.

The shape of the spectrum is largely governed by the non-linear transfer, and is relatively insensitive to the precise form of the input and dissipative source functions. The sharp peak is a self-generated, stable feature of the non-linear energy transfer which migrates slowly towards lower frequencies. To maintain a stable spectral shape, sufficient atmospheric input must be present near the peak to balance the negative lobe of the source function S_{nl}. The spectrum to the right of the peak adjusts rather rapidly to a level such that the input is balanced by the non-linear transfer (plus possible additional dissipation). On the forward face of the

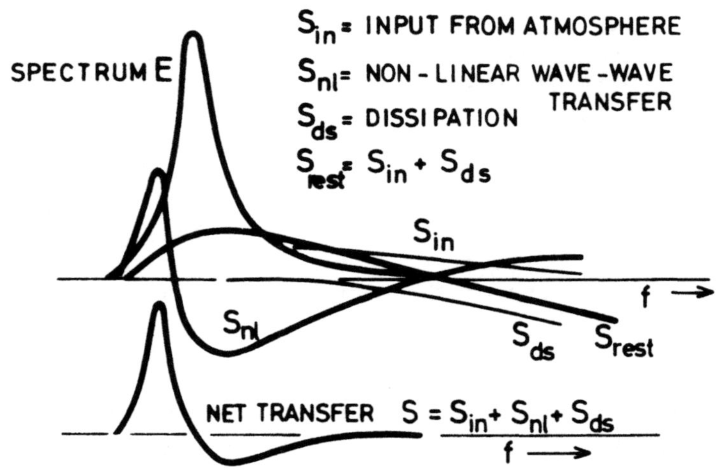

Fig. 1 Schematic energy balance for a growing wind-sea (from Hasselmann et al., 1976). The source function S_{rest} = $S - S_{nl}$ follows from S (measured) and S_{nl} (computed). In the quasi-equilibrium approximation the (uncertain) decomposition of S_{rest} into S_{in} and S_{ds} is irrelevant for wave prediction.

spectrum, however, the narrow positive lobe of S_{nl} cannot be balanced by the more smoothly varying input and dissipative source functions: theories suggest that both source functions are given by the product of the spectrum times a smoothly varying factor, thus they cannot simultaneously balance both the positive and negative lobes of S_{nl} near the peak. Consequently, the spectrum grows on the forward face, making the peak migrate towards lower frequencies. The characteristic time scale of this migration is found to be an order of magnitude greater than the time scale for the adjustment to a quasi-equilibrium on the high frequency side of the peak. The rapid establishment of an equilibrium in the main part of the spectrum and the generation and migration of the peak by internal non-linear interactions implies that the angular distribution of the spreading function should also adjust to a universal quasi-equilibrium distribution (including a possible frequency dependence) which is oriented relative to the local wind direction.

Since the spectral shape is quasi-universal, the spectrum can be characterised by two scale parameters, e.g. the non-dimensional peak frequency

$$\nu = Uf_m/g$$

where U is the wind speed at some reference level (taken here as 10 m), f_m is the peak frequency, and g is the acceleration of gravity; and a second scale parameter characterising the energy level, such as Phillips' constant α or the non-dimensional energy

$$\varepsilon = \frac{<\zeta^2>\cdot g^2}{U^4}$$

where ζ is the surface displacement.

These parameters are in fact not independent, since ε adjusts to a quasi-equilibrium value, dependent on ν, on a time scale small compared with the time scale of changes in ν. In the ε-ν-plane, the equilibrium curve is given by the relation (Hasselmann et al., 1976)

$$\varepsilon = C\nu^{-10/3} \tag{2.2}$$

where C = (5.1) 10^{-6} (Fig. 2). As the waves grow, the equilibrium state migrates to the left along the curve towards lower frequencies and higher energies. A displacement from the equilibrium line, induced, for example, by a rapid change in the wind field, gives rise to restoring terms producing a return to equilibrium along the curves indicated.

The relation (2.2) was deduced from non-linear transfer calculations and an empirical calibration using fetch-limited data for ideal, uniform wind cases, but should be valid generally for growing wind seas and not too rapidly varying wind fields (Hasselmann et al., 1976; data supposedly violating this relation recently presented by Regier and Davis, 1977, were not non-dimensionalised). However, the concept of a quasi-equilibrium spectrum can be regarded only as a limiting approximation which must break down if the time and space scales of the wind field and sea state become too small. The model will also cease to be valid for fully developed spectra, for which the input source function near the peak becomes too small to balance the negative lobe of the non-linear transfer function. In this case a new equilibrium shape is probably established with a flatter peak in accordance with the Pierson-Moskowitz (1964) form. This yields a low-frequency positive lobe of S_{nl} directly at the peak (so that the peak no longer migrates) and a negative lobe at higher frequencies, where it can be balanced by S_{in}.

To discuss the limits of validity of the quasi-equilibrium model it is convenient to consider non-dimensional transfer rates

$$\beta^q_{pr} = \frac{1}{qf_m} \left| \left(\frac{\partial q}{\partial t}\right)_{pr} \right|$$

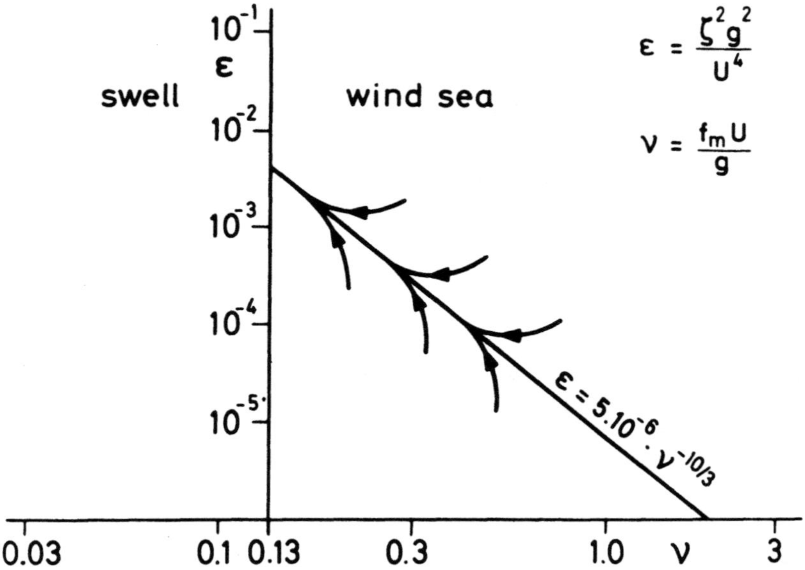

Fig. 2 Evolution of a wind sea in the ε-ν-phase plane (after Hasselmann et al., 1976).

defined with respect to the rate of change $(\partial q/\partial t)_{pr}$ of an appropriate spectral parameter q (or the spectral density itself) due to a particular process pr.

The role of the non-linear interactions can be summarised by two transfer rates:

$$\beta_{nl}^{\alpha} = \frac{1}{\alpha f_m} \left| \left(\frac{\partial \alpha}{\partial t}\right)_{nl} \right|$$

characterises the rate of change of the spectrum due to the negative lobe of the non-linear transfer to the right of the peak (where α is defined here by the spectral level in the band to the right of the peak for which S_{nl} is negative) and the rate of migration of the peak frequency

$$\beta_{nl}^{\nu} = \frac{1}{\nu f_m} \left| \left(\frac{\partial \nu}{\partial t}\right)_{nl} \right|$$

due to the non-linear transfer on the forward face. Both transfer rates exhibit the same functional dependence on the scale parameters ε and ν (as can be seen by dimensional analysis of the Boltzmann scattering integral), but the numerical computations yield an order of magnitude greater transfer rates for the adjustment of the spectral level to the right of the peak than for the peak migration (Hasselmann et al., 1976)

$$\beta_{nl}^{\alpha} = (5.0)\alpha^2 \tag{2.3}$$

$$\beta_{nl}^{\nu} = (0.54)\alpha^2 \tag{2.4}$$

Also of interest are the non-linear transfer rates characteristic for the establishment of the spectral shape. The functional form of β_{nl}^{s}, where s denotes a shape parameter, is again given by (2.3), (2.4), the numerical computations yielding coefficients for β_{nl}^{s} of the same order or larger than for β_{nl}^{α} - as required for the quasi-equilibrium concept.

For the present discussion the transfer rates of the input and dissipation source functions need not be considered separately. To first order the sum $S_{rest} = S_{in} + S_{ds}$ must balance the non-linear energy transfer in the quasi-equilibrium range of the spectrum: $\beta_{rest}^{\alpha} \approx \beta_{nl}^{\alpha}$ for ε, ν on the equilibrium line. The behaviour of β_{rest}^{α} away from the equilibrium line is not critical, as long as it differs sufficiently from β_{nl}^{α} to yield stable net restoring forces. For example, if dissipation is neglected (in the relevant quasi-equilibrium band) and S_{in} is assumed proportional to the spectrum, as suggested by theory, one finds

$$\beta_{rest}^{\alpha} = (5.1)\cdot 10^{-3}\nu^{4/3} \tag{2.5}$$

which yields the required stability. Inclusion of dissipation would yield a still more stable system.

Finally, the advective transfer rate can be characterised by the quantity

$$\beta_{adv} = \frac{v_i}{F f_m}\frac{\partial F}{\partial x_i} = O(L^{*-1}\nu^{-2}) \tag{2.6}$$

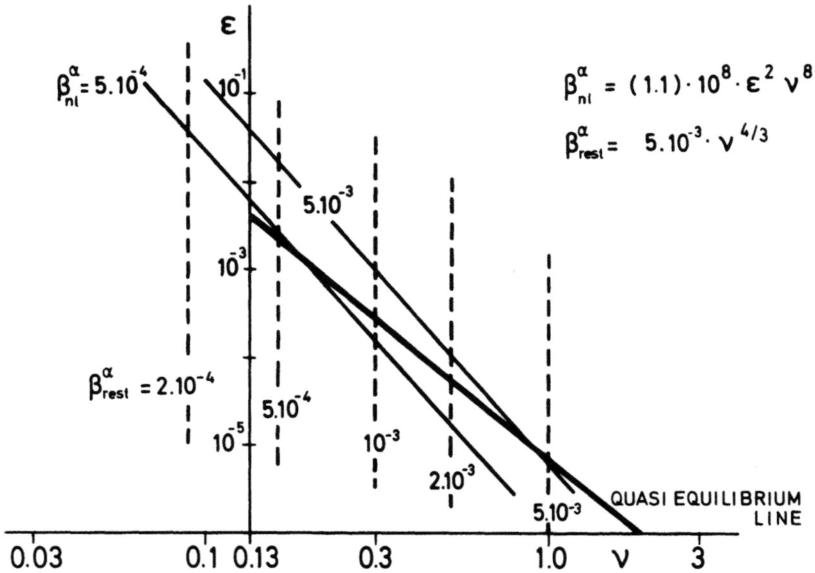

Fig. 3 Nondimensional transfer rates β^{α}_{rest} (assuming $\beta^{rest}_{\alpha} \approx \beta^{in}_{\alpha}$) and β^{α}_{nl}. Along the equilibrium line, $\beta^{\alpha}_{rest} \approx \beta^{\alpha}_{nl}$. The difference $\beta^{\alpha}_{nl} - \beta^{\alpha}_{rest}$ off the equilibrium line yields stabile restoring forces.

where $L^* = Lg/U^2$ is a non-dimensional space scale of the wave field. For simplicity, the parameter index has been suppressed in (2.6). In a more detailed analysis, the space scale should be defined in a reference frame moving with the characteristic propagation velocity of the parameter concerned.

The quasi-equilibrium concept requires both the transfer rate β^{ν}_{nl} for the migration of the peak and the advective transfer rate β_{adv} to be small compared with the transfer rate β^{α}_{nl} which governs the rate of approach to equilibrium and the shape-maintaining transfer rates β^{s}_{nl}. The various transfer rates are summarised in Figs 3 and 4. Since $\beta^{\nu}_{nl} << \beta^{\alpha}_{nl} \lesssim \beta^{s}_{nl}$, we need compare in the following only β_{adv} and β^{α}_{nl}.

In the ε-ν-plane, curves of constant β_{adv} appear as vertical straight lines, changes in the non-dimensional length scale L^* displacing the lines laterally. Examples of β_{adv} values are given

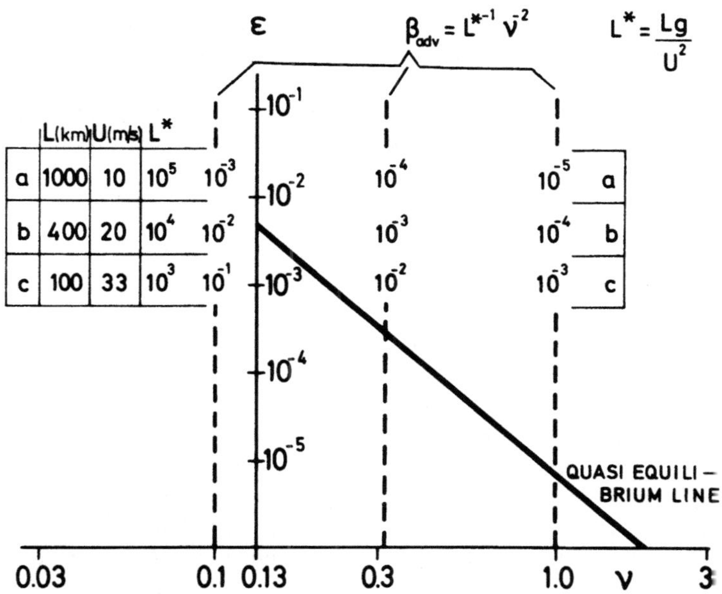

Fig. 4 Nondimensional advection β_{adv} for three generation cases corresponding to (a) large scale, (b) synoptic scale, and (c) hurricane scale wind fields. The quasi-equilibrium conditions are reasonably well satisfied for cases (a), (b), but not for case (c).

in Fig. 4 for (a) a large-scale wind region, (b) a typical mid-latitude cyclonic disturbance, and (c) a hurricane-scale system. For cases (a) and (b) the quasi-equilibrium requirement $\beta_{adv} << \beta_{nl}^{\alpha}$ is well satisfied in the wind-sea range. However, case (c) satisfies this condition only for large ν. Small values of L^* of the same order as in case (c) may be encountered also during frontal passages or for off-shore generation situations with strongly slanting fetches.

In general, the condition $\beta_{adv} << \beta_{nl}^{\alpha}$ becomes more difficult to satisfy as one approaches the fully developed state: as ν decreases along the equilibrium line, β_{nl}^{α} decreases, whereas β_{adv} increases. Close to the fully developed state itself the equilibrium model must break down independently of these considerations since the input distribution can no longer maintain the shape of the spectrum.

A fully developed Pierson-Moskowitz spectrum corresponds in our representation to a non-dimensional peak frequency $\nu = 0.13$ (with U defined at 10 m). For $\nu < 0.13$ the sea state corresponds to swell and the quasi-equilibrium model is not applicable. The transition from the wind-sea to the swell regime occurs through a decrease in wind speed U. A change ΔU in U yields changes

$$\frac{\Delta\nu}{\nu} = \frac{\Delta U}{U}$$

$$\frac{\Delta\varepsilon}{\varepsilon} = -4 \cdot \frac{\Delta U}{U}$$

in the non-dimensional parameters ε and ν. Thus for negative ΔU the wave state in the ε-ν-phase plane is moved upwards and to the left along lines (in a log-log plot) parallel to $\varepsilon = \text{const}\,\nu^{-4}$. Once the wave state lies within the swell region $\nu < 0.13$, the input source function is negligible, and the non-linear transfer then rapidly decreases ε to a level where advection and the non-linear transfer become comparable. Advective dispersion then normally decreases the energy level further, leaving advection as the remaining significant term in the transport equation.

3. THE TRANSITION REGIME

A realistic treatment of the transition regime between the small-mean-path and large-mean-path approximation is important for the transformation of wind sea to swell and for the generation of waves by small-scale, intense wind fields. In the transition regime the redistribution of energy by the nonlinear transfer remains important, but is not sufficiently rapid to smooth out the past history of individual wave groups coming from different regions of the generation region. Consequently, a wide variety of spectral distributions, reflecting the diversity of possible space-time structures of wind fields, can occur.

Two numerical approaches have been considered for the general case: the traditional discretisation of the transport equation in terms of a large number of frequency-directional spectral bands, with appropriate parameterisation of the non-linear transfer source function; and the application of the general dynamical parametrical method, in which the full transport equation is projected on to a parametrical subspace. The parameter space should contain enough degrees of freedom to define a sufficiently broad class of spectra, and is best taken as the same space as is used to parameterise the non-linear transfer. Neither technique has yet been incorporated fully in a numerical wave model, mainly because the parameterisation of the non-linear transfer integral, which is required for both approaches, has not yet been carried out adequately. Earlier parametrical expressions used by Ewing (1971) and Barnett (1968) were based on computations for a fully developed spectrum and did not model essential features such as the generation and migration of the peak of a growing wind-sea spectrum. These effects are included in the recent models of Günther and Rosenthal (1977), Weare and Worthington (1977) and Günther et al. (1978), but a general parameterisation should include also more complex cases such as directionally skewed spectra, cross seas, and multiple peaks.

Also important for the construction of a physically realistic wave model in the transition regime is the form of the input and dissipative source functions. In contrast to the quasi-equilibrium case, for which the details of S_{in} or S_{ds} turned out to be rather irrelevant, the transition regime depends on the structures of all source functions. If S_{in} were known (and some encouraging progress is evident in the papers presented at this meeting), the source function S_{ds} could possibly be determined using Hasselmann's (1974) relation

$$S_{ds} = -\gamma f^2 E$$

for the dissipation due to white capping and residual energy-balance calculations to infer the unknown dependence of the functional γ on the sea state.

4. CONCLUSIONS

Recent studies of the energy balance of wind waves have not yet been fully incorporated into numerical wave-prediction models. The results of these investigations can be translated rather easily into a numerical prediction scheme for growing wind seas under normal wind conditions. However, wave growth for strongly variable wind fields and the transition from wind sea to swell present numerical modelling problems which have yet to be resolved. In the intermediate regime between weakly and strongly coupled wave trajectories, in which advection and the non-linear transfer are comparable, a wide variety of spectral distributions can occur, and the representation of the non-linear transfer then poses a difficult parametrisation problem. An extensive program of numerical experimentation, involving exact integrations of the Boltzmann transfer integral for many types of spectra, will probably be needed to resolve this question satisfactorily.

The most efficient numerical treatment of the transport equation when one of the key source functions has to be parameterised needs further investigation. It can be argued that the projection of the complete transport equation on to the parameter space used to represent the non-linear source function is the most economical and consistent approach. On the other hand, the traditional discretisation in terms of a large number of frequency-directional wave bands, although requiring considerably more computer time, is simpler conceptually and algebraically, and yields a more straightforward treatment of the transition to decoupled swell propagation.

REFERENCES

Barnett, T.P., 1968: On the generation, dissipation and prediction of ocean waves. J.Geophys.Res. 73, 513-530.

Ewing, J.A., 1971: A numerical wave prediction method for the North Atlantic Ocean. Deut.Hydrogr.Z. 24, 241-161.

Günther, H., W. Rosenthal, K. Richter, 1977: Parametrical numerical wave prediction tested in wind situations varying in time and space. (published in these Proceedings).

Günther, H., W. Rosenthal, T.J. Weare, B.A. Worthington, K. Hasselmann, and J.A. Ewing (1978): A hybrid parametrical wave prediction model. (to be published)

Hasselmann, K., 1964: On the spectral dissipation of ocean waves due to white-capping. Bound.-Layer Meteor. 6, 107-127.

Hasselmann, K., T.P. Barnett, E. Bouws, H. Carlson, D.E. Cartwright, K. Enke, J.A. Ewing, H. Gienapp, D.E. Hasselmann, P. Kruseman, A. Meerburg, P. Müller, D.J. Olbers, K. Richter, W. Sell, and H. Walden, 1973: Measurements of wind-wave growth and swell decay during the Joint North Sea Wave Project (JONSWAP). Deut.Hydrogr.Z., Suppl. A, 8, No. 12.

Hasselmann, K., D.B. Ross, P. Müller, and W. Sell, 1976: A parametric wave prediction model. J.Phys.Oceanogr. 6, 200-228.

Mitsuyasu, H., 1966: Interaction between water waves and winds (I). Rept.Res.Inst.Appl.Mech., Kyushu Univ., No. 14, 67-89.

Mitsuyasu, H., 1968: On the growth of the spectrum of wind-generated waves. 1. Rept.Res.Inst.Appl.Mech., Kyushu Univ., No. 16, 459-465.

Mitsuyasu, H., 1969: On the growth of the spectrum of wind-generated waves. 2. Rept.Res.Inst.Appl.Mech., Kyushu Univ., No. 17, 235-243.

Mitsuyasu, H., 1973: The one-dimensional wave spectra at limited fetch. Rept.Res.Inst.Appl.Mech., Kyushu Univ., No. 20, 37-53.

Pierson, W.J., Jr., and L. Moskowitz, 1964: A proposed spectral form for fully developed wind seas based on the similarity theory of S.A. Kitaigorodskii. J.Geophys.Res. 69, 5181-5190.

Regier, L.A., and R.E. Davis, 1977: Observations of the power and directional spectrum of ocean surface waves. J.Marine Res. 35, 433-451.

Weare, T.J., and B.A. Worthington, 1977: A numerical model hindcast of severe wave conditions for the North Sea. (published in these Proceedings).

Wu, H.-Y., E.Y. Hsu, and R.L. Street, 1977: The energy transfer due to air-input, non-linear wave-wave interaction and white-cap dissipation associated with wind-generated waves. Stanford University Techn.Rep. No. 207.

DISCUSSION

D.E. Hasselmann:

It is physically not really entirely obvious that a spectrum remains self-similar if nonlinear transfer and input dissipation only balance approximately. It is hard to believe that if one doubled the input at a specific medium frequency band, that then the spectrum would remain self-similar when integrated over a longer period of time.

K. Hasselmann:

Self-similarity holds if the input and dissipation are smoothly varying functions of frequency - for example, if they are proportional to the spectrum, with slowly varying proportionality factors, as suggested by some theories. Also, there has to be some input in the neighbourhood of the peak. The self-similarity is then a consequence of the very much more rapidly changing nonlinear transfer source function S_{nl} - with its pronounced plus-minus-plus signature in the neighbourhood of the peak - and the fact that the precise distribution of S_{nl} is very sensitive to the sharpness of the peak. Self-similarity would, of course, not hold for your particular case of a spike in the input source function. The spectral shape also changes to a flatter form for a fully developed spectrum, when the input near the peak is probably insufficient to balance the negative lobe of S_{nl} near the peak. For a flatter spectrum, the negative lobe of S_{nl} then shifts to higher frequencies, where it can again be balanced by the input.

D.J. Webb:

1) Your self-similar spectra depend on parts of the spectra being controlled by wave generation and wave breaking. The self-similar assumption has been checked over short fetches in the JONSWAP experiment. On what evidence do you base your parameters for large fetches.

2) Do you think that it is physically reasonable to take the wave dissipation rate as being proportional to the amount by which the spectrum exceeds the f^{-5} part of a Pierson-Moskowitz spectrum.

K. Hasselmann:

1) We have checked the self-similarity concept for a large number of open ocean cases, including non-uniform wind fields, in our paper on the parametrical wave model (J.Phys.Ocean. 6, 200-228, 1976). It appears to be valid generally for growing wind seas. However, as pointed out in my answer to D. Hasselmann's question, it is no longer applicable for a fully developed sea, when the input source function at the peak vanishes.

2) No. My own best guess (Boundary-Layer Met. 6, 102-127, 1974) of the dissipation due to wave breaking is that it is proportional to the spectrum times the square frequency, with a numerical factor which is an (unknown) nonlinear functional of the complete spectrum.

O.M. Phillips:

One must be extremely careful in drawing any analogy between energy loss by wave breaking and viscous dissipation at high wave-numbers. Wave breaking is sporadic and local in physical space; the energy loss is therefore broad band in Fourier space, rather than being concentrated at high wave-numbers.

K. Hasselmann:

I agree that there is no grounds for drawing any analogy between dissipation by viscous stresses and wave breaking. In fact, your point that wave breaking is essentially local in physical space was the basis of my estimate of the source function for this process mentioned in my reply to Webb.

E. Bouws:

Conditions in North Sea can be such that $L^* = Lg/U^2$ becomes much lower than in the example shown, due to high U, limited fetches and strong decay, particularly in southern areas, which causes a much shorter effective scale length due to smaller amount of advection over longer distances. For example, U = 25 m/s, depth h = 30 m (south of Dogger Bank), $L_{effective} \approx 300$ km at most, also for NW gales. It then becomes nearly comparable with the tropical hurricane example.

K. Hasselmann:

I agree that the tropical hurricane is not the only case in which the quasi-equilibrium conditions required for the one-parameter model break down. In your example the length scale is not governed by the wind field, but by the smaller scale of the wave field in a rapidly decaying area. In these situations one would need a multi-parameter model or a traditional spectral model.

B. Golding:

How will the nonlinear transfer mechanism work in an oscillatory wind field such as is found under cumulus convection.

K. Hasselmann:

I suspect that the average shape of the spectrum will still remain more or less the same, but the level of the spectrum would fluctuate rather rapidly as the spectrum responds to the rapidly changing input. This would cause different rates of shift of the peak frequency, and the advection of these different spectra on rather small scales would cause fluctuations also in the peak enhancement and width parameters. In fact, we think much of the variability of the spectral parameters we observed during "ideal" fetch-limited conditions in JONSWAP may be attributed to small scale variability of the wind field.

A TWENTY YEAR NORTHERN HEMISPHERE WAVE SPECTRAL CLIMATOLOGY

Sheldon M. Lazanoff and Norman M. Stevenson

Metoptics Corp., P.O. Box 117 Carmel Valley, Ca. 93924
Fleet Numerical Weather Central, Monterey, Ca. 93940

INTRODUCTION

A Northern Hemisphere twenty year (1956-1975) deep water wave spectral climatology is being developed at Fleet Numerical Weather Central (FLENUMWEACEN), Monterey, California, at the request of a Seakeeping Workshop held in June 1975 under the sponsorship of the U.S. Naval Sea Systems Command. A modified version of the FLENUMWEACEN operational Spectral Ocean Wave Model (SOWM), originally developed under the direction of Professor Willard J. Pierson (Pierson, Tick and Baer, 1966), is being used to compute the wave spectra. The SOWM has been operational since 1974 and provides real time semi-daily analyses and forecasts out to 72 hours for the Navy's routine use. In addition, analyses and forecasts have been provided to special projects such as the Marineland and West Coast experiments sponsored by the Jet Propulsion Laboratory.

The SOWM computes two-dimensional (15 frequencies by 12 directions) wave spectra, using a modified Phillips-Miles growth mechanism. Growth is limited by the Pierson-Moskowitz fully developed wave spectrum for the given wind velocity. The wave energy is spread directionally by the technique developed in the Stereo Wave Observation Program (SWOP). Wave energy is propagated across approximately 2000 grid points throughout the North Pacific and North Atlantic Oceans on an Icosahedral-Gnomonic projection with grid spacing ranging from 160 km to 300 km. The Ocosahedral-Gnomonic projection, which allows great circles to be represented by straight lines within the triangles, is shown in Figure 1. The North Pacific and North Atlantic Oceans are covered by 7 and 6 triangles respectively. Wave energy dissipation is a function of

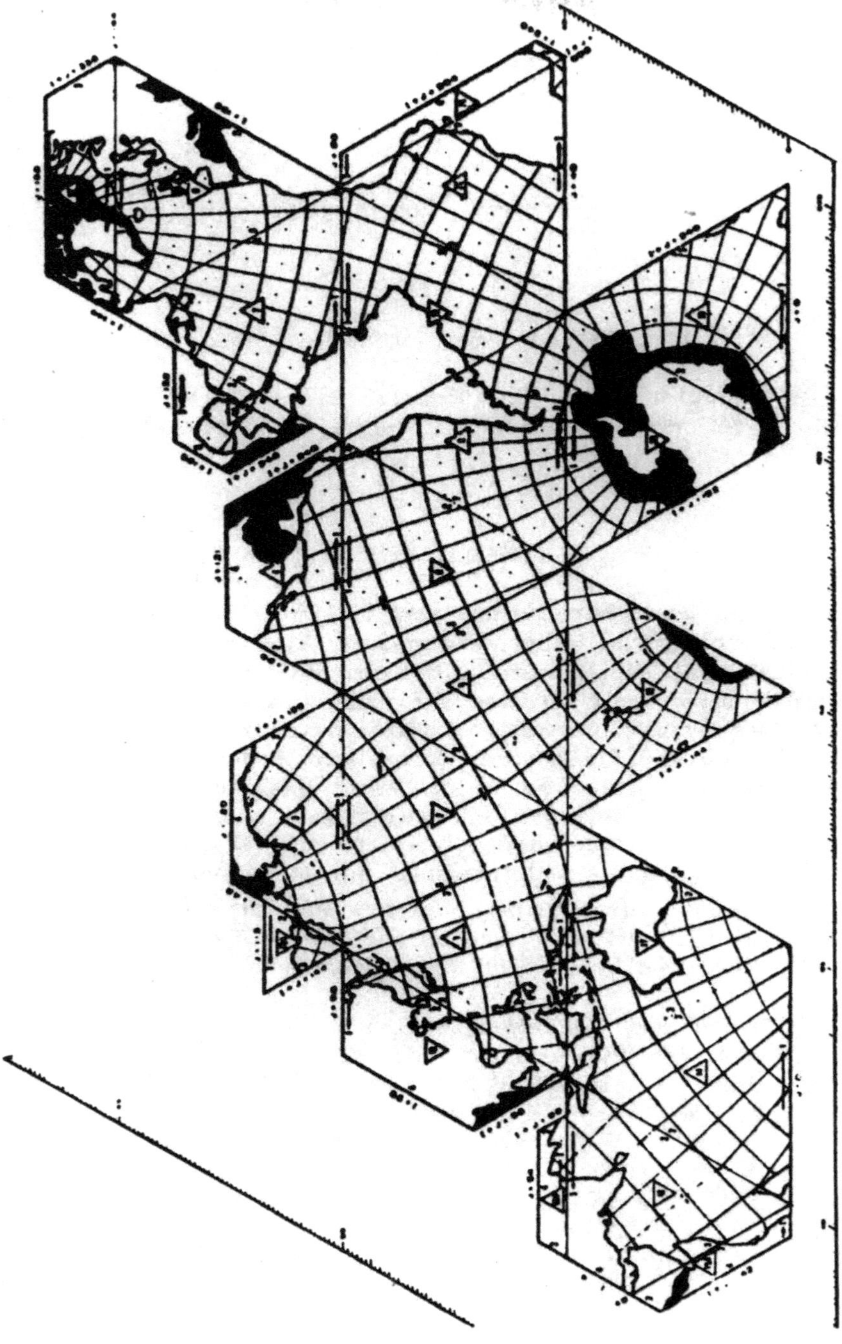

Figure 1. Map of the world ocean for wave studies.

wind direction (wind direction 180° to the wave energy direction causes the most dissipation), frequency to the fourth power, and energy in the wind sea. The model uses a three hour time step. The SOWM does not contain any non-linear effects.

The model SOWM has been evaluated against calculated one-dimensional (frequency) wave spectra obtained from wave data measured by instruments (Lazanoff and Stevenson, 1975). The measured devices include Ocean Station Vessel PAPA (OSV P) Tucker wave recorder, airborne lasers and NOAA Data Buoy Office (NDBO) wave buoys. Special projects such as GATE and Marineland as well as people involved in Optimum Track Ship Routing (OTSR) also provide verification data. Routine visual ship observations for a number of reasons which won't be discussed in this paper, are not considered reliable enough to aid in the verification. Although there has been a definite lack of accurately measured wave data available for verification, the model, generally speaking, produces rather good results on a routine basis. Most of the errors have been traced primarily to inaccurate wind velocity input and secondarily to poor land-sea dissemination in certain areas because of the large mesh length. Unfortunately, there hasn't been enough continuous measured wave spectra data available to FLENUMWEACEN to determine if the SOWM itself has definite weaknesses during various climatic conditions. Hopefully, with the addition of a satellite wave data and NDBO wave spectra, a complete evaluation of the SOWM can be made.

WIND VELOCITY INPUT

The most significant problem that had to be overcome in the production of the twenty year climatology was the calculation of the best wind velocity input for the SOWM on a continuous basis over the twenty year period. First a contract was let to a private company to produce Northern Hemisphere surface pressure analyses every six hours for twenty years. The surface pressure analyses were derived by using previously computed surface pressure analyses as a first guess and upgrading the analyses with surface pressure and wind velocity observations from the U.S. National Climatic Center land and ship data file. The wind velocity data were used to calculate surface pressure gradients. When original surface pressure analyses were not available, the observations were used to produce new surface pressure analyses. The new surface pressure analyses were computed backward in time and were compared to the original surface pressure analyses which were computed forward in time. Examples of the first guess and the reanalyzed surface pressure fields are shown in Figures 2 and 3 respectively. The barbed arrows indicate wind velocity reports in knots. For this report the winds connected with the low pressure system whose center of

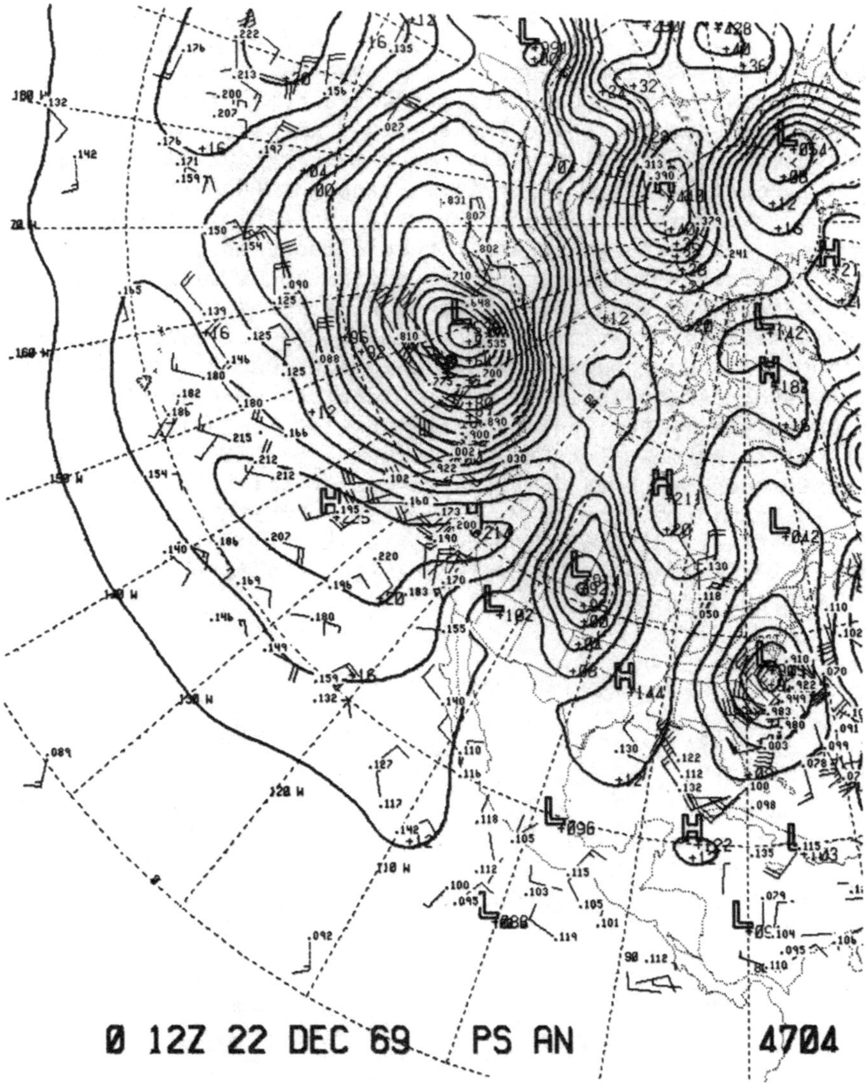

Figure 2

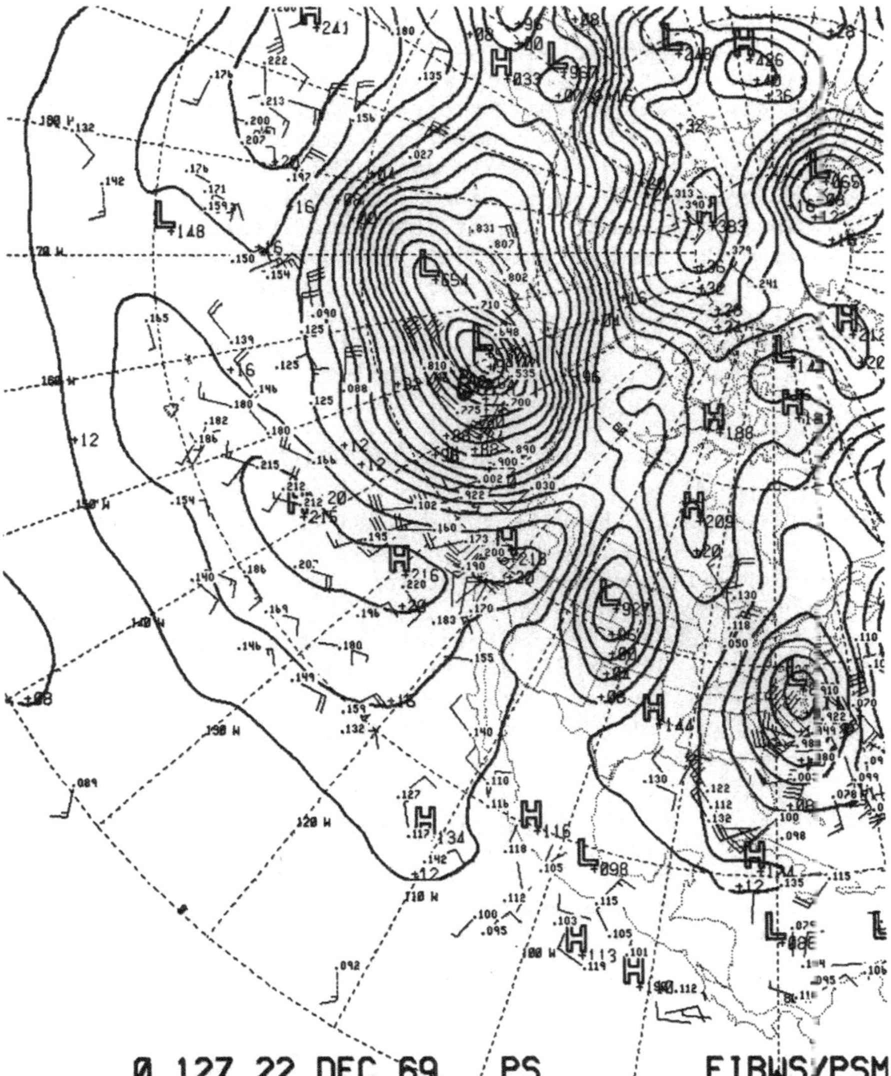

Figure 3

953 mb is located at approximately 55°N, 148°W in Figure 2 will be discussed in a later section. It can be seen in Figure 3 that the reanalysis has altered the shape of the low pressure system and tightened the pressure gradient contours.

An algorithm was developed which computes modified gradient wind velocities and accounts for air-sea temperature differences for all grid points. The winds were adjusted to an elevation of 19.5 meters above sea level. The algorithm supposedly was calibrated by comparing its wind velocities with wind velocities computed by the FLENUMWEA-CEN operational Planetary Boundary Layer (PBL) model, originally developed by Professor Vincent Cardone (Cardone, 1969), for the same period; however, it appears, as will be shown later, that the algorithm seems to underestimate high wind speeds. The cause of the errors is not known at this time since the problem was only determined prior to this symposium and is currently under investigation at FLE-NUMWEACEN. The wind field (in knots) computed for the surface pressure field in Figure 3 is shown in Figure 4.

PROCESSING OF THE CLIMATOLOGY

The processing of the climatology began in January 1977 and this point has been completed from 1 September 1964 to the present for the North Pacific Ocean. The processing is being done on two computers, the CDC 6500 and the CDC Cyber 170. Three days of wave spectra can be processed in one hour of computer processing time and 2.5 hours of input/output time on the CDC 6500 computer. Seven days of wave spectra can be processed in 0.5 hours of computer processing time and 2.5 hours of input/output time on the Cyber 170 computer. Four years of climatology are being processed simultaneously in order to make maximum use of the available computer resources at FLENUMWEACEN. In order to merge years in the most effective manner, it was decided to begin each climatological year on August 15 of one year, since, generally speaking, the Northern Hemisphere oceans have the least amount of energy during late summer periods, and terminate the climatological year on August 31 of the next calendar year. The wave spectra for the first seventeen days of computation are not saved.

MODEL VERIFICATION

In order to be able to use the climatology with some degree of confidence, the climatological wave spectra have to be compared to computed wave spectra from instruments which accurately measure waves. As mentioned earlier in this paper, ship observations are not considered reliable enough to aid in the verification. Unfortunately, there is a scarcity of good measured data. It had

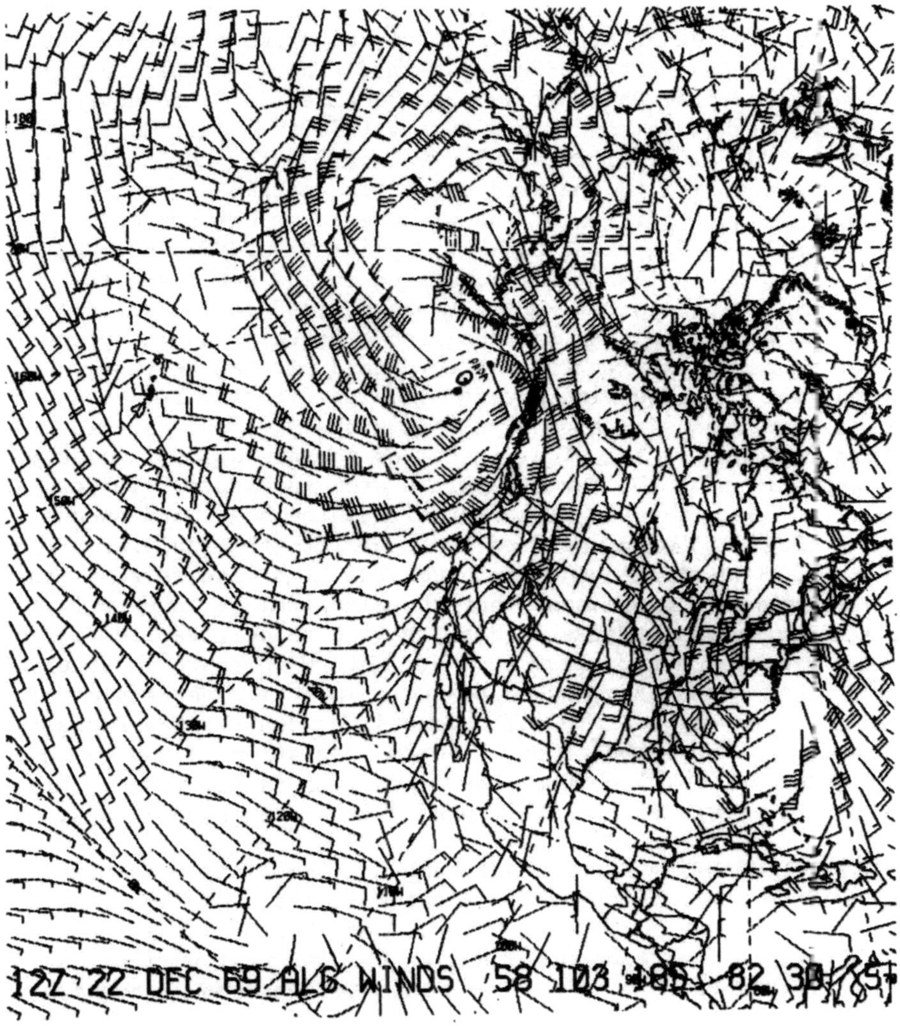

Figure 4

been hoped that wave spectra from the NOAA EB-03 buoy, which is located at 56°N, 145°W could be used; however, it appears that the wave measuring device may have had some hardware problems and it was decided not to use these data until it can be determined if the problems can be overcome. The only other North Pacific data available to us at this time are wave spectra obtained from the Canadian Ocean Station Vessel PAPA (OSV P) Tucker wave recorder. OSV P is located at 50°N, 145°W. Seventy-eight wave spectra were computed using the Blackman-Tuckey method. The wave spectra span the time period from Debember 1968 to December 1970 and cover most of the high wave conditions during this time period. In addition the SOWM climatological wind velocities for the same time period were compared to the wind velocities obtained from the OSV P ship logs. A portion of the wind velocity comparisons are shown in Table 1. The three SOWM grid points, 50.9°N, 145.6°W, 50.2°N, 150.5°W, 48.5°N, 142.5°W, listed in the table are the closest grid points to OSV PAPA. It was decided to look at all the surrounding points in case the analyzed surface pressure systems were not propagating with the correct speed. It can be easily observed that the SOWM wind speeds are significantly less than the measured wind speeds in high wind conditions. Two such cases are outlined in Table 1. The wind directions compare quite favorably which suggests that the low pressure system in Figure 3 is positioned correctly. Thus, it would seem that either the high wind speeds logged by OSV PAPA were not used as input into the analyses or the algorithm did not accurately reproduce the high wind speeds. From past experience it would seem that the algorithm is incorrect. A scatter diagram comparing all (82) the wind velocities from OSV P with the wind velocities from 50.9°N, 145.6°W is shown in Figure 5. The plot reinforces what was observed in Table 1. The mean or bias, 3 m/s, is the average of the differences between the OSV P observations and the SOWM calculations. The RMS deviation, 7 m/s, between the differences indicates that the scatter is rather large. This is particularly true when the OSV P wind speeds are above 15 m/s.

The accompanying significant heights ($\overline{H}_{1/3}$) and modal periods obtained from the wave spectra are shown in Table 2. The significant wave height of 13.9 m for 0900 GMT, 22 December 1969 was the highest wave height obtained from OSV P during the two year period. The scatter diagram for the significant wave height comparisons are shown in Figure 6. Surprisingly, the mean difference 0.1 m is much better than would be expected from the mean wind speed difference. It is difficult to say at this time why the significant wave height mean difference is much better than the wind speed mean difference; however, one possible explanation is that since the significant wave height represents both wind wave and swell energies, the swell energy has less error. The significant wave height RMS deviation is 2.3 meters which is also rather large.

TABLE 1
COMPARISON BETWEEN WIND VELOCITIES (M/SEC) FOR
OSV PAPA (50°N, 145°W) AND SOWM GRID POINTS

DTG	OSV P SPD/DIR	50.9°N, 145.6°W SPD/DIR	50.2°N, 145.5°W SPD/DIR	48.5°N, 152.5°W SPD/DIR
69110121	5/020°	10/165° -	7/207°	9/164°
69110203	16/310°	calm	7/280°	7/233°
69110209	16/260°	12/236°	13/257°	13/233°
69110215	16/240°	15/236°	14/260°	14/232°
69122103	2/140°	8/120°	11/136°	4/174°
69122109	7/150°	9/134°	15/134°	8/199°
69122115	19/110°	15/144°	17/130°	16/185°
69122121	16-20/130°	16/134°	9/100°	22/175°
69122203	24-30/220°	10/160°	4/270°	18/207°
69122209	25-30/220°	13/194°	9/274°	22/211°
69122215	20-25/230°	16/198°	12/245°	18/211°
69122609	9/220°	10/232°	14/206°	9/256°
69122615	16/190°	14/206°	23/192°	10/237°
69122621	20/200°	21/204°	24/202°	16/2[illegible]6°
69122703	17/220°	21/216°	20/212°	20/221°
69122709	17/220°	18/216°	19/206°	21/224°
69122715	15/210°	18/219°	21/210°	19/230°
69122721	12/200°	20/215°	22/205°	18/223°

TABLE 2
COMPARISON BETWEEN $\bar{H}_{1/3}$ (METERS) AND MODAL PERIODS FOR OSV PAPA (50°N, 145°W) AND SOWM GRID POINTS

DTG	$\bar{H}_{1/3}$	OSV P MODAL PER	50.9°N, 145.6°W $\bar{H}_{1/3}$	50.9°N, 145.6°W MODAL PER	50.2°N,150.5°W $\bar{H}_{1/3}$	50.2°N,150.5°W MODAL PER	48.5°N, 142.5°W $\bar{H}_{1/3}$	48.5°N, 142.5°W MODAL PER
69110121	4.5	10.9	5.7	10.9	5.7	10.9	5.1	10.9
69110203	4.1	10.4	5.1	10.9	3.7	13.9	4.6	10.9
69110209	4.0	11.8	4.8	14.9,12.3	4.9	13.9	4.7	10.9
69110215	7.8	12.3	5.0	10.9	5.3	13.9,16.4,10.9	4.5	10.9
69122103	4.9	11.4	4.4	14.9	4.6	12.3,13.8	5.2	14.9
69122109	4.9	11.8	4.5	13.8	6.2	13.9	4.9	14.9
69122115	5.3	12.3	7.1	17.8	8.9	13.9	6.2	14.9
69122121	8.7	8.9	8.1	14.9	8.3	16.4,13.8	10.2	14.9
69122203	9.7	11.1	8.9	14.9	7.3	16.4	10.0	14.9
69122209	13.9	14.2	8.7	14.9	7.1	16.4	10.6	16.4
69122215	10.1	12.7	8.7	14.9	8.5	16.4	10.6	14.9
69122609	3.5	11.8	3.9	10.9	5.0	10.9	3.0	12.3
69122615	3.4	10.3	5.2	10.9	9.5	12.3	2.9	12.3
69122621	missing data		9.2	14.9	11.2	14.9	4.8	13.9,16.4
69122703	7.5	9.8,11.8	10.1	16.4	11.4	17.9,14.9	6.0	17.9
69122709	7.6	12.4	10.1	17.8	11.6	17.9,14.5	6.5	14.9,17.9
69122715	8.6	14.2	10.0	17.8	11.3	17.9,14.9	6.5	14.9,17.9
69122721	7.3	19.2	9.6	17.8	10.6	14.9,17.9	5.9	12.9,14.9

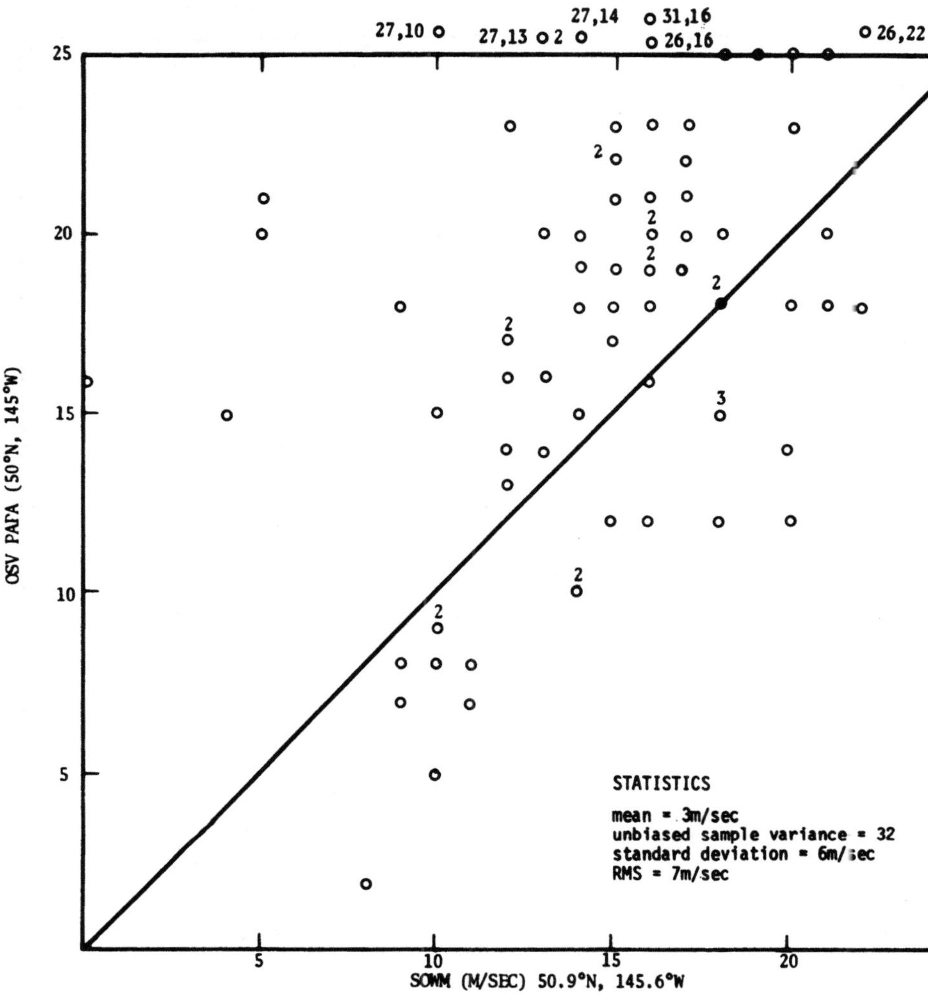

Figure 5. Wind speed (M/S) comparison OSV P, 50.9° N, 145.6° W.

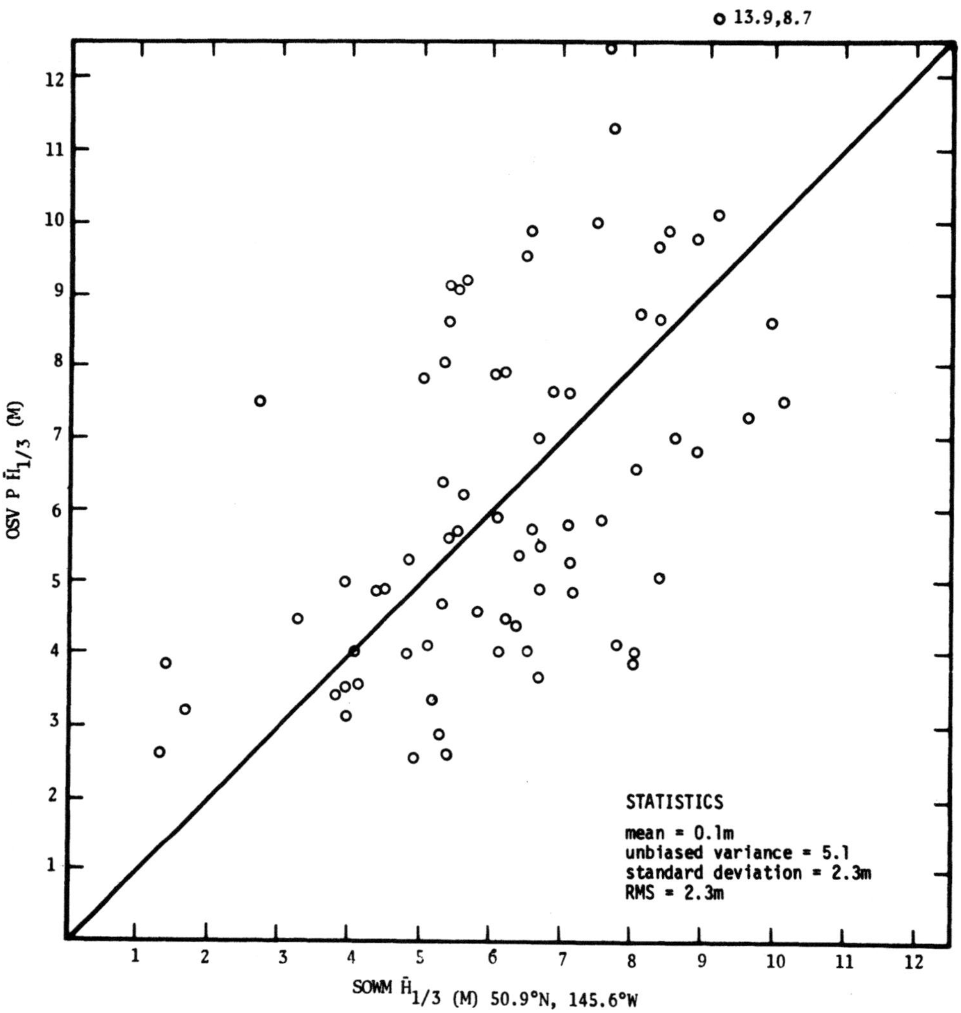

Figure 6. $\overline{H}_{1/3}$ (M) comparisons - OSV P, 50.9° N, 145.6° W.

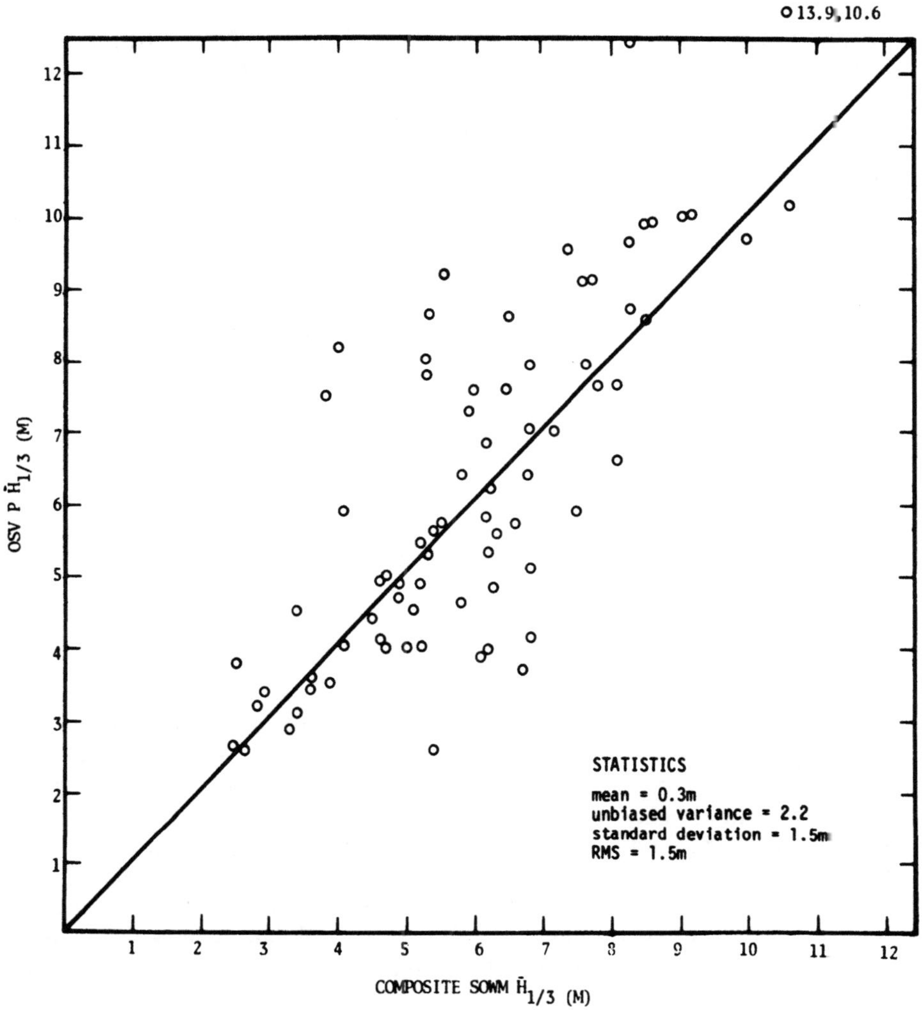

Figure 7. $\overline{H}_{1/3}$ (M) Comparisons - OSV P, composite sown $\overline{H}_{1/3}$.

TABLE 3
SUMMARY OF STATISTICS
OSV P - SOWM GRID POINT

WIND		$\bar{H}_{1/3}$
	50.9°N, 145.6°W	
MEAN = 3 m/s RMS = 7 m/2		MEAN = 0.1 m RMS = 2.3 m
	48.5°N, 142.5°W	
MEAN = 3 m/s RMS = 6 m/2		MEAN = 0.4 m RMS = 2.3 m
	50.2°N, 150.5°W	
MEAN = 3 m/s RMS = 7 m/s		MEAN = 0.3 m RMS = 2.5 m
	COMPOSITE $\bar{H}_{1/3}$	
		MEAN = 0.5 m RMS = 1.5 m

It was decided to make a composite significant wave height scatter diagram containing the best comparisons between all of the grid points and OSV P. The results are shown in Figure 7. In this case, although the mean difference increased to 0.5 m, the RMS deviation decreased to 1.5 m which is considerable improvement. A summary of the mean difference and RMS deviations is shown in Table 3 for all three grid points. It appears that although the SOWM wind speed errors in all cases are bias toward the low side, the significant height errors seem to be significantly less bias which is rather curious.

CONCLUSION

This report demonstrates the need for accurate wind velocity input for the calculation of wave spectra. No matter which wave model is used, if the wind velocity input is inaccurate, the wave spectral models will be inaccurate. In this case either the surface pressure analyses which were used as input to the wind algorithm or the wind algorithm itself seemed to be incorrect. Since higher wind speeds had been calculated in the past from the original pressure analyses, it would seem that the algorithm is probably incorrect. In either case, the errors will be corrected and the wave spectral climatology will be redone.

One significant point to note is that although the wind speed input to the SOWM was biased toward the low side, the significant wave height errors tended to be random. For climatological wave studies such as ship design studies or environmental impact reports it is more important that the long term errors are random than if the analyses are precisely accurate for a given day. A RMS error of 1.5 m is also desirable.

REFERENCES

Cardone, V.J., Specification of the wind distribution in the marine boundary layer for wave forecasting, New York University Department of Meteorology and Oceanography, TR 69-1, 131 pp., 1969.

Lazanoff, S.M. and N.M. Stevenson, An Evaluation of a Hemispheric Operational Wave Spectral Model, Fleet Numerical Weather Central TN 75-3, 103 pp., June 1975.

Pierson, W.J., L.J. Tick and L. Baer, Computer based procedures for preparing global wave forecasts and wind field analyses capable of using wave data obtained by a spacecraft, Proceedings of the Sixth Naval Hydrodynamics Symposium, 1966.

DISCUSSION

J. A. Businger:

The algorithm for the planetary boundary layer developed by Cardone, which was presented by Pierson, should give a reasonable approximation of the windspeed, although the model may be somewhat improved.

S. M. Lanzanoff:

I agree with this statement; however, the input needed for the planetary boundary layer model is difficult to obtain on a climatological basis. Therefore, it was decided to use a simpler algorithm which needed less sophisticated input. It was hoped that even though there could be significant error at any given time, the errors over a twenty year period would be random.

For climatological studies this would be satisfactory.

C. A. Friehe:

Were the comparisons of the analysed wind fields to OWS PAPA independent? That is, were the PAPA data used as input in the analysis?

S. M. Lazanoff:

The OWS PAPA wind velocities were supposed to have been part of the input but it is not known at this time whether they actually were. FNWC personnel are investigating this point. If the OWS PAPA data were not part of the analyses, then they should be added; however, if the data were part of the analyses, then this would seem to indicate that the wind velocity algorithm was not correct.

K. Hasselmann:

For engineering applications of wave climate statistics one really needs more than the significant wave height or significant wave period. One would like the probability distribution of wave spectra. This involves constructing joint probability distributions of a number of characteristic spectral parameters simultaneously. Have you considered how this difficult problem could be tackled?

S. M. Lazanoff:

We have discussed this type of approach but the present studies which are being done in the Gulf of Alaska will not contain this type of information. It is hoped that more time will be allowed to ponder this problem later in the project.

W. Rosenthal:

You have a wave model for the whole earth. Did you have the opportunity to compare with other sea truth data than from weather ship PAPA?

S.M. Lazanoff:

The operational analysis forecast model is compared to other data such as wave spectra from the NOAA Data Buoy Office buoys in the North Atlantic Ocean and aircraft laser measurements in both the North Pacific and North Atlantic. Since, at this time, the climatological wave spectral model has only been run for the North

Pacific Ocean, we only have the OWS PAPA wave spectral data for comparison. More data will be available for comparison when the climatological model is run for the North Atlantic.

I should point out that the comparisons made with the operational model, generally speaking, have been quite satisfactory and most of the errors have been caused by inaccurate wind velocity input rather than from wave model failures.

SEVEN YEARS OF ROUTINE NUMERICAL WAVE PREDICTION WITH THE DSA 5 MODEL

R. Gelci, P. Chavy

Centre de Meteorologie Maritime,
B.P. 337, Plouzane, 29273 Brest Cedex - France

INTRODUCTION

DSA models are spectral models based on three source terms: the theoretical advection term, a dissipation term, and a generation term, the last two being empirically fitted to reproduce the observed fetch and duration dependence.

The first model was in operation in 1956 at the meteorological center of Casablanca. It provided predictions at a single point, using the technique of propagation diagramms of J. DARBYSHIRE. The space-time wind fields were manually plotted from wind ship observations, or, where these were not available, from the geostrophic wind, multiplied by a correction factor of 0.7. Spectral growth was "simultaneous", all components growing independently to their asymptotic equilibrium value given by the PIERSON-NEUMANN spectrum with a cosin-square spreading function. This technique was rather tedious.

In 1960, the introduction of computers permitted the development of a numerical grid point model with an improved damping term, using a formal extension of a result of LAMB. Hereby it was possible to simulate the rapid decrease of the waves after leaving a generation area, the much slower decay of dispersed swell, and the lull preceding the onset of a new period of generation.

The model has not been modified since its application in routine predictions for the North Atlantic Ocean. The details of the model have been described elsewhere, so that we mention here only the main operational characteristics.

The computational grid of 30 x 50 = 1,500 points, with a mesh size of 90 nautical miles, covers the major part of the North Atlantic Ocean (Fig. 1). The plane of the grid is tangent to the sphere at 45°N, 40°W.

There are 96 spectral components, each representing an angular interval of 22.5 degrees and a period interval of 3 seconds.

The time step is 3 hours. The computation time is 22 seconds per prediction day, on a CDC CYBER computer.

Four wind field analyses are used daily at 0000, 0600, 1200 and 1800 GMT, the wind being taken as constant during the 6 hour intervals.

Until 1972, all wind fields were analysed and digitised manually. Subsequently, automatic analyses replaced the manual analyses, first at 1200, then at 0000 and 1800 GMT. Since August 1974, only the 0600 GMT analysis is still made manually.

Manual (i.e. subjective) analyses are based mainly on wind observations, or, if these are lacking or obviously in error, on the gradient wind (multiplied by 0.7). Manual analyses permit good data control, good continuity of successive fields, and reproduce correctly areas of strong gradient.

Automatic (i.e. objective) analysis take into account both pressure and wind observations (but not the wind height).

The model output is presented in two forms: listings of two dimensional spectra for several points of the grid, and synoptic charts of the one-tenth highest wave heights, including arrows of variable length representing the direction and the period of the principal components.

A hindcast (analysis) chart and two forecast charts for 24 and 36 hours are broadcast.

SOME RESULTS

Lack of synoptic sea truth makes the evaluation of a sea state model difficult.

Four different methods of evaluation may be distinguished:

(1) The response of the model to ideal situations, for which the sea state is well known.

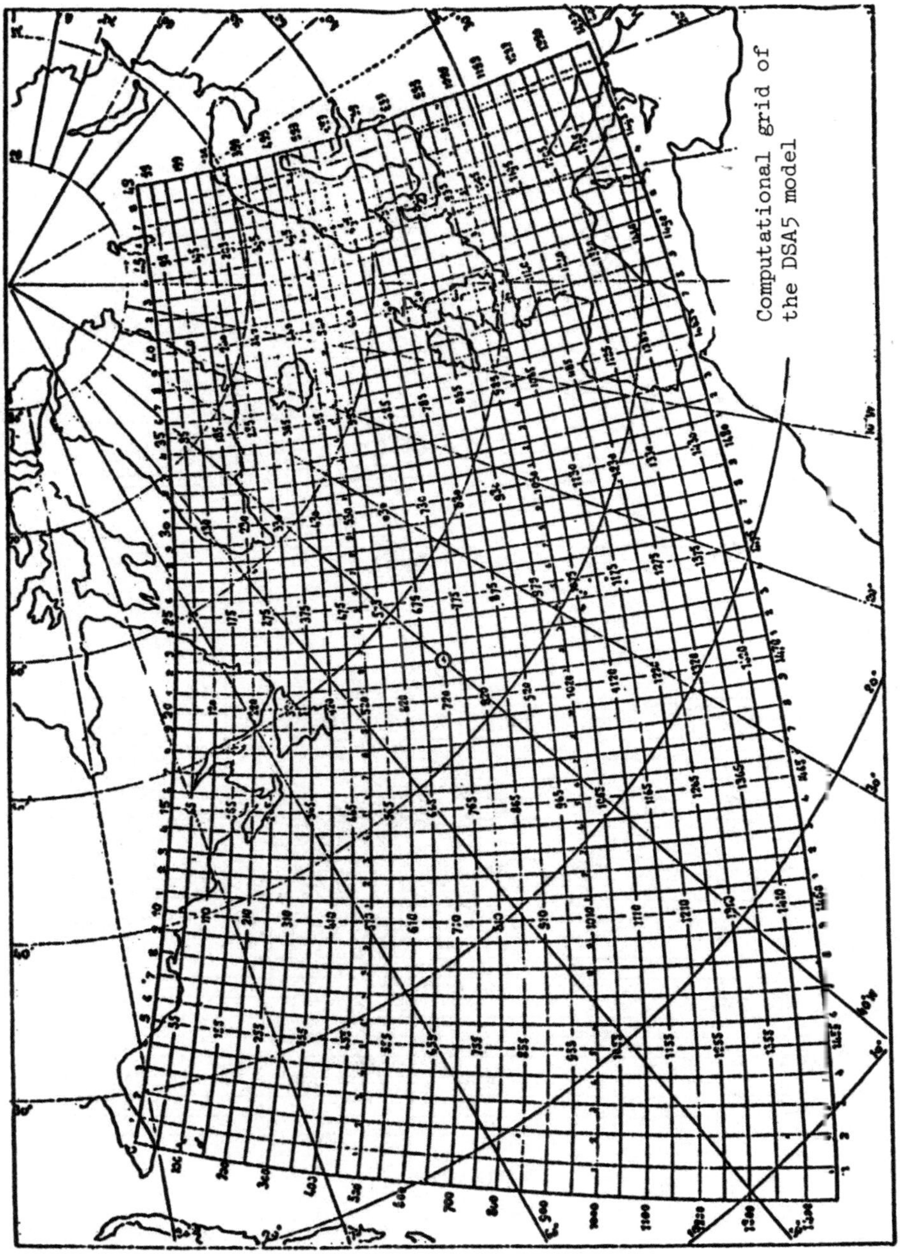

Computational grid of the DSA5 model

Figure 1

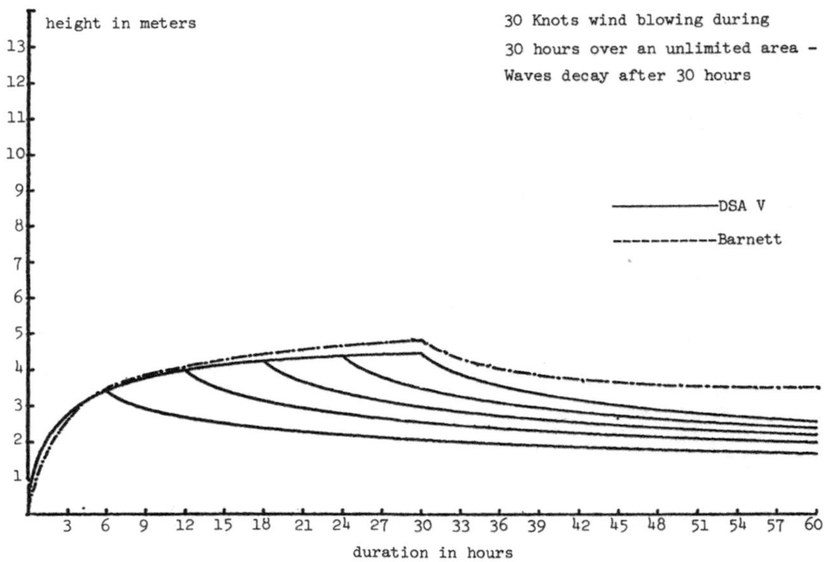
height in meters
30 Knots wind blowing during
30 hours over an unlimited area -
Waves decay after 30 hours
DSA V
Barnett
1
2
3
4
5
6
7
8
9
10
11
12
13
3 6 9 12 15 18 21 24 27 30 33 36 39 42 45 48 51 54 57 60
duration in hours

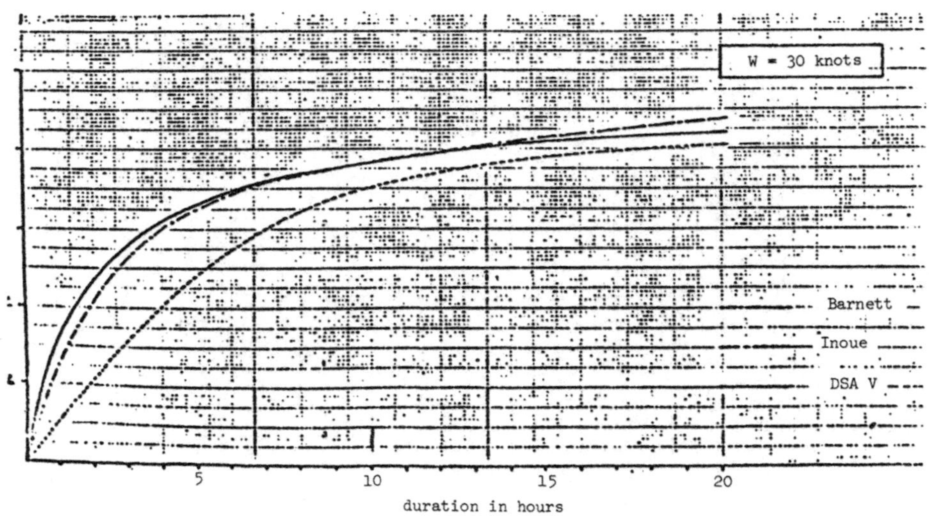
W = 30 knots
Barnett
Inoue
DSA V
5 10 15 20
duration in hours

Figure 2

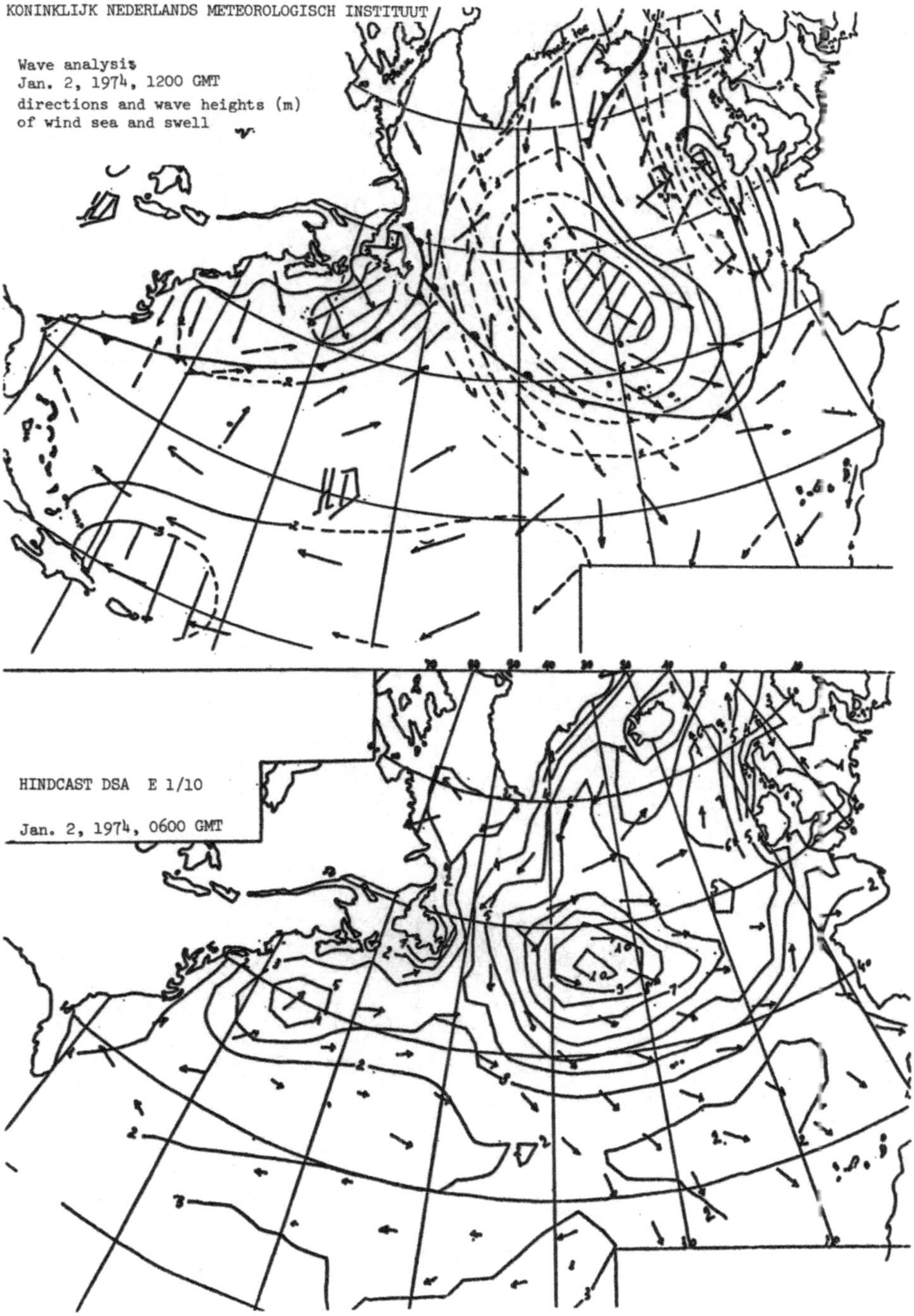
KONINKLIJK NEDERLANDS METEOROLOGISCH INSTITUUT
Wave analysis
Jan. 2, 1974, 1200 GMT
directions and wave heights (m)
of wind sea and swell
HINDCAST DSA E 1/10
Jan. 2, 1974, 0600 GMT

Figure 3

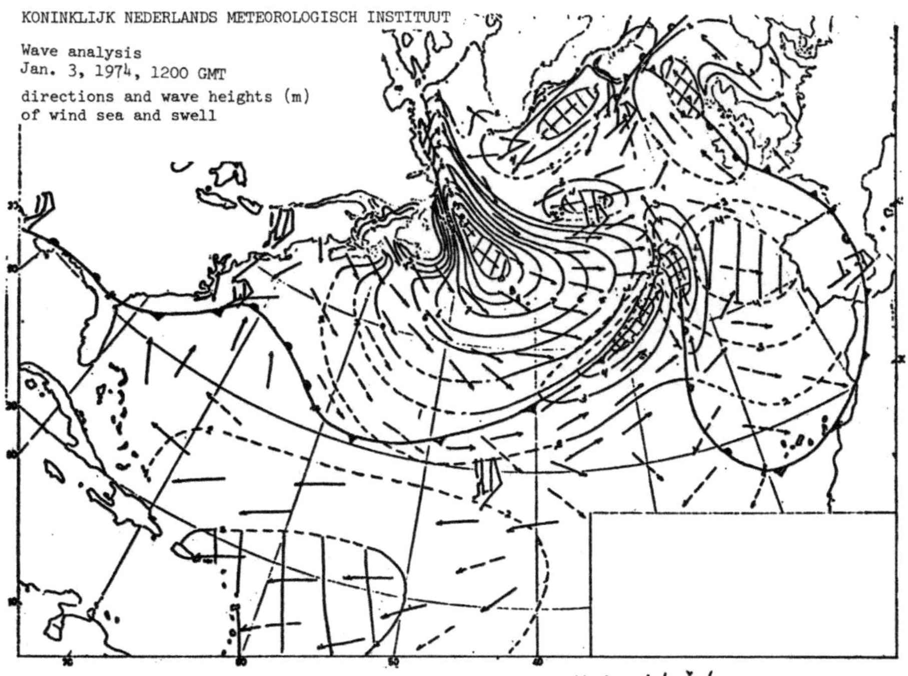
KONINKLIJK NEDERLANDS METEOROLOGISCH INSTITUUT
Wave analysis
Jan. 3, 1974, 1200 GMT
directions and wave heights (m)
of wind sea and swell

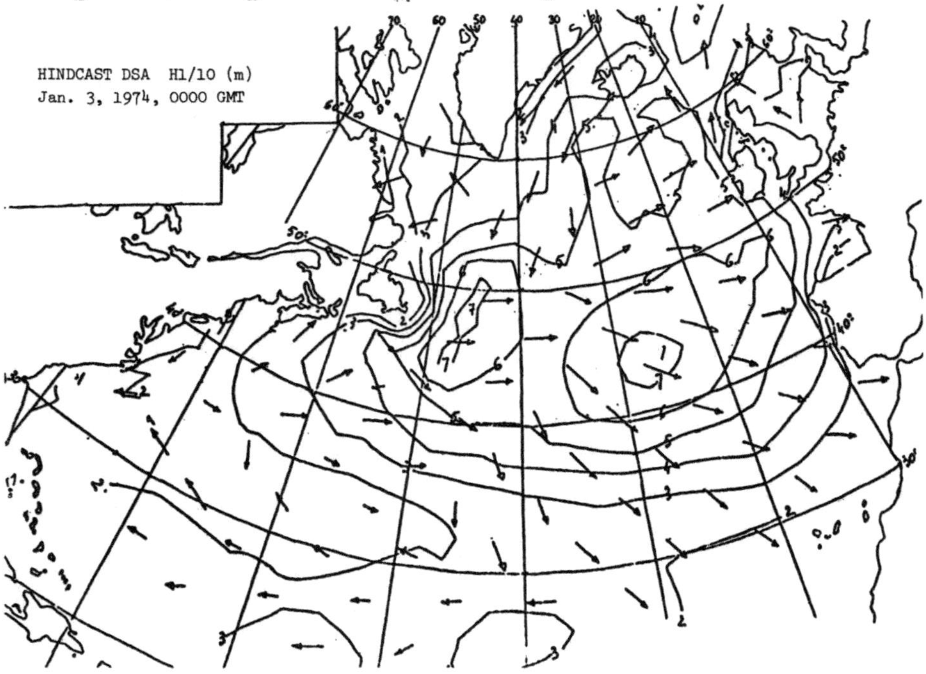
HINDCAST DSA H1/10 (m)
Jan. 3, 1974, 0000 GMT

Figure 4

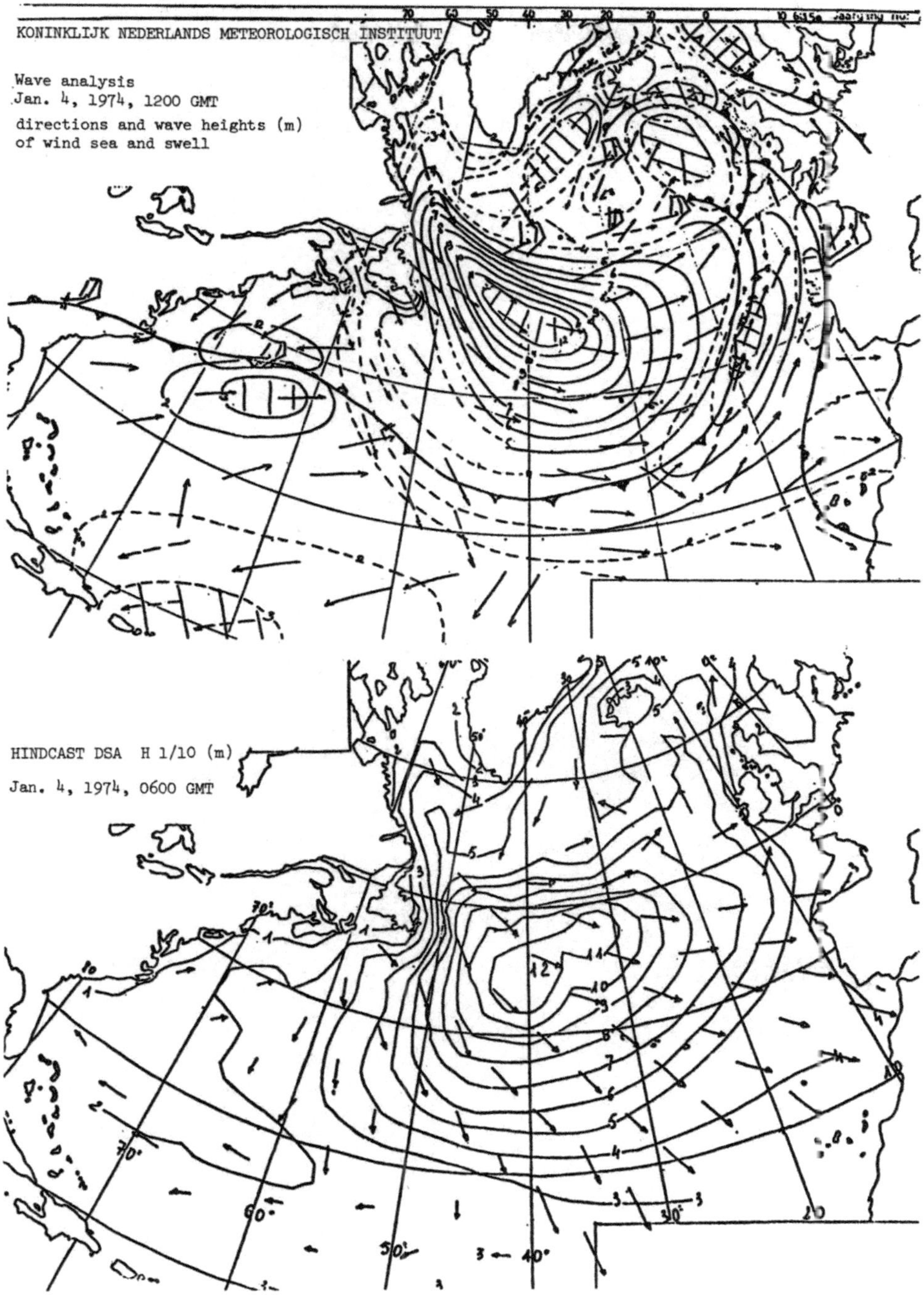
KONINKLIJK NEDERLANDS METEOROLOGISCH INSTITUUT
Wave analysis
Jan. 4, 1974, 1200 GMT
directions and wave heights (m)
of wind sea and swell
HINDCAST DSA H 1/10 (m)
Jan. 4, 1974, 0600 GMT

Figure 5

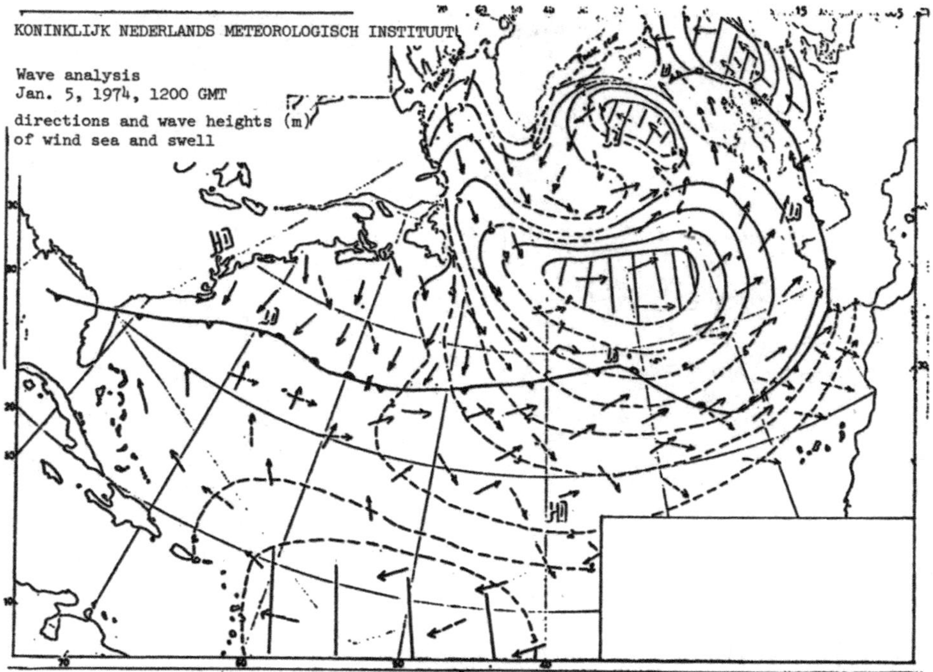
KONINKLIJK NEDERLANDS METEOROLOGISCH INSTITUUT
Wave analysis
Jan. 5, 1974, 1200 GMT
directions and wave heights (m)
of wind sea and swell

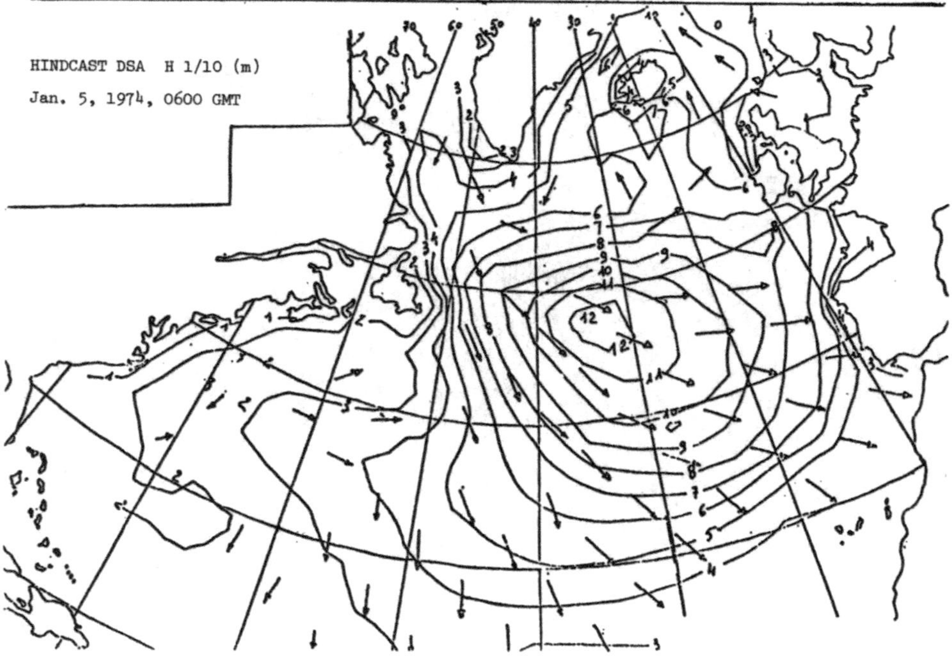
HINDCAST DSA H 1/10 (m)
Jan. 5, 1974, 0600 GMT

Figure 6

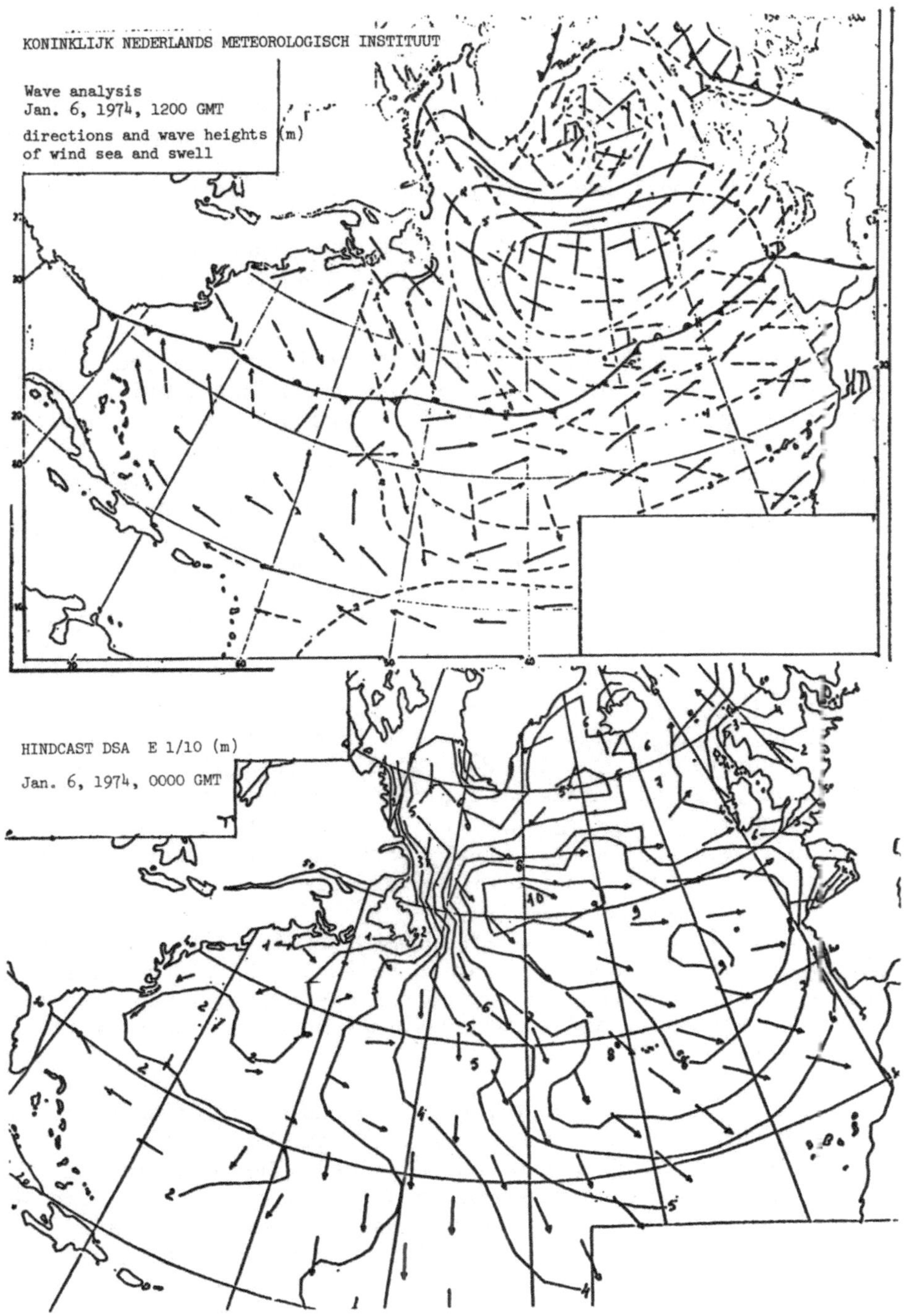
KONINKLIJK NEDERLANDS METEOROLOGISCH INSTITUUT
Wave analysis
Jan. 6, 1974, 1200 GMT
directions and wave heights (m)
of wind sea and swell
HINDCAST DSA E 1/10 (m)
Jan. 6, 1974, 0000 GMT

Figure 7

(2) The comparison with synoptic analyses derived from visual sea state observations of meteorological ships.

(3) The comparison with instrumental wave recordings at individual positions.

(4) The "reasonableness" of hindcast wave climatology.

With respect to the first test, the model response does not differ essentially from that of other well tested models, cf. Fig. 2 for the case of thirty knot winds.

The second type of test was carried out in a comparison with analyses of the Koninklijk Nederlands Meteorologisch Instituut. Five such comparisons are represented in Figs 3 to 7. Close agreement between the observations and the model cannot generally be expected, because of the distinction in the observations between sea and swell. Nevertheless, strong generation areas generally coincided (allowing for the different times of the analyses), and where the distinction between sea and swell does not introduce any ambiguity overall agreement between hindcast heights and visually observed heights exists, if the latter are multiplied by 1.30.

A third test was made by comparing the model against a TUCKER wave recorder on the weather ship FRANCE II at weather station KILO during the period from the 28th of December, 1973, to the 22nd of January, 1974. During this period, the measurements indicated three maxima exceeding 7 meters, separated by lulls. Fig. 8 shows a comparison of the wave height throughout the period. Figs 9 to 19 show the hindcast and observed spectra.

Other comparisons made at station KILO yield similar agreement, except for a limited number of discrepancies due to instrumental errors. Particularly we note a systematic overestimation of the medium wave components, and an underestimation of the short and long components.

An interesting multi-point comparison was made during a period of more than three months with an array of five buoys, launched off Britanny by the CNEXO (French Oceanic Office). The comparison is summarized in Fig. 20. The measured values of the one tenth highest waves are bracketed by the two curves, the hindcast values are represented by points, and some visual observations by triangles.

Other comparisons made with measurements nearer to shore generally show less satisfactory agreements.

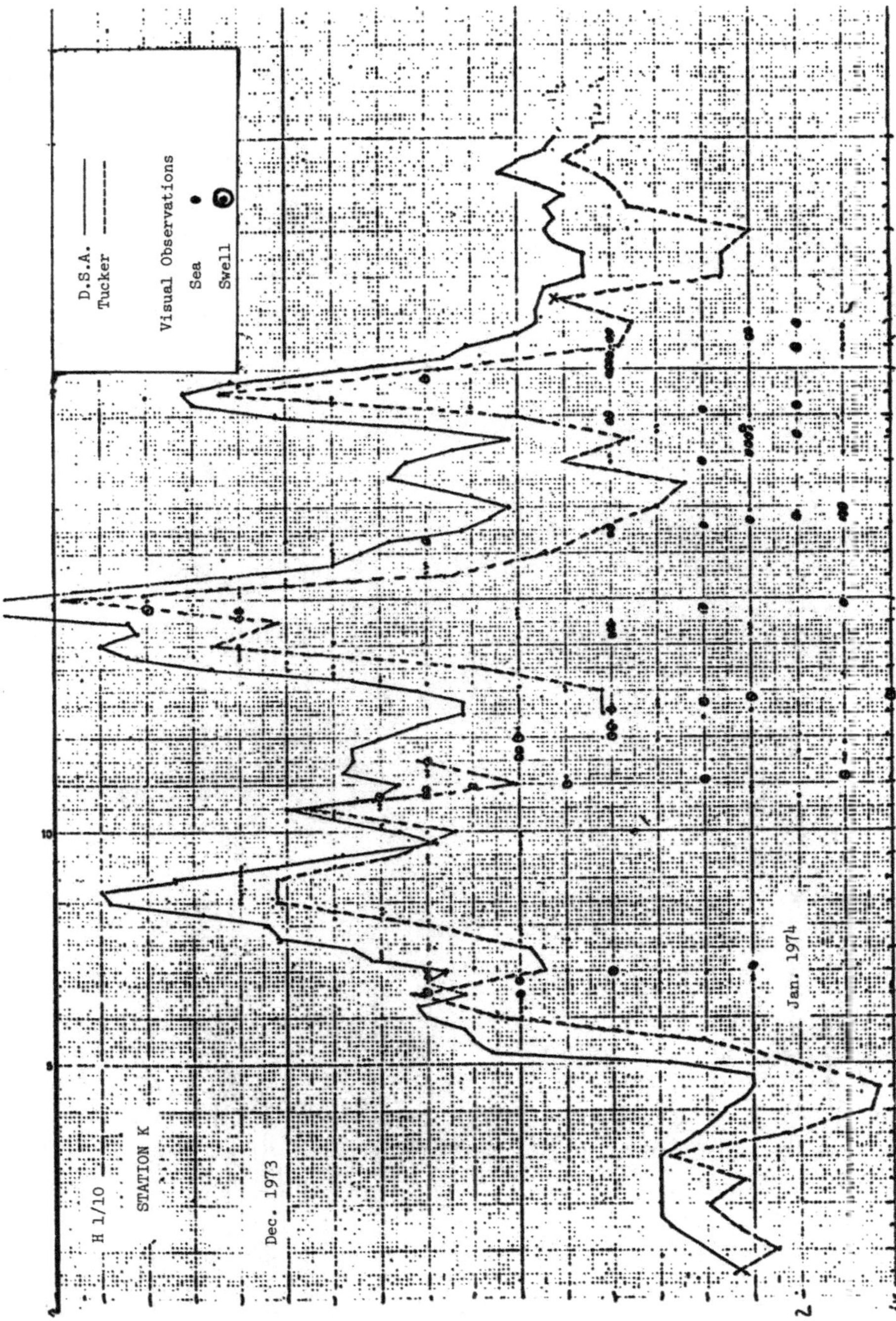
D.S.A.
Tucker
Visual Observations
Sea
Swell
H 1/10
STATION K
Dec. 1973
Jan. 1974
28 29 30 31 1 2 3 4 5 6 7 8 9 10 11 12 13 14 15 16 17 18 19 20 21 22
10
5
2

Figure 8

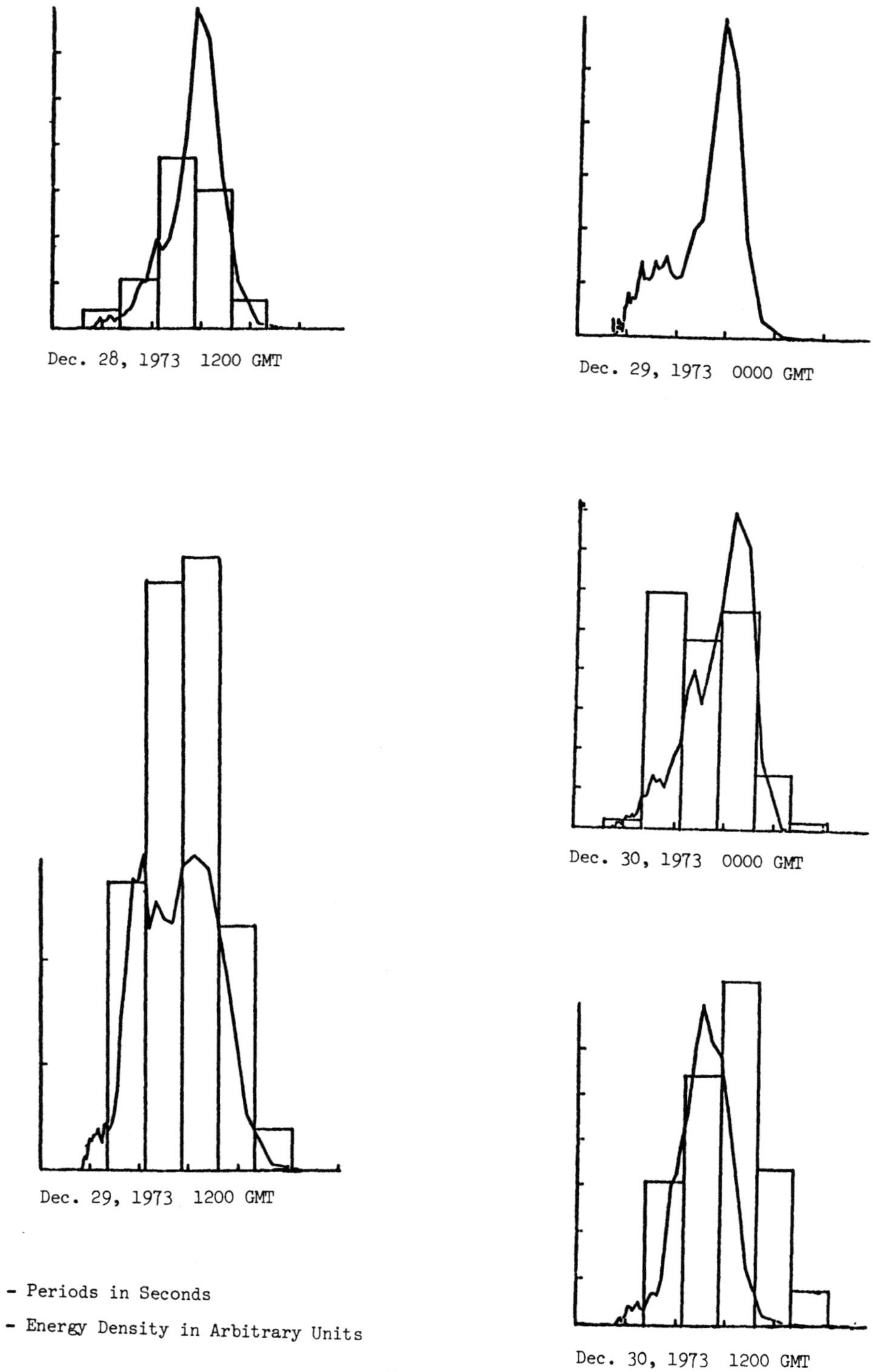
Dec. 28, 1973 1200 GMT
Dec. 29, 1973 0000 GMT
Dec. 30, 1973 0000 GMT
Dec. 29, 1973 1200 GMT
Dec. 30, 1973 1200 GMT
- Periods in Seconds
- Energy Density in Arbitrary Units

Figure 9

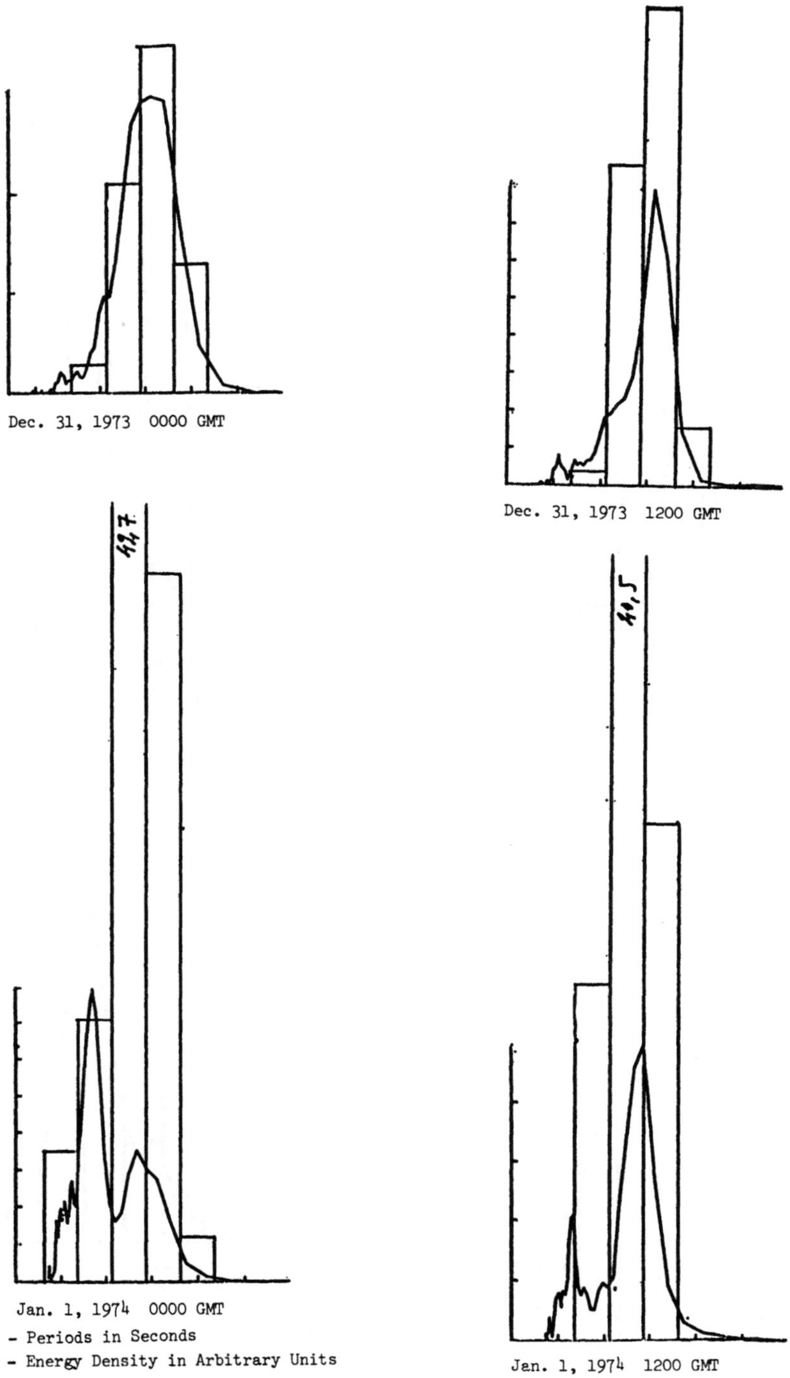
Dec. 31, 1973 0000 GMT
Dec. 31, 1973 1200 GMT
42,7
40,5
Jan. 1, 1974 0000 GMT
- Periods in Seconds
- Energy Density in Arbitrary Units
Jan. 1, 1974 1200 GMT

Figure 10

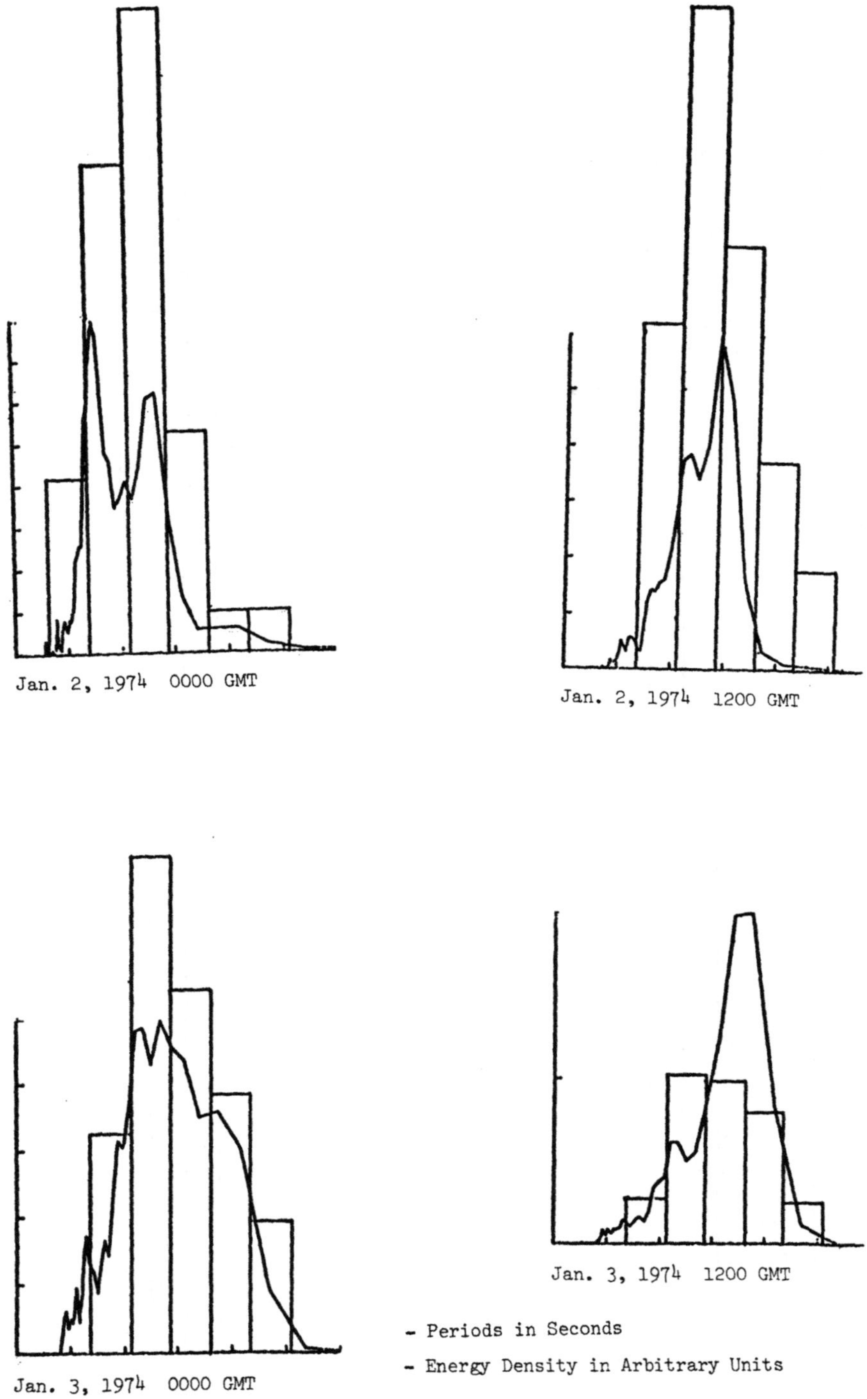
Jan. 2, 1974 0000 GMT
Jan. 2, 1974 1200 GMT
Jan. 3, 1974 0000 GMT
Jan. 3, 1974 1200 GMT
- Periods in Seconds
- Energy Density in Arbitrary Units

Figure 11

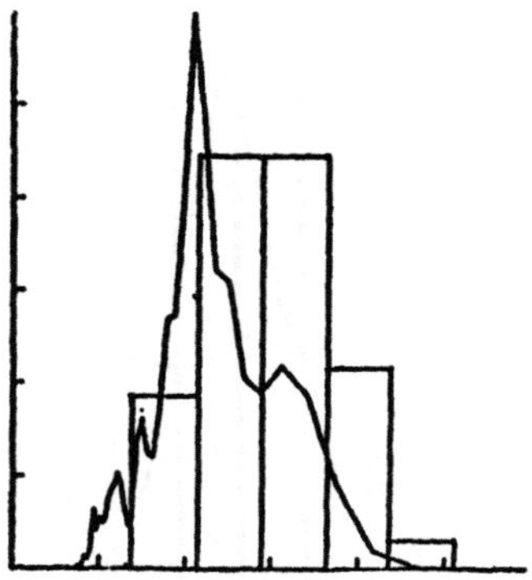

Jan. 4, 1974 0000 GMT

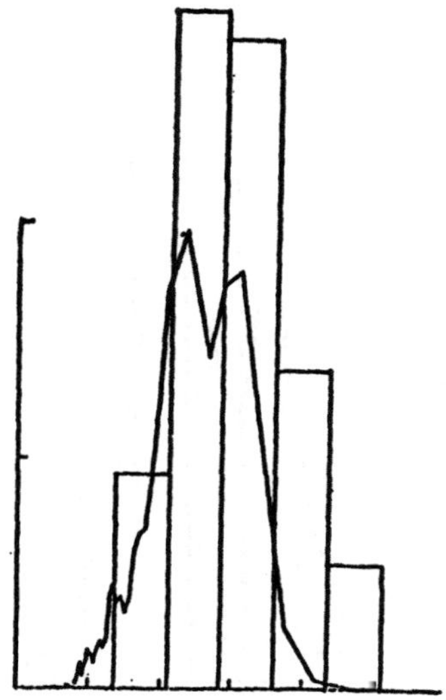

Jan. 4, 1974 1200 GMT

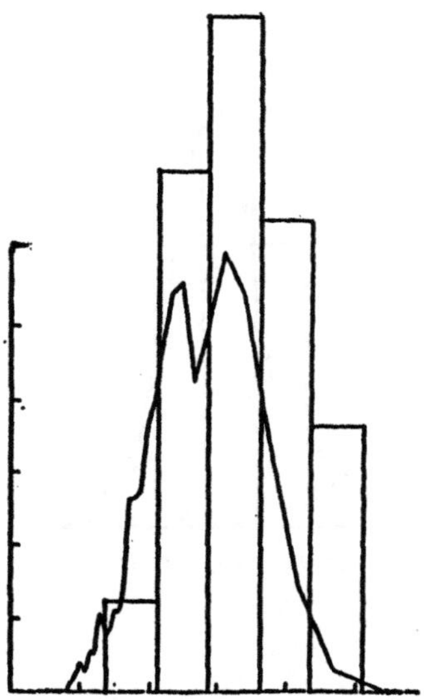

Jan. 5, 1974 0000 GMT

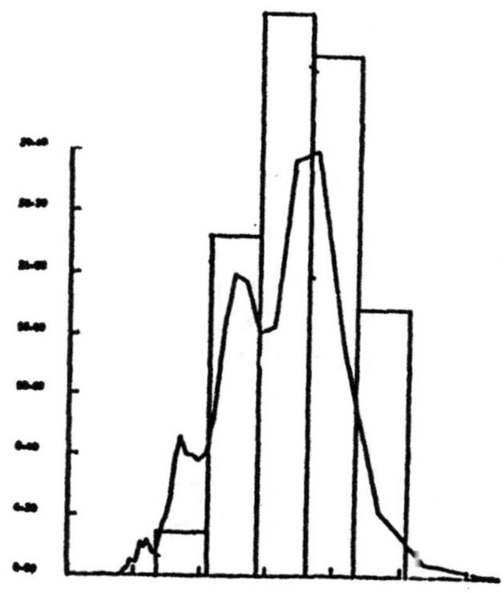

Jan. 5, 1974 1200 GMT

- Periods in Seconds
- Energy Density in Arbitrary Units

Figure 12

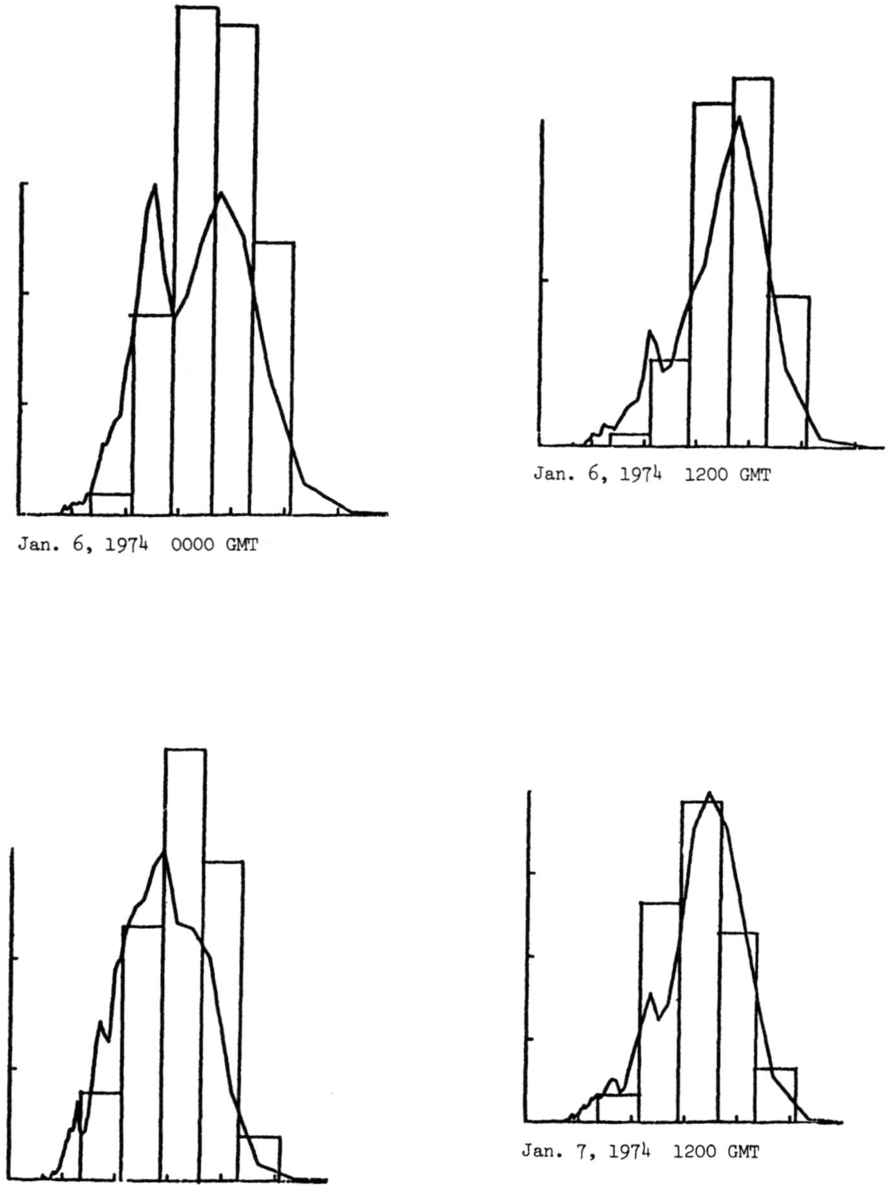

- Periods in Seconds
- Energy Density in Arbitrary Units

Figure 13

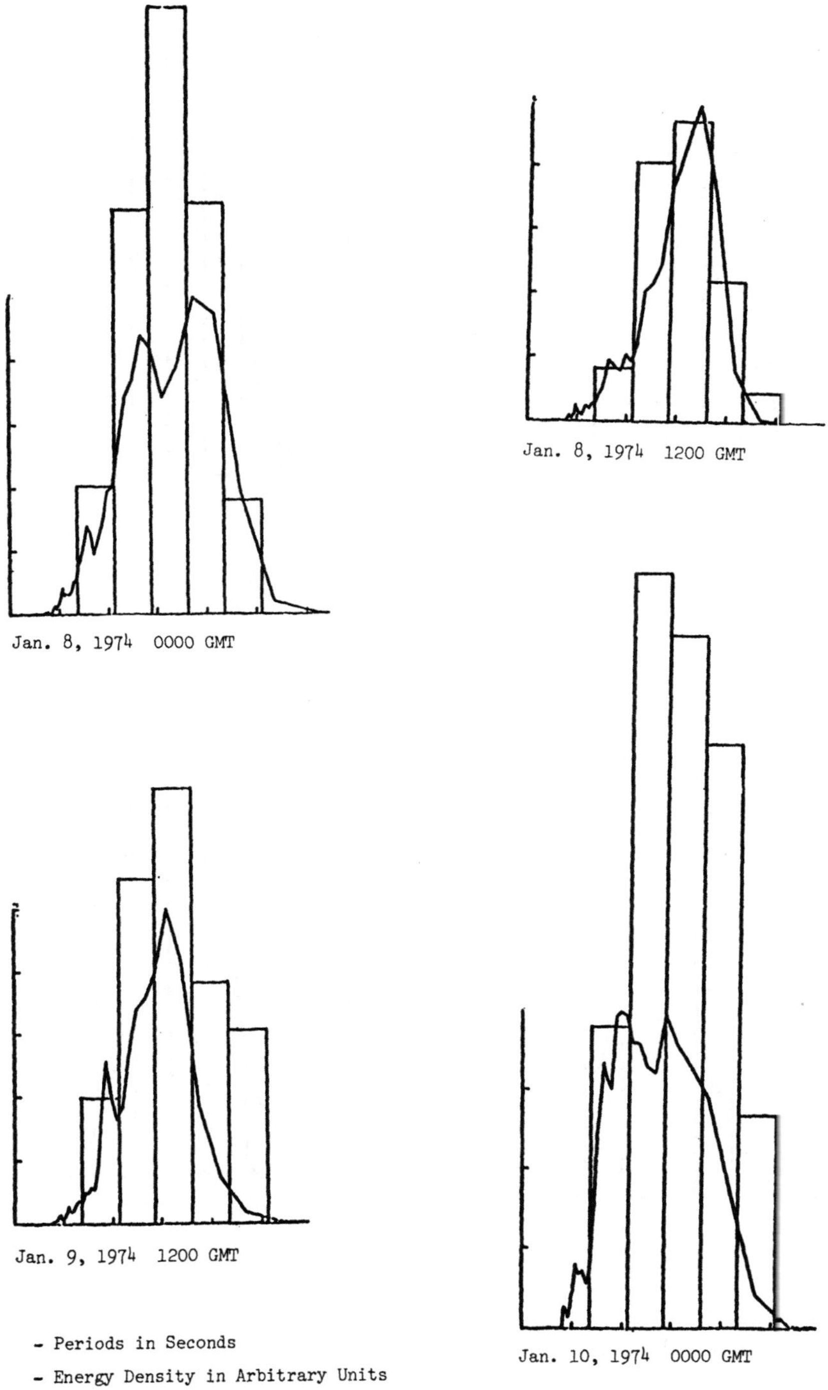
Jan. 8, 1974 0000 GMT
Jan. 8, 1974 1200 GMT
Jan. 9, 1974 1200 GMT
Jan. 10, 1974 0000 GMT
- Periods in Seconds
- Energy Density in Arbitrary Units

Figure 14

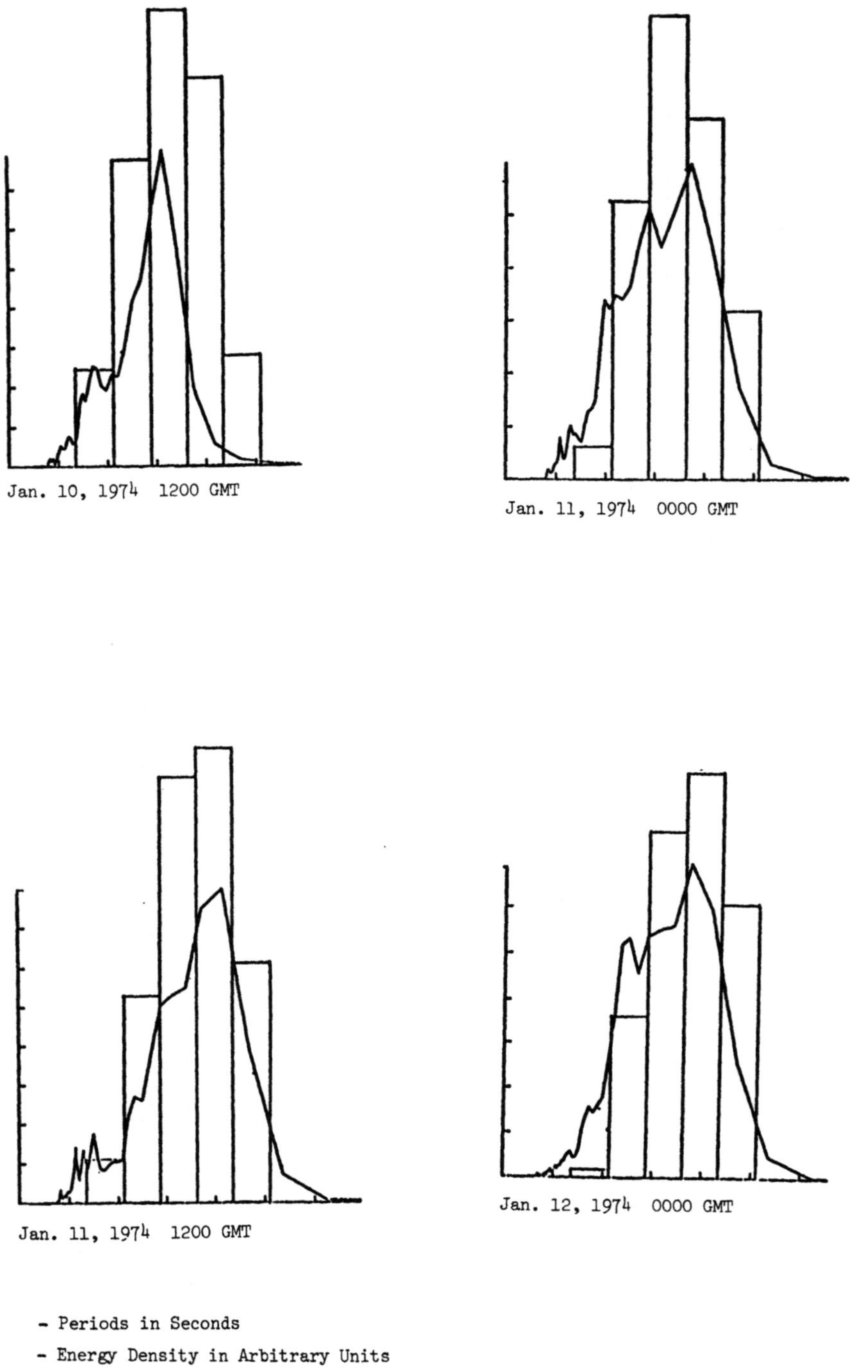
Jan. 10, 1974 1200 GMT
Jan. 11, 1974 0000 GMT
Jan. 11, 1974 1200 GMT
Jan. 12, 1974 0000 GMT
- Periods in Seconds
- Energy Density in Arbitrary Units

Figure 15

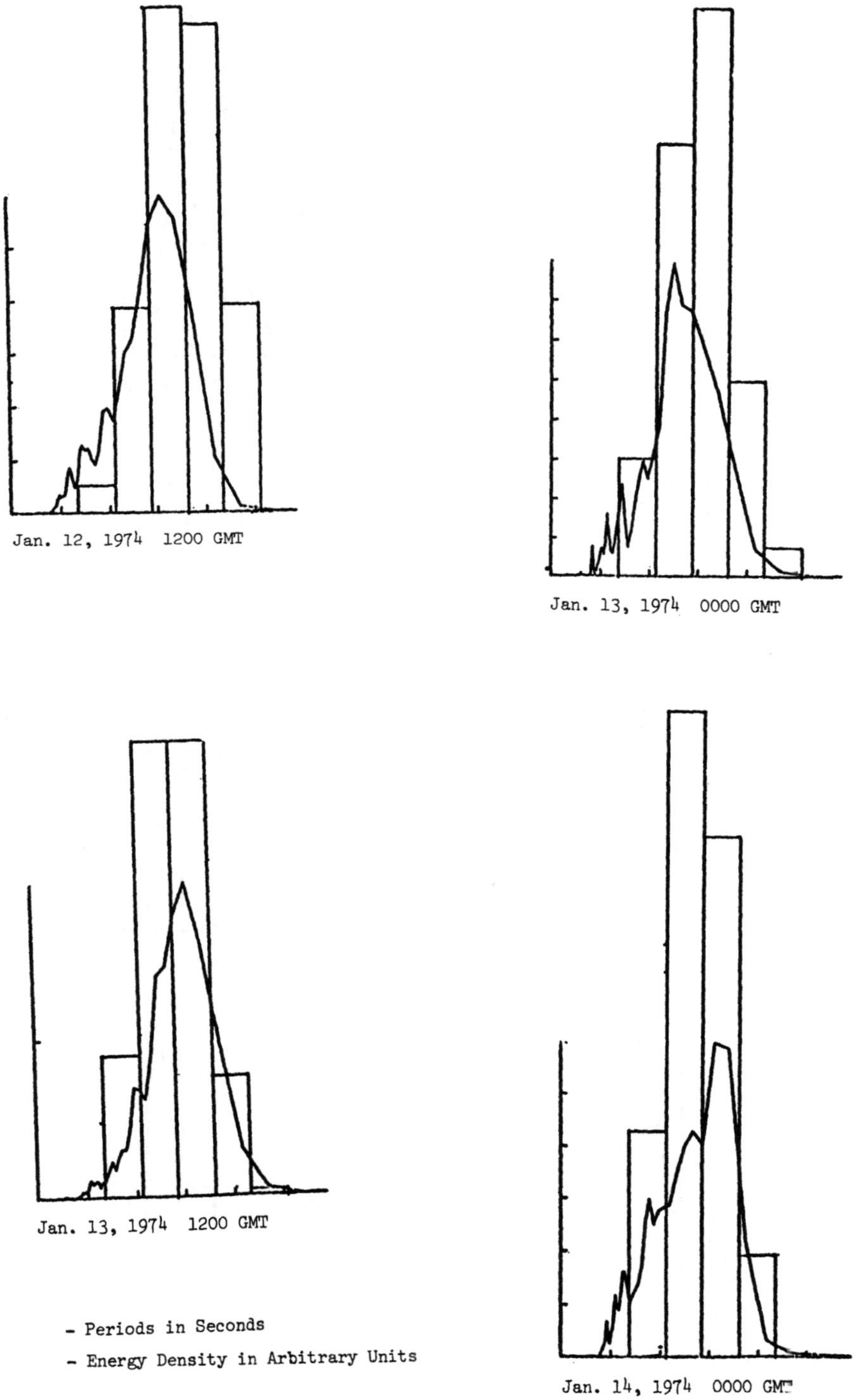
Jan. 12, 1974 1200 GMT
Jan. 13, 1974 0000 GMT
Jan. 13, 1974 1200 GMT
Jan. 14, 1974 0000 GMT
- Periods in Seconds
- Energy Density in Arbitrary Units

Figure 16

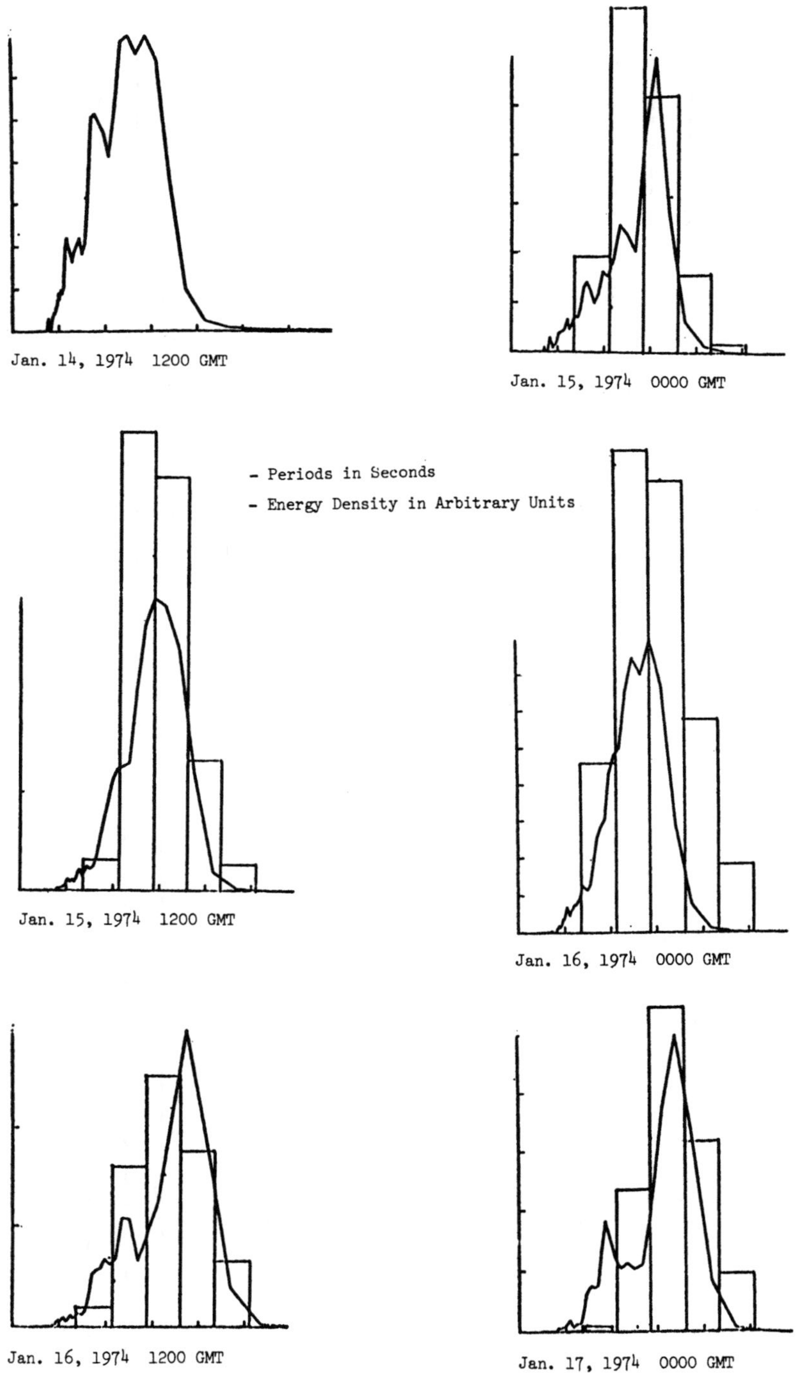
Jan. 14, 1974 1200 GMT
Jan. 15, 1974 0000 GMT
- Periods in Seconds
- Energy Density in Arbitrary Units
Jan. 15, 1974 1200 GMT
Jan. 16, 1974 0000 GMT
Jan. 16, 1974 1200 GMT
Jan. 17, 1974 0000 GMT

Figure 17

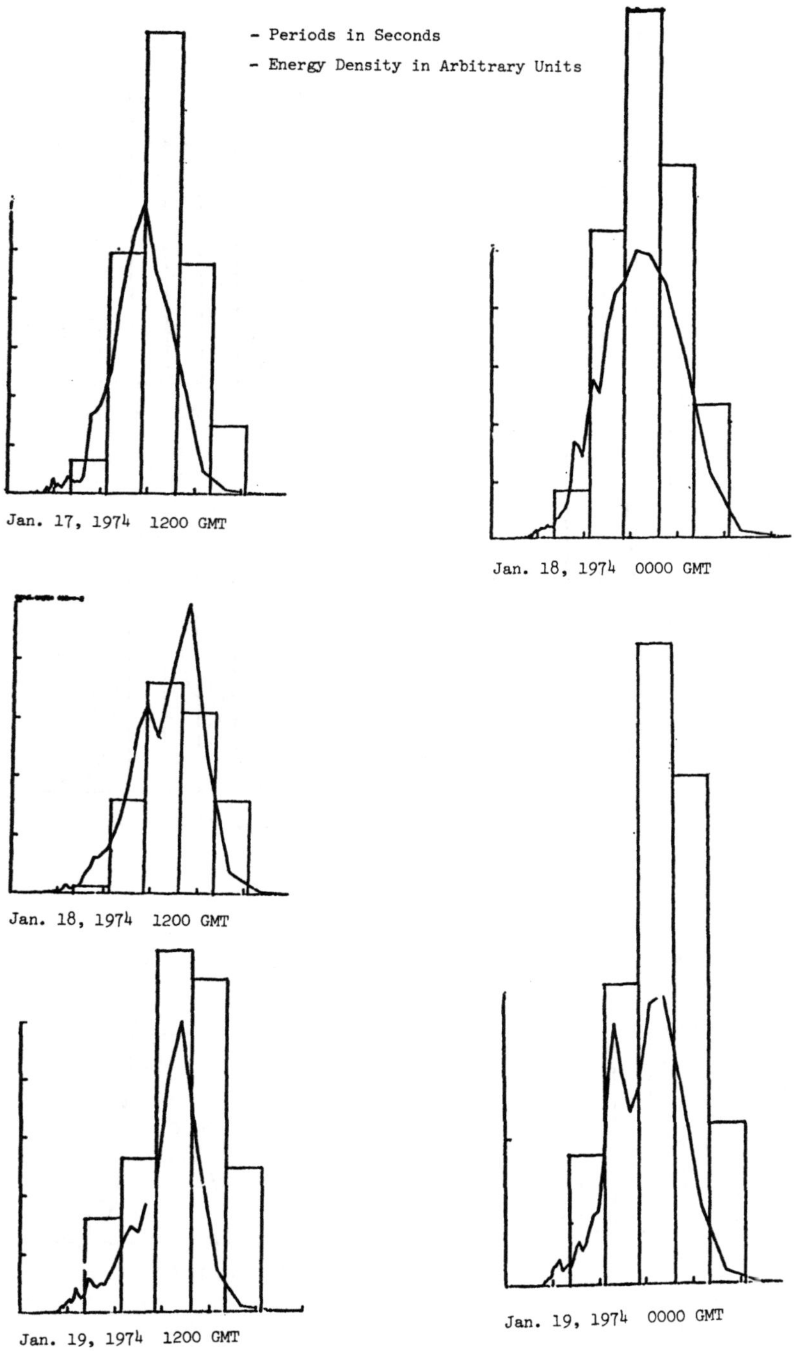
- Periods in Seconds
- Energy Density in Arbitrary Units
Jan. 17, 1974 1200 GMT
Jan. 18, 1974 0000 GMT
Jan. 18, 1974 1200 GMT
Jan. 19, 1974 1200 GMT
Jan. 19, 1974 0000 GMT

Figure 18

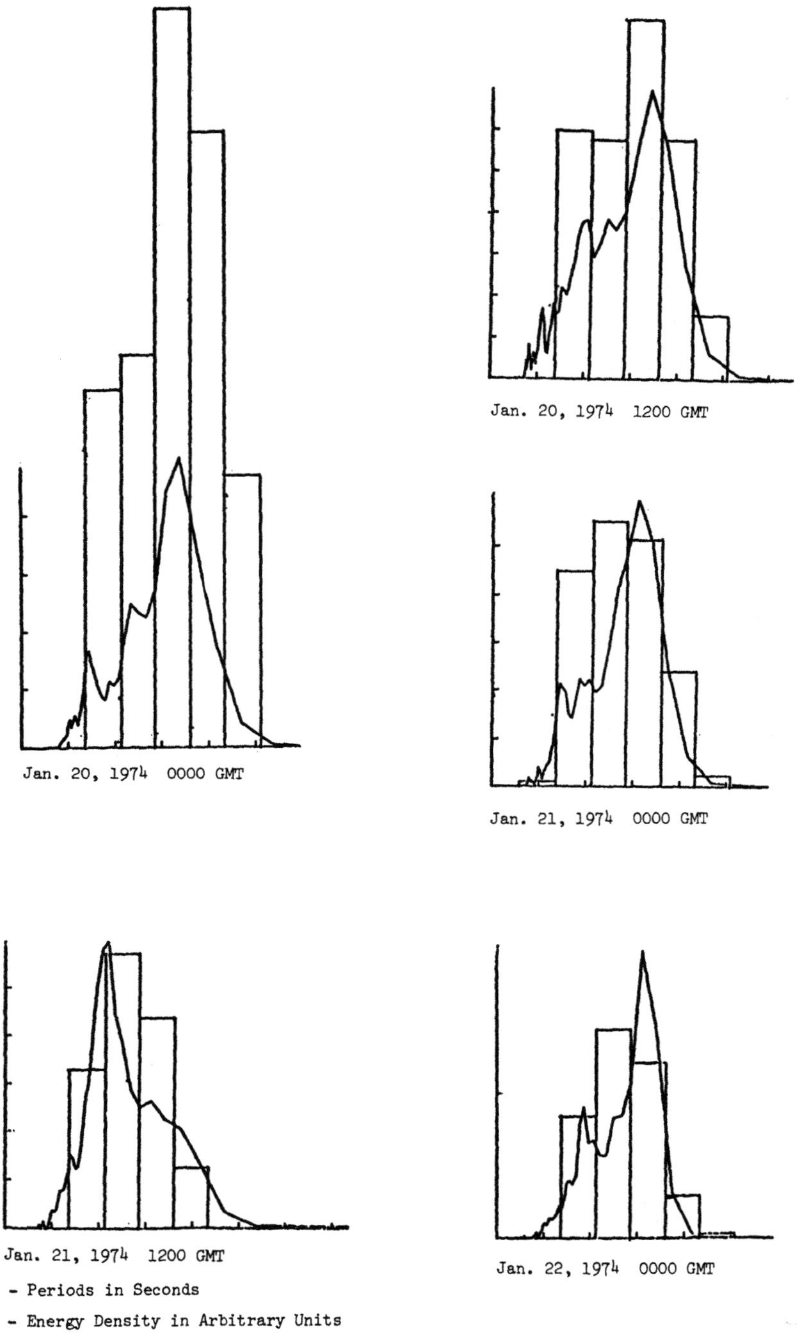
Jan. 20, 1974 0000 GMT
Jan. 20, 1974 1200 GMT
Jan. 21, 1974 0000 GMT
Jan. 21, 1974 1200 GMT
Jan. 22, 1974 0000 GMT
- Periods in Seconds
- Energy Density in Arbitrary Units

Figure 19

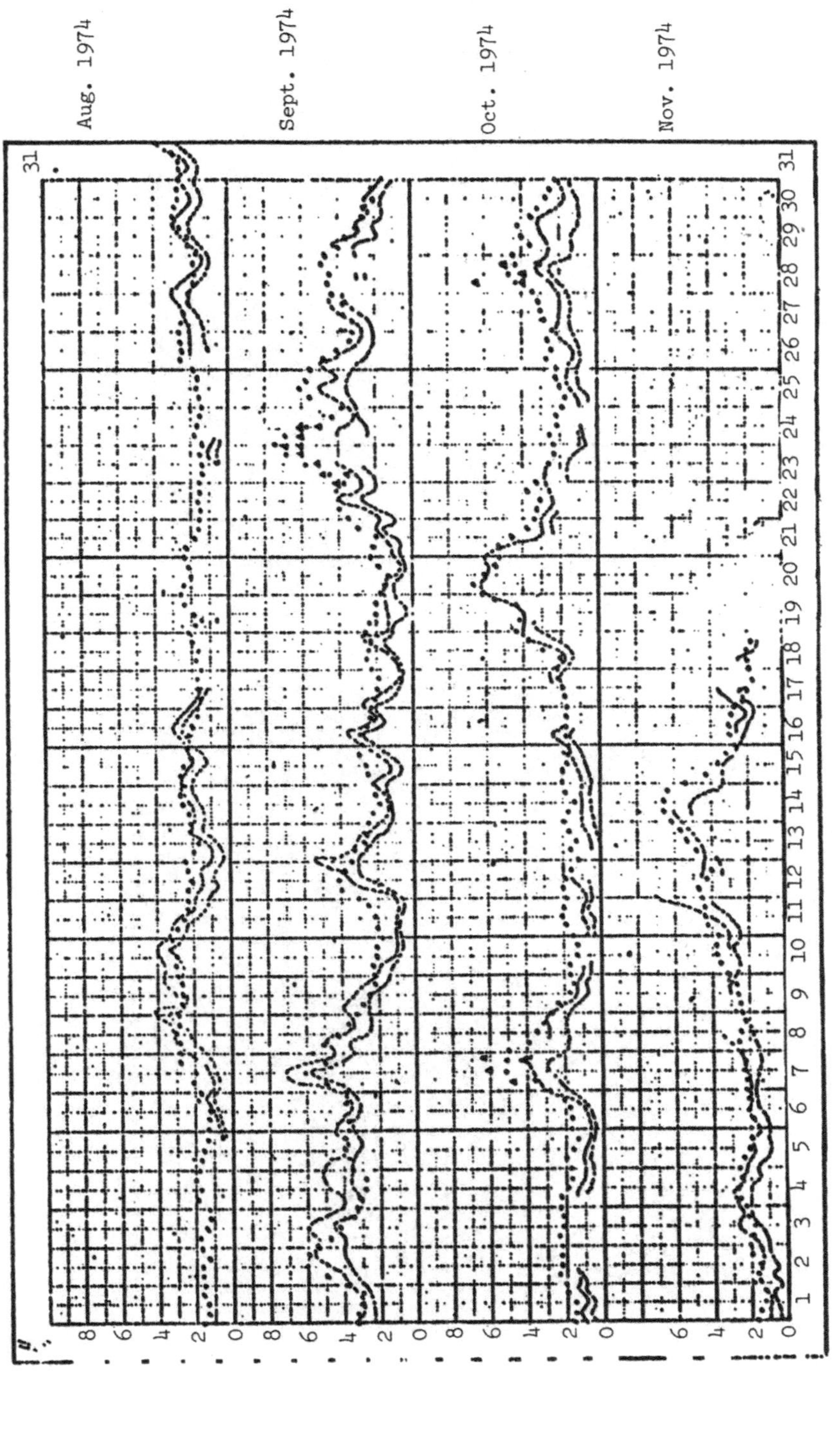

Comparison with Buoys (● Hindcast Δ Ships)

Figure 20

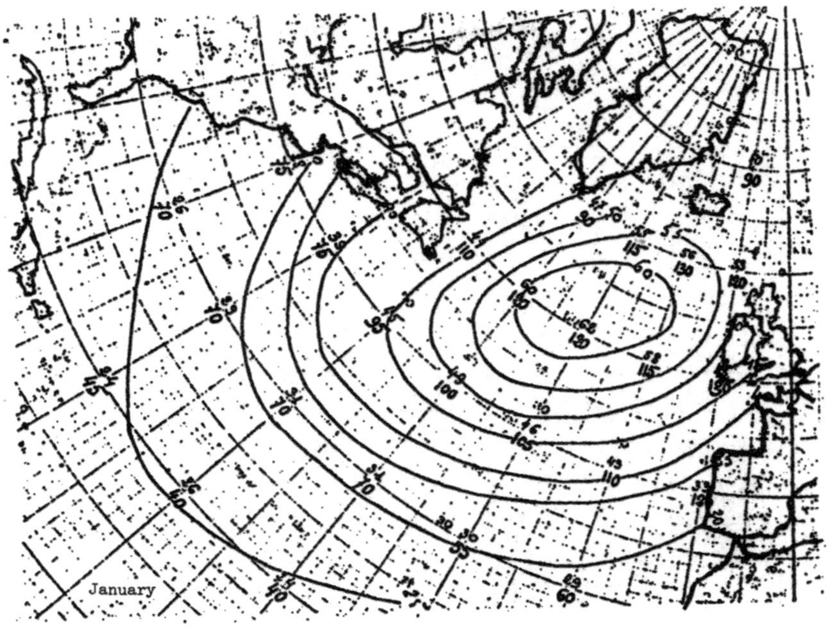

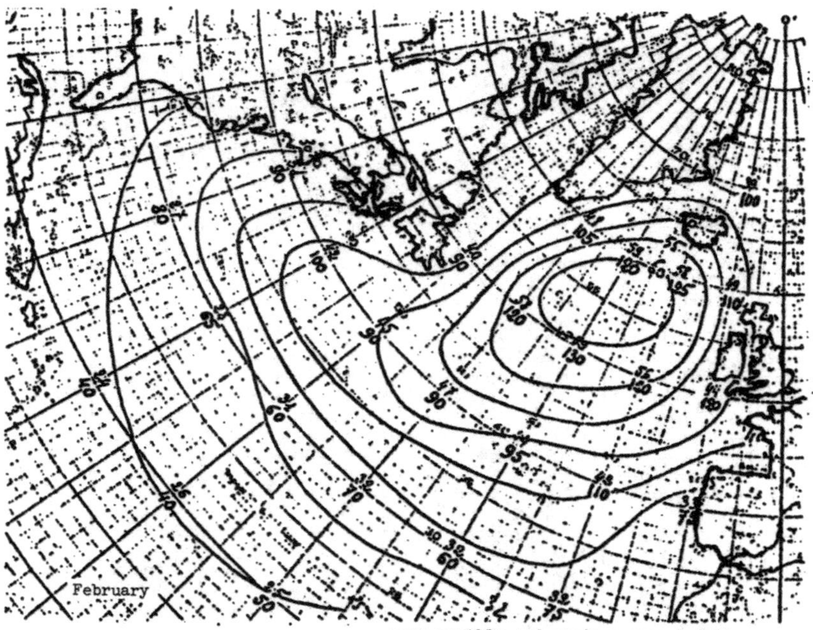

mean monthly and maximum wave heights
H 1/10 in decimeter.

Figure 21

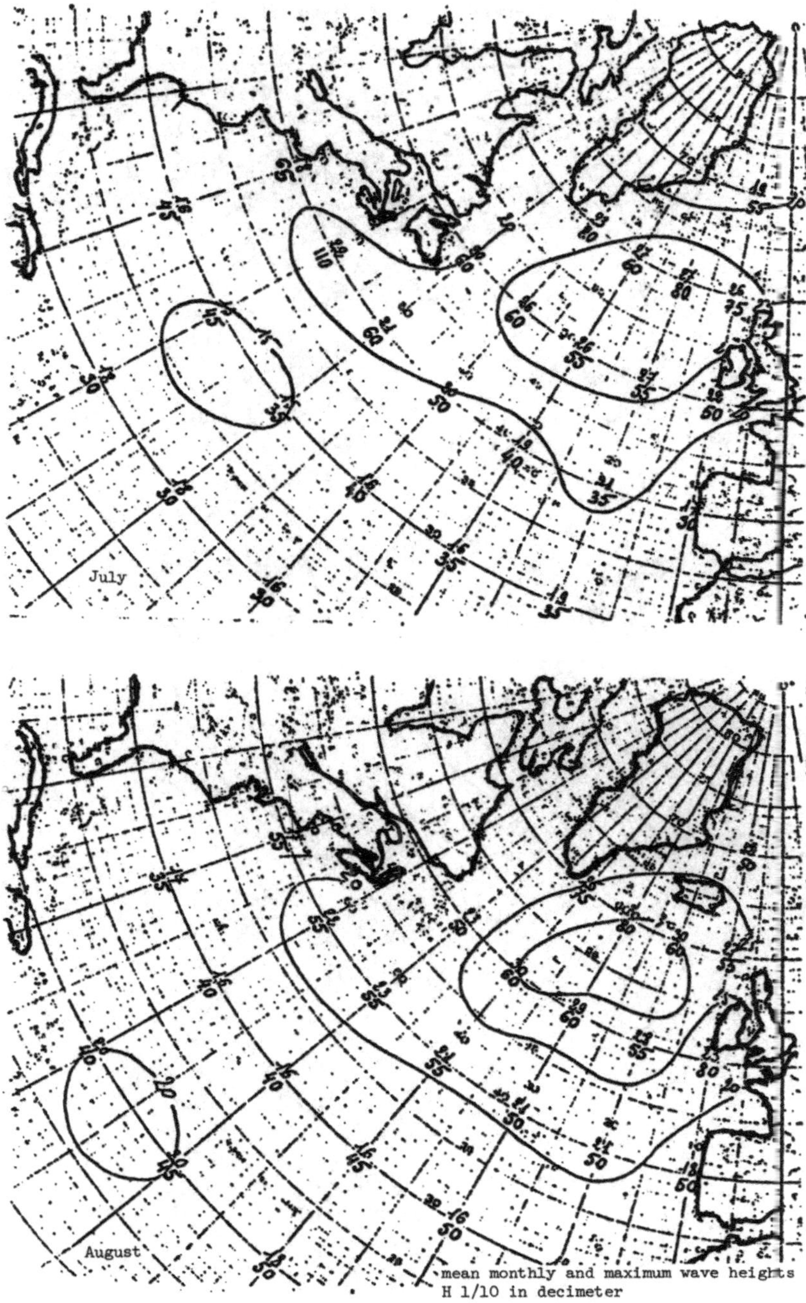
July
August
mean monthly and maximum wave heights
H 1/10 in decimeter

Figure 22

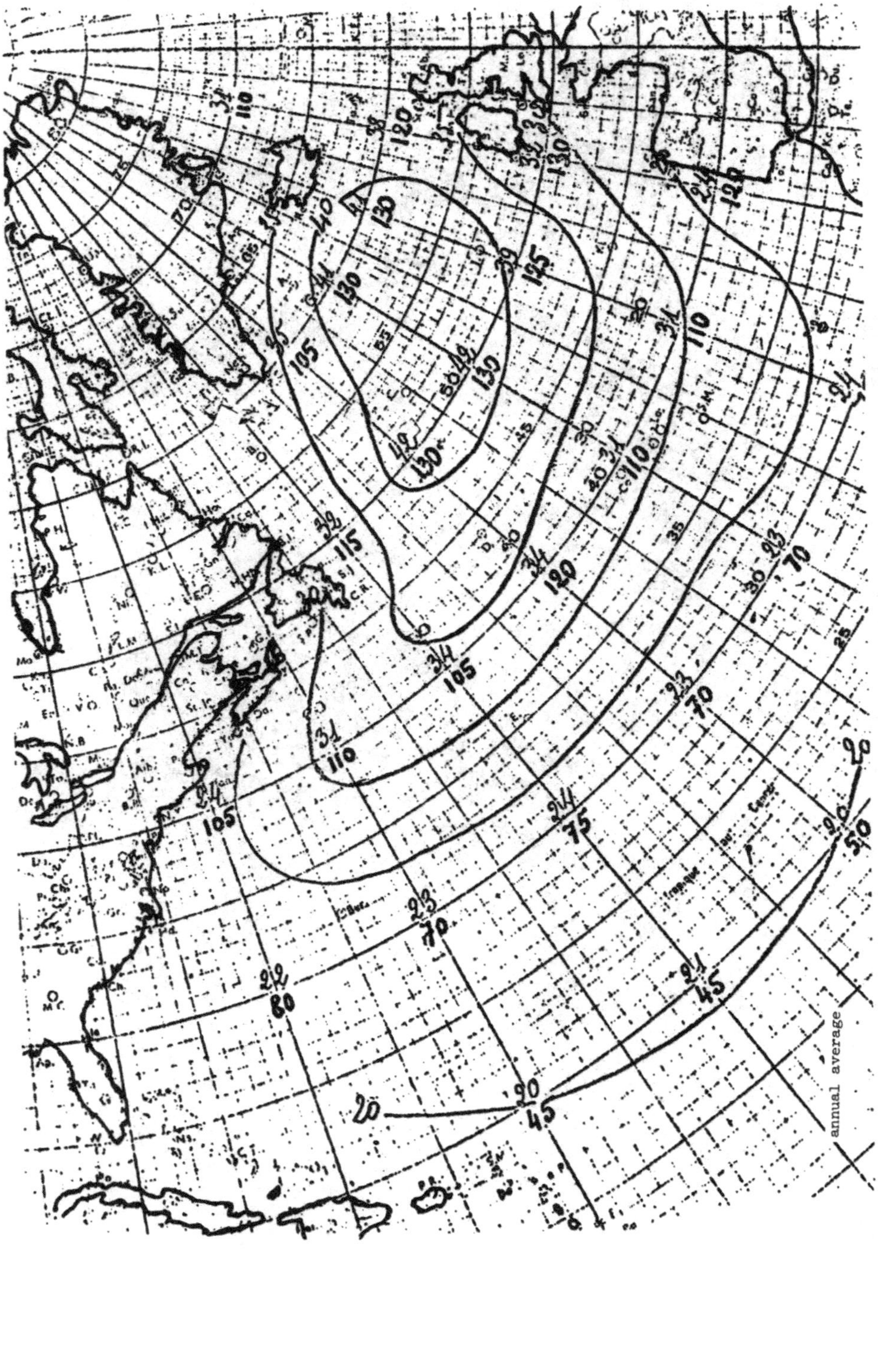

Figure 23

Lastly, in a climatological application of the model, mean monthly wave height data have been computed for seven years. Four monthly charts and an annual distribution are shown in Figs 21 to 23. The maximum values computed are also indicated in bold-faced types, below the mean values. All charts were constructed without modification from the routine analysis as broadcasted daily.

NEW DEVELOPMENTS

The model could be improved by statistically fitting the constants, and by a more sophisticated treatment of the energy transport equation.

However, it is evident that at present, for any sea state model, inaccurate wind fields are the main cause of errors. In the next years, SEASAT A analysis will probably change this situation and will affect not only the degree of sophistication appropriate for an operational sea-state model but also the method of constructing initial conditions by combining hindcast and observed analysis.

To improve the satellite analysis, more sea-truth data will be needed. As a contribution to this task the French Meteorology is developing the construction of MARISONDE buoys to measure sea-state and other parameters. At the same time efforts are being made to improve the quality of ship observations. Thus it may be possible to compute three hourly pressure fields using tendency pressure data and mean wind velocity measurements from ships.

Lastly, a wave model is being developed which attempts to model the wave generation process in greater detail as a function of wave slope and the wind profile.

ACKNOWLEDGEMENTS

We have been assisted in this work by Mr. Hervé TOLHOAT and Mr. Albert SCHAEFFER for the acquisition and processing of the data, and Mr. Jean QUENET for the presentation.

DISCUSSION

W. Rosenthal: Did you look at the statistics between observations and modelling?

R. Gelci: We have not computed such statistics, and prefer to pay attention to numerous special case histories.

A DEPTH-DEPENDENT WAVE MODEL FOR OPERATIONAL FORECASTING

Brian W. Golding

Meteorological Office
London Road, Bracknell, Berkshire, England

1. Introduction

The wave forecast system at present under development for routine wave forecasting in the United Kingdom Meteorological Office is designed to give detailed forecasts for the continental shelf areas around the UK up to 36 hours ahead using wind forecasts interpolated from the fine mesh (100 Km gridlength) version of the Met Office's 10-level numerical atmospheric model (Burridge and Gadd 1977). A spectral representation of the wave field is used and the forecast equation is a version of the energy transport equation. Boundary conditions are supplied from a coarse mesh version of the model which is already in routine use for forecasts up to 48 hours over the North Atlantic and North Pacific Oceans.

2. The Forecast Model

The version of the energy transport equation used in the model is the following

$$\frac{\partial E}{\partial t} = -\nabla \cdot (\underline{c}_g E) - c_g \frac{\partial E}{\partial \theta} \frac{\partial \theta}{\partial s} + G + D \qquad (2.1)$$

Here $E = E(f,\theta;\underline{x},t)$ is the energy density at frequency f propagating in direction θ and located at position $\underline{x}$ and time t. The depth dependent group velocity $\underline{c}_g$ is that derived from linear theory.

$$\underline{c}_g = \frac{1}{2}\sqrt{\frac{g}{k}\tanh kH}\left(1 + \frac{2kH}{\sinh 2kH}\right)\hat{\underline{k}} \qquad (2.2)$$

where the wavenumber k is given by

$$f^2 = gk \tanh kH \qquad (2.3)$$

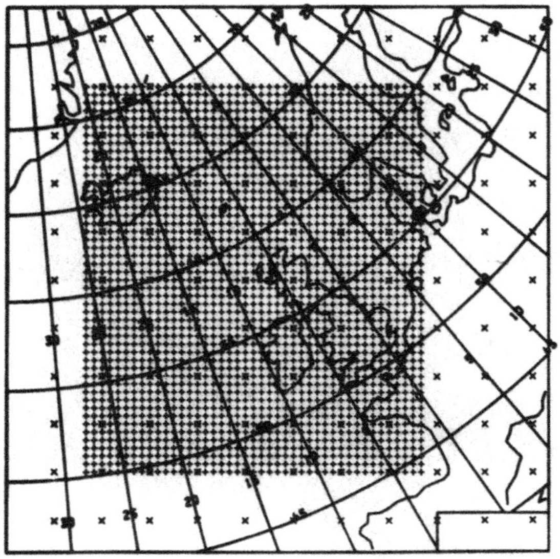

Fig. 1 Grid Points used in Fine-Mesh Wave Forecast.

The terms on the right hand side of the energy transport equation (2.1) are: first, the propagation term modified to take account of variations in group velocity; second, the refraction term which has been expressed in terms of the rate of change of direction along the wave path 's'; and finally, the source and sink terms G and D respectively which **parameterise** the effects of wind energy input, wave breaking etc.

The energy spectrum used in the model consists of 6 frequency components from 0.05 Hz to 0.15 Hz in steps of 0.02 Hz, and of 12 direction components spaced at 30 degree intervals. Spatially, a 50 Km grid length is used over a rectangular area covering the North East Atlantic from Iceland in the northwest to Biscay in the southeast. The grid is shown in Fig 1 along with some of the coarse mesh grid points which provide boundary values. A polar stereographic map projection is used. The timestep is half an hour, as required for the computational stability of the integration scheme.

The different terms in equation (2.1) are each integrated separately. The most accurately treated is the basic propagation equation

$$\frac{\partial E}{\partial t} = -\nabla.(\underline{c}_g E) \qquad (2.4)$$

This is integrated using a modified Lax-Wendroff scheme developed by Gadd at the UK Met Office. The phase and amplitude properties of this scheme are compared with a fourth order leapfrog scheme and

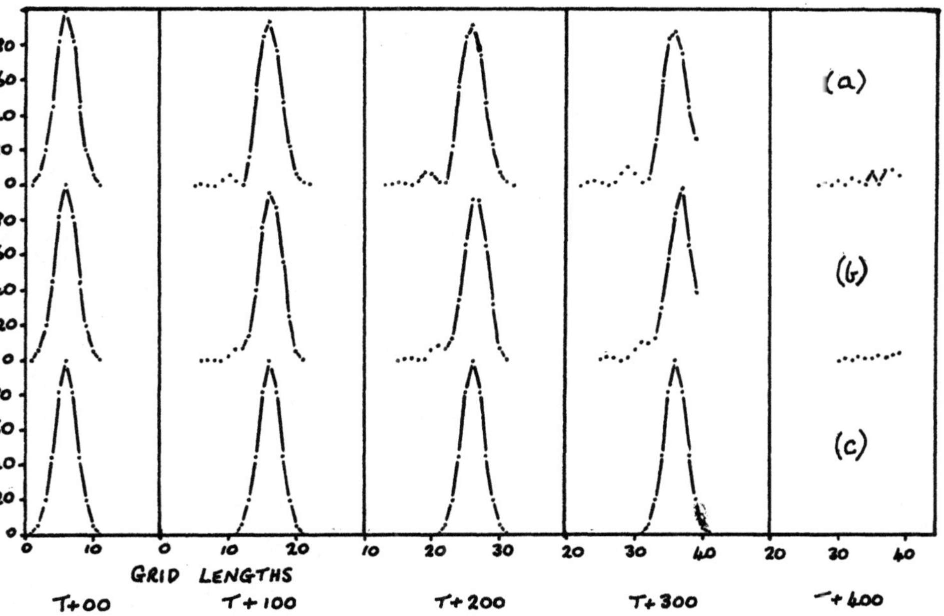

Fig. 2 Propagation of a Gaussian Profile of Wave Energy Travelling at 0.1 Grid lengths/Timestep
(a) 4th Order Leapfrog
(b) Modified Lax-Wendroff
(c) Analytic

the analytic solution in Fig 2 where a Gaussian profile of wave energy is moving at 0.1 gridlengths per timestep. The method may be illustrated by applying it to the one-dimensional propagation equation

$$\frac{\partial E}{\partial t} = -c\,\frac{\partial E}{\partial x} \tag{2.5}$$

The solution then consists of the following two steps

$$E_{j+\frac{1}{2}}^{n+\frac{1}{2}} = \tfrac{1}{2}\left(E_j^n + E_{j+1}^n\right) - \tfrac{1}{2}\mu\left(E_{j+1}^n - E_j^n\right)$$

$$E_j^{n+1} = E_j^n - \mu\left\{(1+a)\left(E_{j+\frac{1}{2}}^{n+\frac{1}{2}} - E_{j-\frac{1}{2}}^{n+\frac{1}{2}}\right) - \frac{a}{3}\left(E_{j+\frac{3}{2}}^{n+\frac{1}{2}} - E_{j-\frac{3}{2}}^{n+\frac{1}{2}}\right)\right\} \tag{2.6}$$

where $\mu = c\,\dfrac{\Delta t}{\Delta x}$ and $a = \frac{3}{4}\left(1-\mu^2\right)$

j signifies spatial position and n the time level.
The method is readily extended to two dimensions. Near the grid edge and internal coasts, not all of the required points are available. To deal with this, points one grid length from the edge use the ordinary Lax-Wendroff integration scheme. Along the edge, upstream differences are used for outflow components while inflow values are specified - either zero along land coasts, or by interpolation from the coarse mesh model at open sea boundaries. In the simplified case used above, if J is the right hand edge

point then the integration at points J, J-1 may be written

$$E^{n+1/2}_{J-1/2} = \frac{1}{2}\left(E^n_{J-1} + E^n_J\right) - \frac{1}{2}\mu\left(E^n_J - E^n_{J-1}\right)$$

$$E^{n+1}_{J-1} = E^n_{J-1} - \mu\left(E^{n+1/2}_{J-1/2} - E^{n+1/2}_{J-3/2}\right) \qquad (2.7)$$

and

$$E^{n+1}_J = \begin{cases} E^n_J - \mu\left(E^n_J - E^n_{J-1}\right) & \text{for right going components} \\ 0 \text{ or specified} & \text{for left going components} \end{cases} \qquad (2.8)$$

The refraction term in (2.1) has to be treated rather differently because of the large, 30°, spacing between angular components. The term is calculated as a correction to the values obtained above from the propagation equation. For each component $E(f,\theta)$, the angular turning in a timestep is first calculated

$$\Delta\theta = c_g \frac{\partial\theta}{\partial s}\Delta t \qquad (2.9)$$

$\frac{\partial\theta}{\partial s}$ may be obtained from Snell's law for optical refraction.

$$k\sin\alpha = \text{Const.} \qquad (2.10)$$

where α is the angle between the direction of propagation and the normal to the depth gradient, and k is the wave number. In terms of available quantities this gives

$$\frac{\partial\theta}{\partial s} = -\frac{1}{k}\,\frac{\partial H/\partial x \tan\theta - \partial H/\partial y}{\partial H/\partial x + \partial H/\partial y \tan\theta}\,\frac{\partial k}{\partial s} \qquad (2.11)$$

where H is the sea depth, and from (2.3) with frequency held constant

$$\frac{\partial k}{\partial s} = \frac{-k^2 \operatorname{sech}^2 kH}{\tanh kH + kH \operatorname{sech}^2 kH}\,\frac{\partial H}{\partial s} \qquad (2.12)$$

For each component, having calculated $\Delta\theta$, the energy is then partitioned between the original direction and the next component direction according to the ratio of $\Delta\theta$ to 30°, the component separation. $\Delta\theta$ is not permitted to exceed 30° in a single timestep. The ability of this formulation to reproduce results usually obtained by ray diagram methods is illustrated in Fig 3 where a swell front advancing over half the grid has been refracted by a sea mount with minimum depth 3 metres at the centre. A reduction of wave height can be seen to the left while over the mount heights increase. The refraction term is not calculated within two grid lengths of the boundary in order to avoid numerical instability.

The energy source and sink terms G and D are treated empirically. The fully developed spectrum described by Pierson and Moskowitz (1964) with an angular spreading factor is taken as the limiting sea state in both processes

$$E_{PM}(f,\theta) = \begin{cases} \dfrac{0.0081\,g^2}{(2\pi)^4 f^5}\exp\left(-0.74\left(\dfrac{g}{2\pi V}\right)^4\right)\cdot\dfrac{2}{\pi}\cos^2(\theta-\Psi) & |\theta-\Psi| < \pi/2 \\ 0 & |\theta-\Psi| \geq \pi/2 \end{cases} \qquad (2.13)$$

where V is the wind speed and Ψ its direction.

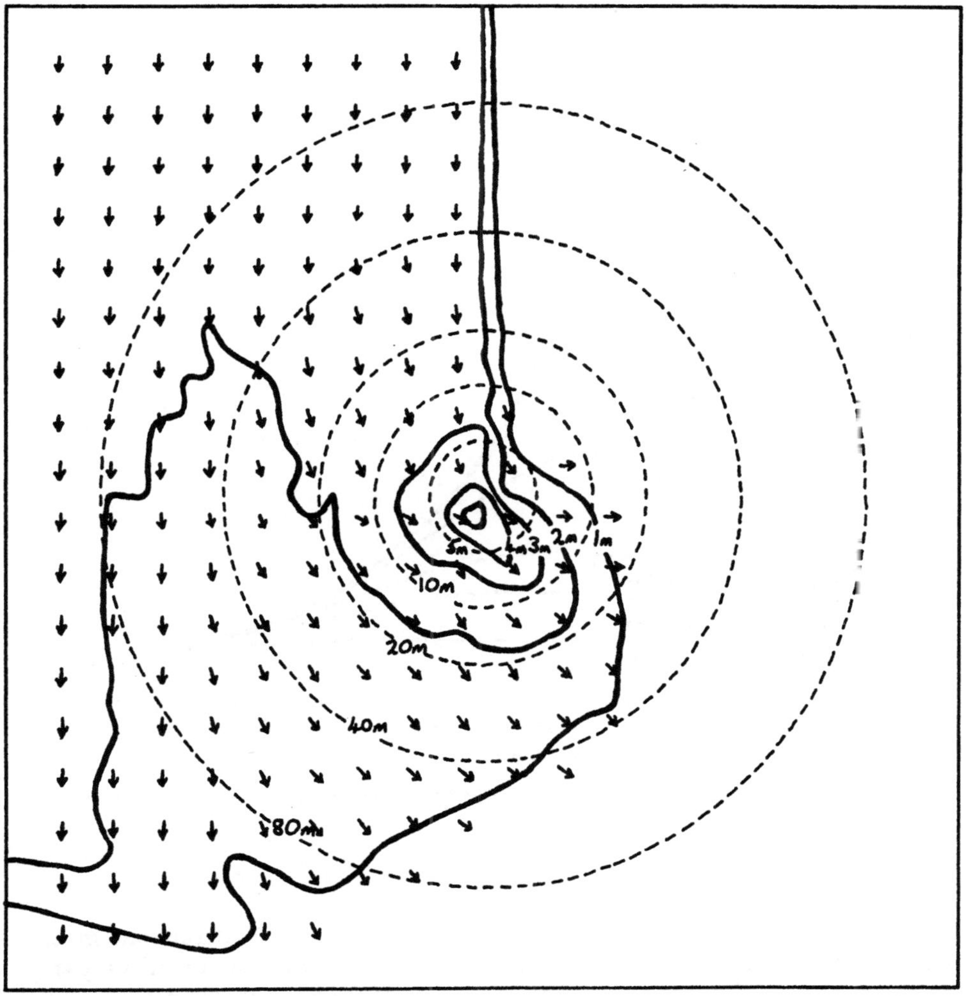

Fig. 3 Swell Front after Crossing Sea Mount (Dashed lines show sea depth in metres; Solid lines show wave height in metres)

The terms G and D are applied as a correction after the two processes already mentioned have been dealt with. If the energy of a component at this stage is E_{old} then if this is less than E_{PM}, energy is input according to the formula

$$E_{new}(f,\theta) = E_{PM} - (E_{PM} - E_{old})(1-(7f)^4)^{\Delta t} \qquad (2.14)$$

The resultant growth of the wave energy spectrum in a constant 20 m/s wind with unlimited fetch is shown in Fig 4.

Similarly if E_{old} is greater than E_{PM}, energy is dissipated according to the formula

$$E_{new}(f,\theta) = E_{PM} - (E_{PM} - E_{old})(1-(3f)^4)^{\Delta t} \qquad (2.15)$$

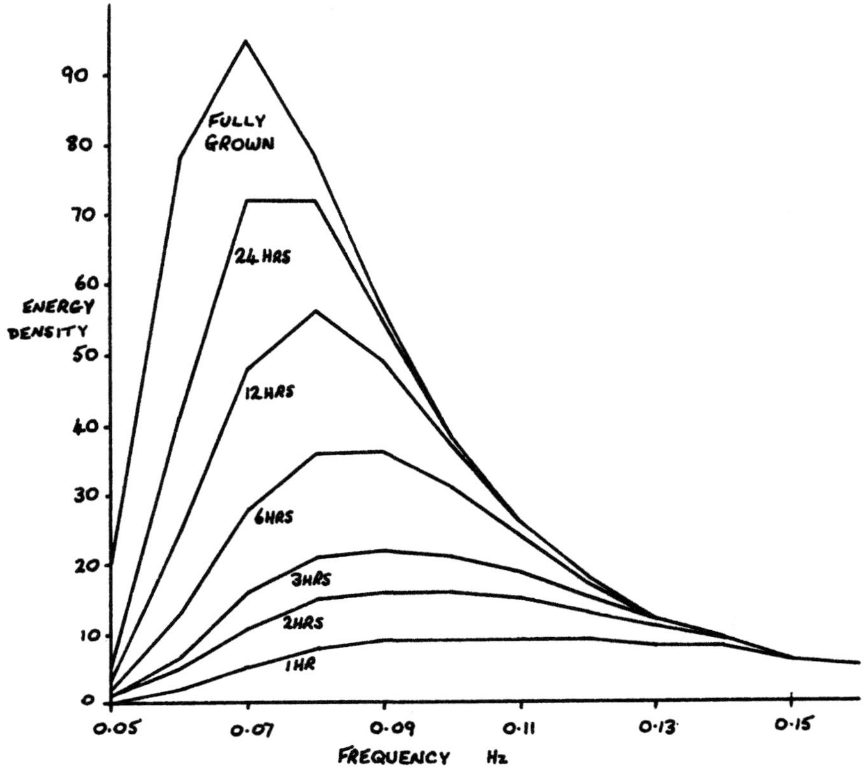

Fig. 4 Development of the Wave Energy Spectrum from Rest in a Steady 20m/s Wind.

The effect of this formula is illustrated in Fig 5 which shows the time taken for half of the energy in a wave train to be dissipated if E_{PM} is zero. The dissipation at frequencies less than 0.09 Hz is negligible, in agreement with observations of low frequency swell travelling over very large distances.

The energy input process has a built in limit to wave growth. However the refraction mechanism does not have this limit, and in reality may produce such features as caustics where energy is lost to turbulence. In order to take account of this situation, an absolute limit is placed on the energy in any component which corresponds to the Saturation Energy of Phillips (1958)

$$E_{PK}(f,\theta) = \frac{\alpha g^2}{(2\pi)^4 f^5} \quad \text{with } \alpha = 0.0081 \qquad (2.16)$$

This spectrum was derived on dimensional grounds from the assumption that waves cannot grow beyond the point where the downward acceleration at the crest is equal to the acceleration of gravity. In the model a 1% overshoot is permitted in order that this state

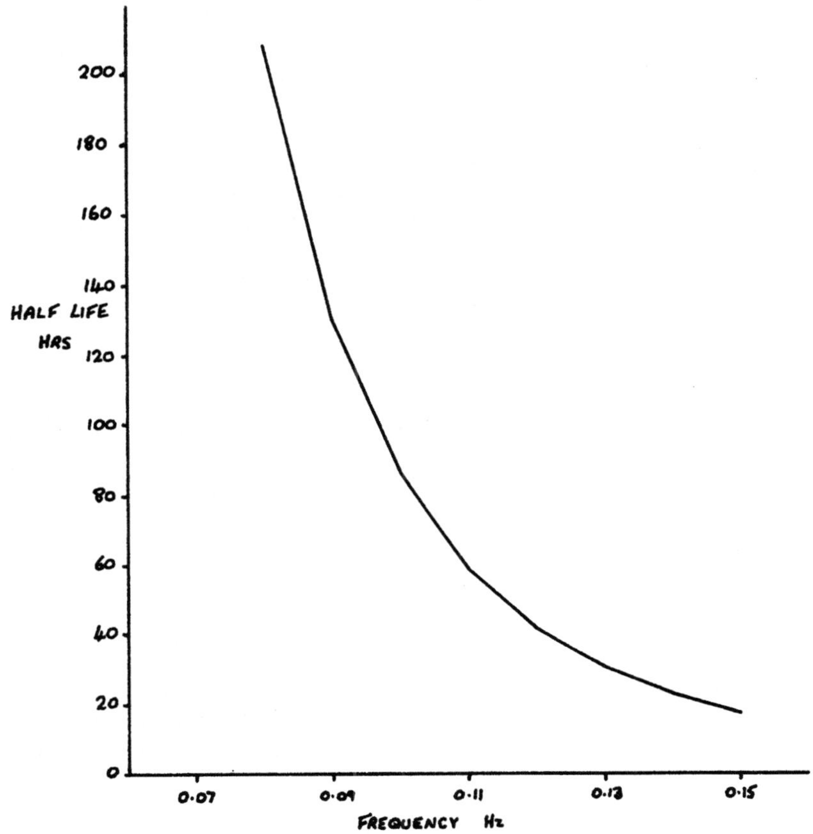

Fig. 5 Half Life of Swell Energy in Zero or Opposing Winds.

can be recognised in output programs. In the frequency ranges likely to become saturated this overshoot will have a negligible effect.

3. Wind Input

The forecast is run in two modes: first, a twelve hour hindcast using analysed wind fields; then, using this as a starting point, a thirty-six hour forecast. This cycle is repeated every twelve hours, the first part providing a continual update of the initial field. The winds for the forecast are interpolated from hourly fields stored in the output dataset of the fine mesh version of the 10-level atmospheric model. For the hindcast, both the previous cycle's forecast winds and those derived from the current height analysis are used. This is done to reduce the large errors introduced by direct interpolation from twelve-hourly datasets. The method used is to calculate the difference between the twelve hour forecast and the analysis, and then to adjust each of the intermediate steps of the forecast by a proportion of this error.

In both cases, the surface winds are derived from the 900 mb atmospheric model winds by empirical formulae based on the work of Findlater et al (1966). These formulae give a speed reduction factor and an angular turning which depend on the mean lapse rate from 900 mb to the surface, and on the wind speed.

The relationships are expressed in the form

$$V_0/V_{900} = a V_{900}^2 + b V_{900} + c \qquad (3.1)$$

$$\alpha = d \qquad (3.2)$$

where a, b, c, d are constants which have five values depending on the class of lapse rate; V_0, V_{900} are the surface, 900 mb wind speeds respectively; and α is the angular backing of the surface wind. The lapse rate is calculated from the mean temperature T_{950} of the lowest 100 mb layer in the atmospheric model, and a surface temperature T_0 calculated in the following manner

$$T_0 = \begin{cases} T_{950} + 0.5 h'_{950} \gamma_d & \text{if } T_0 < T_s \\ 0.1(T_{950} + 0.5 h'_{950} \gamma_{900}) + 0.9 T_s & \text{if } T_0 > T_s \\ T_s & \end{cases} \qquad (3.3)$$

where T_s is the analysed sea surface temperature, otherwise, h'_{950} is the lowest layer thickness; γ_d is the dry adiabatic lapse rate, and γ_{900} is the lapse rate defined by the thicknesses of the lowest layers h'_{850} and h'_{950}.

4. Experimental Hindcasts

Experiments have been conducted to investigate the sensitivity of the predicted wave field to the presence of bottom topography in the North Sea. It should be noted that no account of bottom friction has yet been taken in the model. The refractive effects are greatest at large wavelength and large amplitude so that the most dramatic differences are to be looked for in storms. For this reason, the comparisons shown are for the storm of the 2nd and 3rd of January 1976. The actual accuracy of the predictions is reduced because of the interpolation from twelve hourly wind fields as mentioned before, but the case can nonetheless be used to test the model's sensitivity to bottom topography. The bottom topography used in the depth-dependent runs is shown in Fig 6.

Fig 7 shows the synoptic pattern as the storm crossed the British Isles and the North Sea. The hindcast was started from a state of rest at midday on Jan 1st and run with fixed boundary values. Figs 8, 9 show the predicted wave field at midday on Jan 3rd at the height of the storm in the North Sea for runs with and without bottom topography. Originally, runs with bottom

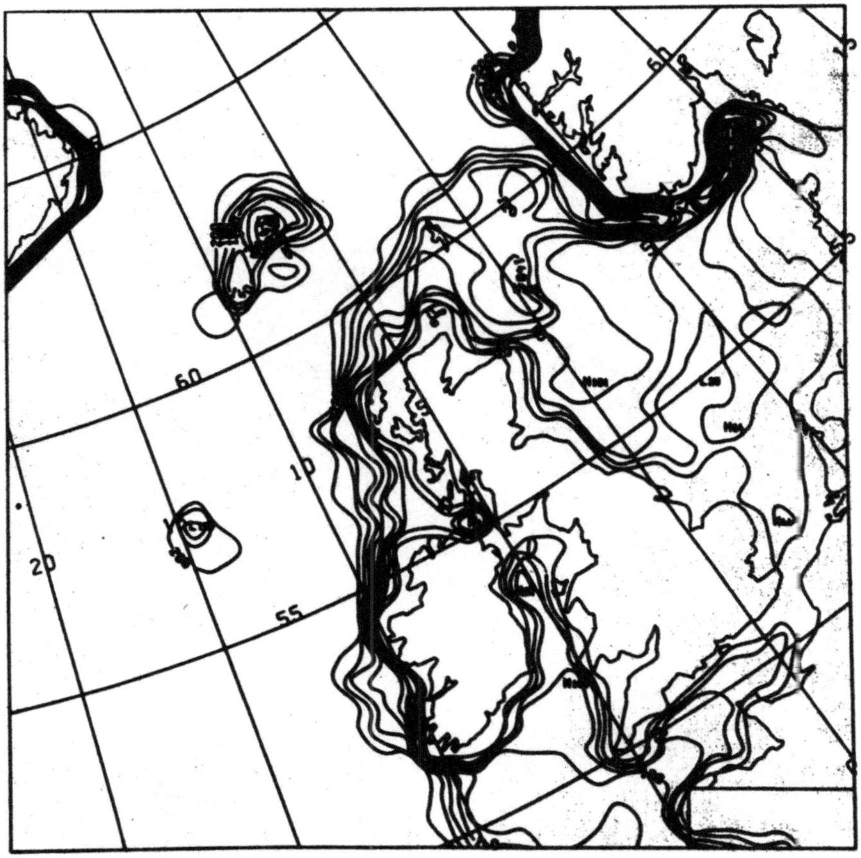

Fig. 6 Depth Field Used in Fine-Mesh Wave Forecast (Contour Interval 20m).

topography predicted a maximum of 11 metres near the east coast of England. This was due to excessive accumulations of high frequency refracted energy near a sand bank. The introduction of the Phillips Saturation Limit (2.16) removed this spurious feature. Comparison of the two forecasts shows only small differences in predicted wave heights at the height of the storm. Figs 10, 11 show the same two forecasts twenty four hours later when the storm has moved away over continental Europe. Here the differences in the swell remaining are more marked particularly in the southern part of the North Sea where heights have been decreased by the introduction of bottom topography.

The small differences between the forecasts may be due to two features of this forecast. First, the inclusion of bottom friction should lead to a substantial reduction in wave heights

Fig. 7 Surface Pressure Analysis at Daily Intervals for the Period of the Hindcast.

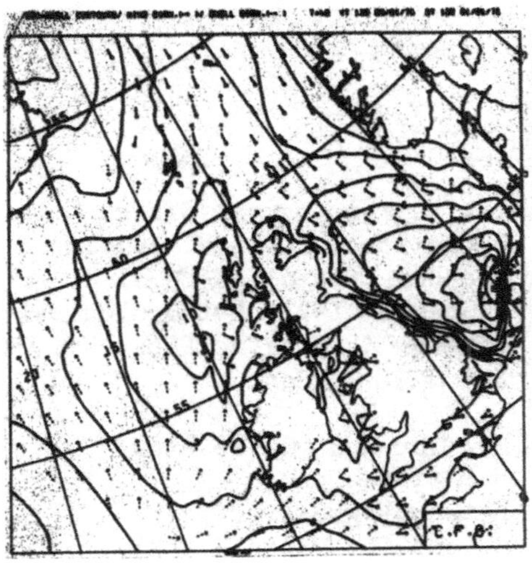

Fig. 8 T + 48 With Bottom Topography

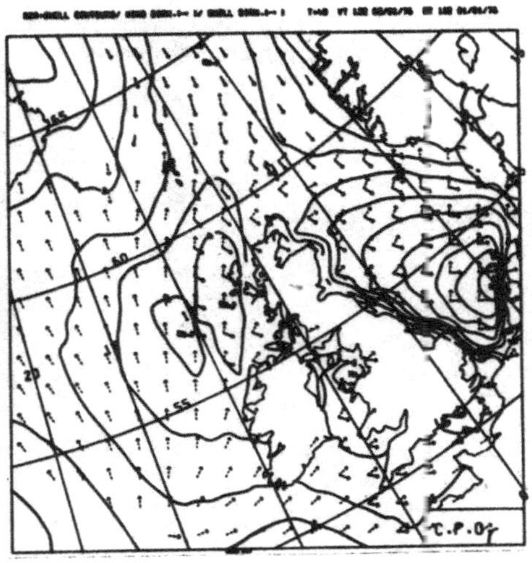

Fig. 9 T + 48 Without Bottom Topography

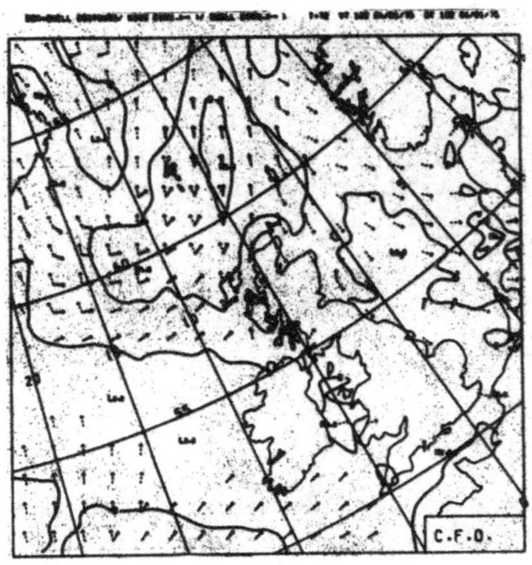

Fig. 10 T + 72 With Bottom Topography

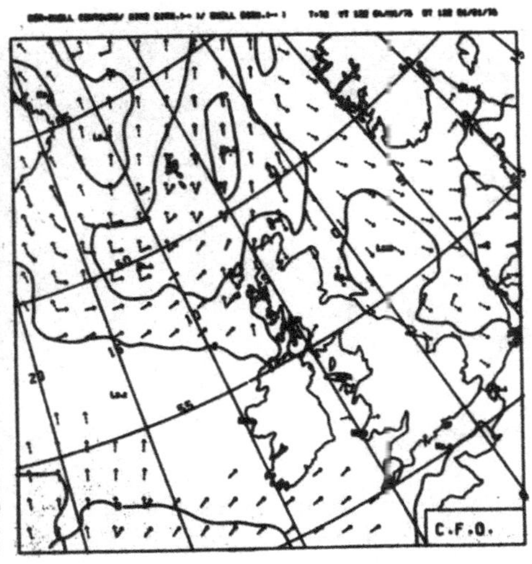

Fig. 11 T + 72 Without Bottom Topography

Figs. 8-11 Hindcasts of the Jan 1st - 4th 1976 Storm

south of the Dogger Bank. Also, the passage of the storm through the southern North Sea masked refraction effects on low frequency components.

REFERENCES

BURRIDGE D M and GADD A J	1977	The Meteorological Office Operational 10-level Numerical Weather Prediction Model (December 1975) Meteorological Office, Scientific Paper No. 34 H.M.S.O. London.
FINDLATER J HARROWER T N S HOWKINS G A and WRIGHT H L	1966	Surface and 900 mb Wind Relationships. Meteorological Office, Scientific Paper No. 23, H.M.S.O. London.
PHILLIPS O M	1958	The Equilibrium Range in the Spectrum of Wind-Generated Waves. Journal of Fluid Mechanics 4 p 426.
PIERSON W J and MOSKOWITZ L	1964	A Proposed Spectral Form for Fully Developed Wind Seas Based on the Similarity Theory of S A Kitaigorodskii. Journal of Geophysical Research 69 p 5181

DISCUSSION

R. B. Long:

Your model does not include the kind of depth dependent dissipation which was discussed by Prof. Shemdin earlier. Is it not likely that incorporating this effect will eliminate the factor-of-two overestimate of significant wave height produced by your model at the Dutch coast?

B. Golding:

I expect that this effect explains most of the discrepancy. I hope shortly to include a bottom friction term in the model. I shall also look closely at the possibility of incorporating the other mechanisms described by Prof. Shemdin.

F. Dobson:

If swell attenuation is a function of depth as suggested by R. Long's question, then you must include storm surge information in your model.

B. Golding:

When the IOS storm surge model for the North Sea becomes available for real-time use it will be an interesting experiment to try including these effects.

E. Bouws:

At KNMI we used for a wave climate study hourly waverider measurements taken at 3rd January, 1976, near lightship Texel (06220), obtained from Rijkswaterstaat. They showed a remarkably constant significant wave height, scattered around a mean value of 6.8 m, in agreement with the equivalent number of degrees of freedom. The spectral shapes were roughly similar to the shape of the JONSWAP spectrum. Apparently almost no wave energy has been advected from the area north of Dogger Bank to the lightship Texel area (average depth about 30 m). Refraction will probably be of minor importance in this case, also due to this. It appeared from the data that only locally generated waves with fetches of 200 - 300 km occurred, despite the large size of the wind field.

B. Golding:

These data will be usefull for verifying the effect of any bottom friction mechanism put into my model.

K. Hasselmann:

I liked your treatment of refraction, as it introduced a numerical angular diffusion which we believe, from our analysis of JONSWAP swell cases, to be a real effect - it is due to the small-angle scattering of rays by sub-grid scale bottom depth irregularities. A small mathematical point: the group velocity should be taken outside the ∇ operator in the transport equation as you have used it. If it is placed behind the ∇ operator, you should also take the refraction term k_i behind the differential operator in the form $\frac{\delta}{\delta k_i}(k_i F)$. The two forms are equivalent because of the incompressibility of the flow in $\underline{x}$. $\underline{k}$ phase space:

$$\frac{\delta}{\delta x_i} (\dot{x}_i) + \frac{\delta}{\delta k_i} (\dot{k}_i) = 0.$$

B. Golding:

In f, θ space the form I have integrated is correct. However, in writing the transport equation I should have made the refraction term $\frac{\delta}{\delta\theta}(Ec_g \frac{\delta\theta}{\delta s})$.

Thank you for pointing this out. The results are not affected since the numerical scheme already used this form.

V. Cardone:

Your duration limited growth curves for a stationary wind field do not show any of the properties of the JONSWAP spectral evolutions - that is, migration of the spectral peak with the stage of wave development and sharply peaked spectra in the initial stages of generation. However, your model may exhibit those properties in the fetch-limited domain as is shown for example, in other discrete models of the Pierson type. There is some evidence in nature, for example Giovanangeli's results at this symposium, that duration limited growing spectra do not obey the shape invariance characteristic of idealized fetch growth spectra. Have you simulated simple fetch limited spectral evolutions with your model?

B. Golding:

I have not compared the duration and fetch limited spectral growths but I expect them to be very similar. My spectral growth formula is, of course, very crude but does include a shift of the peak frequency as well as the total energy growth.

D. Ross:

While it is true that the growth of the amplitude of the peak for duration limited seas without a frequency shift has been reported by d'Allonnes, JONSWAP 73 and 75, and the Marineland 75 experiment do exhibit situations where a change in wind direction is accompanied by a change in direction of the highest frequency components creating a high-frequency peak which then migrates toward lower frequencies.

PARAMETRICAL NUMERICAL WAVE PREDICTION TESTED IN WIND SITUATIONS VARYING IN SPACE AND TIME

H. Günther[1], W. Rosenthal[1] and K. Richter[2]

[1]Max-Planck-Institut Für Meteorologie, Hamburg and Institut Für Geophysik, Universität Hamburg

[2]Deutsches Hydrographisches Institut, Hamburg

I. INTRODUCTION

In the last years a hybrid parametric wave prediction model has been developed in the "Sonderforschungsbereich 94" of the University of Hamburg. This project was carried out jointly by the Max Planck Institut for Meteorology, the Institute of Geophysics of the University and the German Hydrographic Institute. It was part of the international North Sea Wave Prediction Model (NORSWAM) project. More results from NORSWAM are mentioned in the paper of T.J. Weare and B.A. Worthington.

The first part of this paper describes the fundamental principles of the model. In the second part we present results of two test runs using JONSWAP 73 data.

II. HYBRID MODEL

Like most of the surface wave prediction models developed in the last years, our model is based on the transport equation:

$$\frac{\delta F}{\delta t} + v\frac{\delta F}{\delta r} = T = T^{in} + T^{dis} + T^{nl} \qquad (1)$$

where $F = F(f,\theta,r,t)$ is the energy density of the frequency f and the direction θ at location r at time t. v is the group velocity of the frequency f and T is the source function, which is the sum of the wind dependent input term T^{in}, the dissipation term T^{dis} and the nonlinear interaction T^{nl}. The transport equation is treated differently for the windsea part of the spectrum F, with frequencies greater than $f_{PM} = 0.13\ g/U$, and the swell part, with frequencies less than f_{PM}. U denotes the wind velocity at 10 m height.

Windsea Model

Following Hasselmann et al. (1973, 1976) we assume that it is possible to describe a developing windsea by the five parameters $(f_m, \alpha, \gamma, \sigma_a, \sigma_b)$, where f_m is the peak frequency, α the Phillips' "constant", γ the peak enhancement factor, and σ_a, σ_b the left and right peak width of a JONSWAP spectrum (Fig. 1).

$$E(f) = \alpha\, g^2\, (2\pi)^{-4}\, f^{-5} \exp\left(-\frac{5}{4}\left(\frac{f}{f_m}\right)^{-4} + \ln\gamma \exp\left[-\frac{(f-f_m)^2}{2\sigma^2 f_m^2}\right]\right) \qquad (2)$$

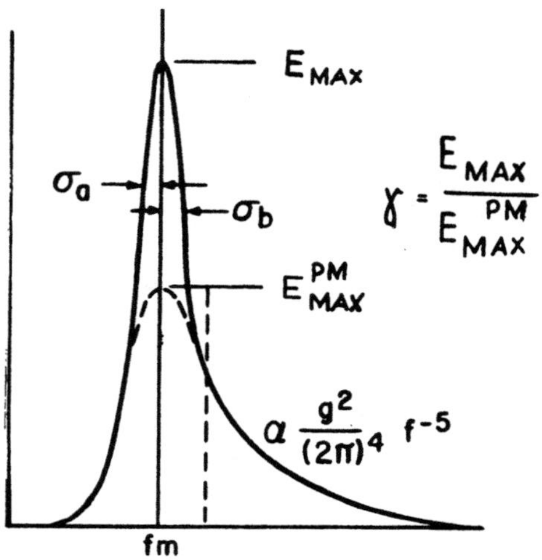

Fig. 1: Definition of the JONSWAP parameters of a windsea spectrum

Assuming a frequency independent cosine-square directional distribution, the set of five parametric transport equations

$$\frac{\delta a_i}{\delta t} + D_{ijk}\frac{\delta a_j}{\delta x_k} = S_i = S_i^{in} + S_i^{dis} + S_i^{nl} \qquad (3)$$

$$i,j = 1,\ldots,5 \qquad k = 1,2$$

is calculated from the equation (1) using the functional mapping concept of Hasselmann et al. (1976). D_{ijk} is the propagation matrix and S_i is the source function, both dependent on the parameter vector $a = (f_m, \alpha, \gamma, \sigma_a, \sigma_b)$. The determination of the source function is basically the same as in the above references. A further assumption is that the wind field does not change its direction very rapidly, so that the wave spectrum can be regarded as propagating in the local wind direction.

In our model we fixed $\sigma_a = 0.07$ and $\sigma_b = 0.09$ (mean JONSWAP values), and therefore our windsea model (3) consists of three coupled differential equations, which are solved by a second order finite difference integration scheme (predictor-corrector method of R.W. MacCormack 1971). It can be argued that it is inconsistent to consider a three parameter model without free directional parameters: the quasi-equilibrium concept invoked to justify the neglect of prognostic directional parameters leads to the one-parameter model described in Hasselmann et al. (1976). However, the model presented here should be regarded as a first step towards a more general multi-parameter hybrid model, including free directional parameters.

Swell Propagation and Energy Exchange between Windsea and Swell

In the swell band ($f < f_{PM}$) the spectra are no longer of the self-stabilized JONSWAP shape, and the energy has to be propagated with its group velocity along prescribed characteristics. In the model, energy is transferred from the windsea to swell characteristics when the peak frequency f_m becomes less than f_{PM}. Energy is transferred in the opposite direction when swell energies appear with frequencies greater than the peak frequency of the windsea. The transfer algorithm is based of the requirement of overall energy conservation and a JONSWAP shape for the new windsea spectrum. If energy occurs in the swell part of the model for frequencies between f_{PM} and f_m, decoupled propagation along characteristics is assumed, but a linear Snyder-Cox source function is included.

III. TESTS

The hybrid model was tested against two JONSWAP 73 cases, each of 24 hours duration. Fig. 2 shows the experimental array west of the island of Sylt. The integration area was 44 x 44 km, with a 2 km grid. The time step was 5 minutes. The resolution for swell propagation was 24 frequency-bins from 0.078 to 0.45 Hz and 8 direction-bins. Measurements for comparison exist for Stations 4, 5N, 5, 8N, 8S, 9 and 10. Boundary values were taken from the

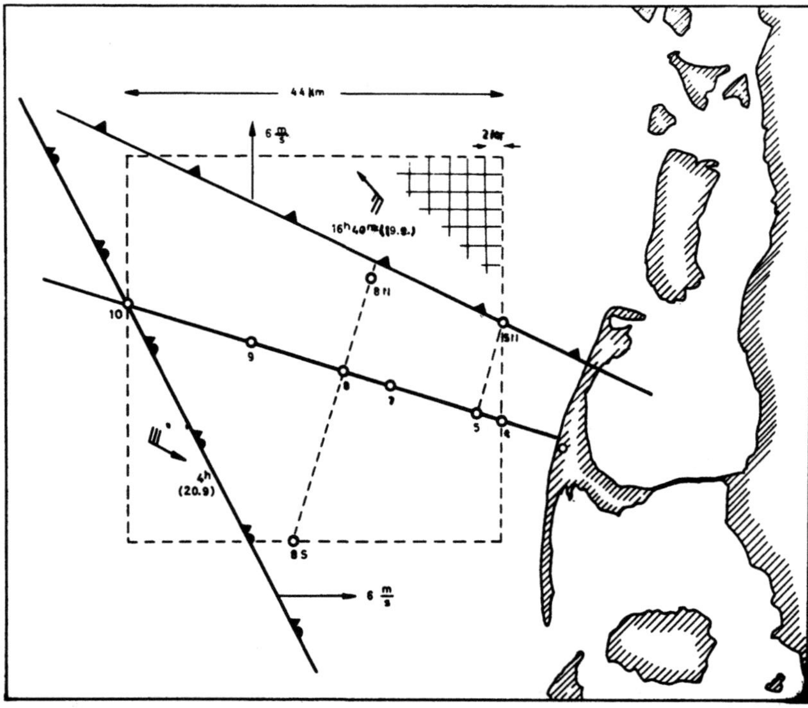

Fig. 2: The JONSWAP 1973 array. Also shown is the prediction model grid of both test cases and the two fronts of the first test case at different times.

observed wave data at Station 8S for the south, and Station 10 for the west and north boundary. The values for the east boundary were calculated from the JONSWAP fetch laws. The initial values were obtained by linear interpolation of the measured values.

First Case: Sept. 19(12:00) - Sept. 20(12:00)

The wind field of the first case from Sept. 19 (12:00) - Sept. 20 (12:00) is marked by a frontal system which moved over the JONSWAP array in the northeast direction. Fig. 1 shows the first front at 16:40 h on the 19th of Sept. and the second front at 4:00 h on the 20th of Sept.

Before the first front arrived the wind was south-east at about 10 m/sec. Between the fronts the wind slowed down and shifted to the South and South-West. After the transition of the second front the wind turned to the North-West and increased to about 12 m/sec.

Figs. 3 - 6 show the results of our calculations at Stations 7 and 9, Figs. 3 and 4 the wind-history parameters and the significant wave height, and Fig. 5 and 6 a comparison of the spectra. The results are generally satisfactory. The differences arising during the last 2 hours are due to a numerical overshoot which will be remedied.

Second Case: Sept. 20 (12:00) - Sept. 21 (12:00)

During this period the wind was spatially homogeneous over the area but changed direction from North-West through South to North-East. The wind strength also varied considerably.

Most of the hindcast period is characterized by considerable wave energy in the swell domain. The wind-sea and swell regimes are not well separated, and it is therefore difficult to identify the wind-sea parameters used in the numerical model in the measurements (Fig. 7 and 8). Thus, the power spectra themselves are more appropriate for a comparison of measured and hindcasted sea state than the spectral parameter (Figs. 9 - 11).

The swell energy at Station 7 (Fig. 9) is higher in the model at the beginning, but compares well later on. The differences in the first phase may be attributed to incomplete knowledge of boundary values; it was assumed that the northern boundary had the same energy-flux as measured at the western boundary at Station 10. This probably yielded an overestimate of the energy coming across the northern border, because the shallow water north of the prediction area effectively shadows swell coming from North-West. The error becomes smaller later, when the swell came in from more westerly directions. The discrepancies are smaller at Station 9 (Fig. 8 and 10), which is further from the coast and less influenced by refractive shadowing.

Fig. 11 shows a comparison of the spectra at Station 9 on Sept. 21st. The hindcast windsea developed several hours later than the observed windsea. The reasons for this are not clear, but it is possible that the source functions for the windsea might be influenced by the presence of large swell running in a direction opposite to the windsea, or -- more likely -- the windsea development for changing wind direction was overestimated by treating the wave energy as propagating in the local wind direction, without incorporating a relaxation time.

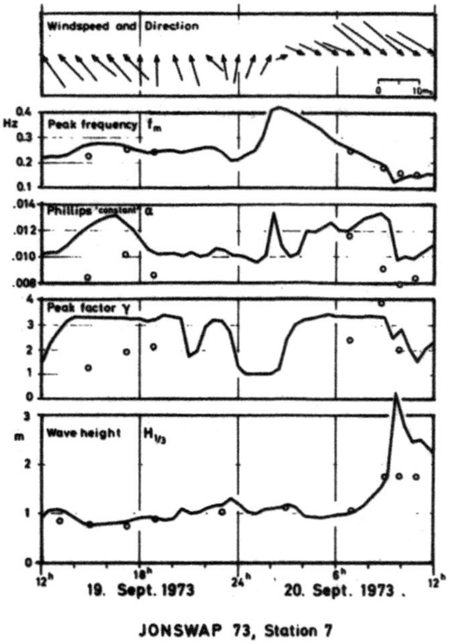

Fig. 3

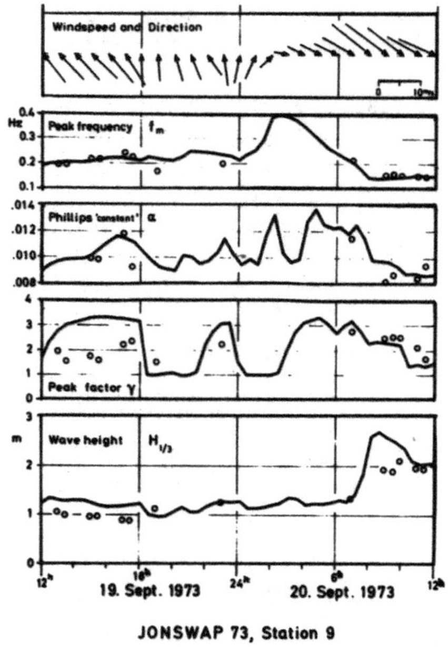

Fig. 4

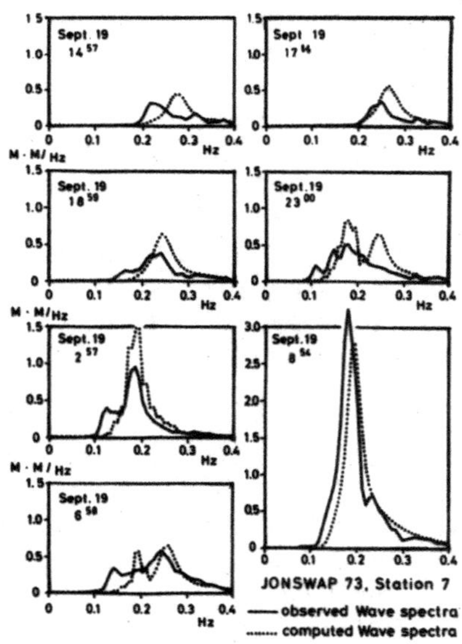

Fig. 5

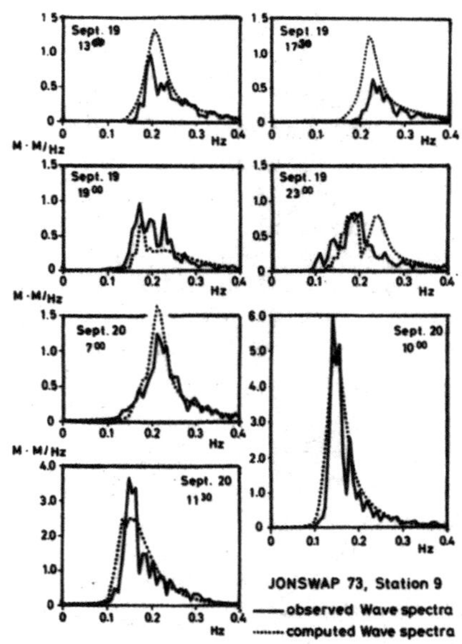

Fig. 6

Fig. 3 - 6: Model results and measurements for the first test case. Sept. 19 (12:00) - Sept. 20 (12:00) at Station 7 and 9.

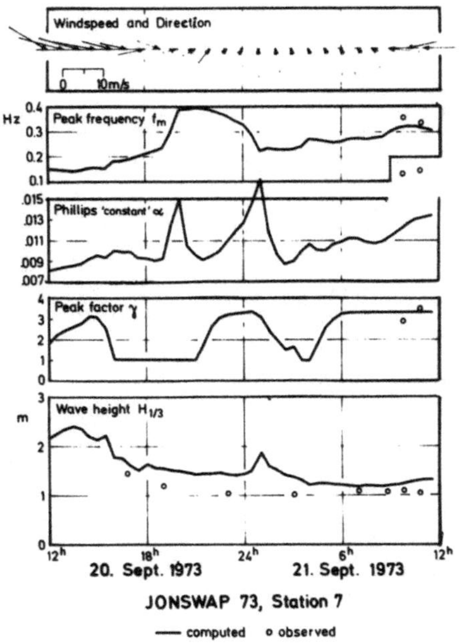

Fig. 7

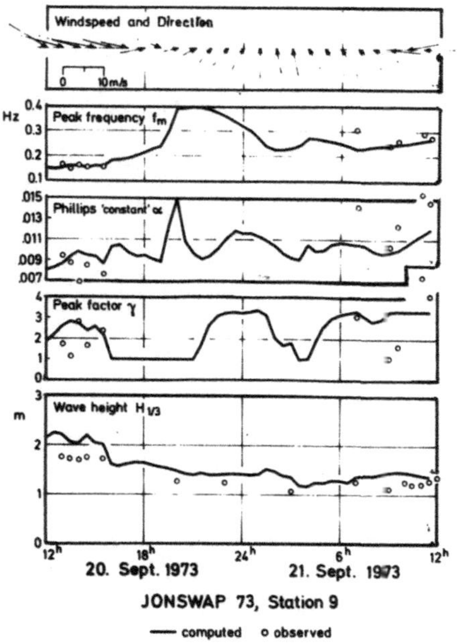

Fig. 8

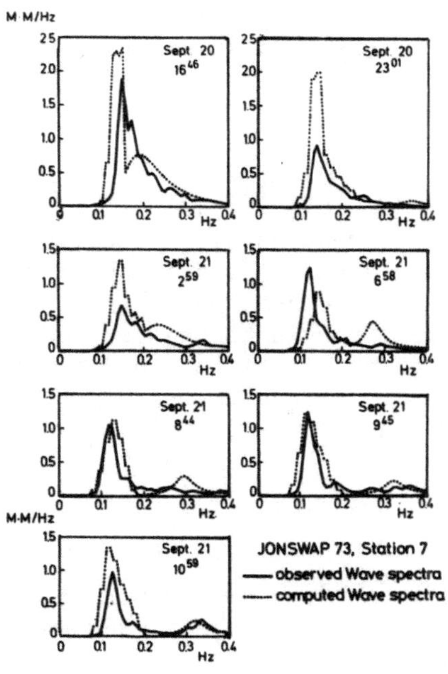

Fig. 9

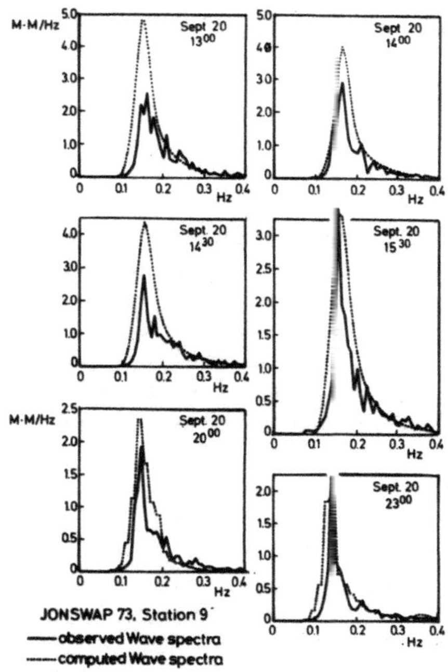

Fig. 10

Fig. 7-10: Model results and measurements for the second test case. Sept. 20 (12:00) - Sept. 21 (12:00) at Station 7 and 9.

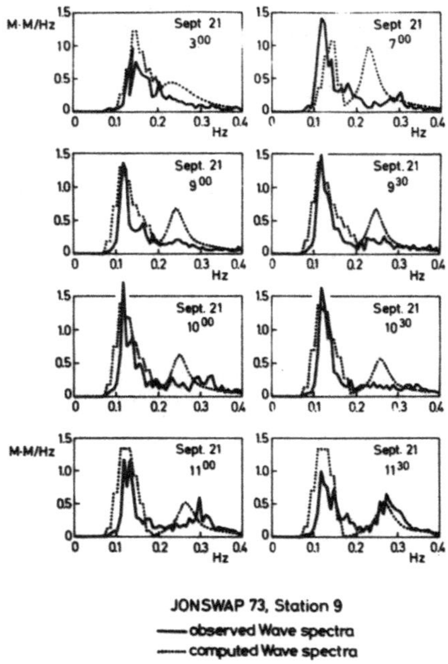

Fig. 11: Model results and measurements for the spectra of the second test case at Station 9 on Sept. 21.

CONCLUSION

The model is able to predict deep water waves with reasonable accuracy. The existing shortcomings should be removed in the next test phase. For wave prediction over specific shallow water regions, the southern part of the North Sea, for instance, the characteristics for swell propagation have to be calculated and implemented. Shallow water effects of dissipation and possibly modified generation have to be included in the source function.

REFERENCES

Hasselmann, K. et al. 1973. Measurement of wind-wave growth and swell decay during the Joint North Sea Wave Project (JONSWAP), Deutsche Hdr. Zeit. Reihe A (8°), Nr. 12

Hasselmann, K. et al. 1976. A parameterical wave prediction model. J. phys. Oceanogr., 6(2), 200-228

MacCormack, R.W. 1971. Numerical solution of the interaction of a shock with a laminar boundary layer. Proc. Second. Intern. of Numerical Methods in Fluid Dynamics, M. Holt, Ed. Springer-Verlag, 515-163

DISCUSSION

H. Mitsuyasu:

Did you check your prediction method for a case in which the wind waves were growing very rapidly (duration-limited case). According to our recent laboratory study the growth rate of spectral components is much larger for the duration-limited case than for the fetch-limited case. However, these data are limited to the case of laboratory waves. We would like to have similar information in the ocean.

If the laboratory results are satisfied also in the ocean, prediction of the duration-limited wave -, which uses the same growth rate as obtained for the fetsch-limited wave, will underestimate the results.

H. Günther:

We did not have the opportunity to run our wave model for a well defined situation described by Dr. Mitsuyasu. Beside very homogeneous and stationary winds, with associated wave fields, we also have windfields in our data bank changing in direction and strength on various time scales.

In these cases we face the problem of the non-linear coupling of waves travelling into different directions. This problem is currently being investigated. It appears that the present model overestimates the growth of the wind sea in situations of turning winds.

K. Hasselmann:

It will, of course, be difficult to find situations corresponding to ideal duration limited wave growth (i.e. a spatially uniform wind field with a step-function onset) outside the laboratory. One will presumably have to consider mixed space-time dependences , which normally involve also angular direction changes, as discussed by Günther.

However, it may be mentioned that according to the quasi-equilibrium concept the evolution of the wave field for the fetch- and duration-limited cases should be governed by different power laws (cf. Hasselmann et al., Journ. Phys. Ocean., 6, P. 200-228, 1977). In the first case the decrease of the non-dimensional peak frequency ν is proportional to the - 3/10 power of the non-dimensional fetch gx/u^2, in the latter case to the - 3/7 power of the non-dimensional duration gt/u (i.e. the waves grow faster.

D. Hasselmann:

I would like to comment that looking at the response time of the sea to a change in wind direction I arrive at response times in the order of 1 - 2 hrs at about .4 Hz,

W. Pierson:

How far from wave spectral forecast point were the estimated spectra used as boundary values for the model?

H. Günther:

I did not mention this to save time, but as pointed out in the paper, we used the measured values of St. 8S for the south, of St. 10 for the west and north boundary. The values of the east boundary are fixed by the fetch laws.

The prediction area was 44 x 44 km. The area was of the same scale or larger than the inhomogeneity of the wind field, so that the interior solution was not simply given by the boundary values.

D. J. Webb:

On September 21 when the model predicted a wind sea which did not develop - what was the angle between the predicted wind sea and the swell present. I ask because I would expect the swell to have most effect on the development of the wind sea when the angle between the two was typically 45°.

H. Günther:

In the last hours of the hindcast swell is running from NW to SE and the windsea from E to W. The angle therefore is about 180°.

A NUMERICAL MODEL HINDCAST OF SEVERE WAVE CONDITIONS FOR THE NORTH SEA

T. J. Weare and B. Worthington

Hydraulics Research Station
Wallingford, United Kingdom

1. INTRODUCTION

The sharp increase of activity in the North Sea in recent years has led to a need for better long-term extreme wave statistics to assist in the design of off-shore structures. In view of the necessarily long period required to gather a valid statistical base of observations and the geographical extent of the area of interest, a numerical model offers the only solution. For this reason the NORSWAM model has been developed to calculate wave conditions from a retrospective analysis of the windfield for a selected number of severe storms over the past 10 years.

The project has been jointly financed by the UK Departments of Energy and Industry, the Hydraulics Research Station (UK), and the UK Offshore Operators Association; it was administered by a tripartite executive committee representing the three funding agencies. Technical guidance has been provided throughout the project by a specially constituted international Technical Advisory Group under the chairmanship of J. A. Ewing (Institute of Oceanographic Sciences, UK).

The model is based on the parametrical wind-sea model of Hasselmann et al. (1976). The analysis of the windfield input and the long-term statistics has been performed by the Institute of Oceanographic Sciences. The development and calibration of the model was carried out by the Hydraulics Research Station and it is principally this work which is to be described in the following.

2. Theoretical basis

The fundamental principle on which the model is based is the energy balance equation for the frequency spectrum, ie

$$\frac{\partial e}{\partial t} + \underline{v} \cdot \underline{\nabla} e = S \qquad (1)$$

where e = the energy density for wave frequency f and direction Θ at x,y,t.

$\underline{v}$ = the wave group velocity

$= \frac{g}{4\pi f}$ in the direction Θ

The source term S represents all physical processes which add energy to or remove energy from the spectrum except, of course, for wave propagation. Equation 1 is treated in two distinct ways depending on whether the sea state is a wind-sea or a swell condition.

WIND-SEA Following Hasselmann et al. (1973; 1976), we assume that for a developing wind-sea the input of energy from the atmosphere and the transfer of energy within the spectrum due to the non-linear wave-wave interactions conspire to give a stable spectral shape. In particular we assume that the wind-sea frequency spectrum is described by the JONSWAP parametrical form, viz.

$$E(f) = \alpha \frac{g^2}{(2\pi)^4} f^{-5} \exp \{- \frac{5}{4} (\frac{f}{f_m})^{-4}$$

$$+ \ln\gamma . \exp \left[(\frac{f-f_m}{f_m})^2 \frac{1}{2\sigma^2} \right] \} \qquad (2)$$

$$\sigma = \sigma_a \; f<f_m \; ; \; \sigma_b \; f>f_m$$

Of the five parameters introduced $(\alpha, f_m, \gamma, \sigma_a, \sigma_b)$, only the energy level parameter α, the peak frequency f_m, and the peak enhancement factor γ will be treated as free; σ_a and σ_b being fixed at 0.07 and 0.09 respectively, these being their JONSWAP mean values (Hasselmann et al., 1973).

Provided the windfield does not suffer excessively rapid changes in space or time, the stabilising effect of the wave-wave interactions leads to the further simplification that the angular distribution of the wave spectrum is centred on the local wind direction. Furthermore, it is assumend that the angular distribution of the wave spectrum can be represented by a cosine^2 form. Thus the two dimensional wind-sea energy spectrum is given by

$$e(f,\theta) = \frac{2}{\pi}\cos^2(\theta-\phi).\ E(f) \qquad (3)$$

where ϕ = the direction of the wind and E(f) is given by equation 2.

One of the advantages of assuming a parametrical form for the spectrum is that the energy balance equation (equn 1) can be transformed into a set of coupled equations describing the evolution of the five parameters. This is much more manageable computationally than the large set of coupled equations which results when equation 1 is discretised directly (ie discrete f and θ). In addition, the contribution of the wave-wave interaction to the source term is more readily described in the parametrical form. When transformed to parametrical space the energy balance equation becomes (Hasselmann et al., 1976)

$$\frac{\partial a_i}{\partial t} + D_{ijk}\frac{\partial a_j}{\partial x_k} = S_i \qquad (4)$$

D_{ijk} is a generalised propagation velocity; S_i is the transformed source term for parameter a_i ($a_i = \alpha, f_m, \gamma$ etc.). A detailed discussion of the derivation and form of the coupling coefficients D_{ijk} would be inappropriate here. However, to understand the calibration procedure for the model, a description of the main features of the source terms S_i is useful.

Only three free parameters are employed in the model: α, f_m and γ, and it is the equation governing the peak frequency f_m that is the most important. The source term for f_m is dominated by the non-linear wave-wave interactions and consequently is prescribed on theoretical grounds. It dictates the rate of evolution of the spectrum in space and time. In the present model we take

$$S_{f_m} = -0.5861\ \alpha^2 f_m^2\ \left(\frac{\gamma-1}{2.3}\right) \quad \text{for} \quad \gamma>1$$

$$= 0 \quad \text{for} \quad \gamma<1 \qquad (5)$$

The source term for α involves important contributions from wave-wave interactions and from other sources (denoted S_{rest}), principally the atmospheric input. S_{rest} is not well defined theoretically. We assume the following form for S_α

$$S_\alpha = \underbrace{-5.0\ \alpha^3 f_m}_{\text{wave-wave}} + \underbrace{a\nu^p\ \alpha f_m}_{S_{rest}} \qquad (6)$$

where $\nu = \frac{U}{g} f_m$ and U = wind speed.

The coefficient a and exponent p are the principal free parameters in the model and are calibrated by comparison with observed wave conditions. Except with very rapidly varying wind the contributions from the wave-wave and the rest source terms are in balance so that a relationship between α and the dimensionless peak frequency ν is implied, viz.

$$\alpha = \left(\frac{a}{5.}\right)^{1/2} \nu^{p/2} \qquad (7)$$

This, of course, is the main motivation for the assumed form of S_{rest}.

The source term for γ is dominated by the wave-wave interaction term and is taken to be

$$S_\gamma = -16.0\ \alpha^2 f_m\ (\gamma - \gamma_o) \qquad (8)$$

where $\gamma_o = 3.2787 \quad \text{for} \quad \nu \geq 0.16$

$= -14.95 + 113.94\nu \quad 0.16 \geq \nu \geq 0.14$

$= 1 \quad \nu \leq 0.14$

The numerical coefficients in S_α, S_{f_m} and S_γ are based on computations of the non-linear wave-wave source term for the mean JONSWAP spectrum ($\gamma = 3.3$, $\sigma_a = 0.07$, $\sigma_b = 0.09$). The dependence of S_γ on ν and S_{f_m} on γ is chosen to ensure transition to a fully developed state ($\gamma \to 1$) as $\nu \to 0.14$. ($\nu = 0.14$ gives a value of f_m corresponding with the Pierson-Moskowitz spectrum).

<u>SWELL</u> The parametrical approach can only be applied to the wind-sea region of the energy spectrum; for swell the atmospheric input is too low to support the spectrum at a level at which wave-wave interactions are effective. A shape stabilising dynamic balance does not exist and swell has to be treated as a freely propagating wave field. However, situations will arise in which swell exists at frequencies such that energy is received from the atmosphere although the swell is outside the range of the non-linear wave-wave interactions and is not, therefore, absorbed into any existing wind-sea. It is assumed that swell at a frequency f will not be absorbed into a wind-sea with a peak frequency f_m if $f<0.9f_m$. On the other hand, energy will be received from the atmosphere if $f>f_o = \frac{g}{2\pi U\cos\theta}$. Thus if $f_o<f<0.9f_m$ it is appropriate that the swell energy should grow and an atmospheric input source function is assumed of the following form (Snyder, 1974)

$$S(f,\theta) = \beta(f,\theta)\, e(f,\theta) \qquad (9)$$

where
$$\beta = 2\pi f\, \{\frac{f}{f_o} - 1\}\, \frac{\rho_a}{\rho_w}\, c \qquad f>f_o$$
$$= 0 \qquad f>f_o$$

(ρ_a, ρ_w = density of air, water)

The constant c is taken as 0.05, a value consistent with the parametrical wind sea source function S_α.

HYBRID MODEL: SWELL + WIND SEA

The NORSWAM model is a hybrid, combining a parametrical treatment of wind-sea with a classical swell propagation scheme. Clearly dynamical criteria are required to decide when energy is to be transferred between the swell and wind-sea domains. Two simplifying assumptions are made:

(i) energy is conserved overall

(ii) the wave-wave interaction is such that the coupling or de-coupling of wind-sea and swell is rapid when it occurs.

The criteria for wind-sea ↔ swell transfers outlined below are somewhat arbitrary. However, as the NORSWAM model is intended for extreme wave conditions swell does not generally play a major role and this deficiency is not considered serious.

(a) Swell to wind-sea.

If swell exists at a frequency greater than $0.9f_m$ (f_m = peak frequency of the wind-sea), it is assumed to be rapidely absorbed into the wind-sea irrespective of its direction. Energy is conserved by adjusting the wind-sea peak f_m, keeping α and γ fixed, ie

$$f_m)_{new} = f_m)_{old} \quad \{1 + 5 \frac{(2\pi)^4}{\alpha g^2} f_m^{\ 4} \frac{\delta E_{swell}}{c(\gamma)}\}^{-1/4} \qquad (10)$$

The correction factor $c(\gamma)$ is necessary because the JONSWAP spectrum is not integrable for $\gamma = 1$. To a good approximation $c(\gamma) = 1 + (\gamma-1)\ 0.224$ for σ_a, $\sigma_b = 0.08$

(b) Wind-sea to swell.

It is assumed that, when the value of the dimensionless peak frequency falls below the fully developed PM limit of 0.14, ie $f_m < \frac{0.14g}{U} = f_{PM}$, energy will be shed from the wind-sea to the swell domain. Energy is conserved overall by adjusting f_m to f_{PM} and by dumping to swell all energy at frequencies less than a cut off frequency f_c defined by

$$f_c = (\tfrac{5}{4})^{1/4} \ f_m \ [-\{\ln \ 1 - (\frac{f_m}{f_{PM}})^4 \}]^{-1/4} \qquad (11)$$

A cosine^2 spreading factor is assumed when distributing the swell energy.

3. Numerical and computational techniques

Limitations of space preclude a detailed description of the numerical techniques and computer program. A brief outline of the salient features is given in the following two sections; further details are available elsewhere (Hydraulics Research reports, EX775 and EX776).

WIND-SEA The parametrical wind-sea component of the model is based on a computer program originally developed at the Max-Planck-Institut für Meteorologie, Hamburg. The set of coupled partial differential equations (equn 4) governing the evolution of the spectral parameters is solved by means of a finite difference integration scheme. A second order Lax-Wendroff method is employed to integrate the equations on a uniform space-time grid (a first order accurate predictor-corrector scheme originally used proved to be too inaccurate). Although programming the equations is reasonably straightforward, care must be taken with the logic near boundaries so that when one sided spatial differences are necessary they are taken in an upwind direction. Also, to avoid an instability arising when the change in the parameters due to the source term is large, it is necessary to reduce the timestep so that the maximum change at each sub timestep is less than some pre-set limit (say 10 %).

SWELL The swell field is represented on a set of ray paths which, in the absence of refraction, follow great circles. Ray tracks are very nearly straight lines on the stereographic projection used by NORSWAM. The energy density of the swell for a particular frequency f and direction θ is defined at points uniformly spaced along each ray in the θ direction with an interval

$= \Delta t \ \frac{g}{4\pi f}$. ($\Delta t$ = integration timestep). With all rays stored

in a single one-dimensional array the entire swell field can be propagated one timestep by a simple array or input-output process.

Transformation between the characteristic ray grid and the cartesian swell grid is necessary when swell-wind sea transfers occur. In the NORSWAM model ten frequency bins between 0.05 and 0.15 Hz and eight direction bins (45° interval) are used.

4. Application to the North Sea

The model grid selected for the wave calculations is shown in figure 1. It is the same as the grid used by the British Meteorological Office atmospheric model. Based on a stereographic projection, it has an average grid spacing of 100 km. Since the primary interest of the study is in the northern North Sea the model boundaries extend well to the north and west into the Atlantic.

The analysis of the wind field was carried out by the Institute of Oceanographic Sciences. From the 200 or more gale occurences during 1966-76 a sub set of 42 storms was selected for analysis. The selection was made preserving the same proportion of storms according to type and the monthly/yearly frequency. Synoptic weather charts were reanalysed to include late ship reports and oil-rig observations. Corrections were made for atmospheric stability and for markedly cyclonic flows. The windfield was specified every 3 hours throughout a storm at the grid points indicated in figure 1. Linear interpolation in space and time provided windfield values on the 100 km grid and 1 1/2 hour integration step used by the model.

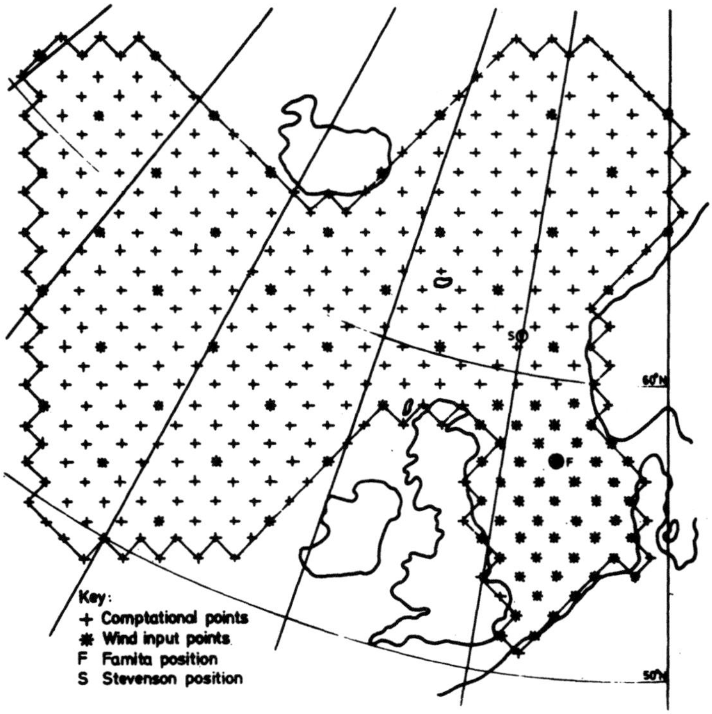

Fig. 1. Model grid.

Calibration of the model was possible by comparison with observed wave recorder data for 14 storms at a number of sites (see figure 1). Adjustment of the model was made through the α source term (equn 6) to minimise any bias in the maximum significant wave heights predicted by the model. Various combinations of the values of the coefficient a and exponent p in the α rest source term were tested, and the model and observed values for the significant wave heigth H_s and mean zero-up-crossing period T_z compared. Values of a = 0.00512 and p = 1.314 were selected to minimise the bias in H_s at positions FAMITA and STEVENSON.

This corresponds to an α-ν relationship of

$$\alpha = 0.032 \quad \nu^{0.657} \tag{12}$$

which is very close to that obtained in the JONSWAP experiment

$\alpha = 0.033 \quad \nu^{0.667}$ (Hasselmann, 1973).

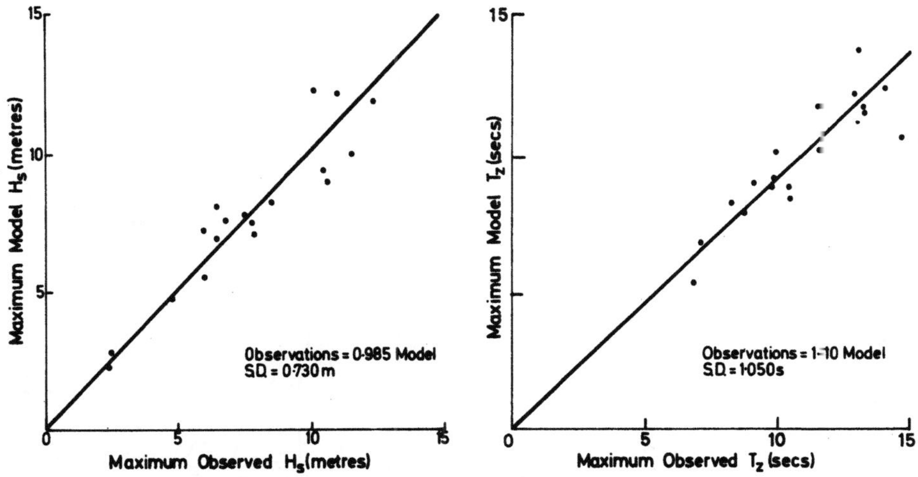

Fig. 2a. Correlation of observed and model peak values.

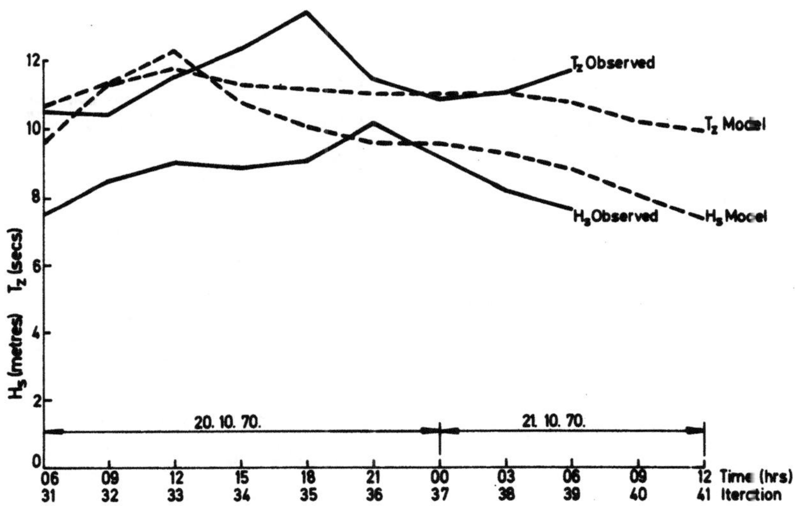

Fig. 2b) Storm 1 FAMITA position.

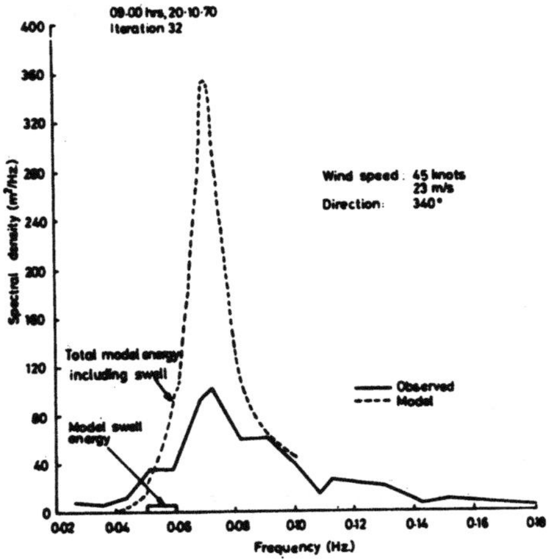

Fig. 3a. Storm 1 - FAMITA position.

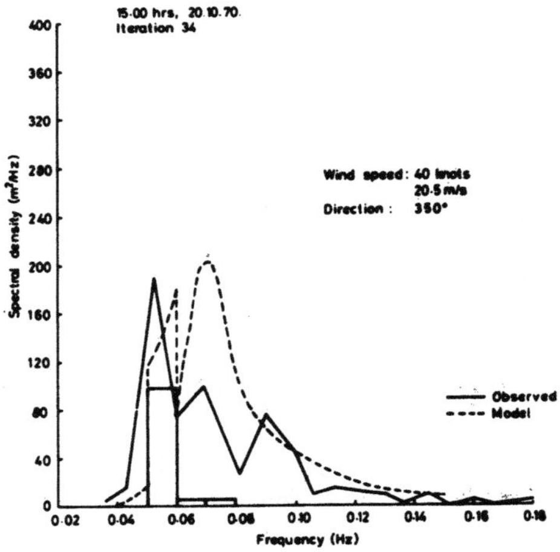

Fig. 3b. Storm 1 - FAMITA position.

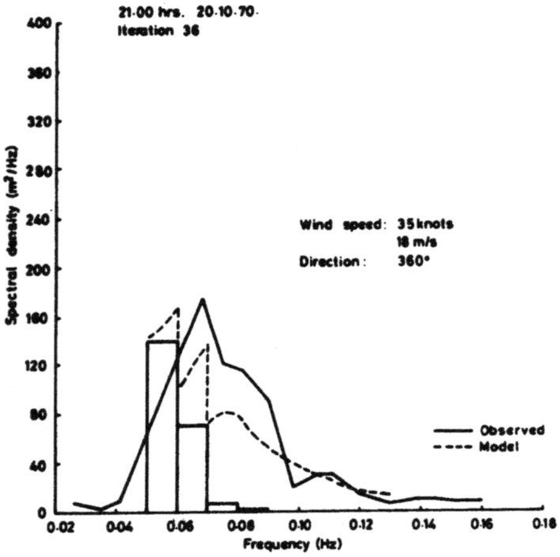

Fig. 3c. Storm 1 - FAMITA position.

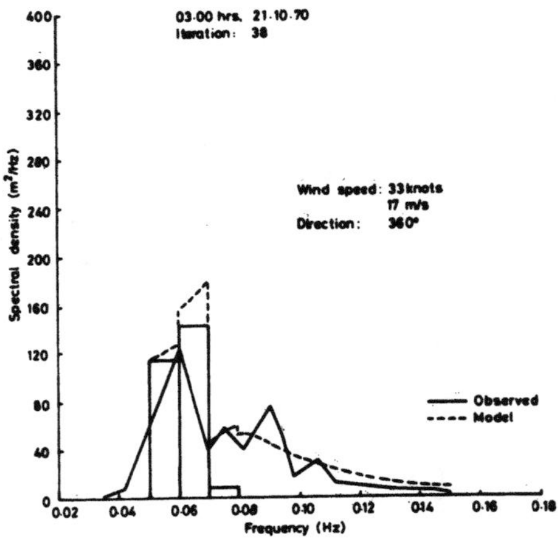

Fig. 3d. Storm 1 - FAMITA position.

Although equation 12 differs somewhat from the fit obtained for an enlarged data set (data set J; Hasselmann, 1976),

$\alpha = 0.036\ \nu^{0.87}$, it is well within the scatter of the data.

Figure 2a shows the correlation between predicted and observed maxima H_s for FAMITA and STEVENSON positions. A weighted least squares fit yields

$$H_s)_{Observed} = 0.985\ H_s)_{Model} \qquad (13)$$

with a standard deviation from equation 13 of 0.7 m (= 10 % of the mean). A comparison of model and observed H_s and T_z for one storm at FAMITA is presented in figure 2b, and in figure 3 spectra for the same storm are compared.

When all the observed data, ie not only the storm peaks, are taken into account the scatter between predicted and observed H_s and T_z rises to around 20 % of the mean. This is probably as good an agreement as can realistically be expected from the model. The composite data set J (Hasselmann, 1976) referred to earlier has scatter in the α-ν plane with a standard deviation of 38 %. Since $H_s \propto \alpha^{1/2}$ this implies an error in H_s of ~ 19 %. In addition to this presumably inherent variability, there is also the uncertainty in the windfield estimate in the model. The overall agreement between model and wave recorder data is therefore considered acceptable and long term wave statistics have been deduced from the model output.

5. Future development

Inevitably, in the course of a major development program of this nature a number of new questions are raised, some purely practical, others of a more fundamental nature. Undoubtedly, one major area requiring further research concerns the swell - wind sea exchanges. Although adequate for the purpose of this project, the somewhat arbitrary criteria adopted here may be less than satisfactory for a forecasting mode in which swell conditions are more significant.

Transition to a fully developed sea state has been allowed in the model by modifying the f_m and γ source terms. At present there is very little theoretical or experimental evidence in this area and further research is needed.

The assumption that the angular distribution of the wind-sea follows the local wind direction is not expected to hold for rapidly varying conditions. Some way of relaxing this constraint must be found if situations involving restricted fetches are to be handled satisfactorily.

For an operational forecasting model of the North Sea a finer grid is necessary to resolve important details such as the Orkney and Shetland Isles. To extend the model to the southern North Sea will require the inclusion of refraction and other shallow water effects which will be particularly important for swell. Of course, as greater computational emphasis is placed on the treatment of swell the economic advantage of the parametrical wind-sea component is reduced. In its existing form already more than 50 % of the computational effort is devoted to swell.

The accuracy of the wave predictions is controlled to a large extent by the accuracy of the wind input. For this reason it was felt that the extra effort involved in the subjective manual windfield analysis adopted for NORSWAM was worthwhile in terms of the improvement in accuracy. In spite of this, a number of cases have been found in which significant errrors in the wave predictions were attributed to deviations of the input winds from those observed. Considerable importance must be attached to the development of a reliable numerical windfield model. Indeed, it is likely that future improvements in the accuracy of wave model predictions await further advances in atmospheric modelling.

Acknowledgements

The authors are indebted to all the members of the NORSWAM Technical Advisory Group and, in particular, to J A Ewing and K Hasselmann for their constant help and encouragement throughout the project. The work reported in this paper is published with the permission of the Director of Hydraulics Research, UK.

References

Hasselmann. K, et al, 1973; "Measurement of Wind-Wave Growth and Swell Decay during the Joint North Sea Wave Project (JONSWAP)", Deutsche Hydr. Zeit. Reihe A (8°), Nr. 12.

Hasselmann. K, et al, 1976; "A Parametrical Wave Prediction Model", Journal of Physical Oceanography, Vol 6 (2) pp 200-228.

Snyder. R. L, 1974; "A Field Study of Wave-Induced Pressure Fluctuations above surface Gravity Waves", Journal of Marine Research, 32, pp 497-531.

A WIND WAVES PREDICTION MODEL IN THE ADRIATIC SEA

L. Cavaleri, P. Malanotte Rizzoli

L.D.G.M., Consiglio Nazionale delle Ricerche
San Polo 1364, Venice, Italy

ABSTRACT

A numerical wave prediction model for shallow water has been developed with specific application to the Northern Adriatic Sea. Refraction patterns from a single prediction point are evaluated for a given number of frequencies and directions. The energy balance equation is successively integrated along the characteristics providing the wave spectra at the reference point and at a given time. Predicted and experimental results are compared and discussed.

INTRODUCTION

One of the most important problems affecting Venice and its environment is the wind wave field produced in the Adriatic sea under particular meteorological conditions, which has been responsible for severe destructions along the coastlines of the three islands protecting the lagoon of Venice from the open Adriatic. In the most known example, the famous storm surge of November 1966, which caused the flooding of the city of Venice, the associated wind wave field completely destroyed the defensive structures along vast portions of the above mentioned islands and the Northern Adriatic coastline.

Apart from these extreme situations, this coastline shows moreover extensive processes of erosion and sedimentation, obviously related to the Northern Adriatic wave regime. It is therefore of fundamental importance to be able to predict the wind wave field generated in the Adriatic under any kind of meteorological situation. In the past, wave spectra forecasts were based on a number of empirical spectral formulas not always applicable to every kind of basin and often giving different results. We refer to the now classical methods of Sverdrup-Munk-Bretschneider and Neumann and Pierson.

More recently, methods of wave forecasting were based on the integration of the energy balance equation, combining in a synthesis theories of wave dynamics and empirical functions derived by existing wave growth data (Barnett, 1968 - Ewing, 1971).

Since then, extensive wave measurements have been made, such as during the JONSWAP experiment (Hasselmann et al., 1973). They have not only clearified some of the main mechanisms controlling the evolution of the wave spectrum but also changed the outlook and the interpretation of the wave growth process. In particular, the importance of resonant wave-wave interactions has been enhanced (Hasselmann, 1962-1963 a, b), as producing a nonlinear energy transfer which is, in particular, responsible for the growth of the newly developing waves on the forward face of the spectrum.

As it has been recently pointed out by Hasselmann et al. (1975), a new start seems necessary for the wind wave prediction problem, in the light of these new findings. Nevertheless, for what the wave prediction in shallow basins is concerned, the simplest and most logical first approach is still the construction of a wave ray model coupling the traditional energy balance equation with those modelling the interaction with the bottom topography, which strongly modifies the propagating waves. The Adriatic sea, infact, in its northern half is entirely on the continental shelf, with a depth which gradually increases from 0 to 100 m, abruptly falls to about 250 m and finally forms the deep southern basin (1200 m) (Fig. 1). In our ray model, the integration of the transport equation is done straightforwardly along characteristics, following therefore the change in energy of the individual wave packets propagating along the rays. Even if, after JONSWAP, the implicit limitations of a decoupled ray model are evident and well known, still this method is particularly simple to handle as it deals with a single prediction point. Moreover, that method is particularly efficient for what numerical construction and computer time are concerned. Our reference point is a tower belonging to our laboratory, situated outside the Venice lagoon, about 7 miles offshore, on a depth of 16 m.

For what this area is concerned, two characteristic meteorological patterns are dominant. The first has a typical generation area in the southern-central part of the basin, and originates a South-East wind, which blows parallel to the longitudinal axis of the Adriatic, with very long fetch and moderate intensity. The

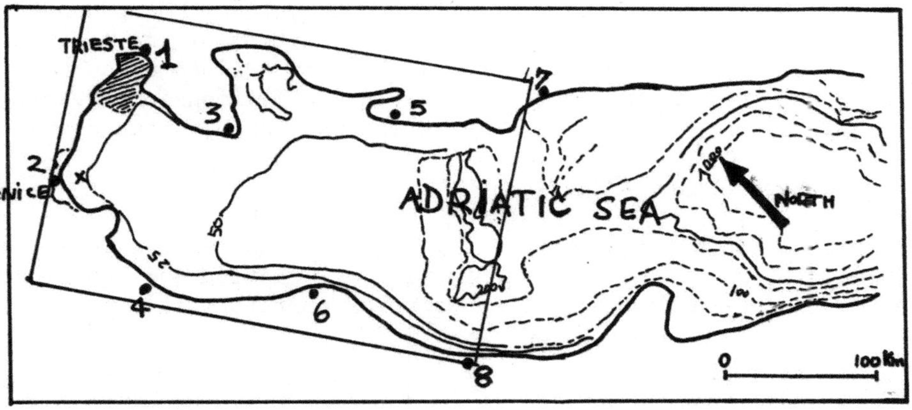

Fig. 1 - The upper Adriatic sea is covered with a 30x52 points grid. Numbers 1 to 8 are the meteorological stations. "x" is the position of the tower used for the measurements.

typical waves thus originated are swell waves with periods in the range from 8 to 10 sec. The second meteorological pattern is characterized by a generation area involving the whole northern most part of the Adriatic, and produces a North-East wind which blows parallel to the cross-basin axis in the gulf of Trieste and Venice (Fig. 1). The wave field thus originated is that of a typical wind sea, with periods in the range from 4 to 8 sec. The two experimental series here considered belong to this second meteorological situation. The influence of refraction on the South-East wave field was analyzed in a previous study by Malanotte Rizzoli and Battiston (1973).

THEORY

The sea state is described through the use of the spectral density function $E(\vec{k})$ defined per unit area per unit wave number space. This can be considered as the description of the sea near a point in space (x, y). In general, $E(\vec{k})$ is defined as a function of (x, y) and time t, being therefore a 5-dimensional variable $E(x, y, t, k_x, k_y)$ if k_x, k_y are the wave number components along cartesian axes (x, y).

Within the spectrum, the energy changes can be considered as due to local rate of change, convection, refraction and shoaling. The considered energy balance equation is therefore:

$$\frac{dE}{dt} = \frac{\partial E}{\partial t} + \frac{\partial E}{\partial x_i}\frac{dx_i}{dt} + \frac{\partial E}{\partial k_i}\frac{dk_i}{dt} = Q \qquad (1)$$

where:

$$x_i \equiv (x_1, x_2) \equiv (x, y) \quad ; \quad k_i \equiv (k_1, k_2) \equiv (k_x, k_y)$$

and the summation symbolism over equal subscripts is considered.

The source function Q will be constructed by the superposition of processes governing the possible sources or sinks of wave energy. For the case of no dissipation or generation, Q=0 and $E(\vec{k})$ = constant (Longuet-Higgins, 1957).

The energy equation (1) is coupled with the Hamiltonian equations of the wave field (Phillips, 1966).

$$\frac{dx_i}{dt} = v_i = \frac{\partial w}{\partial k_i} \quad ; \quad \frac{dk_i}{dt} = -\frac{\partial w}{\partial x_i} \qquad (2)$$

where: $\omega = \omega(\vec{k}, \vec{x}) = g\,\kappa \tanh(k\,H)$, with g = gravity acceleration and H = depth, is the finite depth dispersion relationship for wind gravity waves. $\upsilon_i = \partial\omega/\partial\kappa_i$ with i = 1, 2 are the (x, y) components of the group velocity $V = \partial\omega/\partial\kappa$. For computational purposes, the spectrum is usually represented as a density $E\ (f,\ \emptyset)$ with respect to frequency and propagation direction, through the transformation $E(\vec{\kappa})\ V/2\pi\kappa \cdot E(f,\ \emptyset)$ and the frequency spectrum is then evaluated as $E(f) = \int_0^{2\pi} E(f, \phi)\, d\phi$. For the specification of the source function Q the following processes have been taken into account.

a. Phillips reasonant mechanism. This mechanism was first proposed by Phillips (1957) and governs the wave growth in the early phases of generation. The term does not contribute significantly to the source-function, but it is included to initiate the evolution of the wave spectrum. Phillips' basic definition has been taken

$$\alpha = \frac{\pi}{\rho_w^2}\, c^2 \cdot \tanh^2 (\kappa H) \cdot P(\vec{\kappa}, \omega) \qquad (3)$$

valid for water of any depth H, where $c = \frac{\omega}{\kappa}$ is the wave celerity and $P(\vec{\kappa}, \omega)$ is the turbulent pressure spectrum.

Since we have found the previously proposed expressions for $P(\vec{\kappa},\omega)$ (Barnett, 1968) unsuitable for the shallow water prediction problem, we have used instead the approximate expression of Phillips (1966), giving the mean spectral density $P(\vec{\kappa},\omega)$ as proportional to the 4th power of the wind speed. A more precise expression for α is, anyway, unnecessary, as its main purpose is to "start" the sea from zero initial conditions as the coastline.

b. Miles instability mechanism. Once the water surface is disturbed, it in turn disturbs the air flow through a linear feedback mechanism (Miles 1958). It has recently been shown (Hasselmann 1974) that the empirical representation used in the early wave models are not sufficiently representative because of the dominant contribution of non-linear interaction to wave growth on the forward phase of the spectrum, not recognized in the earlier wave growth studies. In the same reference, Hasselmann proposes a new expression for this source function input, based on the JONSWAP data, which is several factors smaller than the previous ones.

As, anyway, non-linear interactions are not taken into account in the present model, we had to resort to another expression for the function which, at least partially, compensates for the lack of non-linear energy transfer. We use the expression given by Barnett (1968)

$$\beta = 5\, s\, f\ (W \cos\delta/c - 0.90) \qquad (4)$$

where ρ = ratio of densities of air to water, f = wave frequency in hertz, W = wind speed in m/sec., δ = angle between wind and wave vectors.

c. Bottom friction dissipation. Two theories have been proposed to explain the energy loss, in particular of swell, in shallow water: bottom friction (Hasselmann and Collins, 1968) and bottom scattering (Long, 1974). The swell attenuation of JONSWAP data is not consistent with the theory of Hasselmann and Collins, even if it might be explained by Long's theory. However, in the northern part of the Adriatic sea, for the North-East wind situation, waves run essentially parallel to the isobaths, and the back scattering caused by the irregularities of the bottom topography is probably not the dominant effect. We have therefore modelled the bottom frictional dissipations following Hasselmann and Collins (1968) in which a quadratic relationship is assumed for the bottom friction stress. The following approximate treatment has been used in the uncoupled case (Collins, 1972) for the dissipation function:

$$\Phi_{b\ friction} = - \frac{c_f\ g\ \kappa V}{2\pi\omega^2 \cosh^2(\kappa H)} E(f,\phi)\cdot\langle u\rangle \qquad (5)$$

with the usual meaning of symbols, where < > denotes ensemble average for the wave orbital velocity at the bed of a wave field in water of constant depth H, in the absence of any mean current velocity. $\langle u\rangle$ is related to the frequency energy spectrum $E(f)$ which, therefore, must be known for all frequencies at every point along each ray. Differently from Collins, we do not have an initial value to give to $E(f)$ in the integration of the energy equation along the characteristics. Infact, the great majority of the initial points of rays - and surely for all the rays of significance for a North-East wind- the initial points is at the coastline, where the initial energy is obviously zero. Being, moreover, under typical generation conditions, at each point of every ray we have assumed a parametrical representation of $E(f)$ as given by the "mean" JONSWAP spectrum, computed according to the wind component along the specific ray.

d. Whitecaps and breaking. It is a well observed fact that waves grow to some maximum amplitude before the onset of breaking or whitecaps will limit their energy. A general form of the whitecapping source function has recently been proposed by Hasselmann (1974) giving for the dissipation function

$$\Phi_{wave\ breaking} = - \eta\ \omega^2\ E(\vec{\kappa}) \qquad (6)$$

The coefficient η is again estimated on the basis of the experimental results of JONSWAP. In (6) anyway, η has been determined from energy balance including both non linear interaction - not present in our model - and a source function input as given by Hasselmann in the same reference (1974), which is, as previously mentioned, several factors smaller in respect to the β - expression proposed by Barnett which we use. We have therefore added a further constraint to the energy growth in the saturation range, adopting the traditionally used criterion for which the energy density cannot overcome the value of the saturation energy, the expression of which was first proposed by Phillips (1958). We have, in particular, used for $E(f,\emptyset)_{saturated}$ the expression reported in Collins (1972),

NUMERICAL PROCEDURE

The northern half of the Adriatic sea has been covered with a network of 30 x 52 grid points, respectively on x and y direction. The axis coincides approximately with the longitudinal axis of the basin. The grid size is 7500 m. The bottom topography field is taken from a 1:750,000 map, with the same network of points. We first integrate the refraction equations (2) starting from the tower (Fig. 1) using a 4th order Runge-Kutta scheme. The integration is done for 20 frequencies and, for each of these, for a number of rays spaced at 10° direction interval, thus covering the whole basin under consideration. The direction spacing is reduced to 5° interval for longer rays. The integration is stopped when the ray reaches the coastline or the southern end of the grid. The numerical procedure has been checked repeating the integration procedure in the backward direction assuming as initial point the final one on the ray. The maximum difference between the corresponding points of the two processes was in the order of the 5th significant digit in grid units.

All ray patterns were stored on computer tape. Using the memorized data, the energy balance equation (1) is integrated backwards for each frequency ray by ray. From the directional spectrum thus obtained, the frequency spectrum is evaluated by weighted summation over all the directions. The computer time for the evaluation of a spectrum is of the order of 8 minutes on an IBM computer 370/155.

WIND EVALUATION

The wind field knowledge is one of the basic points required for an accurate estimate of the wave spectrum. Apart from a few exceptions, like in JONSWAP, the lack of wind data has been a strong limitation to the further improvement of the wave prediction models. Findlater et al. (1966) on the basis of long time series, have analyzed the general dependence among the wind speed and direction at a height of 19.5 m, the 900 m bar geostrophic wind and the temperature lapse rate. On the average the surface wind speed over the sea has been found to be 80 % of the 900 mbar geostrophic wind and turned 8° towards the low pressure zone in respect to it. These values have been used by Ewing (1971) for his model of the North Atlantic Ocean.

This procedure, assuming the correctness of the meteorological map, fails when we go into small details, strongly influenced by the local orography, that cannot be represented on a map drawn on an European basis. This has been clearly put in evidence for the northern Adriatic sea by Cozzi. His numerical model takes into account the surface friction and the local topography, but it is not directly applicable because of the computer time required. In our case we have used a simpler model developed by Tomasin (Robinson et al., 1972) for the storm surge forecast in the Adriatic. In it, 3 hour surface pressure data at several stations along the Adriatic coast were least square (LSQ) fitted to a linear polynomial in the two horizontal grid coordinates, with each station weighted proportionally to its inverse square distance from the point at which an estimate was to be obtained. The pressure gradient and corresponding geostrophic wind were obtained by differentiating the LSQ polynomial. The geostrophic wind was attenuated by 50 % and rotated by 17° to estimate anemometer height winds (Robinson et al., 1972). These values are much larger than those we have previously quoted from Findlater et al. (1966) and may be justified by the average different conditions present in the Adriatic region. This procedure fails when the local topography leads to local winds definitely different from those present in the general area under consideration. In this case the wind field is smoothed out and the local effects disappear. In the attempt to improve the local estimate in the Northern Adriatic and particularly in the Trieste gulf, we have here reduced the number of coastal meteorological stations to those strictly surrounding the area.

Fig. 1 shows the overall area covered by the grid and the stations numbered in sequence. As previously pointed out, in this paper we deal only with North-East wind situations. This wind, called "bora", practically blows from Trieste towards Venice; hence for what concerns a test for the wind evaluation we will refer only to the Northern Adriatic. We have estimated the wind field considering the stations a) 1 to 8, b) 1 to 6, c) 1 to 4, and finally in a combined system, d) 8 stations with the exception of the Trieste gulf where only numbers 1 to 4 have been taken into account (the corresponding area has been shadowed on Fig. 1). The results of the tests are shown in Fig. 2, comparing the wind recorded on the tower with that evaluated with the various methods. The

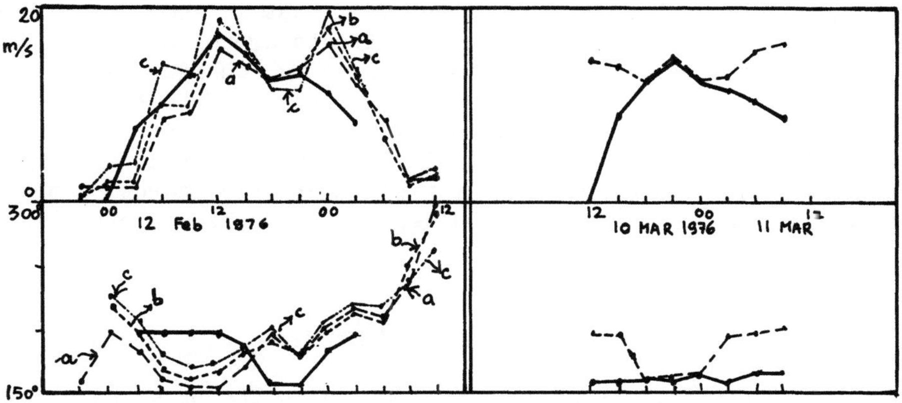

Fig. 2 - Measured (continuous line) and computed (broken line) wind speed and direction at the tower, (fig. 1). Computation with stations a) 1 to 8, b) 1 to 6, c) 1 to 4.

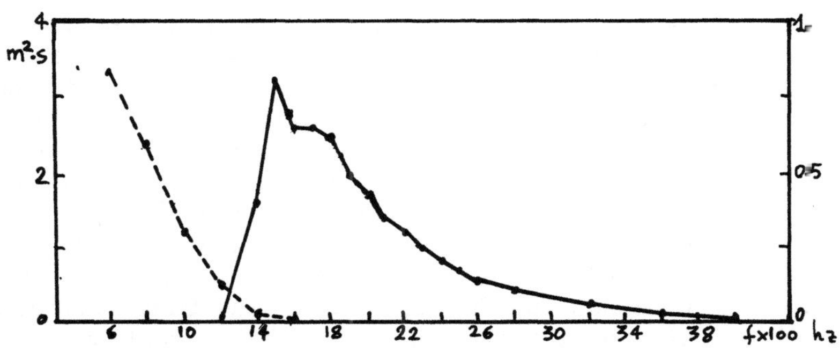

Fig. 3 - Two spectra a) with, b) without bottom friction, coincident in the continuous line. The broken line is the percentage of energy dissipated by bottom friction.

methods a) and d) provide here the same results, d) simply enhancing the fact that the local effects are stronger in the Trieste gulf.

From Fig. 2 we see that the wind speeds are modelled well enough while stronger differences may be present in the estimate of the direction. This is due to the fact that in some meteorological situations the local orographic effects are of a smaller scale than those which can be resolved even by considering only the local surrounding stations. The consequence of this is not simply a shift of the estimated main wave direction in respect to the actual one. From Fig. 1 it is in fact clear that a slight rotation of the wind direction may imply a drastic reduction of the available fetch. This must be kept in mind when interpreting the results of the model.

EXPERIMENTAL EQUIPMENT

The experimental measurements reported in the paper have been obtained from the oceanographic tower of C.N.R., placed 13 km off the coast of Venice in the Adriatic sea. The average water depth at the tower is 16 m; this increases offshore with a regular bottom slope of 1:1000. The main isobaths are sketched in Fig. 1. All the reported records have been taken under the direct continuous supervision of the on-board personnel.

The wave elevation was measured with a resistance wave-staff developed at C.N.R. Its overall accuracy (Cavaleri, 1973) was found to be better than 2 %. The output from this instrument provided the data for the wave spectrum evaluation. The further information required for the directional spectrum estimate was obtained from an electromagnetic current meter developed by I.O.S. (Tucker et al., 1970), measuring the two horizontal components of the water particles velocity. Its accuracy is better than 1 % up to 1 m/s with a resolution of 0.001 m/s. Calibration of the currentmeter was carried out in a towing tank. The outputs of the instruments were recorded on a magnetic tape in digital form at 0.25 s interval (Clayson and Smith, 1970). The Fast Fourier Transform algorithm was used for the analysis. Each spectrum is provided at 1/128 hz interval with 28 degrees of freedom. Further information on the instrumental set-up and analysis procedure are reported by Cavaleri, Ewing, and Smith (1977), whose paper is part of the same series of lectures.

APPLICATION, DISCUSSION, AND CONCLUSIONS

The influence of bottom friction dissipation can be displayed by comparing two computational spectra under the same wind input, respectively with and without the bottom friction. A typical result is shown in Fig. 3. The spectrum corresponds to both the considered cases, the difference between the two being inappreciable in the scale of the drawing. The broken line gives for each frequency the percentage of energy dissipated in respect to the energy value computed without bottom friction. It is evident that in the range of frequencies with significant contribution to the total energy, the effect of the process is negligible, even if in the low frequency end of the spectrum its influence is quite considerable.

Two sequences of tests have been run, to be compared with the corresponding two experimental series taken on the tower.

a) February 12, 1976. In Fig. 4 six spectra are shown out of eleven, at approximately 2 hours interval, (continuous line experiment, broken line model). The general feature which can be recognized in the model spectra is the absence of the sharp energy peak often present in the experimental ones. On the other side, the model predicts energy values generally greater than the experimental ones in the range of frequencies higher than the peak frequency. Both facts are evidently due to the absence of non linear interactions, the main effect of which is to produce a leftward transfer of energy from the right lobe to the forward face of the spectrum. Both facts, moreover, partially compensate each other, so that the total energies (model and experiment) are approximately the same. This is evident when comparing experimental and theoretical significant heights H_s, reported in Fig. 5. In this case, one can also notice that the computed wave heights are on the average over-estimated of about 15 % in respect to the experimental ones. In Fig. 6 the computed and actual main wave directions for each frequency are reported. The effect of refraction is quite evident in the model through the gradual shift from the low frequency range of waves, strongly refracted, to the high frequency waves aligned in the direction of the estimated wind. In this latter range, the difference between the computed and observed wave directions is due to the corresponding difference between the estimated and actual wind direction at the tower.

b) March 10-11, 1976. As for the previous series, wave spectra (six out of twenty, at approximately 2.5 hours interval), significant wave heights H_s, and main wave directions for each frequency are respectively reported in Figs. 7, 8, 9. The same general considerations discussed for the previous series of spectra can be repeated also for this second sequence. Anyway, for this period it must be put in evidence that meteorological data from coastal Adriatic stations were missing for the hours 18-21 of March 10 and 00 of March 11. In this period of time the wind input to the model was supplied by the actual wind data recorded on the tower. Strong differences between theoretical and experimental

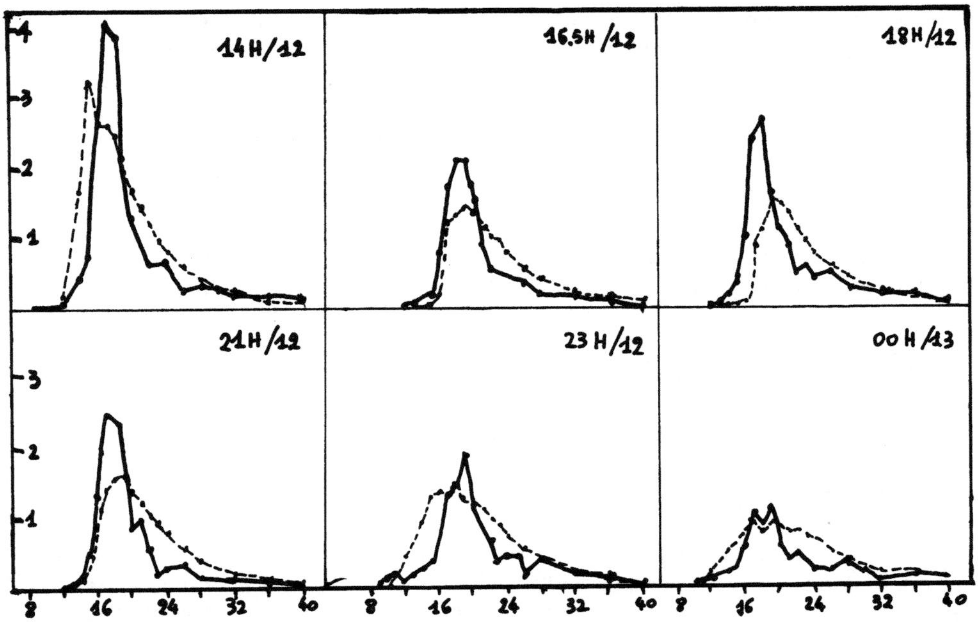

Fig. 4 - Observed (continuous line) and computed (dashed line) spectra for February 12, 1976 - Vertical scale unit is $m^2 \cdot sec$, horizontal scale unit is f·100 in hertz.

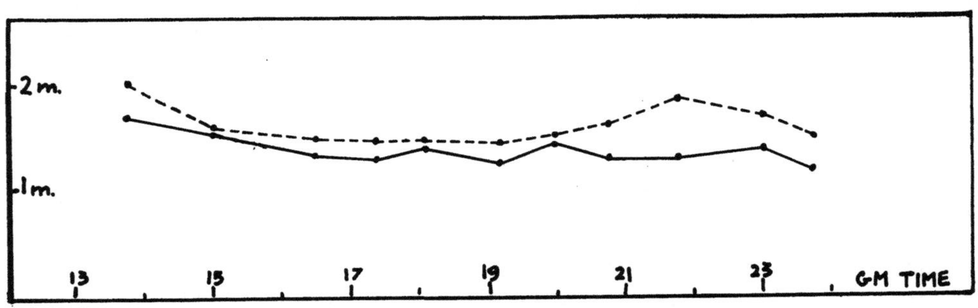

Fig. 5- Observed (continuous line) and computed (dashed line) significant wave heights for records of February 12, 1976.

Fig. 6 - Observed (continuous line) and computed (dashed line) main directions for each frequency - vertical scale unit is degrees CCW in respect to grid x-axis.

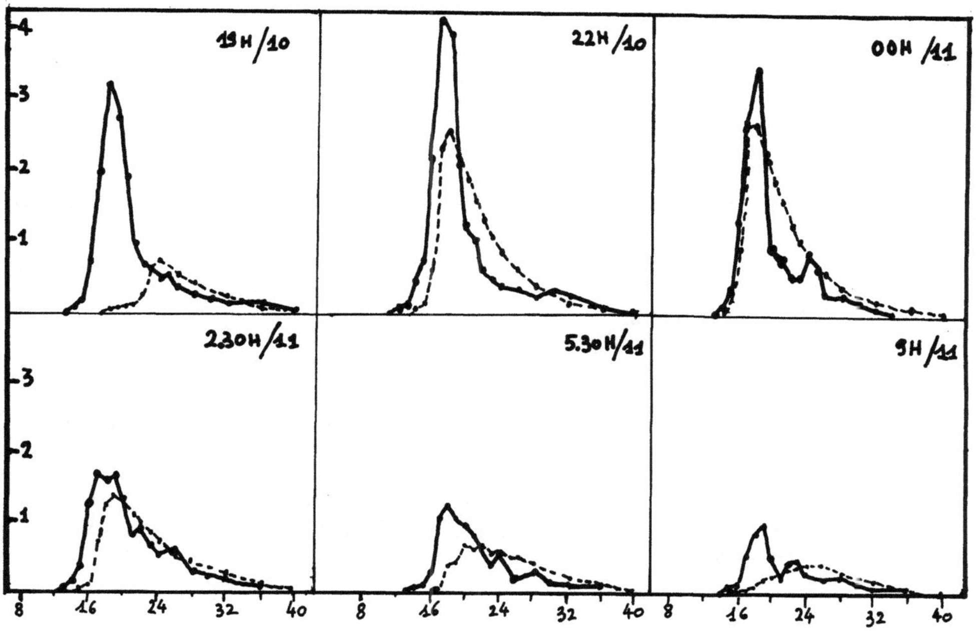

Fig. 7 - Observed (continuous line) and computed (dashed line) spectra for March 10-11, 1976 - Vertical scale unit is $m^2 \cdot sec$, horizontal scale unit is f·100 hertz.

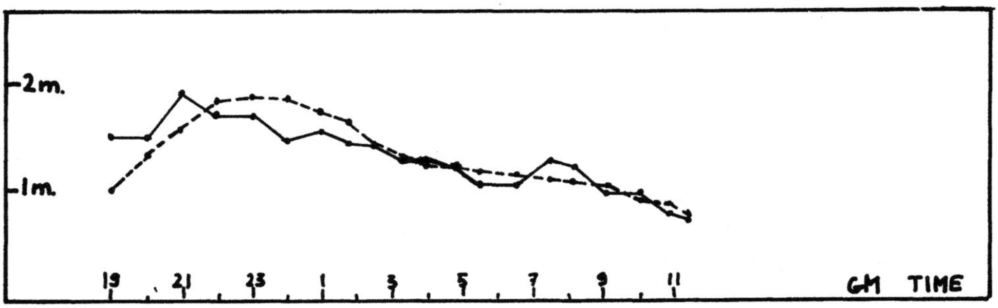

Fig. 8 - Observed(continuous line) and computed (dashed line) significant wave heights for records of March 10-11, 1976.

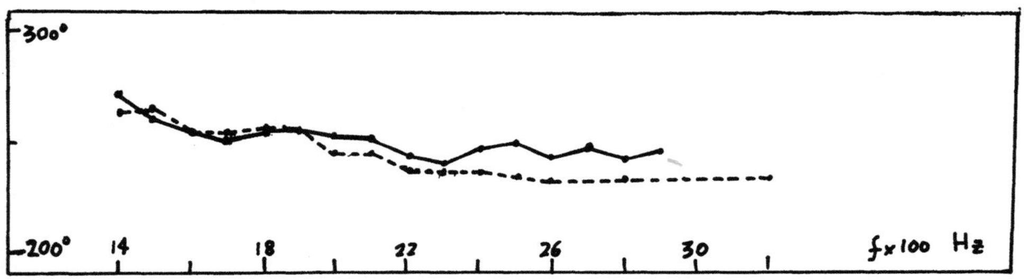

Fig. 9 - Observed (continuous line) and computed (dashed line) main directions for each frequency - vertical scale unit is degrees CCW in respect to grid x-axis.

results are present in the early estimates of spectra and H_s. This is due to the shortness of the fetch because of the 30° degrees difference between computed and actual wind directions. On the other side, the spectra and H_s evaluated with the real wind at the tower show a remarkably good agreement, as well as the last estimates of H_s, when the wind input was again interpolated from coastal data. In Fig. 9, again, the effect of refraction is quite evident in the model. This time, the agreement between observed and computed main directions is quite good.

The fair agreement of most of the results is encouraging for future applications. The main problem to get significantly better model estimates is to improve the wind prediction. As the collection of experimental series on the tower is going on, with new sequences of data at our disposal, hopefully different and better defined meteorological situations, the model will be further extensively tested.

ACKNOWLEDGEMENTS

All the computations were performed on the computer of the IBM Scientific Research Center, Venice. We wish to thank Dr. L. Lippi, director of the center, for having put the necessary facilities at our disposal.

REFERENCES

Barnett, T. P. "On the generation, dissipation and prediction of ocean wind waves", J. Geophys. Res., 73, 513-530, 1968.

Cavaleri, L., J. A. Ewing and N. D. Smith, "Measurements of the pressure and velocity field below surface waves", NATO Symposium 1977, Plenum Press, New York.

Cavaleri, L. "Ondametro a resistenza-critica, miglioramenti e precisione ottenibile", C.N.R. Tech. Report no. 78, 1973-b.

Clayson, C. H. and N. D. Smith. "Recent advances in wave buoy techniques at the N.I.O.", Electronic Engineering in Ocean technology, London: Institution of Electronic and Radio Engineers, 1970.

Collins, J. I. "Prediction of shallow-water spectra", J. Geophys., 77, 2693-2707, 1972.

Cozzi, R. "A detailed analysis of surface winds for oceanographic applications" unpublished manuscript.

Ewing, J. A. "A numerical wave prediction method for the North Atlantic Ocean", Deutsche Hydrog. Z., 24, 241-261, 1971.

Findlater, J. et al. "Surface and 900 mbar wind relationships", Scient. Pap. Met. Off., London, no. 23

Hasselmann,K. and J. I. Collins. "Spectral dissipation of finite depth gravity waves due to bottom friction", Mar. Res., 26, 1-12, 1968.

Hasselmann, K. et al. "Measurements of wind-wave growth and swell decay during the Joint North Sea Wave Project (Jonswap)", Deutsche Hydrogr. Z., no. 12, 1973.

Hasselmann, K. "On the nonlinear energy transfer in a gravity wave spectrum", Part 1, J. Fluid Mech., 12, 481-500, 1962; Part 2, J. Fluid Mech., 15, 273-81, 1963.

Hasselmann, K. "On the spectral dissipation of ocean waves due to whitecapping", Boundary-layer Meteo., 6, 107-127, 1974.

Hasselmann,K. et al. "A parametric wave prediction model", J.Phys. Oceanography, 6, 200-228, 1976.

Long, R. B. "Scattering of surface waves by an irregular bottom", J. Geophys. Res., 78, 7861-7870, 1973.

Longuet-Higgins, M. S. "On the transformation of a continuous spectrum by refraction", Proc. Camb. Phil. Soc., 53, 226-229, 1957.

Malanotte Rizzoli, P. and L. Battiston. "The influence of refraction due to bottom topography on the distribution of wave energy along the coastline of the lagoon of Venice", CNR T.R. no. 34, 1973.

Miles, J.". "On the generation of surface waves by shear flows", J. Fluid Mech., 3, 186-204, 1957.

Phillips, O.M. "On the generation of waves by turbulent wind", J. Fluid Mech., 2, 417-445, 1957.

Phillips, O.M. "The equilibrium range in the spectrum of wind-generated waves", J. Fluid Mech., 4, 426-434, 1958.

Phillips, O.M. "The Dynamics of the upper ocean", Cambridge University Press, London, 1966.

Robinson, A.,A. Tomasin and A. Artegiani. "Flooding of Venice;phenomenology and prediction of the Atlantic storm surge",Quart. J. Royal Meteo. Soc., 99, 686-692, 1972.

DISCUSSION

M. A. Donelan:

I would like to complement you on a nice piece of work and to express the hope that Venice rises as quickly as the progress of wave prediction models indicated by this morning's paper. I have the following comment regarding the specification of wind input in this and other models. The basis of the form of the wind input is in the experiment of Snyder and Cox (J. Mar. Res., 1966) and Barnett and Wilkerson (Barnett, J. Geophys. Res., 1967). However, these experiments succeeded in measuring, not the wind input to a particular wavenumber or frequency, but the balance of total wind input to the spectrum, dissipation and non-linear transfers as reflected in the growth of the forward face --- the only appreciably growing part of the spectrum. Therefore to use the results of these experiments to yield the overall spectral dependence of wind input, and further to compute dissipation and non-linear transfers separately, is inconsistent.

L. Cavaleri:

We do agree. In our model, anyway, the dissipation source function, as given by Hasselmann (1974) is unessential as we do impose as limiting shape to the spectrum the ω^{-5} dependence (Phillips) with the angular spreading factor proposed by Barnett. With the future introduction of the non-linear transfer source function (JONSWAP), we'll have to shift to the β-function proposed by Hasselmann (1974).

A COMPARISON OF PARAMETRIC AND SPECTRAL HURRICANE WAVE PREDICTION PRODUCTS

Duncan Ross* and Vincent Cardone**

*National Oceanic and Atmospheric Administration, Atlantic Oceanographic and Meteorological Laboratories, Miami, Florida, 33149
**City University of New York, New York, 10031

INTRODUCTION

Statistics of extreme wave heights created by tropical cyclones are urgently needed for the design of harbor breakwaters, ships, and offshore structures to prevent accidents from underdesign and to avoid the untenable costs of overdesign. Wave data alone cannot provide the needed statistics due to the long periods required to collect statistically reliable data sets and the inability to deploy enough systems to obtain measurements of maximum conditions. Needed statistics, therefore, are usually extracted from models which attempt to specify the wave environment associated with particular storm systems. These models are either largely untested against measurements, or are validated over a limited range of storm conditions. One such model is that of Cardone et al. (1976a) which was developed and calibrated using a data set obtained in hurricane Camille, a severe Gulf of Mexico storm which struck the coast of the U. S. in August 1969. Another is an empirically-based parametric model proposed by Ross (1976), based upon forty measurements of one-dimensional frequency spectra obtained in Camille, Pacific hurricane Ava (Ross, 1975), and Gulf of Mexico hurricane Eloise (Withee and Johnson, 1975).

A brief comparison of the Cardone et al. (1976a) model and the Ross (1976) parametric model, including data obtained in Atlantic hurricane Belle, has been conducted by Cardone et al. (1976b) and Cardone and Ross (1977). That study shows that both models produced comparable results for maximum wave conditions for storms moving at a forward velocity within the range of the calibration data sets (20-40 km/hr) but that the Cardone et al. model produces higher waves at lower frequencies than the parametric model for storms moving at higher forward velocities.

In this paper, the parametric and spectral models are intercompared systematically through a sensitivity study involving several hindcasts of a hypothetical hurricane, with the forward velocity varying from 0 to 80 km/hr. The differences between the spectral and parametric hindcasts are interpreted in terms of the alternate parameterizations of the physics of wave generation in the two methods.

The divergent results documented in the sensitivity study emphasize the limitations of state-of-art spectral and parametric models which are quasi-empirical and which depend upon observations of net wave growth for calibration.

Tropical cyclones provide a powerful test of wave prediction models since they exercise model parameterizations of growth and dissipative processes over a broad range. Fast moving tropical cyclones are associated with "moving fetches" which emphasize the differences between wave growth algorithms which are coupled and decoupled in frequency space. Wave measurements under such conditions would enable further testing of existing models and possibly provide insight into the nature of the spectral energy balance at large fetch.

PARAMETRIC AND SPECTRAL WAVE PREDICTION MODELS

The spectral model (Cardone et al., 1976a) is based upon the numerical integration of the wave energy balance equation

$$\frac{\partial S(f,\theta)}{\partial t} = -V_g(f,\theta)\cdot\nabla S(f,\theta)-F \tag{1}$$

where S is the directional wave energy spectrum, f is wave frequency, θ propagation direction, V_g is the group velocity, and F represents a source function which accounts for all processes that transfer energy to, from, and within the wave spectrum. Equation (1) is solved by the successive simulation of the processes of wave growth, dissipation, and propagation at each time step, on a hexagonal grid of 1694 points nominally spaced 35 km apart on a gnomonic map projection. At each grid point, the spectrum is resolved into fourteen frequency and twenty-four direction bands. During each one-hour time step, equation (1) is solved for each of the 336 spectral elements. The source function F is applied only to spectral components propagating within $\pm 90^{o}$ of the local wind direction according to

$$F=\left[A(f,\theta,\vec{V})+B(f,\theta,\vec{V})\cdot S(f,\theta)\right] \times \left\{1-\left[\frac{S(f,\theta)}{S_{\infty}(f,\theta,\vec{V})}\right]^2\right\} \quad (2)$$

where A represents the linear growth rate (Phillips, 1957), B is the exponential growth term of the form suggested by Miles (1957), and $S_{\infty}(f,\theta,V)$ is the Pierson-Moskowitz (1964) fully-developed form of the spectrum with a windspeed-dependent high-frequency equilibrium range (Phillips, 1958) as suggested by Stacy (1973). Spectral components travelling against the wind are attenuated at a rate proportional to the fourth power of the wave frequency and the local energy of the wind sea. A frequency-dependent functional form for the directional spread of wave energy is prescribed, and in regions of rapidly shifting wind direction, an angular redistribution of spectral energy is imposed to preserve the known tendency for a growing spectrum to redistribute directionally, probably through non-linear interactions about the local wind direction. Propagation is accomplished by a Lagrangian scheme whereby spectral components jump to adjacent grid points in the appropriate direction after a sufficient number of time steps have elapsed to account for the displacement according to that component's deep water group velocity.

The parametric model is a simplification of the parametric approach to wave prediction proposed by Hasselmann et al. (1976). The method is considered to be applicable only to growing wind seas. The method follows from the data collected in the JONSWAP study, the analysis of which (Hasselmann et al., 1973) suggested that fetch-limited wave spectra can be normalized in a way such that there is an approximate invariance of normalized spectral shape with fetch. It was proposed that non-linear resonant wave-wave interactions were responsible for the shape invariance and that it characterized not only idealized fetch-limited siutations but growing wind seas in general.

The parametric approach exploits the shape invariance by representing the spectrum in terms of a small number of shape and scale parameters (e.g., five in the JONSWAP spectrum) and projecting equation (1) onto a set of prognostic equations for the parameters. The wave prediction problem then becomes computationally much simpler than the explicit application of equation (1) to a discrete two-dimensional spectrum.

Evidence of the shape-invariance of hurricane-generated wave spectra has been provided from spectra measured in hurricane Ava from an aircraft equipped with a laser altimeter, from wave measurements made in hurricane Camille as part of an oil industry measurement program (cf. Hasselmann et al., 1976), and from a meteorological data buoy during hurricane Eloise (Withee and Johnson, 1976). Those spectra matched well the JONSWAP spectral form, and Ross (1976) proposed that the advecting wave field in a hurricane might be determined by the local wind and a fetch determined by the local radius of curvature of the wind field, which is proportional to the radial distance to the eye, and derived a special parametric solution applicable only to hurricane generated seas.

The original analysis of JONSWAP data followed the procedure proposed by Kitaigorodskii (1961) such that for ideal fetch-limited wind seas, all wave variables, when non-dimensionalized in terms of g, gravity, and the windspeed, U, should be functions only of the single non-dimensional fetch, $\xi=gX/U^2$, where X is the fetch. In Hasselmann et al. (1976), the non-dimensional peak frequency $\nu=Uf_m/g$, where f_m is the peak frequency, was found to be described well by the relation

$$\nu=2.84\ \xi^{-0.3} \tag{3}$$

The non-dimensional total energy $\varepsilon=Eg^2/U^4$ where $E=\int S(f,\theta)d\theta df$ showed a linear fetch dependence

$$\varepsilon=1.6 \times 10^{-7}\xi \tag{4}$$

Similar power-law fits were obtained for γ, σ_a, σ_b, and α (see Hasselmann et al., 1976, for a further discussion of these parameters).

In the Ross (1975) model, the fetch, X, is replaced by r, the radial distance to the eye of the hurricane, which should parameterize fetch or duration. The data set from hurricane Ava, Camille, and Eloise were then combined to produce the following power-law fits:

$$\left.\begin{aligned} \nu&=.97\ \xi'^{-.21} && (5)\\ \varepsilon&=2.25 \times 10^{-5}\xi' && (6)\\ \gamma&=4.7\ \xi'^{-.13} && (7)\\ \alpha&=.035\ \nu^{.82} && (8) \end{aligned}\right\}\quad \xi'\leq 3.0 \times 10^{+4}$$

where $\xi'=gr/U^2$. σ_a and σ_b were found to scatter somewhat about an average of approximately .1 but showed no consistent dependence on ξ'.

The wind field used to initialize both models is an adaptation of the theoretical model of horizontal air flow in the boundary layer of a moving vortex developed by Chow (1971). The model is based upon the numerical integration of the vertically averaged equations of motion that govern a boundary layer subject to horizontal and vertical shear stresses on the rotating earth. The model was calibrated against wind data collected in hurricane Camille, after the windspeed measurements were adjusted from measurement altitude to 20 meters using the boundary layer theory of Cardone (1969). For a more detailed description of the model, including comparison of predicted to observed winds, the reader is referred to Cardone et al. (1976a) and Forristall, Hamilton, and Cardone (1977).

MODEL VALIDATION

The spectral model, though calibrated largely on the basis of hurricane Camille wave data, has been rather extensively applied both in hindcast studies of historical hurricanes and in detailed hindcast studies of more recent storms for which wave data in one form or another exist. Cardone et al. (1976a) compared hindcast and measured or estimated maximum wave height in five separate Gulf of Mexico hurricanes. In eighteen cases, with measured maximum heights ranging from 6 to 24 meters, the RMS difference was 1.5 meters. The hindcast spectra of maximum conditions agreed well with a more limited number of available measured frequency spectra.

More recently, the deployment of automatic recording buoys equipped with wave recorders, by the National Oceanic and Atmospheric Administration National Data Buoy Office (NDBO), has resulted in the acquisition of unique wave data sets in hurricane Eloise (1975), in the Gulf of Mexico, and hurricane Belle (1976) on the U. S. East Coast. Cardone and Ross (1977) compared the spectral and parametric hindcasts of NOAA NDBO buoy EB10, which experienced eye passage and maximum wind and wave conditions in Eloise. Figures 1 and 2 from that study compare observed to hindcast time histories of wind and wave conditions during hurricane Eloise. It should be noted that wave data in Eloise from this buoy were used in the calibration of the parametric method.

Since it has only recently been proposed, the parametric method has not been extensively tested on storms independent of those which provided a basis for model tuning. Using both models, Cardone et al. (1976b) made a real-time wave forecast of hurricane Belle (1976), a moderately intense U. S. East Coast storm which passed directly over buoys EB15 and EB41, as shown in Figure 8. Errors in the wave forecasts were found to be caused mainly by errors in the official forecasts of storm intensity and the track

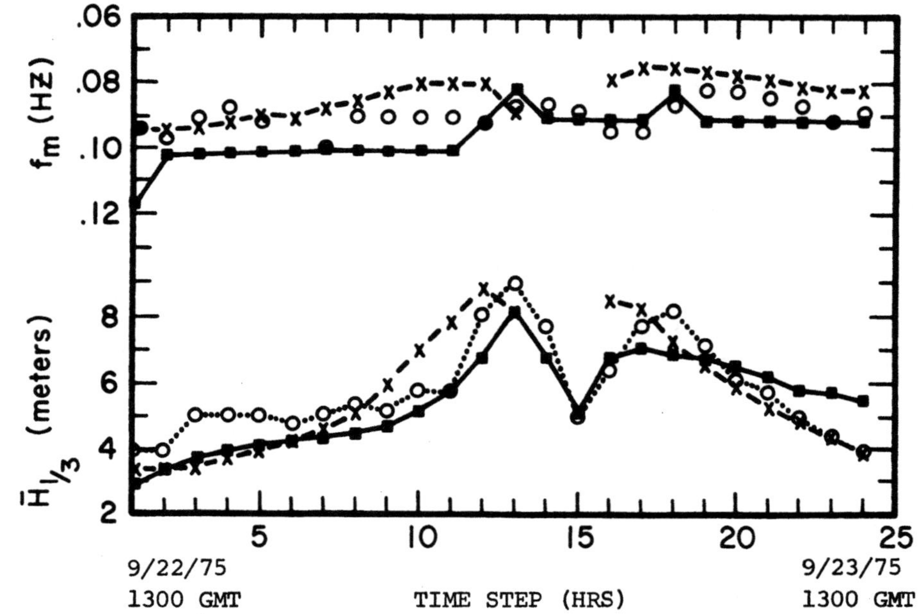

Figure 1. Comparison of measured (O), and hindcast significant wave height and spectral peak frequency from the parametric (X), and spectral (■) models at EB10 in hurricane Eloise.

used to calculate the forecast wind field. In the same study, the parametric and spectral models were used to hindcast two severe historical East Coast storms, Donna (1960) and a September, 1944, storm. Significant differences in maximum hindcast wave height and peak frequency were found, with the spectral model producing much higher sea states and much lower peak frequencies. This was a surprising result and was related at the time to the major difference between East Coast and Gulf hurricanes - the fast forward motion (65-85 km/hr) of the former.

It was hoped that the relatively fast forward motion (40 km/hr) of Belle along the upper coast and the availability of frequency spectra from EB41 would provide a basis for resolving the discrepancy noted in the historical hindcasts. Thus a detailed hindcast study of Belle was performed. It was found possible to specify the wind field quite accurately throughout its history, as verified by observed wind time histories at the buoys. The hindcast time histories of significant wave height are compared to the measurements in Figures 3 and 4. Hindcast frequency spectra associated with maximum sea states when the eye was near EB15 and 41 are compared in Figures 5 and 6. The sampling interval of three hours at the buoy apparently was not sufficient to resolve maximum conditions at EB15, as the measured hourly winds and both

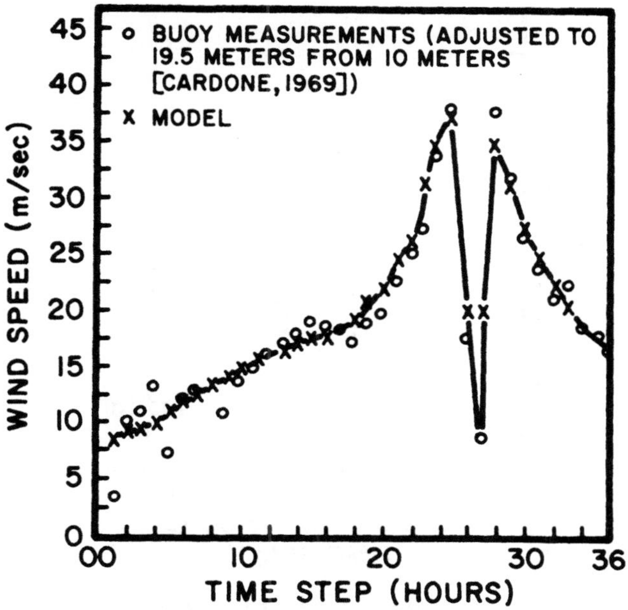

Figure 2. Comparison of measured and modeled average windspeed time history at EB10.

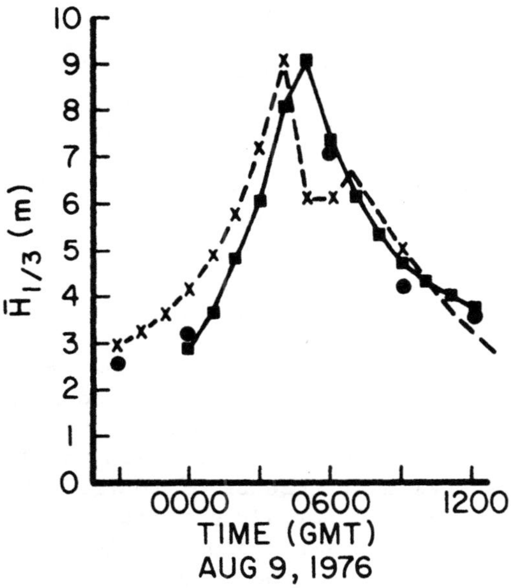

Figure 3. Comparison of measured and hindcast significant wave height at EB15 during hurricane Belle: ● - measured; X - parametric model; ■ - spectral model.

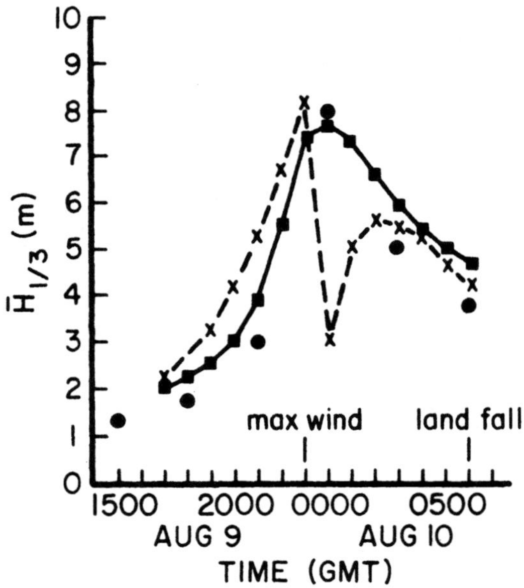

Figure 4. Comparison of measured and hindcast significant wave height at EB41 during hurricane Belle: ● - measured; X - parametric; ▪ - spectral.

hindcast time histories suggest that peak sea states occurred just prior to the 0600 GMT buoy wave measurement. At EB41, however, peak conditions were more likely measured, and it is possible to compare the hindcasts and the measurement. Differences in the hindcasts for this storm are small, though the spectral model appears to simulate better the observed time history of wave height at the buoy locations. Belle, therefore, provided further validation of both models but failed to provide information on the divergence of the solutions for faster moving storms.

SENSITIVITY STUDY

To study the effect of storm motion on the relative differences between the parametric and spectral hindcast models in a more controlled fashion, a sensitivity study was conducted that involved four hindcasts. In that study, the wind field was specified for a hypothetical hurricane with attributes of central pressure (975 mb), peripheral pressure (1013 mb), pressure radius (75 km), ambient geostrophic flow (9 m/s), and forward velocity (10 m/s) that are close to the median values observed in the population of important hurricanes that have affected the U. S. East Coast this century. A similar sensitivity study has been reported recently by Uji (1975), but

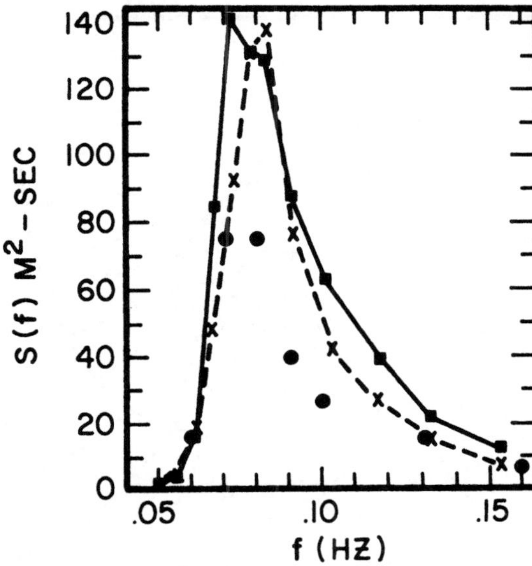

Figure 5. Maximum predicted spectra for hurricane Belle: X - parametric model, $\overline{H}_{1/3}$=9.0 m; ■ - spectral model, $\overline{H}_{1/3}$=10.1 m. Also shown is the maximum observed spectrum (●), $\overline{H}_{1/3}$=7.1 m, at EB15.

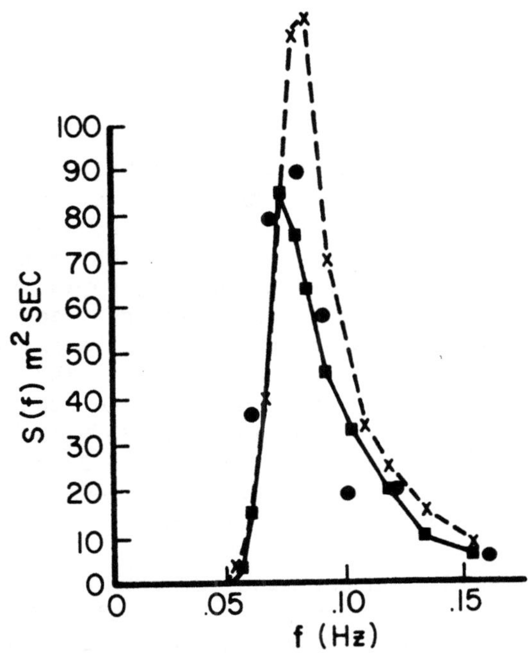

Figure 6. Maximum predicted spectra for hurricane Belle at EB41: X - parametric model, $\overline{H}_{1/3}$=8.2 m; ■ - spectral model, $\overline{H}_{1/3}$=7.8 m. Also shown is the maximum observed spectrum (●), $\overline{H}_{1/3}$=7.8 m.

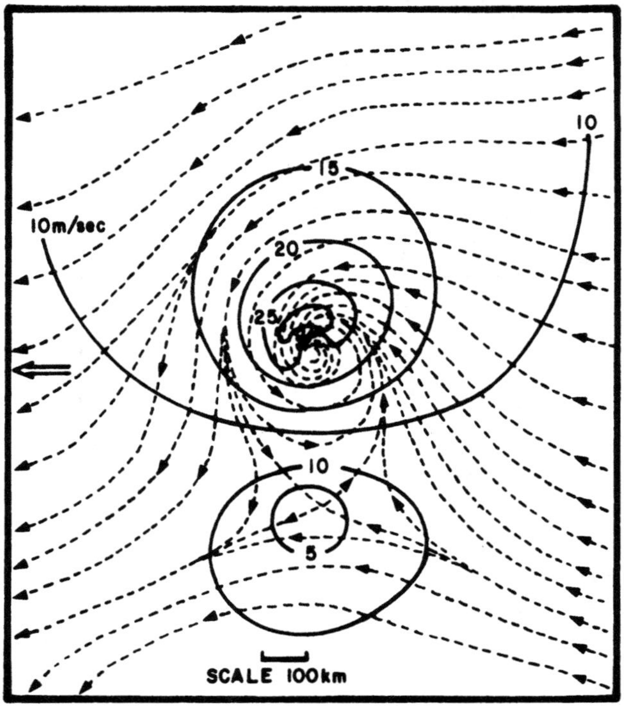

Figure 7. Hindcast wind field analysis for hurricane Belle.

no evidence is presented on the accuracy of the hindcast model when applied to actual tropical cyclones.

The computed wind field is shown in Figure 7. The maximum average 19.5 meter windspeed is 37 m/s at a distance of 50 km from the pressure eye. The wind field shown is very similar to that specified for hurricane Belle near EB41, though at that time in its history, the Belle wind field was undergoing rapid changes. In this sensitivity study, the wind field was assumed steady state and was translated across the hindcast grid to reflect storm forward velocities of 0, 27 km/hr, 55 km/hr and 80 km/hr. The location of the grid system used is shown in Figure 8. For the stationary storm simulation, the hurricane eye was placed at 35°N, 74°W. The moving storms arrived at 40°N, 70°W at the end of the 21 time step hindcasts. The origin of the track simulated for all three of the moving storm cases is also shown in Figure 8.

Since all spectral hindcast runs are started from initially calm conditions, there is a spinup period required before the distribution of simulated directional spectra reflects an equilibrium relationship with the driving input steady state wind field. The parametric model provides its equilibrium solution directly from the wind field and is independent of storm motion variations. Also,

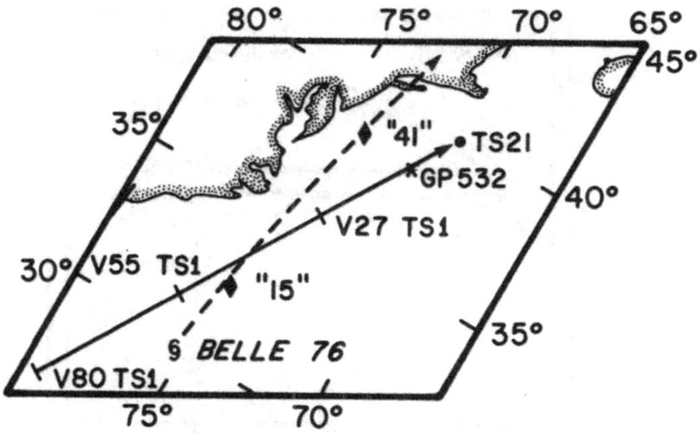

Figure 8. Domain of grid system for "Belle" hindcast and sensitivity study: ◆ - location of buoys EB15 and 41. Solid line shows track of hypothetical storms; ticks indicate initial position for indicated storm velocity in km/hr.

because of the time required for swell generated in the areas of highest winds around the eye to propagate away from the storm, the spinup period for the spectral hindcasts increases with increasing distance from the storm center.

The behavior of the solution near the eye for the spectral hindcasts is shown in Figure 9. The solution shown is for the grid point which exhibits maximum energy each time step. This grid point is near, but not always coincident with, the grid point in a given time step with the highest wind. It is evident that steady state maximum wave conditions are very nearly reached within the 21 time step run, and that there is a significant dependence of the maximum sea state hindcast on storm forward speed. The significant wave height and peak frequency hindcast by the parametric model for maximum conditions are 9.2 m and .078 Hz, respectively, values which match closely the spectral predictions for the 27 km/hr run.

In Figure 10 the one-dimensional spectra predicted for equilibrium maximum sea states by the spectral model are compared to the frequency spectrum constructed from the scale and shape parameters predicted by the parametric model. Again, the solutions are most similar for the 27 km/hr run but are quite different for slower or faster moving storms. The difference is especially pronounced for the 50 km/hr run.

Figure 11 (a-e), displays the distribution of significant wave height about the storm. The parametric solution and the spectral

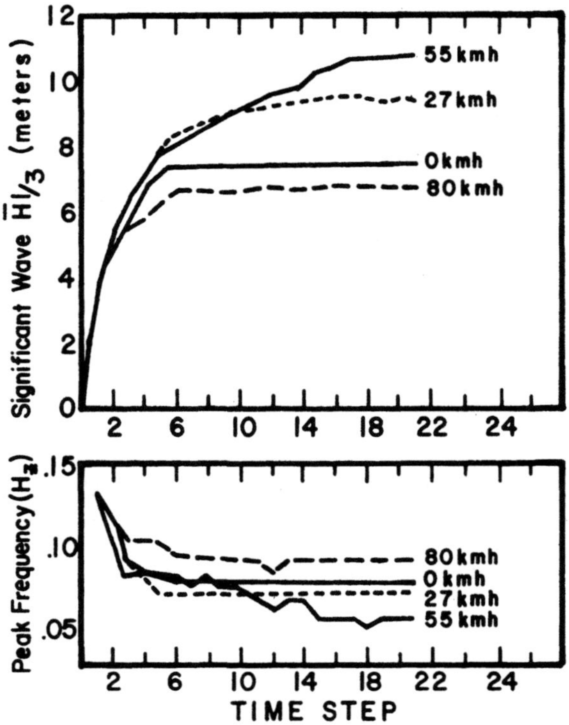

Figure 9. Time history of significant wave height and peak frequency for grid point 532.

solutions for the stationary and 27 km/hr cases show patterns of wave height resembling the isotach pattern. For faster moving storms, however, the wave height distribution predicted by the spectral model is affected by the storm motion, as wave heights decrease in the forward quadrants and increase in the rear quadrants, relative to the slow-moving case. The 50 km/hr case exhibits not only the highest waves, but the greatest areal coverage of high sea states. The solution is apparently very sensitive to storm speeds near and above 50 km/hr.

DISCUSSION

In a stationary cyclone, the stage of wave development, as gauged by the non-dimensional peak frequency, is limited in the area of strong winds by the physical fetch limitation imposed by the highly curved flow. In a moving cyclone, however, the effective fetch is determined by both the vector wind pattern and the propagation vector of storm motion.

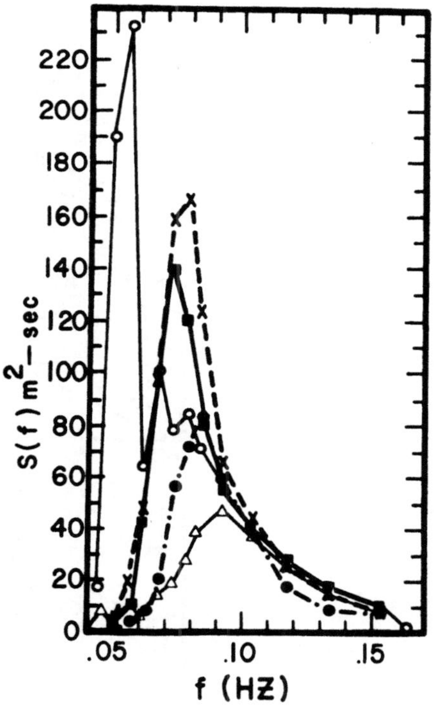

Figure 10. Maximum conditions for storms of sensitivity study. Spectral model: ● - 0 km/hr; O - 27 km/hr; ■ - 55 km/hr; Δ - 80 km/hr; X - parametric solution.

In the spectral growth model used here, wave growth, in the absence of swell, is essentially uncoupled in frequency, and each spectral component travelling within $\pm 90^{\circ}$ of the local wind direction responds independently to the linear and exponential growth mechanisms for the local windspeed. For simply defined linear fetches, such a spectral growth model is in reasonably good agreement with observed fetch-limited spectra (e.g., Pierson, 1977). However, in more complicated wind systems the method can produce spectral growth patterns which do not obey, even approximately, the shape invariance theorem upon which the parametric method is based.

Moving tropical cyclones appear to be especially capable of differentiating between the two approaches. If growth can be considered a process decoupled in frequency, then moving hurricanes provide, over limited regions of wave number space, infinite fetch conditions over small regions of the storm circulation. Thus, if in those regions of the circulation, the local windspeed is suffi-

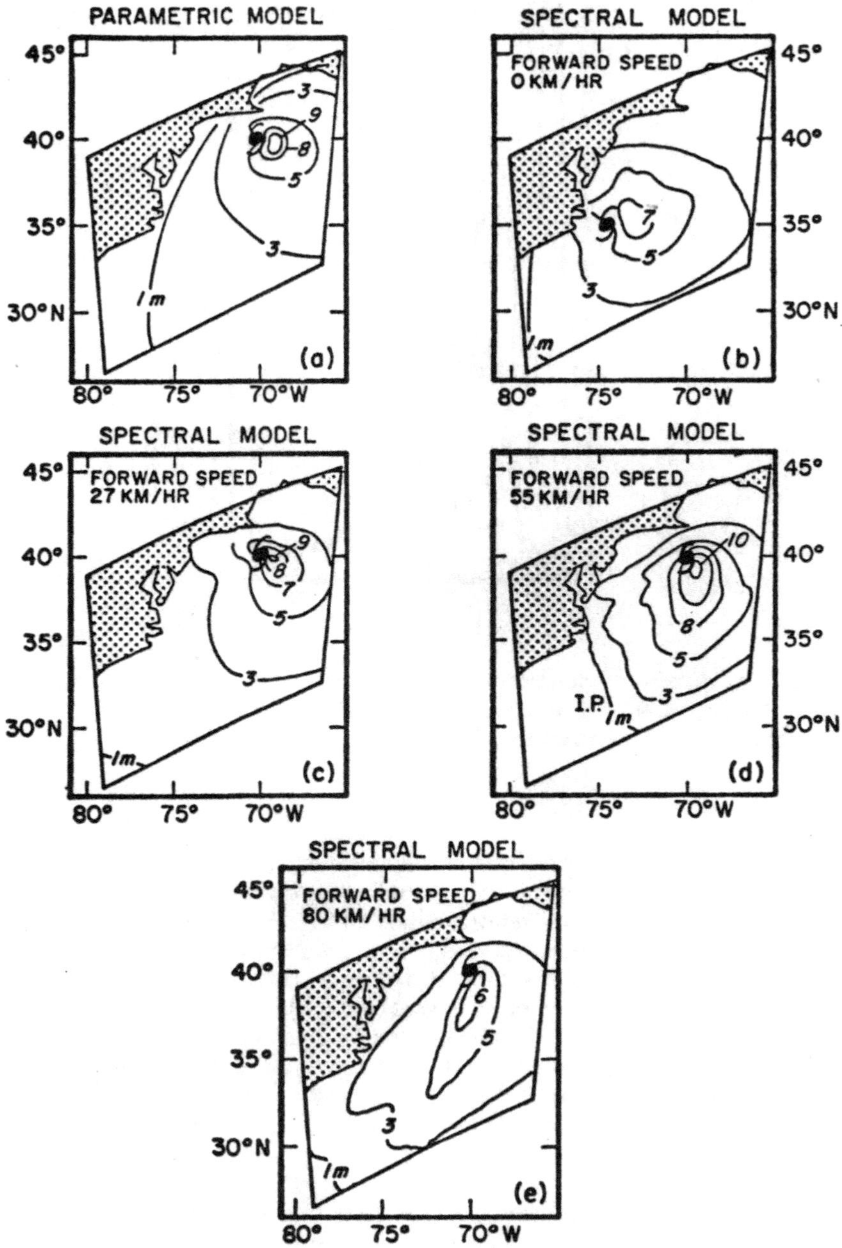

Figure 11 (a-e). Contour diagrams of significant wave height for storms of the sensitivity study.

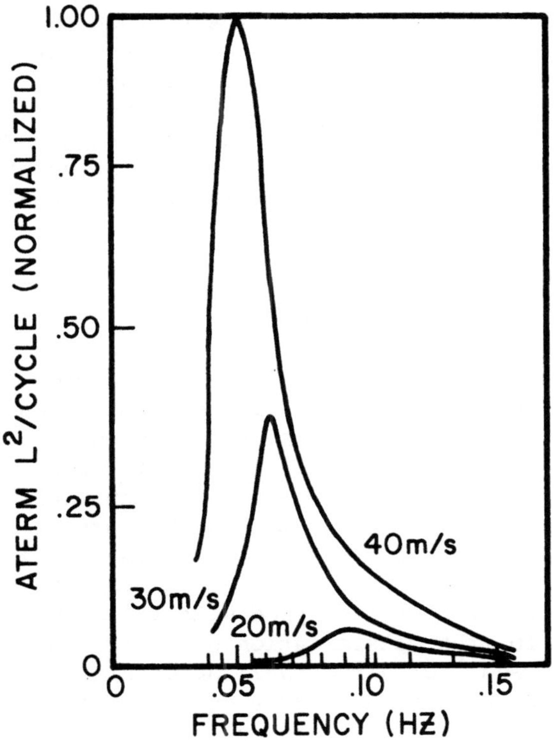

Figure 12. Characteristic growth due to resonance mechanism.

cient to provide excitation by means of the linear growth mechanism, or A term in equation (2), then spectral components are able to grow exponentially to saturation levels determined by the Pierson-Moskowitz fully-developed spectrum, which at hurricane windspeeds can support enormous energies. The dependence of the resonance mechanism growth rate (A term) on frequency and windspeed is shown in Figure 12. It can be inferred that, for the spectral growth method, wave generation in tropical cyclones will be maximized if the windspeed in that portion of the circulation where the wind direction matches the storm direction (the right rear quadrant) is such that the peak frequency in the resonance mechanism parameterization is at a frequency associated with gravity wave group velocities that match the propagation speed of the storm. In the sensitivity study described above, this condition was met for the 55 km/hr storm.

On the other hand, the parametric method derives from a picture of the spectral energy balance involving wave-wave interactions as well as processes uncoupled in frequency. The non-linear component of the source function is in fact a function of the position of a spectral component with respect to the peak in the spectrum, which itself migrates from high to low frequency more slowly as the stage of wave development increases. Growth of low-frequency energy in fast-moving cyclones therefore may not be expected to be significantly different from that in slow-moving storms of the same meteorological nature, although detailed integrations with full multi-parameter models would be required to quantify the differences.

SUMMARY

Given a proper specification of the wind field within tropical cyclones, it has been shown that accurate predictions of the generated wave field can be produced by both the spectral and parametric methods. Nevertheless, significant differences do occur for the unusual case of very slow and very fast moving storms. While these storms are rare, they may be the most significant from the standpoint of engineering designs. The measurements required to verify the differences probably can only be obtained in the near future by means of aircraft (equipped with air-droppable buoys or laser altimeters for the determination of energy spectra and real aperture imaging radar for directional spectral information) which are capable of seeking out the proper quadrant of a fast-moving storm. The aircraft techniques are well within the state-of-the-art and are not costly. It is imperative, therefore, that a measurement program be undertaken to validate the model predictions.

ACKNOWLEDGMENTS

The development of the spectral model employed in this study was supported by a consortium of U. S. oil companies as part of the Ocean Data Gathering Program (ODGP). We thank the administrator of the ODGP program, Shell Development Company, for supporting, jointly with NOAA, the adaptation of the model to the NOAA AOML computer facility. Dr. Cardone's participation in the study reported herein and in the NATO Symposium was supported by NOAA under Grant 04-7-022-44014.

The authors thank Merlin Ahrens of the National Oceanic and Atmospheric Administration's Atlantic Oceanographic and Meteorological Laboratories for his diligent computer programming assistance.

REFERENCES

Cardone, V. J., 1969: Specification of the wind distribution in the marine boundary layer for wave forecasting. Rept. TR 69-1, Geophys. Sci. Lab., New York University NTIS No. AD702490 .

———, W. J. Pierson, and E. G. Ward, 1976a: Hindcasting the directional spectra of hurricane generated waves. J. Petrol. Technol., 28, 385-394.

———, D. Ross, M. Ahrens, J. A. Greenwood, C. Greenwood, and R. E. Salfi, 1976b: Forecasting hurricane winds and waves - A pilot study. Final report to NOAA Sea-Air Interaction Laboratory and Shell Development Company. CUNY Institute of Marine and Atmospheric Sciences at the City College, New York.

———, and D. B. Ross, 1977: State of art wave predictions and data requirements. Paper presented at the NOAA Climate Symposium, Washington, D. C., July 12-14, 1977. Proceedings to be published by Plenum Company in 1978.

Chow, S., 1971: A study of the wind field in the planetary boundary layer of a moving cyclone. M.S. Thesis, Department of Meteorology and Oceanography, New York University.

Forristall, G. Z., R. C. Hamilton, and V. J. Cardone, 1977: Continental shelf currents in tropical storm Delia: observations and theory. J. Phys. Oceanogr., 7, 532-546.

Hasselmann, K., T. T. Barnett, E. Bouws, H. Carlson, D. E. Cartwright, K. Enke, J. A. Ewing, H. Gienapp, D. E. Hasselmann, P. Kruseman, A. Meerburg, P. Müller, D. J. Olbers, K. Richter, W. Sell, and H. Walden, 1973: Measurements of wind-wave growth and swell decay during the Joint North Sea Wave Project (JONSWAP). Deut. Hydrogr. Z., Suppl. A 8: #12.

———, D. B. Ross, P. Müller, and W. Sell, 1976: A parametric wave prediction model. J. Phys. Oceanogr., 6, 200-228.

Kitaigorodskii, S. A., 1961: Applications of the theory of similarity to the analysis of wind-generated wave motion as a stochastic process. Izv. Akad. Nauk SSSR, Ser. Geofiz., No. 1, 73-80.

Miles, J. W., 1957: On the generation of surface waves by shear flows, J. Fluid Mech., 3, 185-204 233 .

Phillips, O. M., 1957: On the generation of waves by turbulent wind. J. Fluid Mech., 2, 417-445.

———, 1958: The equilibrium range in the spectrum of wind generated waves. J. Fluid Mech., 4, 426-434.

Pierson, W. J., Jr., and L. Moskowitz, 1964: A proposed spectral form for fully developed wind seas based on the similarity theory of S. A. Kitaigorodskii. J. Geophys. Res., 69, 5181-5190.

———, 1977: Comments on "A parametric wave prediction model". J. Phys. Oceanogr., 7, 127-134.

Ross, D. B., 1975: A comparison of SKYLAB S-193 and aircraft views of surface roughness and a look toward Seasat. Proceedings of the NASA Earth Resources Survey Symposium, Lyndon B. Johnson Space Center, Houston, Texas, June 1975, 1911-1920.

Ross, D. B., 1976: A simplified model for forecasting hurricane generated waves (Abstract). Bull. A.M.S. January, 113. Presented at conferences on Atmospheric and Oceanic Waves, Seattle, Washington, March 29-April 2. American Meteorological Society.

Stacey, R. A., 1973: High frequency gravity waves in hurricanes. M.S. Thesis, Department of Meteorology and Oceanography, New York University.

Uji, T., 1975: Numerical estimation of the sea wave in a typhoon area. Papers in Meteorology and Geophysics, 26, 199-217.

Withee, G. W., and A. Johnson, 1975: Buoy observations during hurricane Eloise. Data Buoy Office, National Oceanic and Atmospheric Administration, Bay St. Louis, Mississippi.

DISCUSSION

W. Rosenthal:

Can you give the functional dependence and the parameters for the directional coupling of different frequencies that you mentioned.

V. Cardone:

As in other parts of the model, the effects of non-linear interactions are simulated in a highly implicit way with regard to directional coupling. In the present model, the algorithm operates only in developing or fully developed sea situations, and has the effect, in rapidly shifting wind direction zones, of "turning" the directional spectrum gradually so that the maximum energy is located in the wind direction. However the adjustment time is long, as spectral components that become overdeveloped with respect to the local wind direction and the SWOP spreading function, are attenuated at a rate proportional to the degree of overdevelopment. In hurricanes, this algorithm is responsible in part for the simulation of the rather flat directional spread of energy observed in recent measurements.

K. Hasselmann:

I agree it would be very valuable to use hurricane data to test the different physics of models with and without significant non-linear coupling between different rays. This would be a very valuable exercise. In this context I should perhaps mention as a point of clarification that the parametric hurricane model discussed here - which is based on dimensional argument without use of the transport equation - should be distinguished from the dynamical parametrical models discussed in the previous papers, which are based on the projection of the full transport equation on the parametric space. These latter models are not different in principle from the usual spectral models, they use only different numerical techniques.

INDEX

GPSR Compliance

The European Union's (EU) General Product Safety Regulation (GPSR) is a set of rules that requires consumer products to be safe and our obligations to ensure this.

If you have any concerns about our products, you can contact us on ProductSafety@springernature.com

In case Publisher is established outside the EU, the EU authorized representative is:

Springer Nature Customer Service Center GmbH
Europaplatz 3
69115 Heidelberg, Germany

Batch number: 10683525

Printed by Printforce, the Netherlands